Conservation Planning

Shaping the Future

Edited by F. Lance Craighead and Charles L. Convis Jr.

Esri Press
REDLANDS | CALIFORNIA

Esri Press, 380 New York Street, Redlands, California 92373-8100
 First edition 2013
17 16 15 14 13   1 2 3 4 5 6 7 8 9 10

Printed in the United States of America

Library of Congress Cataloging-in-Publication Data
Conservation planning : shaping the future / F. Lance Craighead, Charles L. Convis, Jr., editors. — First edition.
pages cm
Includes bibliographical references and index.
ISBN 978-1-58948-263-0 (pbk. : alk. paper) 1. Landscape protection—Planning. 2. Habitat conservation—Planning. 3. Landscape protection—Data processing. 4. Habitat conservation—Data processing. I. Craighead, Lance. II. Convis, Charles L., Jr., 1954–
QH75.C68146 2013
333.95'16—dc23   2012033091

Esri Press titles are distributed to the trade by the following:
In North America:
Ingram Publisher Services
Toll-free telephone: 800-648-3104
Toll-free fax: 800-838-1149
E-mail: customerservice@ingrampublisherservices.com

In the United Kingdom, Europe, Middle East and Africa, Asia, and Australia:
Eurospan Group
3 Henrietta Street
London WC2E 8LU
United Kingdom
Telephone: 44(0) 1767 604972
Fax: 44(0) 1767 601640
E-mail: eurospan@turpin-distribution.com

Contents

Foreword

When Charles Convis and Lance Craighead first approached me on behalf of the Craighead Institute with the idea for a book that would bridge the gap between current ecological research and the best practices of working conservation planners around the world, I was delighted to support them. In the years since we began this collaboration, geographic information systems (GIS) technology has evolved rapidly, especially in the online arena. However, the fundamental concepts and approaches presented in this book are timeless and, we hope, will provide a foundation for conservation planning and landscape design for decades to come.

With increasing population pressure, civil strife, and economic challenges, it's easy to throw up one's hands and despair. That's the mistake people make when they think they are alone and they have no tools and no options. If Esri's 40-plus-year history and hundreds of thousands of users mean anything today, it's to show us that we are never alone and never without skills or tools.

Through the efforts of the Society for Conservation GIS to our annual Esri International User Conference (attended now by thousands of people, representing hundreds of worldwide regions and organizations), we recognize that tough struggles can only be won when like-minded people unite. With this book and others like it, and through our tools, technologies, and our support of communities around the world, we continue to do all we can to empower willing workers and thoughtful activists to improve society in whatever way they can. Knowing how to engage different kinds of people in discussion and pursue agreement on what our communities and landscapes ought to look like in the future is a very powerful technique for determining what the future will be. That's what planning is.

Now go out and design a future that makes sense!

Jack Dangermond
President, Esri

Preface

The need to systematically address many of the problematic issues in conservation science became evident from our experience over several decades in conducting conservation planning throughout large areas in the western United States and Canadian Provinces. These needs coalesced over an eight-year period during which the Craighead Institute held an annual meeting and workshop at the B Bar Ranch near Livingston, Montana, a beautiful working ranch that for a time had been modified as a guest ranch. More recently, the guests have been conservation groups that are given use of the ranch at low cost by Maryanne Mott, the ranch owner. At the B Bar, scientists from an eclectic mix of disciplines, including mathematics, GIS, and biology, met informally and discussed current challenges in conservation. It provided fertile ground for cross-disciplinary innovation and collaboration. Many of the advancements in conservation planning in the Yellowstone to Yukon region have directly resulted from these workshops.

Over the same several decades, the Esri Conservation program provided GIS technology grants and support to thousands of conservation organizations around the world. The Craighead Institute received some of the early grants, and Lance Craighead and Charles Convis, editors of this book, became better acquainted. As the technical and analytical capacity of the Esri Conservation Program grantees grew, and the problems they addressed became more complex, their primary needs shifted from technical capacity to better science guidance. In survey after survey, conservation GIS groups asked for a user's guide to conservation planning, which led the Esri Conservation Program to join the Craighead Institute as the best venue for developing such a book.

Although many useful tools and approaches developed over the years were published in various formats, we felt there was a need to make these advances available to a wider audience of conservation practitioners, particularly those in other regions of North America and around the world who may not have access to original sources. Even at the stylistic level, we have tried to adopt plain-language conventions to make these concepts more accessible, such as our use of "data" as a singular term, in common use but unlike the strict scientific convention of using it as a plural term. We envisioned a book that a scientist or manager could refer to for guidance in developing a conservation plan for any area in the world, from the ground up, using the best available data, hardware, and software. The book would be a starting point that could be augmented and updated in the future as better data, tools, and approaches become available.

To that end, we sought to formalize the workshop environment that had developed over the years and produce a book that captures the current state of knowledge in various conservation approaches. We identified 13 or more areas of focus that were systematically addressed. The results are provided in this book. Here you will find an analytical framework for conservation planning using standardized methods and tools that we feel will improve accuracy, efficiency, and replicability of solutions and strategies developed to conserve ecosystems throughout the world, beginning in local communities.

The approach is based upon the focal species concept, which has been widely adopted by conservation scientists working to protect large landscapes. The primary assumption is that by protecting habitat for key species, we can preserve healthy ecosystems for all or most native wildlife. Fundamentally, this approach involves the modeling of "core" habitats needed to maintain populations of focal species and identifying critical linkage corridors needed to knit the region into a healthy integrated ecosystem, while allowing sustainable human economies and development in areas less critical for biodiversity. In many parts of the world, these core and corridor habitats still exist; they need to be identified, maintained, and protected. In many other areas, these habitats have been degraded or destroyed and will need to be restored for natural processes to again function effectively.

Our experience in conservation planning has taught us that the focal species approach works well at multiple scales, but most project funding cycles do not allow sufficient time for analytical methods to be fully developed and refined, or for results to be finalized and implemented. Lack of standardized analytical frameworks and methods creates inefficiency because of duplicated efforts in research and development among different groups, and because results cannot be compared between programs, making it difficult to integrate results from multiple scientists working in large regions over multiple scales. This book is our first step toward working together with common tools and approaches to achieve a common goal: the maintenance of biodiversity and ecosystem function.

Using the existing resources of the Craighead Institute as the foundation, we decided to more formally address conservation problems and produce this book. Start-up funds over three years were generously provided by the M. J. Murdock Charitable Trust to support staff, provide fellowships for visiting scientists, and purchase computer equipment and software to support program activities. The purpose of the fellowships and workshops was to stimulate creativity and produce innovative ideas in a supportive, interdisciplinary environment.

Subsequent workshops at the B Bar and other venues addressed conservation problems by bringing together some of the brightest minds working in the region and providing state-of-the-art computer technology to address the most pertinent scientific questions and challenges. Our goal is to incorporate existing theory in landscape ecology, mathematics, and other disciplines into research frameworks, methods, and tools with broad consensus among conservation planning practitioners. In a sense, our path converged with the concepts of geodesign being formulated by Carl Steinitz with support from Esri over the same time period.

As the process unfolded, we formed expert teams to develop practical and coherent guidelines for the development, interpretation, and implementation of conservation plans at multiple scales. These guidelines were then applied in real conservation projects in Montana and around the world, and the tools and approaches were modified with feedback from those projects to make them more useful in practical situations. Finally, many of the results were combined with other approaches developed by various authors into chapters of this book to present them to the broader conservation community. Much of the material developed during the latter stages of this project would not fit into the book. In addition, the discipline of conservation planning is changing rapidly, and new and better approaches will be available even by the time this book is published.

In an ever-changing world, we feel that the next phase of this book project will rely on the Internet. A web page has been set up at http://www.esri.com/conservationbook to link to additional materials and to update existing tools and approaches as they become available. It is our hope that the conservation community will take this starting effort and expand it.

As the manuscript was being finalized, Esri began a partnership with the Udall Foundation and US Department of the Interior to support a new National Conservation Planning System. We envision an American network of protected areas and conservation linkages based on a framework of public lands. We hope that this book can provide some guidance toward that effort, but more importantly can empower community-based efforts that can fill in the gaps in any nationwide effort.

Lance Craighead
Charles Convis

Acknowledgments

Bringing together a group of people as diverse as the contributors to this book, and folding their contributions into a single manuscript, was a daunting task. We wouldn't recommend it to anyone. It would not have been possible without the initial support afforded by the M. J. Murdock Charitable Trust, which enabled collaborations with the Craighead Institute, provided expenses and honoraria to contributors, and supplied staff and computer support for several years. We also would like to thank Maryanne Mott and the staff of the B Bar Ranch in Tom Miner Basin (Emigrant, Montana) for providing a place for this project to be formulated and developed. We thank all of the contributors for their hard work and patience with what at times was a slow process. The project was sustained by grants from the New-Land Foundation and Patagonia Inc., and by steady support from a number of individual donors, too many to mention, but exemplified by Carol Tolan and Gilman Ordway.

Many people at Esri made this book happen. Jack Dangermond shared the vision of this project and a personal commitment to the cause of conservation that carried the effort forward. Thanks to Esri employees Aileen Couture and EunKyeong Koh for their work on the book, and also to the Esri Press team.

Kris Ellingsen provided artistic help; John Mitchell and Yvonne Baskin provided invaluable editorial support; and Baden Cross, Mike Rock, Lauren Oeschli, and Rick Tingey assisted with modeling, geographic information systems (GIS) analysis, and cartography over the years on related projects. We especially thank Erica Fleishman, Paul Zwick, and Breece Robertson for reviewing the draft manuscript and helping us find the final form of this book. The Society for Conservation GIS provided a valuable forum for discussions as the book was being developed, and the Global Scholarship Program, directed by Sasha Yumakaev at Esri, has been a continual source of inspiration.

Additional support for the Craighead Institute by the Brainerd Foundation, The Bullitt Foundation, and the Wilburforce Foundation allowed the approaches outlined in this book to be applied in various conservation projects. We are particularly grateful to the Valhalla Wilderness Society, Wayne McCrory, and the late Colleen McCrory for helping apply the results of our conservation plans in the inland rainforest and Chilcotin Plateau of British Columbia. We would also like to thank our friends and colleagues living in the Madison Valley of Montana, where we have practiced and implemented many of the tools and approaches in this book.

And we couldn't have made it through the process at all without the love and support of our wives and families.

Lance Craighead
Charles Convis

1

Introduction: Shaping the future with conservation planning

F. Lance Craighead, PhD; and Charles L. Convis Jr.

This book, *Conservation Planning: Shaping the Future*, provides an introduction and guide for anyone planning for landscape conservation. The goal of a conservation plan is to maintain or enhance biodiversity, ecosystem function, and ecosystem structure, while accommodating human-oriented land uses that may be detrimental to that goal. At its simplest, creating a plan requires the conservation planner to make decisions about the most appropriate land use for a given site at a given time. To make these decisions correctly requires the best available information about the natural systems being conserved, particularly the biological processes that comprise the system. Those processes are represented by the plant and animal communities that an ecosystem supports as well as by the interactions between them. Understanding what these processes are, and how to use them to guide conservation decisions, is the purpose of this book. It outlines an approach using computer-based tools and procedures that are easily accessible via the Internet and the guidelines herein.

These guidelines focus on key natural species as indicators of biological integrity and complexity. They also incorporate best practices for understanding and maintaining ecosystems as integral communities that include human activities but are not dominated by them. Instead of the traditional anthropocentric approach to planning taken by most local governments—a focus on efficient allocation of goods and services for human economies—we explore various pathways to maintain ecosystem health and services as the first priority. The final plan, in most cases, is centered on a map of current or desired land status or use for a given area and scale. Because conservation planning occurs within a nested hierarchy of scales, from the global scale down to the subwatershed or parcel level, this book explores approaches that can be used at all levels. But it focuses primarily on local scales at which conservation plans can be effectively implemented by agencies, organizations, and individuals: for example, county government, forest districts, watershed councils, and local citizens' groups.

This book outlines a virtual journey through all the steps necessary to complete an ideal (at least at this point in time) conservation plan. Viewed through the lens of geodesign (design science based on geographic information systems [GIS]; Steinitz 2010), this approach incorporates several of the nine basic methods that Steinitz described as the basis for modern landscape planning (see sidebar "Nine fundamental geodesign planning approaches" later in this chapter). Developing the plan requires the use of computer-based tools that provide a spatial representation of the biological processes comprising the ecosystem. It is likely that any real-world application will not be able to use every step and tool available. This book should help identify the best tools and approaches available for a conservation planner's specific area of interest. It is important to realize that planning is an iterative process, and each iteration can improve upon the previous version. Do not be deterred by uncertainties: even a very simple conservation plan is better than no plan at all, as long as it is based upon the best available science. The process can be revisited in the future and improved.

Geodesign is one of the fastest growing frontiers in conservation science. According to Steinitz (2010), six fundamental questions must be addressed in any planning or design situation:

1. How should the landscape be described?
2. How does the landscape operate?
3. Is the current landscape working well?
4. How might the landscape be altered?
5. What predictable differences might the changes cause?
6. How should the landscape be changed?

Steinitz also describes nine fundamental methods by which plans are created (see sidebar "Nine fundamental geodesign planning approaches" later in this chapter). This taxonomy is useful because many plans involve a mixture of approaches, each of which has specific strengths and weaknesses as well as specific capabilities for how much or how little it can be participatory, collaborative, or democratic. These are important implications for conservation planners to consider as they go about their work. These methods are not mutually exclusive, and as a conservation plan is developed, elements of several approaches may be incorporated.

The most important step in any journey is the first one. Once that step is firm and solid, take the next one. In this introduction, we briefly outline the topics (steps) that are explored in detail in the chapters to come, and explain how they have developed over time. Each chapter explains why the topic is important and how to apply the tools and procedures as part of a larger plan.

At first, the chapters in this book were planned to go from the general to the specific. We envisioned that readers could begin each topic with a broad overview and continue reading as deep into the subject matter as their interest and understanding allowed. Then they could skip ahead to other chapters, perhaps to return later, with more specific questions to answer. In the end, the chapters did not all fit into an easy formula, which is a testament to the individuality and personal visions of each of the authors. The same approach, however, should be useful: explore with curiosity, and if you get bogged down, skip ahead to more interesting fare and read to the end of the book to get a sense of the subject matter. Then, especially if you have an application in mind, focus on the wealth of detail and related material the authors have provided. In addition, we encourage readers and other conservation planners to draw from, and contribute to, the website we have established at http://www.esri.com/conservationbook. The website is our beginning attempt to address one of the needs that Frank Davis points out in chapter 15: "Most conservation planners lack the resources, formal training, or sufficiently strong incentives to employ best data management practices and to publish their data and information in a way that facilitates reuse. I will simply lodge a plea for allocating more resources to training and open sharing of data and information to continue broadening the collaborative network of conservation planners." As we learned in the eight years of B Bar meetings that led to this book, the other key to successfully navigating a plan is creating a community effort. Planning cannot be done alone; it is a

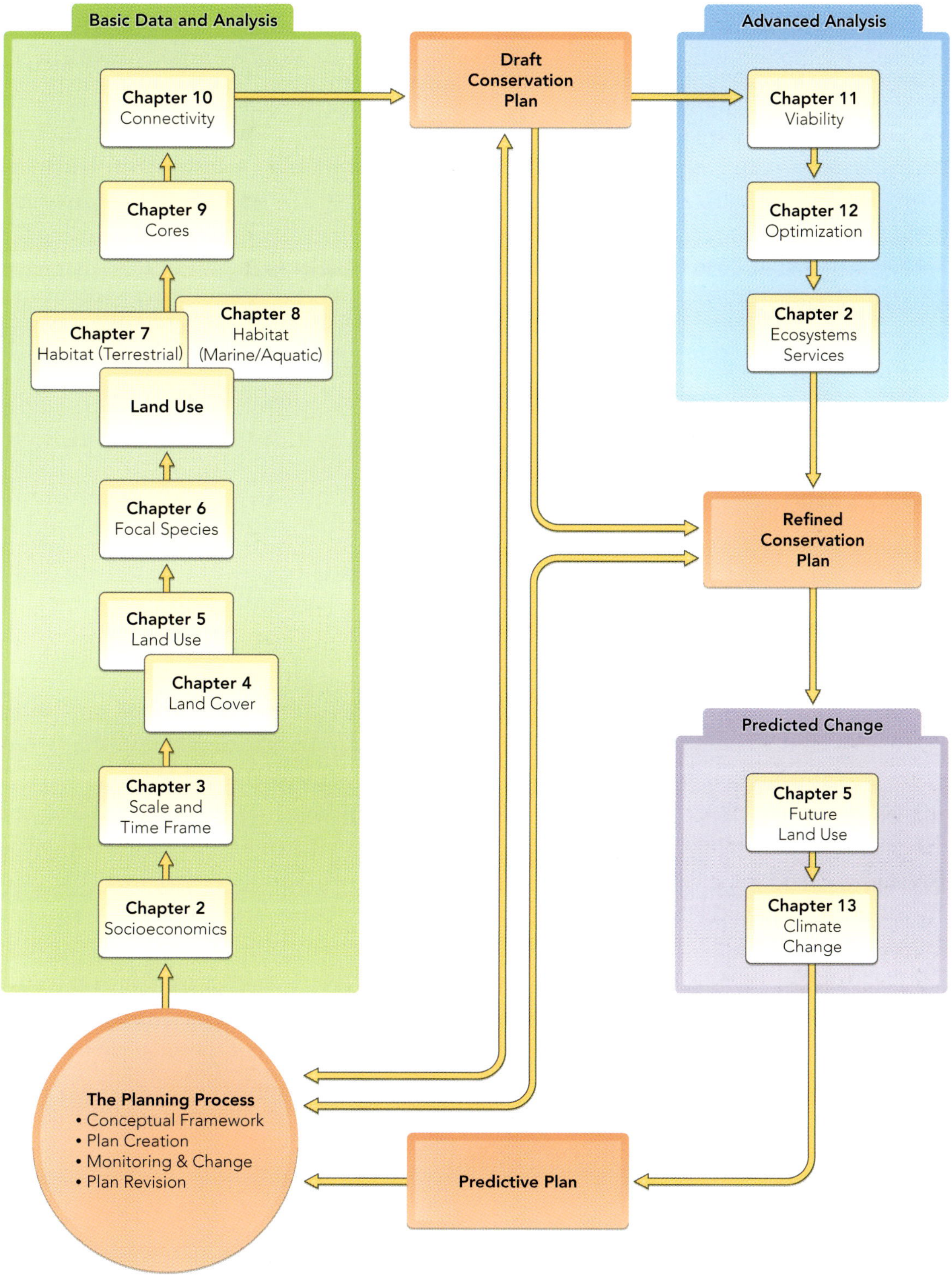

Figure 1.1. A conceptual workflow of the conservation planning process. Arrows lead through a series of topics that correspond to the chapters in the book. The steps to a basic conservation plan are enclosed in the green box. This process can form a conservation plan that can be refined or revisited. Advanced analyses that can improve the basic plan are enclosed in the blue box. Predictive analyses that can help anticipate change in the environment for inclusion in the plan are enclosed in the purple box. Because it is iterative, this flow chart loop is actually a spiral that coils upwards in time, using newer, and hopefully better, data and techniques.
Figure from Lance Craighead and Charles Convis.

community activity requiring a commitment from the stakeholders affected, to the experts and advisors, to decision makers. Our hope is that this book, the website, and additional resources still to come will help to build the larger conservation planning community as a resource for local communities.

This process can be envisioned as a flow chart such as figure 1.1. The process starts within the circle at the bottom as goals and approaches are conceptualized. Part of that step includes reading this book. Some

Figure 1.2. The conservation planning journey. To develop a conservation plan requires data representing the natural world, including humans. The data is organized and manipulated using such tools (represented by the pathway) as GIS and mathematical models into an architecture that becomes more refined and accurate as the steps proceed. The conservation plan (represented by the distant plaza) that is developed is an abstraction of the natural world as it is experienced by key animal species (focal species). Humans experience the world in different ways that are influenced by land use and economies. Spatial units in the plan are assigned values that represent their importance to ecosystems and native species. Once a plan is developed, it can be used to guide land-use choices in the present and to predict how those choices will play out in the future (represented by the pathway in the distance). Kristine J. Ellingsen © 2010.

questions a planner might ask could range from very broad—"What do we want this landscape to look like in 50 years?"—to much more specific—"How can we maintain enough habitat for pygmy rabbits or grizzly bears?" or "Where can we develop wind power without endangering sage grouse?" The question will determine the scale of analysis, the type of data needed, and the methods to be employed. The main focus of this book is on planning for focal species at finer scales of analysis, but the approaches provided are applicable at any scale. As planners proceed step by step through the planning process, they eventually return to the first step when they can refine the questions, use different or newer data, and apply other methods.

Readers can also think of the process of conservation planning as a journey. It is a journey of many steps and can be more artistically envisioned as in figure 1.2. As in the flow chart, it is possible to imagine coming back to the start to begin a new iteration of the process. However, in today's world, irreversible changes are occurring to the landscape so rapidly that a more realistic image may be that a conservation planner has only one opportunity to influence the changes to come.

Regardless of how the conservation planning process is envisioned, it is a journey that can be made by any dedicated group of concerned people. One goal of this book is to level the playing field so that rigorous and scientifically defensible local plans can provide alternatives to top-down approaches that local communities may face from industry or government. Historically, resources have mainly been available for planning from the top down because money and power have been concentrated at the top of social hierarchies, and planning was an expensive proposition. However, in modern societies, the majority of decisions that directly affect landscapes may be influenced from the top but are often made at the bottom, at the level of counties, towns, villages, and landowners.

The book describes a bottom-up approach that has been used successfully by practitioners around the world. Our goal is to provide resources and guidance in one source so that conservation planners do not have to develop their own process by trial and error. Many more tools and approaches exist than can be included in a single book, and they are dynamic: constantly improving as the science progresses. Therefore, the contributors to this book, with support from Esri, have built the website, http://www.esri.com/conservationbook, as an accompaniment where additional, and continually updated, data, information, and tools can be accessed.

This book focuses on the mechanics—the nuts and bolts—of conservation plans. Such planning takes place within broader social contexts, and to be successful, a plan needs to be implemented as it is being designed, through processes such as regulation, legislation, consensus, policy initiatives, land protection, good stewardship, economic incentives, voluntary compliance, and neighborly agreement. The breadth and detail of implementation are beyond the scope of this book, but are outlined by Ban et al. (chapter 8, table 8.1) as described in Pressey and Bottrill (2009), and discussed in more detail by Chan et al. (chapter 2). Implementation of the results of a conservation plan needs to occur concurrently with the development of the plan itself and be driven by stakeholders. Approaches to ensure that the science is applied on the ground, implementation tools, strategies, and tactics, are also discussed in more detail on the accompanying website.

We believe that local stakeholders, particularly people in local communities, have a deeper interest in conservation planning than those in distant positions of authority. They live in, and are more dependent upon, the quality and health of the natural environment around them. We believe that there has been a shortage of guidance and resources to help local residents and governments develop good conservation plans that can best represent their point of view and values. Now that computers and software tools are universally available, and access to data and tools is available through the Internet, it is possible to enable virtually anyone to protect and manage the landscapes they love, in ecologically effective ways as a part of larger, functioning ecosystems. This is the new frontier of conservation planning.

Nine fundamental geodesign planning approaches

Descriptions of these approaches are based on work by Steinitz (2010).

1. *Anticipatory.* Experienced planners use this approach when they are confident in their knowledge and able to create a new plan solely from their own understanding and experience. To implement the plan they typically work backwards from the desired goal to the current conditions to deduce what steps need to happen. This approach is most common on smaller, short-term, local-scale projects.
2. *Participatory.* In this approach, a group of people prepares individual designs, each reflecting the designer's vision, expertise, and knowledge. Then all of the designs are integrated using a GIS to analyze where they agree and differ. The group arrives at one final design by consensus. This approach is common where groups or communities are the beneficiaries of a plan, and the design needs to be done by committee rather than by one individual. Implementation is achieved also by working backwards from the agreed goal to the current conditions in a deductive fashion.
3. *Sequential.* Planners may use this approach when they are not confident about an end design but are certain about the individual steps, elements, or requirements that must be addressed to get from the current conditions to the desired future state. This type of planning frequently occurs when committees cannot visualize a final plan but instead must address each of the critical issues individually, moving on to the next issue only when the prior one has been resolved. This committee-based approach is common on larger plans and often involves 20 to 50 steps, and can take much longer to accomplish.
4. *Combinatorial.* Planners may use this approach when they are not confident in either a design vision or the steps to get there and must account for many options at each design step, creating a huge combination of possible plan alternatives to sift through. To simplify their choices, they use inductive reasoning to simultaneously evaluate the combinations of options that may address the issues best. Often, a starting point is to identify a few extreme cases as the boundaries of the possibilities and then to examine different combinations of the options that create plan alternatives. Each alternative is then evaluated and ranked, and a final candidate is selected. An advantage of this approach is that it requires the evaluation of the most important assumptions early on, which helps avoid serious mistakes further into the planning process. A disadvantage is that it leads to the creation of a large number of alternative designs that must be compared and evaluated.
5. *Constraining.* This approach is similar to the combinatorial one, but it seeks to simplify the planning process by limiting the number of alternative designs. It requires that each major assumption be further evaluated in a Delphi method (expert opinion) process to refine the range of options so that a smaller number of design combinations is possible. This method is useful for enhancing efficiency by limiting the potentially large number of possible combinations when using a committee-based plan that includes direct public participation.
6. *Optimizing.* This approach assumes that the planner knows the major requirements and their relative importance and limits, as in the constraining approach, but goes further by assigning cost

values to each requirement and conducting optimization analysis that identifies which alternatives are the "cheapest" for the outcomes they might produce. It assumes that the planner can assign costs and values effectively to the individual requirements, which turns out to be difficult and depends greatly on the planner's knowledge of the community for which the plan is intended. A key difficulty is ranking and assigning values without first being able to establish their context by studying a proposed comprehensive plan. This approach is commonly used in evaluating potential reserve and corridor designs in terms of their acquisition and management costs, and the relative efficiency by which conservation goals can be achieved for given expenditures of scarce resources, as discussed in chapter 12.

7. *Rule-based.* This approach is similar to the optimizing approach in that planners are not working with a finalized design vision. Many of the major requirements can be expressed only as a set of formal rules or computer algorithms governing the design. Outcomes of algorithms need to be tested and will depend on initial states and inputs. Rules can be applied that currently do not exist, allowing for simulation and testing of different what-if scenarios. One useful technique of rule-based approaches is to use algorithms that translate design outcomes into a common simple metric to facilitate comparisons, such as economic value or replacement cost of ecosystem services. The challenge with rule-based approaches is in designing sequences of rules so that they accurately describe all relevant parts of the system or problem. Once designed, rule-based models are then easy to run and refine, even for very large problems over many different scales and scenarios.
8. *Agent-based.* This is a type of rule-based approach that describes rules not for the whole system, but only for individual computer-generated "actors." The sum of the individual decisions of many actors then defines the system. The goal of this planning exercise is to find the best solution for a number of independent actors carrying out tasks that represent desired activities the final plan will support or enhance. Each independent actor conducts tasks in response to a simulated computer environment, and the combined results of many different actors create an overview of an allocation that best permits them to meet their needs. One strength of this approach is that it can use a relatively small number of reasonably simple rules to model the behavior of a single "actor," and then let the computer randomly generate millions of different tasks and summarize their final patterns in a fairly complex final plan. It is suited to larger, multifaceted planning situations. As chapters 10 and 14 discuss, it is also a commonly used technique for evaluating wildlife movement over the landscape and designing migration corridors.
9. *Mixed agent-based and sequential.* The rules that define a successful agent-based model can themselves arise from other kinds of planning approaches. In this approach, a sequential change model (or a constraining or combinatorial approach) is used to generate rules and evaluation inputs that can drive an agent-based examination of the cumulative effects of many individual actors. This examination can extend through successive time stages to model either historical growth or future projected growth, using a sequential modeling approach again. This overall approach is called "mixed" because it combines elements from several basic modeling approaches.

What is conservation planning?

Recent definitions of conservation planning include that of Pressey et al. (2008): "Conservation planning is the process of locating, configuring, implementing and maintaining areas that are managed to promote the persistence of biodiversity and other natural values." (See also Margules and Pressey [2000] and Ban et al., chapter 8.) Noss et al. (1997) explain that "conservation planning is really as much an art as a science." Groves (2003, 2008) provides two categories of conservation planning: (1) the "scientifically-based, systematic methods for establishing conservation priorities," and (2) "development and implementation of methods for achieving on-the-ground conservation results or in measuring the effectiveness of conservation actions," recognizing the disconnect between developing a plan and actually implementing it. In the sense of this book, which focuses on development, conservation planning can be defined as the use of the best scientific information to ensure that natural systems are conserved as human-induced change takes place.

Conservation planning is closely related to several other ecologically oriented fields and disciplines with similar names, and using similar theories and methods, such as ecological planning, biogeography, landscape architecture, landscape ecology, suitability analysis, range management, and comprehensive land-use planning. It differs, often in subtle ways, in that the primary goal is to maintain populations of native plants and animals (including humans) in functioning ecosystems. Conservation planning addresses patterns (or structure) and processes (or systems) of biodiversity (see also Ban et al., chapter 8). Often there is incomplete knowledge of these elements of biodiversity, and the best science consists of relying on indices or surrogates of the underlying ecological principles.

A good conservation plan must address current economic needs of human societies but also include the economic impacts of damage to vital ecosystem services, such as clean water, in establishing where development can and cannot occur (see Chan et al., chapter 2). Conservation ought to have both biodiversity- and human-oriented objectives because humans are an integral, and extraordinarily influential, component of ecosystems.

Conservation planning is site-specific, scale-specific, data driven, iterative, adaptable, and powered by human desires and understanding. Change is inevitable, but people can direct many of the human dimensions of change: we can guide or control human activities so as to protect or maintain natural systems. Because humans are a part of natural systems and human density varies across the landscape, we are faced with a spectrum of environmental conditions that have been affected to varying degrees by humans. We cannot afford to conserve only intact, pristine landscapes. We also need to consider the bits and pieces of natural systems that still remain within human-altered landscapes. Although this book hopes to provide the best scientific tools and techniques to use with the highest quality base data, we want to emphasize that some approaches can use whatever good data is available. To reiterate: even a very simple science-based conservation plan is better than no plan at all.

Conservation planning is as much a process for balanced landscape decisions as it is a blueprint. That process begins with the best available data and techniques, and then leads to collaborative discussion and development of options leading to working maps, followed by implementation of the plan. As new and better data or techniques become available, or increased interest and funding is developed, the plan can be refined and the process can be modified. The plan evolves or adapts in the sense that it gets progressively better at representing the real world and managing the role of humans in that landscape. In particular, conservation planning has benefited from geographic information systems (GIS). GIS are "computerized systems for the storage, retrieval, manipulation, analysis, and display of geographically referenced data" (Mark et al. 1996).

GIS use has revolutionized the scope and power of conservation planning because of its ability to accommodate, integrate, and analyze physical, biological, cultural, social, economic, and political data.

The goal of conservation planning can be described as shaping the future. Designing a future with intact ecosystems and balanced use of resources is an optimal goal for all people on the planet as well as for its biodiversity. However, strong economic incentives exist for individuals to exploit resources for personal gain at the expense of the common good (Hardin 1968). In modern societies, planning that interferes with profit will often experience resistance from vested interests; conservation planners may need to defend their choices in courts of law and in courts of public opinion. We feel that the strongest defense is to have the best available science supporting decisions. This is another goal of this book: to enable conservationists to apply the best available data and scientific tools to land-use questions, and to answer those questions with defensible results.

The modern science of conservation planning has grown throughout its history as new tools and approaches have been developed. Each new advance has laid a foundation for refinements and new tools to build upon. This history is mirrored in the organization of chapters for this book. If we return to the flowchart of the process (figure 1.1) and the metaphor of conservation planning as a journey of many steps (figure 1.2), we can explore each topic or step as a chapter in this book.

To begin modern conservation planning, the literal first step should be to walk across the landscape so that you can see the environment and talk to local people. This is the best way to learn what the landscape is like. The more direct experience you have with the topography, plants, animals, and human stakeholders, the better. No amount of computerized data can provide the same understanding of a landscape as first-hand knowledge (although in many cases this may not be possible). The next step is to boot up a computer. Within a GIS framework we begin to see the landscape from a variety of perspectives that complement on-the-ground knowledge. Then these perspectives can be combined.

If we focus on North America, one of the first things we might notice is that the environment we have inherited, and which we hope to plan sustainably for the future, is the result of a series of activities, initiatives, policies, and legislation that has produced a patchwork of land uses (figure 1.3). This mosaic of private lands, public lands, protected areas, urban areas, industrial parks, agricultural landscapes, and all other gradations of land use—crisscrossed with a complex network of roads, railroads, transmission lines, pipelines, seismic trails, off-road trails, hiking trails, and myriad other linear infrastructure features—has been so heavily affected by human activities that many of its natural systems no longer function well. Despite this incredible amount of human alteration, though, North America contains many of the healthiest and most intact landscapes left on the planet.

Whom is this book for?

This book is a hands-on guide that describes what the contributors feel are the best scientific methods to be used at each stage of a conservation plan. It should be most valuable for planners in nongovernmental conservation organizations and in city, county, and state agencies; government professionals in land-management

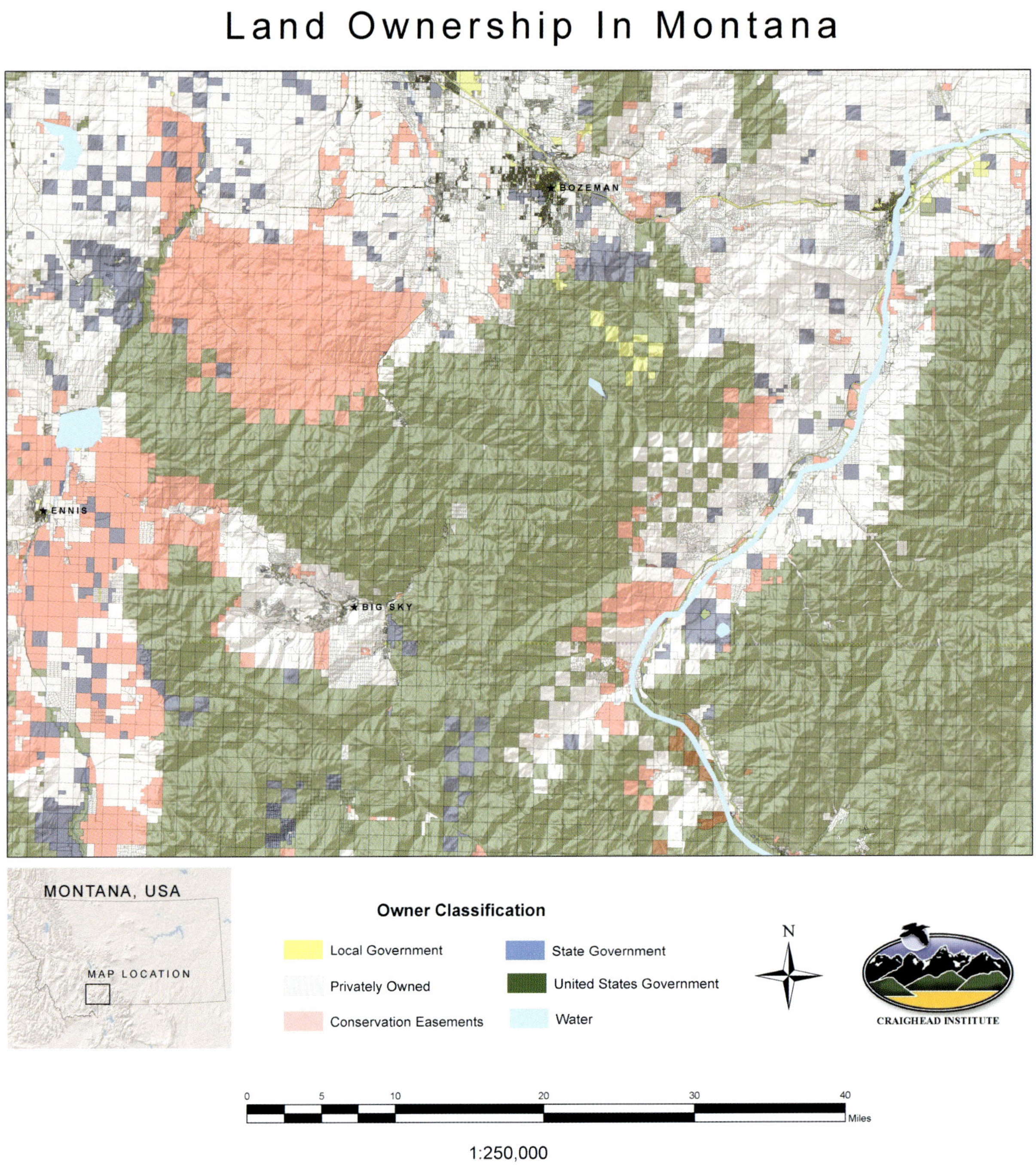

Figure 1.3. The mosaic of land ownership near Yellowstone National Park. Through the Pacific Railway Act of 1862, 10 square miles of land were granted for each mile of railroad laid. The land was distributed in a checkerboard pattern, with every other square mile being granted. The railroad companies received over 175 million acres of public land; more than one-tenth of the entire country. Much of this ownership pattern remains today, and it still influences how conservation efforts proceed. One of the most noteworthy trends in modern conservation is the role of private landowners who voluntarily protect their lands for the greater public good through conservation easements (red areas). Digital elevation model from US Geological Survey; Montana Land Stewardship from MT NRIS and Montana PLSS; Water from State of Montana; Roads/Cities from MT NRIS.

agencies; and students interested in conservation or planning. Although applicable anywhere in the world, this book may be particularly useful in developing countries. It should also be useful for activists concerned about the adequacy of existing land management, policy makers (including at the county, state, and national level), GIS technicians involved in conservation projects, conservation biologists, wildlife biologists, foresters, landscape architects, scientific consultants for real estate developers, or anyone else wishing to understand and actively guide changes in land use. The book is designed primarily for those working at a local scale but within a broader framework of conservation planning.

We hope that this book will be used as a primary or supplemental text for college courses. Suggested courses include landscape ecology, conservation planning, parks management, protected areas management, and wildlife modeling. This book may be the first introduction for many readers into a community that is working together with common tools and approaches to achieve a common goal: the maintenance of biodiversity and ecosystem function.

Implementing the methods outlined in this book requires a good working knowledge of GIS in most cases. However, the book should also be useful for other people, perhaps working as part of a conservation planning team, with expertise in biology, project management, land management, and so on. There are a number of large international conservation organizations employing conservation planning methods at different scales: The Nature Conservancy, Conservation International, World Wildlife Fund, The Wildlands Project, and the Wildlife Conservation Society, to name a few. Noteworthy conservation plans have also been developed by smaller groups like Round River Conservation Studies or the Southern Rockies Ecosystem Project. This book is intended to fill in the gaps that those programs do not cover. If The Nature Conservancy is developing a conservation action plan (see Unnasch and Karl, chapter 3) for your landscape of concern, you may not need this book. However, if you want to help guide conservation for any of the millions of areas that do not yet have scientifically based projects in place, this book is for you.

Where do we go from here?

As concerned citizens of our only planet, we can plan human-impacted landscapes that are healthier than those we have now; and more likely to ensure the clean air and water we will need for survival. In the same way that human beings survived through millennia by saving important resources for later, modern human societies can decide to plan for a rational, sustainable future and not be subject to the whims of chance, self-interest, and greed. We are at a point where we can make conscious decisions about how the land should look now and in the time of our grandchildren. Conservation planners at all levels ask the same fundamental questions: What do we want this landscape to look like in the future? How much do we need to protect to ensure that natural systems and ecosystem services do not collapse? What activities can we undertake without tipping the balance? These are the questions that conservation science and planning has repeatedly addressed for the past four decades or so. These are the questions that we will continue to address, and to answer with progressively greater confidence in the future.

An overview of the process

The modern science of conservation planning encompasses tools and techniques from many disciplines, and it benefits from advances ranging from software to satellites, and genetics to geography. In these sciences, each new advance facilitates new refinements, new processes, and new tools. This interlocking history is mirrored in the progression of chapters of this book.

The tools that modern practitioners have found most useful for conservation planning are presented herein in a logical progression. The book begins with a section containing chapters on foundational tools and data, and explores ways to use these basic concepts to examine complex representations of the real world, including habitat and connectivity. These are the building blocks of a plan and comprise chapters 2 through 10.

The book then discusses ways to include larger issues, such as population dynamics, efficient allocation of space, and human economies, to refine the plan and ensure that the needs of both people and wildlife (and thus biodiversity) are met, but not grossly exceeded. This section includes chapters 11 and 12, along with much of chapter 2.

Finally, dynamic changes in the environment can be factored into a plan that synthesizes what we know and projects those trends into the future. This section includes chapters 13 and 14, along with sections of chapter 5.

The details are presented by a series of different authors who have painstakingly edited their chapters in concert with the rest of the book to present a logical flow of work in a consistent fashion.

This book takes the approach of trying to view the world from the point of view of key animals in ecosystems (focal species). Ecosystems exist because of the relationship of structures, the physical architecture of an ecosystem, with processes, the dynamic flow of interactions, energy, and materials within the ecosystem. Processes happen because physical structures (such as forests, grasslands, or fungi) exist to support them, and structures exist because processes operate to build and maintain them. This distinction is important for considerations of scale, because physical structures have a spatial scale implicit in every description of them, and processes have a temporal scale implicit in measurements or descriptions of their operation.

What geographers and ecologists know is that the understanding of ecological processes and structures at a given scale in the ecosystem may have little or nothing to do with the structures and process that govern that same ecosystem at other scales. In the framework of geodesign, Steinitz (2010) has identified at least three different scales for planning that represent distinctly different methods and dynamics, which any plan must specifically account for to avoid failure:

- 1:5,000—the scale of individual, local projects
- 1:25,000—the scale of municipalities
- 1:250,000—the scale of watersheds and regions

Scales are hierarchical in that processes described at one scale can influence processes occurring at other scales, such as the obvious influence of macroclimate on microclimate. Less obvious are micro-influences that matter at the watershed scale, such as beetle lifecycle biology that allows an outbreak to destroy entire forests. Although this book will focus on local to watershed scales, current conservation plans increasingly are required to account for the global-scale dynamics of climate change (chapter 13).

Developing a basic plan

The basic assumption is that an ecosystem is functioning well if it can continue to support key species that require a comprehensive collection of habitats. The goal of conservation planning is to keep an ecosystem functioning in the face of human-orchestrated change. We begin by trying to understand the social and economic context within which the plan must operate (chapter 2). As we think about which species are most important in a landscape, we begin by focusing on a particular scale or scales (chapter 3). With a scale in mind, we consider land-cover classification and choose datasets that map the areas we want, usually through remote sensing (chapter 4). Similarly, we assemble datasets that capture human activities on the landscape (chapter 5). As we do this, we need to systematically select a suite of species that most efficiently represent the biological systems (chapter 6). Accurate maps of existing vegetation and other land cover are then used to predict the spatial occurrence of habitat for the selected animal species, using habitat models (chapter 7). An analogous approach is used to identify important elements in aquatic and marine systems (chapter 8). Although we have chosen a scale initially to determine what data to use, at each step of analysis there may be additional issues of scale, which must be addressed carefully to ensure that data at the correct resolution is matched with an appropriate analysis scale to produce a meaningful result. Models of predicted habitat are then used as surrogates for species distribution and are used to develop models and maps of "core" habitat (chapter 9). Suitable habitat for one species can be combined with models for other species to delineate important core habitat for a suite of species. Animal movement between core areas can then be assessed and routes can be mapped using connectivity approaches (chapter 10).

At this point in the process, a conservation planner would have a rough blueprint for a network of core areas and linkage corridors that define a conservation area, a nature reserve, or whatever term is desired (the draft conservation plan in figure 1.1).

Refining a conservation plan

Once habitat and core areas have been identified, they can be analyzed in terms of their ability to maintain a given population and allow it to persist, with the tools of population viability analysis (PVA; chapter 11). Multispecies PVAs can help determine the size and configuration of core areas and linkages that can support a suite of focal species. Spatial optimization approaches can be used with a conservation plan as guidelines to determine current minimal areas that meet specific thresholds or targets in terms of number of species protected, percentage of habitat protected, or other metrics (chapter 12). The linkages can also be optimized to identify minimal areas that meet the desired biological thresholds. Similarly, economic goals can be identified, both for the values of ecological processes to humans (ecosystems services valuation), and for extractive or competing uses of space and resources for human economic activities (chapter 2). Any or all of these analyses can result in a more refined conservation plan.

Planning for the future

Finally, reserve designs or conservation areas can be assessed in light of climate change (chapter 13) and land-use trends (chapter 5) to determine what the habitat will look like in the future and what other areas need to be considered to maintain species and ecosystems. Although the landscape in the future, after climate change, is uncertain, a few certainties exist. Focal species in general tend to be sensitive to human activities and developments; providing adequate habitat requires the management of human impacts in areas large enough for population persistence. Even if the landscape comes to support different habitats and focal species in the future, their conservation will still require areas of limited human impact. Thus, the first step in planning for climate change is to identify areas of little human development and to maintain that light human footprint into the future. We can preserve the stage even though the actors may change as the play unfolds.

We can proceed with conservation planning under existing conditions with the understanding that some human aspects of those conditions can, and must, remain somewhat static as the climate changes. We can also predict many of the changes in land use that are likely in the future, in concert with natural changes due to global warming.

Understanding the process

A conservation plan, as we all know, is developed in context. The biophysical context—topography, vegetation, climate, and wildlife—determines the appropriate data and approaches that are used. The social context—public and private lands, government and nongovernment agencies and organizations, stakeholders, economies, values, and beliefs—will determine future anthropogenic changes and also the best approaches to implement the plan.

In chapter 14, an overview of some of the basic scientific modeling approaches is presented in a stepwise process that demonstrates how increasing complexity can be incorporated. The disparity between the modeled environment and the real world is discussed, and the assumptions inherent in common models are articulated. Models are illustrated with examples of elephant and snow leopard conservation that demonstrate many of the challenges faced in parts of the world that do not have the infrastructure and access to information that exists in North America. Chapter 15 summarizes the book as a whole and its context within the changing world of conservation planning.

Implementing conservation on the ground

The social context is an integral part of the scientific approach. Choices that are made in the scientific process need to reflect the fabric of the communities in which the resultant conservation design will be interwoven. We live in a society that views the natural world strongly along a continuum of values, with extreme viewpoints of utilitarianism (human dominion over nature) at one end and protectionism (humans as a destroyer of nature) at the other. Between the two value systems lies an infinite number of viewpoints that see humans as a part of nature: as a steward, an interloper, a bystander, a passive participant, an active recreationist, and so on.

To be successfully realized in a democratic society, a conservation plan must be embraced by the community that it encompasses. It must be adopted by a quorum of stakeholders who will help guide the process and work to make it happen on the ground. To this end, the scientific choices that are made must somehow take into account the social values and beliefs that will determine whether—or not—the science is accepted. A conservation plan must provide something good for almost everybody, and involve the key stakeholders so that they feel a sense of ownership. At the same time, the requirements for maintaining biodiversity must not be compromised.

Modern conservation science has been forged in the crucible of our society and reflects a diversity of interest: currently, many core protected areas with minimal human alteration are juxtaposed with highly urbanized or otherwise altered landscapes. In between are natural habitats with varying degrees of human influence. The challenge is to provide enough habitat to maintain ecosystem processes and functions, vegetation, and animals, even though the habitat is embedded in a matrix of human-altered landscapes. This book should provide the tools to achieve this precarious balance.

In the future, the landscape will be different, and the answers to these questions will also change. We can anticipate some of these changes, but not all, and the overall goals will remain the same. Conservation of nature is based upon four well-accepted tenets: (1) to represent ecosystems across their natural range of variation in strictly protected areas, (2) to maintain viable populations of native species, (3) to sustain ecological and evolutionary processes within an acceptable range of variability, and (4) to build a conservation network resilient to environmental change, particularly climate change (after Noss and Cooperrider 1994). As the climate changes, species will shift in response, and species assemblages will change. Thus, the ecosystems and natural communities themselves will change, but their range of variation should still be represented if protected areas are created with that degree of variation taken into account. We feel that basing reserve designs, and the conservation plans that include them, on focal species rather than primarily on representation of current communities, will be more responsive to the coming changes.

The steps in the journey

The authors of chapters in this book collectively represent centuries of experience in developing and using the tools and processes they describe. Each author brings a unique perspective and voice to the subject. The editors have endeavored to find a balance between presenting the material in a standard and consistent fashion and allowing authors to explain things in the way that they best express themselves.

Chapter 2 Integrating conservation planning with human communities, ecosystem services, and economics

Kai Chan, Natalie Ban, and Robin Naidoo begin by framing conservation planning in the context of social considerations and the importance of stakeholders in implementation. Because people are integral parts of ecosystems, their needs and effects are critically important to consider, even as conservation plans may focus on preservation of other species and the maintenance of systems that humans tend to degrade. The authors explain how the values of ecosystem services can be balanced with the need for human economies that occupy the same landscapes: how to incorporate ecosystem services into conservation planning.

Chapter 3 Scale and conservation planning

Robert Unnasch and Jason Karl discuss the relationship of scale to the ecological factors that are important for conservation. Scale is a fundamental property of how ecological systems, structures, and processes are described and understood. Conservation plans are developed within a nested hierarchy of scales; the appropriate scale depends upon the conservation goals.

Chapter 4 Land-cover data: The foundation for conservation planning

Thomas Olenicki discusses the foundation of the process of building a conservation plan. Once a particular scale (or scales) is determined, the next step is to explore land-cover classification, which is used to map large areas through remote sensing. Many land-cover datasets are available, and this chapter provides guidance in choosing the appropriate data. He outlines a land-cover classification scheme that should elucidate vegetation patterns more relevant to wildlife habitat than current approaches. Additional guidance for use of land cover is provided in chapters 5, 8, and 14.

Chapter 5 Integrating land use and landscape change with conservation planning

David Theobald explains how reserve designs or conservation areas that are planned under existing conditions can be assessed in light of human development. This chapter focuses on how to make use of the tools and approaches that are readily available. He discusses the suite of important land uses, including urbanization, agriculture, energy, recreation. These occur any place human use is dominant. The chapter then addresses the broader view of land-use analysis, which is about application: what is practical to do depends upon the spatial extent and the time frame. Finally, he discusses how predicted changes in land use can be evaluated to guide future conservation decisions.

Chapter 6 Selecting species as targets for conservation planning

Brent Brock and Eric Atkinson explain how we can use species as an index to more complex ecological processes and functions, and how to choose the best suite of species for a particular area. They provide a review of focal species approaches, many of which they have used and experienced in conservation planning processes. Finally, they outline an approach that, compared with other methods, they have found improves the efficiency of analysis without sacrificing objectivity or accuracy.

Chapter 7 Identification of habitat and assessment of habitat quality for conservation of terrestrial animals

Brett Dickson, Steven Sesnie, Erica Fleishman, and David Dobkin explore ways to represent habitat for terrestrial species. They discuss a wide variety of habitat models used to predict the spatial occurrence of habitat for selected animal species based on accurate maps of existing vegetation and other land cover. They then provide an example in which they assess habitat values using an occupancy modeling approach. This approach can be used with any species for which suitable data exists.

Chapter 8 Marine and freshwater conservation planning: From representation to persistence

Natalie Ban and her colleagues, Stephanie Januchowski-Hartley, Jorge Alvarez-Romero, Morena Mills, Bob Pressey, Simon Linke, and Debora de Freitas, explore ways to represent habitat for aquatic and marine species, and discuss a range of issues, from land cover to connectivity, that require different ways of thinking about conservation in much more dynamic systems than on land. Because water-based systems have fundamental differences from terrestrial systems, this chapter revisits many of the topics of other chapters to understand them better from an aquatic or marine perspective. A full exploration of water-based systems is beyond the scope of this book. Because the major focus of the book is terrestrial systems, the authors provide a necessarily broad overview of marine and aquatic approaches with guidance to more in-depth learning.

Chapter 9 Identification and mapping of habitat cores

We return to the terrestrial world, where Richard Church discusses ways to think about and identify which areas of habitat are the most important. “Core” habitat is a commonly used but context-dependent concept, somewhat difficult to define and measure in concrete terms. In most conservation planning, “core” has been used to reflect a sense of intact habitat that is unaffected by human disturbance or other neighboring habitat. This chapter explores various techniques for using models of habitat as surrogates for species distribution that are then used to model, identify, and map core habitat.

Chapter 10 Assessing habitat connectivity

Peter Singleton and Brad McRae explain ways to examine the landscape that lies between intact blocks of good habitat. They explain why connectivity is important and introduce spatial analysis concepts and methods for evaluating habitat connectivity and landscape permeability. The chapter gives an overview of a few of the more common methods of landscape permeability analysis and discusses their strengths and weaknesses. They explain how a team approach can be used to assess connectivity, and they discuss ways to communicate the results—a key final step that should be incorporated into each stage of the conservation planning process.

Chapter 11 Conservation planning to ensure viability of populations and metapopulations

Jessica Stanton and Reşit Akçakaya discuss conservation planning from the perspective of preserving natural populations of species: an important (if not the ultimate) goal of conservation planning is to prevent species extinction. Assessing viability is the most direct way of measuring how close to this goal a conservation plan is, or will be in the future when the planned conservation actions are in place. Viability is the likelihood that a species (or a population) will remain extant in the future. This chapter provides guidance in the use of population viability analysis (PVA): a method for calculating this likelihood, either under current conditions, or under assumed future changes. Multispecies PVAs can help determine the size and configuration of core areas and linkages that can support a suite of focal species.

Chapter 12 Optimization models for reserve site selection and design

Justin Williams brings an engineering approach, operations research, to conservation planning and discusses the tools available for finding optimal spatial solutions. Optimization models are primarily tools for decision support as opposed to decision making. Their purpose is not to dictate a course of action, but to provide guidance, information, and insight for planning and decision making. This chapter provides an overview of what optimization models are and attempts to clarify their role in conservation planning. An example of a simple optimization model used for reserve site selection and design in North America's Inland Temperate Rainforest is provided.

Chapter 13 Integrating conservation planning with projected trends in climate change

Dominique Bachelet discusses how a conservation plan can be assessed in light of climate change. She discusses projected trends in climate, and the resulting simulated climate change impacts, stressing the limitations of current approaches due to model result uncertainties, incomplete understanding of biotic interactions and ecosystem processes, and lack of data. She proposes a list a conservation planning strategies to prepare for future change due to direct and indirect effects of climate change, and provides guidance to other resources.

Chapter 14 Mapping biological processes to the appropriate spatial modeling tools

Kevin Johnston, David Western, and Rodney Jackson begin by explaining how and why we need to view the world from the point of view of animals in ecosystems. Then they outline step by step how this can be done; especially in landscapes where resources and data may be limited. This chapter revisits the entire conservation planning process with a practical perspective to help conservation planners think about what assumptions they are making and why that may be important. They illustrate their approach with case studies where conservation goals were addressed using elephants, snow leopards, and other focal species.

Chapter 15 Summary: Building a broader base for conservation planning

And finally, Frank Davis brings his unique perspective on conservation planning to a discussion of what this book accomplishes, how it compares to traditional planning approaches, and the need to incorporate conservation approaches in the future.

References

Groves, C. R. 2003. *Drafting a Conservation Blueprint: A Practitioner's Guide to Planning for Biodiversity.* Washington, DC: Island Press.

———. 2008. "The Conservation Biologist's Toolbox for Planners: Advances, Challenges, and Opportunities." *Landscape Journal* 27(1): 81–96.

Hardin, G. 1968. "The Tragedy of the Commons." *Science* 162: 1,243–48.

Margules, C. R., and R. L. Pressey. 2000. "Systematic Conservation Planning." *Nature* 405: 243–53. http://courses.washington.edu/esrm200/Margules_Pressey_Conservation_Planning_Nature_2000.pdf.

Mark, D. M., N. Chrisman, A. U. Frank, P. H. McHaffie, and J. Pickles, with contributions from M. Curry, J. Goss, F. Harvey, K. Hillis, R. Miller, E. Sheppard, and D. Varanka. 1996. "The GIS History Project." http://www.ncgia.buffalo.edu/gishist/bar_harbor.html.

Noss, R.F., and A. Y. Cooperrider. 1994. *Saving Nature's Legacy—Protecting and Restoring Biodiversity.* Washington, DC: Island Press.

Noss, R. F., M. A. O'Connell, D. D. Murphy. 1997. *The Science of Conservation Planning. Habitat Conservation under the Endangered Species Act.* Washington, DC: Island Press.

Pressey, R. L., and M. C. Bottrill. 2009. "Approaches to Landscape- and Seascape-Scale Conservation Planning: Convergence, Contrasts, and Challenges." *Oryx* 43: 464–75.

Pressey, R. L., M. Cabeza, M. E. Watts, R. M. Cowling, and K. A. Wilson. 2008. "Conservation Planning in a Changing World." *Trends in Ecology and Evolution* 22 (11): 583–92.

Steinitz, C. 2010. "Ways of Designing." Esri GeoDesign Summit Presentation, January 5, 2010. http://video.esri.com/watch/54/ways-of-designing.

2

Integrating conservation planning with human communities, ecosystem services, and economics

Kai M. A. Chan, PhD; Natalie C. Ban, PhD; and Robin Naidoo, PhD

People are the cause of conservation problems and are fundamental to every conservation solution. Social, economic, and political considerations (hereafter simply "social considerations") are therefore crucial to the success of conservation initiatives (Polasky 2008; Claus et al. 2010). People are also independently important, for moral reasons (Chan and Satterfield, forthcoming) and because the treatment of people affects perceptions of conservation—both individual projects and the larger enterprise. Despite acknowledgement of the importance of such social considerations, conservation planning theory and practice—like much of conservation generally—has been much more focused on ecological considerations than social ones (Knight et al. 2006). Accordingly, much guidance is available for conservation planners on incorporating ecological considerations into conservation planning (see all other chapters in this book), but little guidance exists for incorporating social considerations.

Although the complexity of ecosystems demands considerable attention, the comparative inattention to social factors can be blamed partly for the relatively few and modest impacts that conservation plans have had on conservation on the ground (Knight et al. 2006, 2008). If we assume that the principal purpose of conservation planning is to yield effective protection of ecosystems and their biodiversity, the so-called planning-implementation gap represents missed opportunities or poor allocation of scarce conservation resources. More fully incorporating social considerations into the conservation planning process should improve implementation success by fostering both better plans and better implementation of appropriate plans (Knight et al. 2006).

Increasingly, conservation planners are realizing the importance of social considerations, and particularly of stakeholder involvement. The result is an expansion of conservation-planning approaches. Systematic conservation planning initially was focused mainly on providing a transparent process by which habitat protection would ensure representation of key biodiversity features (species, subspecies, and biological communities) in reserve networks (Margules and Pressey 2000). More recently, Pressey and Bottrill (2009) proposed five additional steps, most of which concern the social, economic, and political context for planning (see also Ban et al., chapter 8). Similarly, Knight et al. (2006) developed an operational framework to facilitate implementation of conservation actions.

In this chapter, we describe several approaches to ensure that social considerations are given their rightful prominent place in conservation planning, through social context analysis, participatory processes, ecosystem service analysis, and inclusion of economic costs in reserve design tools. Social analysis provides a preliminary understanding of the people who affect and are affected by conservation in a particular setting, and provides an entry into participatory processes. Ecosystem service analysis facilitates a more sophisticated understanding of how people might benefit or lose from conservation. Analysis of costs highlights the full suite of costs of implementation. As we describe, such efforts are essential, and they offer opportunities: comprehensive involvement of people can make conservation networks possible, effective, and efficient; and such processes can win new allies and partners for conservation.

Social analysis in conservation planning

The emerging thinking about social-ecological systems applies to social considerations in conservation planning through analyses of social data. Thinking in terms of social-ecological systems recognizes humans as part of the environment, and states that all human-used resources are embedded in complex social-ecological systems (Berkes 2003; Ostrom 2009). Ostrom and colleagues developed a framework to conceptualize social-ecological systems (Ostrom 2009) that is particularly useful for conservation (Ban et al., forthcoming). The framework recognizes that people and ecosystems interact in myriad fundamental and dynamic ways. Accordingly, it is impossible to manage (or protect) ecosystems effectively without accounting for social-ecological relationships through several avenues. This thinking is particularly important in an age of human domination and transformations of the planet (Vitousek et al. 1997). For example, in 1700, about 95% of Earth's ice-free land was wild or seminatural; by 2000, 55% of that land had been transformed into human-dominated landscapes (Ellis et al. 2010).

The social-ecological systems framework evolved in part from research on management of common pool resources. Ostrom's framework is intended to organize findings from several scientific disciplines so that we can identify factors contributing to sustainable management of resources (Ostrom 2009). The framework that Ostrom (2007) presents involves four separate subsystems, and the interactions among them: "(i) a resource system (e.g., fishery, lake, grazing area), (ii) the resource units generated by that system (e.g., fish, water, fodder), (iii) the users or actors in that system, and (iv) the governance system." For example, characteristics of interest regarding users (e.g., stakeholders) might include their number, socioeconomic attributes, history of resource use, and social capital. One of this framework's key contributions has been to identify characteristics affecting the likelihood of users to self-organize to manage their resources. The framework holds similar potential for conservation. In particular, insofar as conservation inevitably involves managing or restricting various human activities ("resource uses"), and the users of those resources are some of the most important and vocal stakeholders of conservation, conservation planning can benefit greatly from an understanding of social-ecological systems dynamics. The social-ecological systems framework is therefore relevant to conservation planning by helping to identify social considerations often neglected in conservation planning (Ban et al., forthcoming).

Common approaches to including social data in conservation planning

Increasingly, conservation planners are incorporating aspects of social considerations in their planning efforts, often through use of social data. Most conservation plans that have used social data have done so through one of three ways: (1) a social assessment, (2) a stakeholder analysis, or (3) inclusion of spatial human-use data in the identification of priority areas for conservation (see Theobald, chapter 5). These important elements of conservation planning are discussed below.

Social assessments

Social assessment in conservation planning, sometimes also called situation analysis, refers to the generation of background documentation that describes the social, cultural, economic, and political context of the study region (e.g., Knight et al. 2006; Conservation Measures Partnership 2007). Social assessments often focus on identifying the key drivers of change that may influence conservation objectives, including direct and indirect threats, opportunities, and enabling conditions (Conservation Measures Partnership 2007). Social assessment is sometimes also used to refer to the analysis of impacts of conservation actions on people (Schreckenberg et al. 2010).

Although social assessment, as commonly conceptualized, is an important start to including social considerations, it often misses important components that the social-ecological systems framework can help illuminate (Ostrom 2009). For example, rarely is discussion of current and potential future governance systems included in social assessments. The social-ecological systems framework outlines components of governance systems, including government and nongovernment organizations; network structure; property-rights systems; operational, collective-choice and constitutional rules; and monitoring and sanctioning processes—all of which would be informative to include in social assessments. Similarly, aspects of resource users that may be important include the number and location of users; their socioeconomic attributes, history of use, leadership, knowledge, and social capital; importance of the resource; and technologies used. Interactions between users and resource systems may include how much is harvested, information sharing among users, mechanisms for deliberation and conflict resolution, and various user activities (investment, lobbying, self-organizing, and networking; Ostrom 2009).

Although the social-ecological systems framework provides a useful list of considerations, the details and application of social assessments vary widely in conservation planning. For example, in a South African study, the social assessment focused on categorization of land-use practices and income (Lombard et al. 2010). The social assessment relied on expert knowledge because it was time- and resource-efficient.

It took only a few days of expert workshops to derive much useful spatially explicit information, while also increasing goodwill among workshop attendees (Lombard et al. 2010). In contrast, another study in South Africa (Pasquini et al. 2010) used a much more detailed data-gathering approach—telephone survey questionnaires to landowners, and detailed semi-structured interviews with selected landowners—to identify the likelihood that stewardship arrangements would be successfully implemented.

Stakeholder analysis

"Stakeholder" refers to anyone, individual or group, who has an interest in or is affected by an existing or proposed conservation action (Kessler 2004; Steffan-Dewenter et al. 2007). A stakeholder analysis—sometimes a component of social assessments—refers to an analysis of groups and individuals who have a stake in conservation planning outcomes. It is an approach to understand a system by identifying stakeholders and their respective interests in that system (Grimble and Wellard 1997). Grimble and Wellard (1997) developed a typology of stakeholders with interest in natural resources, ranging from global to national, regional to local on- and off-site. A key distinction is between stakeholders who affect a decision or action, versus those who are affected by the decision (active and passive stakeholders, respectively). Stakeholders may also be categorized according to their "importance" and "influence" as defined by those in charge of planning (e.g., "importance" may refer to those whose needs and interests are priorities for aid, and "influence" may mean the power of stakeholders over the success of the project; Grimble and Wellard 1997). Stakeholder analysis is therefore a useful tool for understanding the social context for (1) enhancing implementation success (where "influence" is relevant) and (2) evaluating the moral/justice implications of conservation (where "importance" is key for the distribution of effects).

Spatial human-use data

An increasingly common approach to including social considerations in conservation planning is to collate spatial data on human uses, and subsequently use the data as costs within decision support tools (Naidoo et al. 2006; Adams et al. 2010). Decision support tools (e.g., Marxan or C-Plan software) can identify areas of importance for biodiversity conservation (see also Williams, chapter 12). They can do so while minimizing a cost, such as opportunity costs to stakeholders. Hence, conservation planning seeks to identify conservation actions important for biodiversity while minimizing such costs. Other chapters in this book discuss additional ways in which spatial human-use data is integrated into conservation planning. In particular, spatial human-use data is incorporated into land-cover/land-use classifications (Olenicki, chapter 4, and Theobald, chapter 5), habitat characterization (Dickson et al., chapter 7; Ban et al., chapter 8; and Church, chapter 9), and connectivity assessment (Singleton and McRae, chapter 10). Theobald (chapter 5) also discusses the spatial prediction of future human use.

Stakeholder participation

Stakeholder engagement, sometimes also called stakeholder or community or user participation, refers to the involvement of key actors in the planning process (Grimble and Wellard 1997). One of the key purposes of stakeholder analysis (see above) is to identify actors who should then be engaged in the conservation planning process (Mushove and Vogel 2005). It is now generally acknowledged that stakeholder involvement is crucial to generate sustainable conservation outcomes (Grimble and Wellard 1997; Beierle and Konisky 2001; Kessler 2004; Steffan-Dewenter et al. 2007). Such involvement generates a sense of ownership of and goodwill towards proposed conservation actions and enhances the likelihood of successful implementation.

Considerable variation exists in the extent to which stakeholders might be engaged in the conservation planning process. Arnstein (1969) developed the classic typology of levels (or rungs) of citizen participation, ranging from manipulation to citizen control. The bottom two rungs—manipulation and therapy—refer to nonparticipation. The next three levels—informing, consultation, and placation—are classified as "degrees of tokenism" in which the objective is not to enable participation but only to give the appearance of it. The last three rungs—partnership, delegated power, and citizen control—are levels of citizen power with increasing degrees of decision-making responsibilities. The local expectation and importance of true participation in decision making is highly variable culturally (e.g., far more important to stakeholders in North America than in much of mainland China), but such participation has been shown to enhance the sustainability of projects across many contexts.

Examples of participation in conservation planning

Appropriate stakeholder engagement will vary depending on the context of the study region and resources available for the planning process. Several examples of planning for marine protected areas (MPAs) illustrate the range of approaches to gathering information and stakeholder engagement (Osmond et al. 2010). The rezoning of the Great Barrier Reef Marine Park (1999–2004) included 600 public meetings in more than 90 locations, and 21,000 written public submissions. This approach allowed all stakeholders to have input into the planning process if they wished to do so, but it did not give stakeholders direct decision-making power. In contrast to the agency-driven process in Australia, the stakeholder- and consensus-based process in the Channel Islands in California allowed interest groups to have greater control over the process and outcome. Yet another approach was that of the Marine Life Protection Act process in California state waters. After several failed starts, the implementers adopted a transparent process tied to a rigorous timeline, with a 32-member stakeholder group overseeing the public process.

In ideal situations, stakeholder preferences align with ecological goals to maximize conservation benefits, but in reality, this is not always the case (Chan et al. 2007). Thus, linking stakeholder preferences to conservation objectives is a fundamental challenge in conservation planning. A conservation planning process in the Solomon Islands appears to have found a suitable balance between community preferences and ecological goals (Game et al. 2011); this approach has been dubbed "informed opportunism" (Noss et al. 2002; Game et al. 2011). The project uses decision support software (Marxan) collaboratively with communities to assess and guide conservation opportunities (e.g., community preferences for places to protect) as they arise (Game et al. 2011). Similarly, Brock and Atkinson (chapter 6) describe a workshop process that involves local stakeholders in selecting focal species at the beginning of conservation planning. Although generally focusing on stakeholders with biological expertise for the species in question, this process also includes stakeholders who represent social and economic considerations.

Academic studies that explore conservation planning approaches should also consider appropriate stakeholder participation (see sidebar, "Incorporating social elements into marine conservation planning"). For example, a conservation planning research project in British Columbia, Canada, used a three-phase approach for executing the research: building research partnerships, carrying out individual interviews, and holding community discussion sessions (Ban et al. 2008). Building the research partnership (in this example, with indigenous groups) included data-sharing agreements to ensure that the communities involved in the research would have copies of all data used for, and collected during, the project. Interviews built individual relationships with participants, and interview participants were encouraged to express their own goals and issues as well as suggest management actions. Community meetings provided an opportunity for interview participants and others to see the results of the study and provide feedback (Ban et al. 2008). Such an inclusive approach may also be suitable for other research-practice partnerships.

Incorporating social elements into marine conservation planning: An example from British Columbia, Canada

Incorporating social and economic data is a relatively recent practice in systematic marine conservation planning. A literature review of the costs used in marine planning highlighted the predominance of uniform (i.e., where each planning unit has the same cost) and fisheries opportunity costs (figure 2.1; Ban and Klein 2009).

A study in British Columbia, Canada, illustrates a way to incorporate people's preferences directly into decision support tools and the effect of using different costs in one such tool (Marxan). To incorporate people's preferences, Ban et al. developed a partnership with a community, in this case, an indigenous group. They followed a three-phase process of building a research collaboration, carrying out individual interviews, and organizing community feedback sessions (figure 2.2; Ban et al. 2008).

Ultimately, this led to obtaining people's preferences for conservation areas through a map-based interview process. These maps were then digitized, and through community meetings, a set of preferred conservation areas was agreed upon. These preferred areas were then always included as conservation areas in the Marxan software (i.e., they were "locked in"). With these areas selected, Marxan was then used to build on these areas to ensure biodiversity objectives were met (figure 2.3).

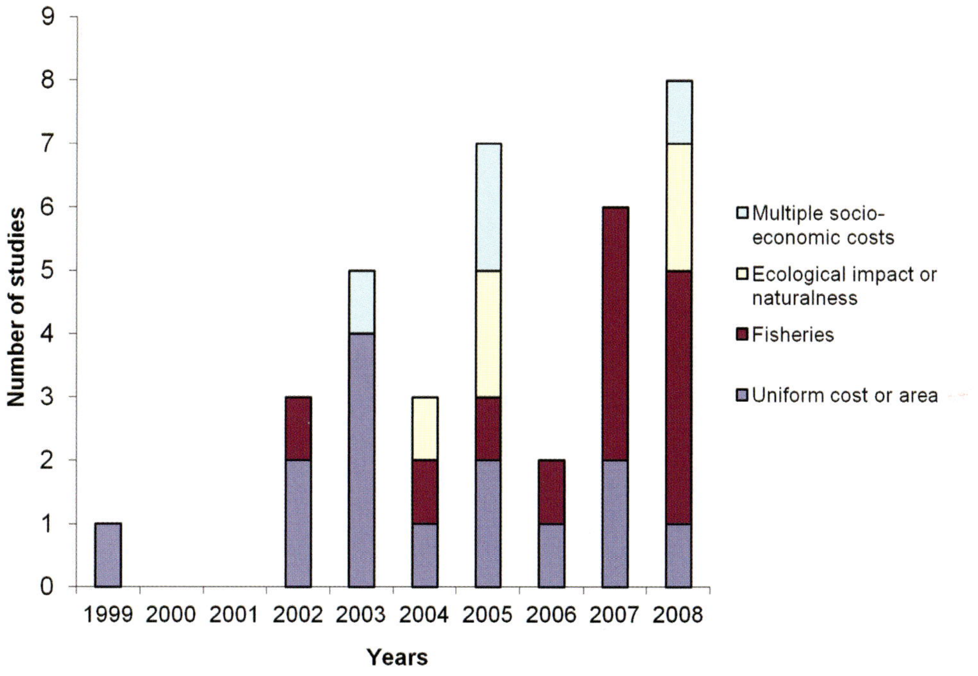

Figure 2.1. Number of studies per publication year encountered in a literature review, by type of socioeconomic cost considered. The number of studies has increased over time, and fisheries as a cost accounted for the largest increase. Figure from Ban and Klein (2009). John Wiley and Sons.

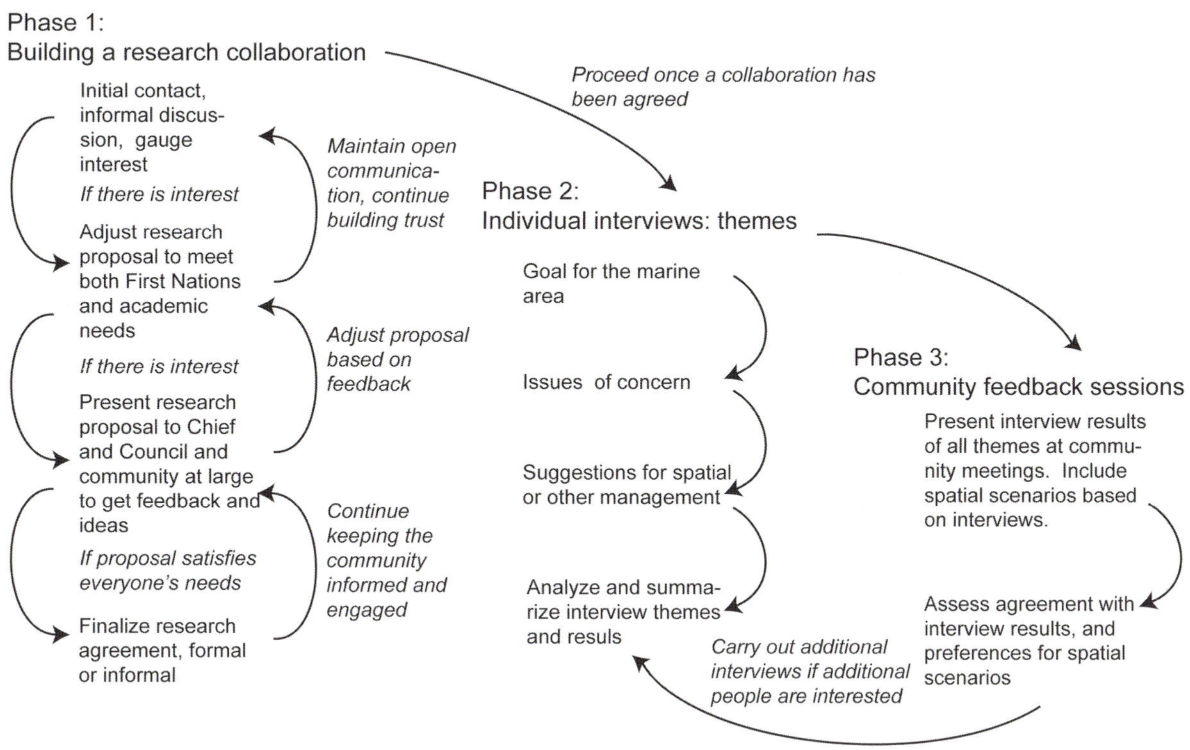

Figure 2.2. Framework for eliciting community preferences for spatial management options. Figure from Ban et al. (2008).

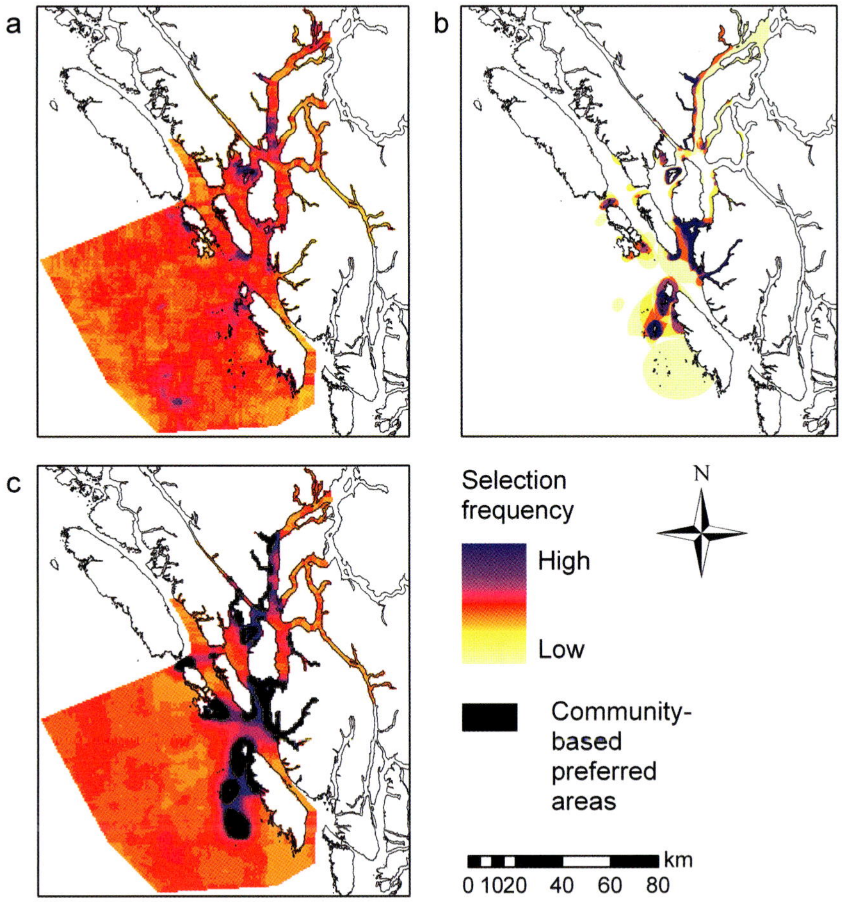

Figure 2.3. Maps of community preferences and Marxan results (selection frequency) for conservation priorities: (a) results of the prioritization with Marxan, showing the scenarios that represent 10% of every feature with area as the cost (same as figure 2.4a), (b) overlays of community preferences from individual interviews, (c) integrated approach with Marxan in which community preferences (black) were locked into the scenarios. Figure adapted from Ban et al. (2009). Basemap provided by the Province of British Columbia under the Open Government License for Government of BC Information v.BC1.0.

Commonly, social data is integrated into conservation planning as costs, which the Marxan algorithm minimizes while meeting biodiversity conservation objectives. To gauge the influence of the cost layer, Ban et al. (2009) contrasted three costs: area, human impacts (all mapped threats, ranging from extractive activities to land-based impacts), and commercial fisheries (figure 2.4). They found that the conservation value of planning units (selection frequency) can vary substantially when different costs are used but biodiversity objectives are held constant. In regions where there is flexibility in achieving conservation targets, selection frequency varies greatly (e.g., offshore regions in figure 2.4). In regions where such flexibility is low (e.g., inshore regions), the selection frequency remains high even when the cost is high (figure 2.4).

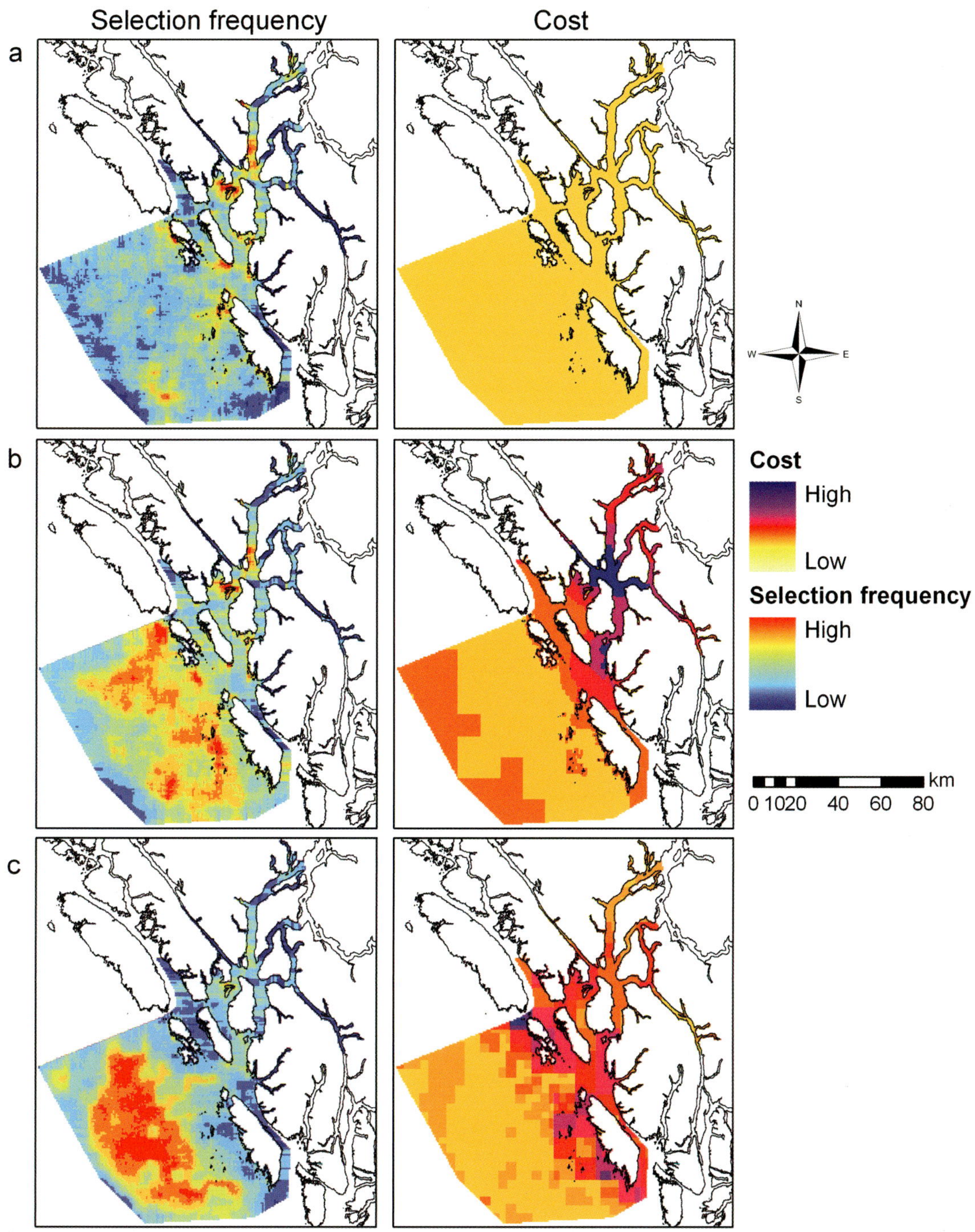

Figure 2.4. Scenarios using different costs for an area on the north coast of British Columbia, Canada. The left column shows the selection frequency of conservation areas under a variety of cost scenarios, and the right column depicts the corresponding costs: (a) area as a cost, (b) human impacts as a cost (including commercial fishing), and (c) commercial fisheries as a cost. The maps depict the scenario that targets 10% of every biodiversity feature. Figure from Ban et al. (2009). Basemap provided by the Province of British Columbia under the Open Government License for Government of BC Information v.BC1.0.

Guidelines for stakeholder participation

Although the most appropriate approach to stakeholder engagement will be context-specific, several guidelines for best practices have been suggested. Based on a literature review, Reed (2008) proposed eight best practices:

1. Stakeholder participation should be built upon a philosophy of empowerment, equity, trust, and learning.
2. Stakeholders should be involved early and throughout the process.
3. Stakeholders need to be analyzed and represented systematically (i.e., a thorough stakeholder analysis should be undertaken).
4. Clear objectives for the participatory process need to be agreed upon by stakeholders from the outset.
5. Context-specific methods should be used, based on the objectives, types of participants, and appropriate level of engagement.
6. Highly skilled facilitation is necessary.
7. Both local and scientific knowledge should be considered.
8. Participation should be institutionalized.

Additional approaches to involving people in conservation planning

Although the common approaches outlined above are important elements for addressing social considerations in conservation planning, many additional tools, techniques, and questions are relevant. Below we outline some questions that conservation planners may want to consider in the planning process (Ban et al., forthcoming), based on Ostrom's (2009) social-ecological systems framework.

- What is the history and context of the study region?
    - What are the governance systems of land/sea use in general and of key resource users?
    - Who are the stakeholders ("users" in Ostrom [2009]), what are their characteristics, and how might they be affected by conservation actions?
    - What are anticipated feedbacks and interactions between the resource system/units and the social system (governance, users)?
- Who is doing the planning, and what are their legal and moral obligations to the public/stakeholders?

Answers to these questions can help inform several aspects of the conservation planning process. First, appropriate stakeholder involvement can be determined by deciding (through a transparent process) who to involve and how to involve them. Second, the specifics of how to involve stakeholders can be integrated into the planning process design. For instance, Arnstein's (1969) ladder of citizen participation can be used to identify an appropriate level of participation, given the context of the planning region. Third, appropriate conservation and management actions can be identified: for instance, whether the best conservation action is to designate protected areas, and if so, whether the areas should be fully closed to human uses or accessible for some level of extraction. Finally, conservation opportunities and win-win scenarios can be identified and used to prioritize actions. For example, stakeholder preferences for conservation areas can be assessed for their conservation value, and supplemented with additional areas if necessary.

Ecosystem services and human values

The concept of ecosystem services (ES)—the processes by which ecosystems benefit people, directly and indirectly—promises to contribute greatly to conservation. Its greatest promise is to render visible numerous benefits associated with conservation that would otherwise be unseen. Given the explicit focus of ES on human benefits, it is no surprise that the concept, developed substantially by Daily (1997), has gained considerable traction within and outside the conservation community (The Economist 2005; Millennium Ecosystem Assessment 2005; World Resources Institute et al. 2008; TEEB 2009; Sodhi and Ehrlich 2010; Kareiva and Marvier 2011). The tools for mapping and valuing ecosystem services have advanced greatly (Kareiva et al. 2011), and the concept is central to current notions of ecosystem-based management (McLeod and Leslie 2009; Lubchenco and Sutley 2010) and sustainability science (Levin and Clark 2010).

One of the most important advances has been the explicit characterization of how changes in ecosystems result in changes to human benefits, through ES "production functions" (Chan and Ruckelshaus 2010; Nelson and Daily 2010). This kind of characterization enables ES to contribute substantially to particular decisions, including conservation plans. But although many conservation planning analyses pay lip service to ES, few such analyses include production functions (Egoh et al. 2007). A literature review shows that the inclusion of ES in conservation planning frequently occurs through coarse proxies (e.g., ecological processes that "support the persistence of biodiversity pattern," or presence or abundance of an ecosystem service provider) (Egoh et al. 2007). Without explicit characterization of production functions, the potential of ES to link conservation to human benefits remains unrealized.

ES include the contribution of forests to hydropower production (Guo et al. 2000; Guo et al. 2007), mitigation of floods by wetlands and other vegetated lands (Bradshaw et al. 2007), contribution of mangrove forests to fish harvests (Aburto-Oropeza et al. 2008), and natural pollination of agricultural crops (Ricketts et al. 2008). Such services, which are often enhanced by conservation actions, can be—and have been—characterized, mapped, and valued, both individually and together (Costanza et al. 1997; Chan et al. 2006; Naidoo and Ricketts 2006; Turner et al. 2007; Naidoo et al. 2008; Anderson et al. 2009; Egoh et al. 2009; Luck et al. 2009; Nelson et al. 2009; Kareiva et al. 2011). Given that debates about conservation areas are frequently dominated by their negative consequences for people, it makes sense also to consider possible positive consequences. Of course, not all ES will be enhanced by conservation—biodiversity protection will preclude realizing benefits of some services (Chan, Hoshizaki, and Klinkenberg 2011). Only a few studies have included multiple ES in a framework for the systematic planning of biodiversity conservation (Chan et al. 2006; Chan, Hoshizaki, and Klinkenberg 2011).

Good practice for including conservation planning in ES would include each of the steps described below.

Identify the most relevant ES and benefits

For an ES analysis to fulfill its potential, it is crucial that the services of greatest relevance be identified and considered. Ideally, identification of priority ES and benefits involves stakeholder interviews and focus groups (Iceland et al. 2008; Chan, Goldstein, et al. 2011; Klain et al., unpublished manuscript); the process can be assisted through interaction with a multi-stakeholder management board (e.g., the Marine Initiative of the Natural Capital Project with West Coast Aquatic on Vancouver Island [Hoag 2010; Guerry et al. 2012]). Furthermore, linking ES to the stated objectives of local governance or management bodies (Espinosa-Romero et al. 2011) can both help prioritize ES for conservation efforts and provide indicators for subsequent monitoring and evaluation.

Characterize relationships between ES, benefits, values, people, and activities

Most ES can be quantified in numerous ways, based on their provision of different kinds of benefits (Chan, Goldstein, et al. 2011; Chan et al. 2012). Stakeholders may not think of these services in the terms used in academic literature (Klain et al., unpublished manuscript). For example, fisheries may be important not only for the gross or net revenue they provide, but also for food security, employment, and so on. Each of these benefits is associated with the ES of the provision of fish for consumption. Given the innumerable possible metrics, it is critical that an ES analysis for conservation planning choose the set that is most meaningful

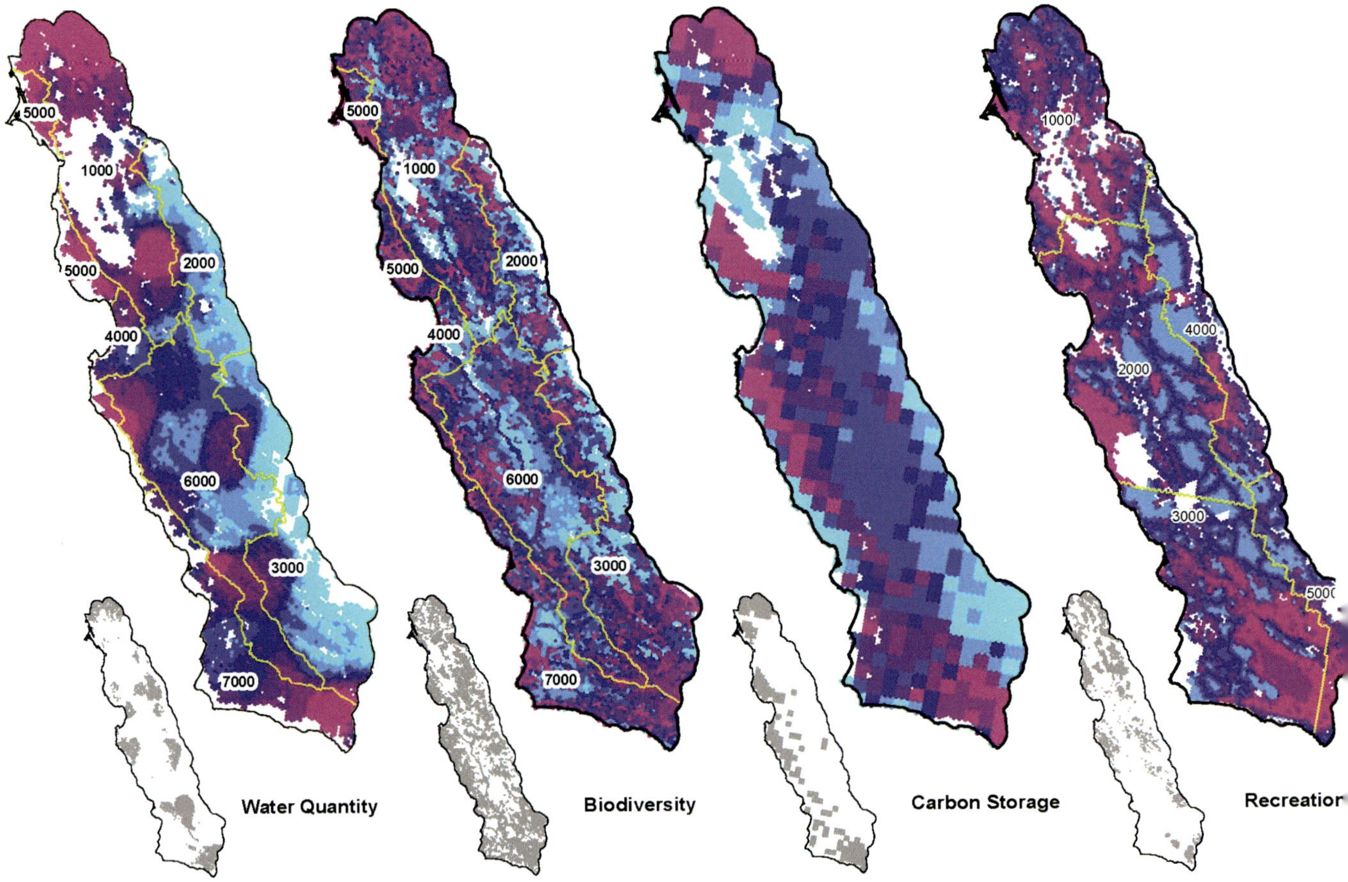

Figure 2.5. Spatial analysis of biodiversity and chosen ecosystem services. The seven benefit functions (feature values) are displayed in color; gray insets show the accompanying best networks of selected planning units. Feature values range from 0 (or locked out; white), to low (light blue), moderate (dark blue), and high (purple). The boundary indicates the ecoregion plus a 10-km buffer. Yellow lines indicate "stratification units," each of which had a separate target (quantitative goal). Numbers in the thousands (e.g., 3,000) are stratification unit labels. Not shown are targets or constraints specific to planning units. Figure from Chan et al. (2006).

not only to local stakeholders but also in light of the possible conservation actions and associated social and ecological changes. Thus, an ES analysis best proceeds by producing—with participation from key stakeholders and local experts—a conceptual map of the major benefits, services, values, people, and human activities associated with an ecosystem (Klain et al., unpublished manuscript; Chan, Guerry, et al. 2012). Such a conceptual map can identify (1) the most relevant services and benefits, based on their connection to conservation actions; (2) appropriate metrics of value for those services and benefits; and (3) appropriate valuation processes, based on the kinds of values at stake (Chan, Satterfield, et al. 2012; Chan, Guerry, et al. 2012).

Select ES strategically and for justice

To avoid the inappropriate dilution of objectives, it is crucial that conservation planning includes only strategically and morally relevant ES. As an example, it is not the role of a conservation organization to maximize agricultural production in a region.

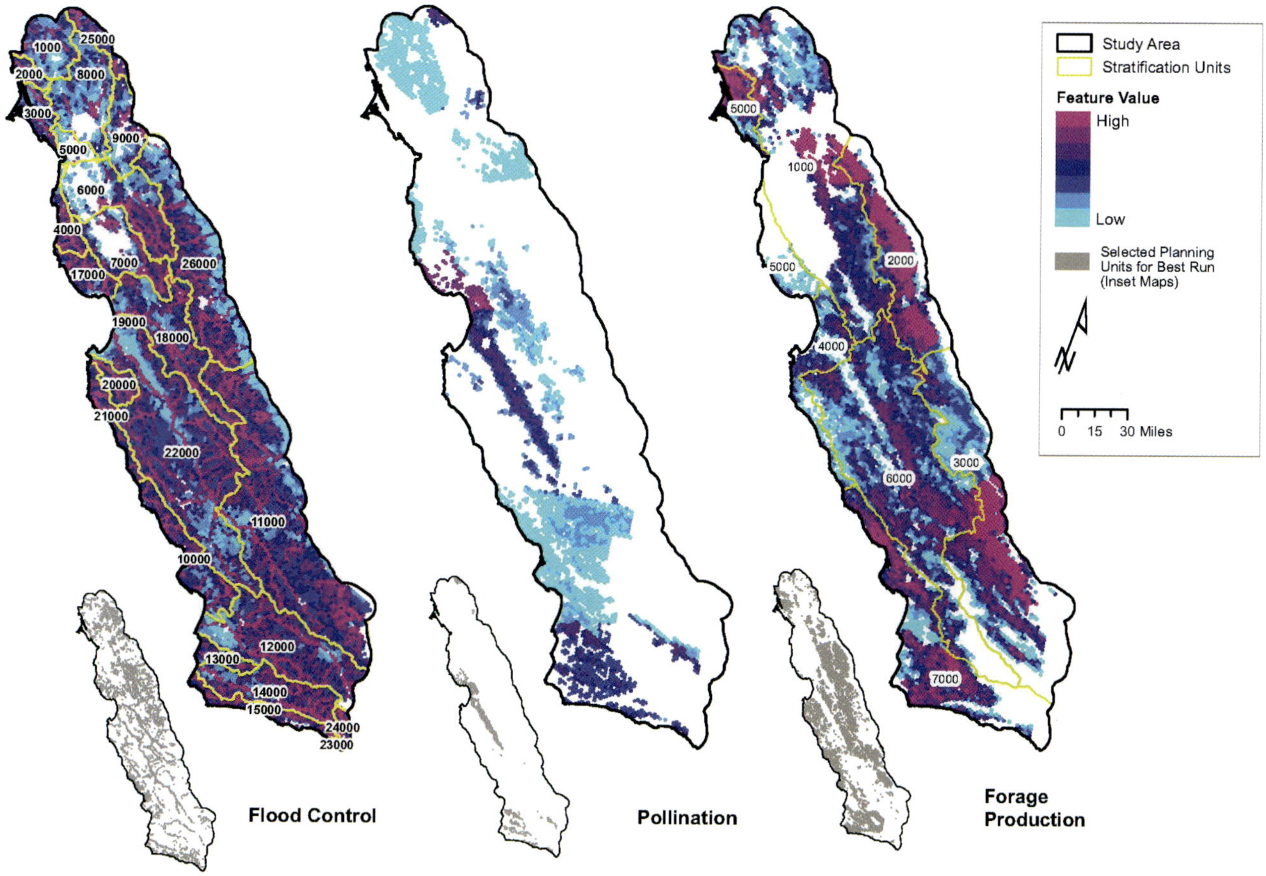

ES should be included if they are associated with opportunity costs (see "Costs" later in this chapter) or if they offer the potential for additional conservation funding or partnerships to enhance the success of implementing a conservation plan. For example, carbon storage in developing nations might garner funding through Reducing Emissions from Deforestation and Forest Degradation (REDD) policies (Naidoo and Ricketts 2006; Venter, Laurance, et al. 2009; Venter, Meijaard, et al. 2009); forage production and pollination services from a reserve network might yield new or strengthened partnerships with ranchers, farmers, and associated organizations (Kremen et al. 2004; Chan et al. 2006); and the continued supply of fishing and logging opportunities might enhance the implementation of protected areas (Ban and Klein 2009; Chan, Hoshizaki, and Klinkenberg 2011).

In conservation planning, consideration of what is "strategic" for ES can continue after analyzing spatial associations between ES and biodiversity priorities. For example, Chan et al. (2006) identified a strategic set of ES by excluding the two services (crop pollination and forage production) whose benefit functions were negatively correlated with biodiversity (figure 2.5).

Another reason for including ES in a conservation planning analysis is for justice considerations. Conservation actions have numerous implications for environmental justice (McShane et al. 2011; Chan and Satterfield, forthcoming). These implications should be explored by understanding the implications (both positive and negative) of planned conservation actions for ES appreciably affected by conservation actions—particularly those ES important to poor, disempowered, and historically disadvantaged groups.

Characterize benefits as affected by conservation—*marginal,* not *total* values—across whole landscapes/seascapes

From a conservation planning perspective, the total value of ES provided in a region or planning unit is not an important factor. What matters is the value of the *difference* that might be expected based on conservation planning alternatives (most simply, whether a parcel is included in a reserve or not). This call for *marginal* changes—realistic changes from the current baseline—is critical for conservation planning generally (Margules and Pressey 2000; Chan and Daily 2008).

Furthermore, as with biodiversity, ES are provided by land/water areas both inside and outside reserves, and the production of ES can be a complex function of spatial pattern of reserves and other land uses, not calculable from consideration of each patch separately. Accordingly, ES objectives should pertain to the change in benefits achievable in whole landscapes or seascapes from realistic changes in land and water use. Accordingly, in characterizing the benefits of conservation for flood protection, Chan et al. (2006) estimated the change in flood risk that might be associated with a change in land cover from its current state to a developed state (the assumed eventual state for land not conserved in the Central Coast of California study region). Such expressions of the ES benefits at stake, as a function of land cover and other variables, comprise the ES production function.

Construct benefit functions using best available science (natural and social)

There is abundant and growing understanding of the production of ES (Kareiva et al. 2011)—an understanding that is budding into a science of ES. Research assists in the construction of benefit functions that express the difference in ES benefits associated with conservation zones, and this research has been integrated into freely available tools, such as the Natural Capital Project's InVEST (Integrated Valuation of Environmental Services and Tradeoffs) software package (Nelson et al. 2009; Kareiva et al. 2011; Guerry et al. 2012). Many ES have been studied extensively outside of the nascent ES field, however. Accordingly, even if formal production functions have not yet been constructed for a given ES, relevant research is likely to be available, although much ecological research does not directly address the ecological characteristics of concern for ES—the "ecological endpoints" (Wainger and Boyd 2009). In the case of Chan, Hoshizaki, and Klinkenberg (2011), the carbon budget model of Kurz et al. (1997) was directly applicable for estimating changes in carbon storage. According to the model, forest types in the central interior ecoregions in British Columbia, Canada, store on average 10% more carbon under natural disturbance cycles than under managed (forestry) cycles. The same carbon budget model is currently used by Canadian Forest Service and has been central to prominent scientific publications (e.g., Kurz et al. 2008).

Align spatial scales and management regimes

Conservation planning exercises generally involve adopting a single spatial resolution for analysis (see Unnasch and Karl, chapter 3), but that resolution may not be the most appropriate one for all ES. Until recently, most reserve design tools also required a single zone for management of uses (but this has changed somewhat with the introduction of Marxan with Zones; Watts et al. 2009). Accordingly, it is important to ensure that ES production functions are appropriate for the spatial scale and management zones of the analysis. For example, given the relatively short flight distance of most insect pollinators (Kremen et al. 2004), the ES contribution of unmanaged pollinators to agricultural production is generally provided *within*, not *across*, standard-sized ecoregional planning units (e.g., 500 hectares). That is, this pollination ES is largely a function of the composition of land-cover types within a planning unit, not the proximity of a planning unit in a large forest to planning units producing insect-pollinated agricultural crops (Chan et al. 2006). The inherent assumption entailed by this pollination ES—that sites with high pollination values will retain agricultural land uses—may contradict other ES or conservation assumptions that conserved sites will be returned to natural land cover.

Classify ES as intrinsically important or as incidental benefits or costs

Individual ES might be targeted as features alongside biodiversity components, or they might contribute to layers representing the incidental effects of conservation (often called the "cost surface"), as either benefits or costs (Chan, Hoshizaki, and Klinkenberg 2011). Critical differences exist between these two options (Chan, Hoshizaki, and Klinkenberg 2011). Specifying targets for an ES suggests that the ES is intrinsically valuable and that there is a meaningful threshold of provision needed (Chan, Hoshizaki, and Klinkenberg 2011). However, including ES as incidental benefits or costs suggests that the benefits of an ES are fungible (freely exchangeable or replaceable) and that there is no necessary minimum level at which a service ought to be provided. Most ES are fungible: for example, flood mitigation might be provided by engineered structures, and reduced atmospheric carbon dioxide concentrations might be achieved either through carbon storage or reduced emissions. When ES are fungible, targets are arbitrary, and the incidental-effect approach is preferable (Chan et al. 2006; Chan, Hoshizaki, and Klinkenberg 2011). Fishing benefits have been included in analyses of costs (as opportunity costs; Ban and Klein 2009), but fishing communities might see fishing opportunities as nonfungible, which would require the target approach (e.g., Ban and Vincent 2009).

Identify relevant constituencies

Any statement of value invites the question, "Value to whom?" Thus, identifying appropriate ES valuation processes in a conservation plan is based on identifying the relevant constituency: those who are affected by inclusion of the ES (see "Select ES strategically and for justice" above). Also relevant is the strategy for aligning the project's objectives with the planners' mission. For example, mitigating climate change has beneficiaries across the planet (in avoided damages), but it can also have local beneficiaries if carbon storage and/or sequestration might yield funding through carbon credits. The choice of which (if any) set of beneficiaries pertains should be determined by (1) whether the planners' mission includes mitigating global social damages associated with climate change, and (2) if the planners wish to and are able to recoup funding through carbon credits (e.g., receiving funding generally requires ownership of land rights and title). It is not appropriate, however, to pursue both avenues, and add the value of mitigated damages for global beneficiaries to that for credit-sellers, because the process of selling the credits involves a transfer of responsibility for the mitigated damages.

Another example involves fishing communities who impede implementation of marine protected areas. If a strategy for enabling implementation involves appeasing such communities (which seems wise), then valuation ought to include fishing wages as benefits, not costs (e.g., by subtracting them from revenues to calculate net revenues), which might be appropriate for ES valuation with fishing-license owners as the only beneficiaries. In this example, the benefit of wages for communities can be added to the benefit of net revenues for license owners, if both are relevant beneficiaries, because both benefits can be achieved simultaneously.

Valuate and evaluate ES benefits in appropriate terms

The incorporation of ES into conservation planning hinges upon appropriate characterization of associated benefits. Both the valuation process and its resulting metrics ought to be determined by the intersection of the kinds of values at stake (see "Characterize relationships between ES, benefits, values, people, and activities" above) and the beneficiaries (see "Identify relevant constituencies" above). For example, the beneficiaries may care about a given benefit based on moral principles or virtues (not just preferences), and based on values characteristic of groups (not just individuals); if so, deliberative valuation processes will be important (Jacobs 1997; Wilson and Howarth 2002; Chan, Goldstein, et al. 2011; Chan, Satterfield, et al. 2012). Where the final benefits associated with ES are market-traded goods, the values can be considered market-mediated (Chan, Satterfield, et al. 2012); in such cases, valuation can consist of mainly market valuation as well as various methods of economic nonmarket valuation (National Research Council 2005; Environmental Protection Agency 2009). For monetary costs and benefits accruing in the future, values should generally be expressed as net present values based on appropriately chosen discount rates (Heal 1998; Portney and Weyant 1999; Goulder and Stavins 2002; Sumaila and Walters 2005). Further explorations of the appropriate choice of valuation processes can be found elsewhere (Environmental Protection Agency 2009; Chan, Satterfield, et al. 2012).

Valuation approaches—whether monetary or not—will not equally reflect all reasons why people care about ecosystems and human uses therein (Klain and Chan, unpublished manuscript; Klain et al., unpublished manuscript). Accordingly, it is important to ensure that appropriately structured participatory processes are associated with later stages of the conservation plan (Chan, Goldstein, et al. 2011; Chan, Satterfield, et al. 2012).

In some cases, rather than expressing ES quantities in metrics of value or importance, it may be useful and appropriate to use biophysical terms (e.g., Chan et al. 2006; Naidoo et al. 2008) or priority scores that include such considerations as local availability of alternatives, local capacity to meet demand, or opportunity to enhance ES (Luck et al. 2009; Satterfield et al. 2011). For example, Chan et al. (2006) adopted a principle-based stand on recreation, assuming that parks and protected areas should be sufficient to provide one day per month of local outdoor recreation to each resident in the ecoregion. Decision makers with such principle-based values may see no need to value recreational experiences in monetary terms. Similarly, Luck et al. (2009) expressed flood mitigation, water provision, and carbon storage through priority scores that integrated criteria for availability of alternatives, opportunities for enhancement, local capacity to meet demand, threat to the service, and (relative) benefit-cost ratios.

Integrate ES into the conservation planning analysis

After appropriate values have been calculated for each ES, they must be incorporated into the conservation planning analysis. For intrinsically important features, targets must be set; for incidental benefits or costs, values must be integrated into the existing cost layer. Targets should be derived from legislation, regulations, or consultation with the relevant constituencies. Incidental benefits and costs must be transformed into terms appropriate for the "cost surface" (i.e., a spatially referenced GIS layer), or the cost surface must be transformed into terms suitable for the incidental benefit or cost values. For example, in the central interior of British Columbia, Canada, Chan, Hoshizaki, and Klinkenberg (2011) calculated monetary values of changes in net benefits of forestry, recreational angling, and carbon storage. They then incorporated the values into the cost surface by transforming the initial surface (a simple function of road density) into approximate dollar terms, and then adding ES cost values and subtracting incidental benefit values (see sidebar, "Transforming a road density suitability index into acquisition costs").

Transforming a road density suitability index into acquisition costs: An example from British Columbia, Canada

Ecosystem services frequently are valuated in dollar terms to compare them with each other and with other economic costs and benefits. To integrate such ecosystem service values into conservation planning often requires converting the preexisting cost surface into dollar terms. In a conservation planning exercise that involved an ecosystem services assessment, the larger analysis applied a cost surface based on a road density suitability index in Marxan, so Chan, Hoshizaki, and Klinkenberg (2011) transformed the road index into a dollar-metric cost surface representing acquisition costs.

To transform the road index into dollar values, Chan, Hoshizaki, and Klinkenberg (2011) used land market values in the study region as a proxy for acquisition costs, effectively assuming that (1) increased road density and proximity to roads implies greater urbanization and correspondingly greater land acquisition costs; (2) land acquisition costs are an important contributor to, and an appropriate proxy for, total costs; and (3) land prices are a parallel proxy for acquisition costs. Land costs for rural areas were provided by the Nature Conservancy of Canada, and costs for urban areas (with high road density) were found on the Multiple Listing Service (MLS) of the Canadian Real Estate Association.

In the absence of a sophisticated analysis relating land prices to road density, Chan, Hoshizaki, and Klinkenberg (2011) used a simple four-part linear transformation to determine acquisition costs for each planning unit. The transformation had four component functions, each based on the interpolation between two of the five anchor points. The anchor points were situated at peaks in the frequency distribution of road index values; at each, we calculated the average land price (figure 2.6).

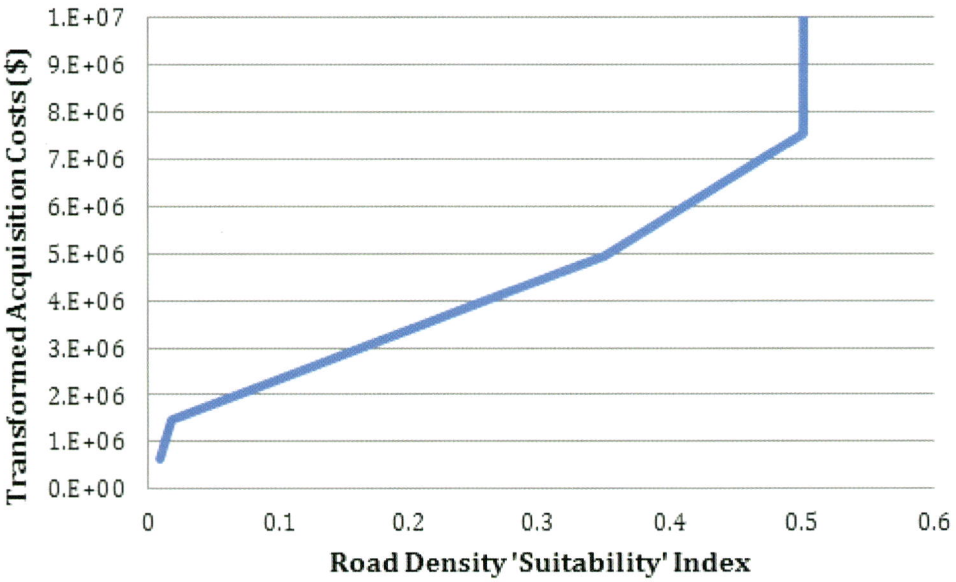

Figure 2.6. The transformation of road density values into approximate acquisition costs (Canadian dollars per 500-ha planning unit). The uppermost point (not shown), representing densely urbanized land, has a cost of $1,606M per 500-ha planning unit at a road density value of 0.7. From Chan, Hoshizaki, and Klinkenberg (2011).

Consider the ramifications of planned conservation actions for stakeholders

Relate the conservation plan, and its expected impacts on ES, to the implications for various stakeholders, particularly those with special moral or legal claims (e.g., indigenous people) and those who are economically, politically, or otherwise disadvantaged. Ecosystem service analyses generally have not characterized distribution of consequences for particular groups, so this is an area of focus for new projects (Chan et al. 2007). Such analyses can draw from a broader literature on environmental justice (Serret and Johnstone 2006; Schlosberg 2007; Chan and Satterfield, forthcoming).

Assess sensitivity to key uncertainties and assumptions

As with many steps in a conservation planning analysis, the incorporation of ES values entails several assumptions and considerable uncertainties. Although prominent conservation planning tools do not provide the means for formal sensitivity analysis, a partial sensitivity analysis can be performed by varying key values and assumptions, and assessing the nature of effects on results (Ardron et al. 2010). Chan, Hoshizaki, and Klinkenberg (2011) used this approach to assess the extent to which differences in quantitative targets for ES (e.g., 40%, 50%, or 60% of total values in the ecoregions) caused differences in spatial correlations between the reserve networks of different scenarios (combinations of ES and biodiversity targets).

Despite the paucity of examples of conservation planning analyses that include multiple ES, there is considerable promise for ES to contribute extensively to conservation planning, and therefore to efficient and effective conservation.

Costs

As with most things in life, the actions we take as part of the conservation planning process are not free. They entail certain costs that must be accounted for, so that in a world of scarce resources, we cannot do everything we might like to do. This obvious fact has nevertheless only been acknowledged relatively recently in the conservation planning literature (Naidoo et al. 2006; Murdoch et al. 2007). Initially, biological considerations were the strong focus. An increasing number of conservation planning studies now explicitly recognize that costs are a crucial part of the planning process, and one that is better dealt with from the outset rather than once the plan has been developed.

As we have argued throughout this chapter, a plan is much more likely to be implemented if socioeconomic concerns are integrated throughout the planning process. Specifically, incorporating costs at the outset into computer planning algorithms allows planners to maximize the benefits (however defined) per unit expenditure (however defined), which translates into getting the greatest "bang for the buck" from a conservation plan. A variety of studies have examined the conditions under which the early integration of socioeconomic concerns is more or less important (Adams et al. 2010; Carwardine et al. 2010), as well as the conditions under which investing in obtaining accurate information on the spatial distribution of costs might actually be more important than investing in the acquisition of spatial data on biodiversity (Ferraro 2003; Naidoo et al. 2006).

Incorporating a comprehensive assessment of costs associated with conservation is also likely to help ensure that a conservation plan is implemented fairly. If the full set of costs borne by others—especially damage and opportunity costs—are assessed at the outset, provisions are more likely to be made for compensation or other measures of accounting for such costs to be implemented along with the conservation measures.

In this section, we focus mainly on the practical issues associated with quantifying the spatial distribution of the economic costs of conservation planning. We leave aside the technical details of how costs enter into mathematical planning algorithms because this is addressed by Ban et al. in chapter 8. At the most basic level, the idea is to construct a cost surface that can then be combined with biodiversity layers to produce a cost-effective conservation plan using the planners' algorithm of choice. The spatial resolution of the cost layer will vary depending on how costs are mapped (see below), but costs may be represented either as a raster (cell-based) layer or as a vector map that corresponds to the planning units.

Types of conservation costs

As reviewed in Naidoo et al. (2006), several types of costs are associated with conservation. Acquisition costs reflect the costs associated with purchasing a parcel of land. Management costs refer to how much it will cost to manage a conservation area. Damage costs reflect the damage that might emanate from a protected area (e.g., damage from wildlife to crops or property). Transaction costs are the costs of identifying, negotiating, and monitoring conservation arrangements with stakeholders. Perhaps the most important category is opportunity costs: in the case of conservation of, say, a parcel of land, these costs are all the benefits associated with the next best use of this land, had it not been put to conservation. Under certain conditions (a well-functioning market with no barriers to sales, well-defined property rights, among other conditions) these may be approximated by acquisition costs, but in instances where these assumptions are not met, opportunity costs must be estimated independently.

Methods to map conservation costs

Acquisition and establishment costs

In countries where private property holdings are widespread, acquisition costs for at least a sample of properties may be available in or near the planning area. Data on acquisition costs may come from registries of actual property sales, assessments of market values (Polasky et al. 2001), formal databases on land prices summarized at various administrative units (Ando et al. 1998), or modeling techniques. The latter is the more complicated because it involves some analysis rather than simply looking up property values and assigning them to the appropriate location. If a representative sample of properties spanning the characteristics of those in the study area can be obtained, a cost surface may be developed by using regression methods to model property costs. The model could be based on biophysical and socioeconomic variables that might already be available as GIS layers (e.g., soil productivity, distance from nearest road, land cover, and so on). Numerous examples exist of this "hedonic" approach (i.e., the prediction of a good's value via a regression model that contains its assumed constituent characteristics) for the reader interested in more detail on methods (Sinden 2004; Chomitz et al. 2005; Nordman and Wagner 2010). In a marine context, researchers have begun to characterize the costs of establishment of marine protected areas (MPA), and how these vary with MPA size, duration of establishment, and so on (McCrea-Strub et al. 2011).

Management costs

The seminal studies on mapping the management costs of protected areas, both terrestrial and marine, come from Balmford and his collaborators (Balmford et al. 2003, 2004; Bruner et al. 2004). For both realms, protected area managers across the globe were surveyed and asked to state the costs of managing their particular area. Using this dataset, a model of management cost per square kilometer was developed based on size of the protected area, economic context (gross domestic product, purchasing power parity, human development index) of the relevant country, and, for marine protected areas, distance from inhabited land. Because the terrestrial and marine models showed high predictive power, and are based on easily obtainable data, this approach to estimating management costs has been used in a variety of conservation planning applications (Moore et al. 2004; Wilson et al. 2006; Underwood et al. 2008; Klein et al. 2009). If management costs can be obtained for protected areas of similar types in more geographically proximate areas, then using these local estimates is preferable; otherwise, the Balmford models provide a straightforward and robust approach to estimate management costs. However, the linear relationship identified by Balmford and colleagues between management cost and the size of marine protected area is unlikely to hold true for very large areas recently designated or proposed (beyond the range of the reviewed protected areas), such as the Coral Sea Conservation Zone in Australia, or the Chagos Island marine reserve in the Indian Ocean (Ban et al. 2011). In other words, costs per unit area cannot continue diminishing linearly, or very large marine protected areas would cost nothing to manage.

Opportunity costs

When acquisition-costs data is unavailable (e.g., in most marine planning efforts) or where they do not accurately reflect the true opportunity costs of conservation, a variety of methods has been used to create spatially explicit surfaces for opportunity costs. Understanding the dominant land uses that exist in a region is fundamental to accurately mapping opportunity costs. Because agriculture is the dominant anthropogenic land use in many areas of the world, terrestrial planning studies often have used the returns to agriculture to estimate the opportunity costs of conservation at a variety of spatial scales and resolutions (Naidoo and Adamowicz 2006; Naidoo and Iwamura 2007; Carwardine et al. 2010; Egoh et al. 2010). A number of conservation planning studies have used the freely available agricultural opportunity cost data of Naidoo and Iwamura (2007), which is a global cost surface of 5-second resolution that may be suitable for conservation planning applications in regions that are large relative to the layer resolution.

In principle, similar approaches could be used to estimate opportunity costs from nonagricultural land uses such as timber harvesting and oil and gas extraction. In the marine realm, opportunity costs of conservation have been estimated as the value of fish from a variety of fisheries stocks (Stewart and Possingham 2005; Richardson et al. 2006; Klein et al. 2008). Given the importance of all of these economic activities, data on opportunity costs is likely to be readily available in some format for the planning region in question; the key issue is to use a sensible method of assigning a cost to each planning unit. This issue becomes more complex as the heterogeneity of the landscape or seascape increases, and as the number of economic activities present grows larger; see Naidoo and Adamowicz (2006) for an approach that addresses both of these complications.

To our knowledge, transaction costs and damage costs have not been mapped in the same way as the categories of costs that we have summarized above. In the case of transaction costs, these are likely to be highest where terrestrial conservation planning involves the development of conservation agreements with a multitude of stakeholders who each occupy small parcels of land in a study area. Rather than explicitly estimating and mapping these costs, it seems rather more productive to consider how transaction costs may affect the overall viability of a conservation project, and what means might exist to abate them. These issues, with specific regard to carbon capture projects under the Clean Development Mechanism, are dealt with by Cacho et al. (2005) and Gong et al. (2010). In the case of damage costs, these could be estimated via surveys of stakeholders in the planning area and then mapped; examples of how to conduct surveys of damage costs and associated perceptions of, for instance, wildlife, include Agarwala et al. (2010) and Naughton-Treves et al. (2003).

We have focused herein on economic costs that can be expressed as monetary values, because they are easily understood by decision makers and conservation planners, and because they can be directly compared to a planner's budget to estimate the feasibility of implementation. Nevertheless, other costs that are not as amenable to economic valuation may also be important considerations in particular contexts; for example, the political costs of protecting contested areas may be difficult to translate into economic terms, but might still have a large impact on the ability to implement a conservation plan. Because of their highly context-specific nature, costs of this type are not covered in this chapter. But the prudent and efficient conservation planner must be aware of and attempt to incorporate all costs, monetary or otherwise, that are likely to effect the successful implementation of a plan.

Discussion and conclusion

Although the integration of approaches discussed in this chapter would go far to improve conservation planning outcomes, several avenues exist for further refinement. First, most social considerations, ecosystem services, and costs are generally integrated into conservation planning only as static elements. In reality, however, they change through time. Therefore, an important area of research is to identify the extent to which social considerations, ecosystem services, and costs are dynamic, and to examine how to integrate such dynamism into the planning process.

Second, an important question to resolve is who in particular will undertake the approaches we have identified. Although the analyses do not necessarily require PhDs in social sciences, they would certainly benefit from social science expertise. Many conservation NGOs and government parks agencies do employ social scientists, but these additional tasks may require additional personnel. Fortunately, resources are available online to assist organizations with such tasks. These resources include the catalog of Conservation Social Science Tools provided by the Social Sciences Working Group of the Society for Conservation Biology (http://www.conbio.org/workinggroups/sswg/catalog/index.cfm), and also the same group's expert database (http://www.conbio.org/WorkingGroups/SSWG/network/dirindex.cfm).

Third, we have suggested many considerations for incorporating social, ecosystem services, and economic elements into conservation planning. Given limited conservation resources, it will not be possible to attempt all of these elements in every plan. Further practice and research may reveal which aspects are most important, or that their relative importance is entirely context-dependent.

For each of the approaches outlined in this chapter, several additional considerations may also be helpful. For social analyses and stakeholder participation, the following points apply:

- An understanding of the social-ecological system may help to identify locally suitable participation strategies. Especially important aspects are governance systems, user characteristics, and users' interactions with the environment, as well as the social and cultural setting.
- Conservation planning is often conceived of as taking place in stages. Social analyses can inform most stages, and ideally, stakeholders should be involved throughout the planning process, from the beginning and at multiple stages throughout revised systematic conservation planning stages (Knight et al. 2006; Pressey and Bottrill 2009).
- Achieving effective stakeholder attendance and participation may require several enabling actions (e.g., compensating for travel as appropriate, building the capacity for stakeholders to engage in planning processes).
- Stakeholder participation processes should contain a clear mechanism for incorporating feedback to change the plan. Therefore, an acknowledgement by participants that any decision will involve trade-offs and conflict is necessary. Explicitly analyzing trade-offs and implications of ES change may help (see "Consider the ramifications of planned conservation actions for stakeholders").
- Discussing the social goals and objectives (in addition to ecological and conservation objectives) is recommended, with consideration of how such social goals might contribute to (or detract from) conservation.

Regarding ecosystem services, it is worth keeping in mind the following:

- ES are not a magic bullet. First, economic markets (and so market values) are fickle friends of conservation: although markets for ES may align with conservation at some times, changing conditions can reverse this and cause market pressures to undermine conservation (McCauley 2006; Chan et al. 2007). Second, ES values are imperfectly correlated with conservation priorities (Chan et al. 2006; Naidoo et al. 2008; Egoh et al. 2009; Luck et al. 2009).
- The ES concept—the idea of ecosystems as service providers—does not include all the kinds of values that people associate with ecosystems, including intrinsic values and many principle-based values (Fairbank and Public Opinion Strategies 2010; Chan, Satterfield, et al. 2012), and the terminology invokes ideas that many people are uncomfortable associating with nature (Fairbank and Public Opinion Strategies 2010). Furthermore, the stock-flow conception of ecology (ecosystems providing ecosystem services) is both an incomplete representation of the dynamism and contingencies of ecosystems (Palmer and Filoso 2009; Norgaard 2010), and a particular metaphor for human-environment relations that may crowd out other such metaphors (Klain et al., unpublished manuscript; Levine et al., unpublished manuscript).

Additional considerations for including economic costs in planning include the following:

- Recent research has highlighted the dynamic nature of planning decisions and suggested that an optimal conservation planning process would include considerations of targets, threats, and costs in a spatially dynamic planning algorithm (Costello and Polasky 2004; Wilson et al. 2006).
- Careful consideration of the appropriate measures of economic costs should be given, because depending on which cost measure is chosen, the results of planning algorithms can vary considerably, as shown by Adams et al. (2010).
- As with ecosystem services, the costs of conservation for various stakeholders may not be expressed effectively using economic costs. Analyses of the values at stake in conservation decisions (see "Characterize relationships between ES, benefits, values, people, and activities" above), could provide valuable additional insight into the planning process.

A crucial question for improving all aspects of social considerations in conservation planning is, "What approaches are most suitable for the context in question?" Documenting successes and failures of various approaches and conservation actions, along with the corresponding social-ecological context, is of paramount importance for improving conservation planning and practice. To enable such comparisons, studies should use a common framework so that multiple case studies can be analyzed to identify common elements to successful conservation actions. The social-ecological systems framework was developed to enable just such a comparison (Ostrom 2009), and conservation planners should consider using it to document lessons learned (Ban et al., forthcoming).

In this chapter, we have discussed several avenues by which conservation planning can be more attentive to the needs and interests of people: social context analysis, participatory processes, ecosystem service analysis, and inclusion of economic costs in reserve design. Comprehensive involvement of people can make conservation networks possible, effective, and efficient, and such processes can win new allies and partners for conservation.

References

Aburto-Oropeza, O., E. Ezcurra, G. Danemann, V. Valdez, J. Murray, and E. Sala. 2008. "Mangroves in the Gulf of California Increase Fishery Yields." *Proceedings of the National Academy of Sciences of the United States of America* 105: 10,456–59.

Adams, V. M., R. L. Pressey, and R. Naidoo. 2010. "Opportunity Costs: Who Really Pays for Conservation?" *Biological Conservation* 143: 439–48.

Agarwala, M., S. Kumar, A. Treves, and L. Naughton-Treves. 2010. "Paying for Wolves in Solapur, India and Wisconsin, USA: Comparing Compensation Rules and Practice to Understand the Goals and Politics of Wolf Conservation." *Biological Conservation* 143: 2,945–55.

Anderson, B. J., P. R. Armsworth, F. Eigenbrod, C. D. Thomas, S. Gillings, A. Heinemeyer, D. B. Roy, and K. J. Gaston. 2009. "Spatial Covariance between Biodiversity and Other Ecosystem Service Priorities." *Journal of Applied Ecology* 46: 888–96.

Ando, A., J. Camm, S. Polasky, and A. Solow. 1998. "Species Distributions, Land Values, and Efficient Conservation." *Science* 279: 2,126–28.

Ardron, J. A., H. P. Possingham, and C. J. Klein, eds. 2010. *Marxan Good Practices Handbook*, version 2. Victoria, BC: Pacific Marine Analysis and Research Association. http://www.uq.edu.au/marxan/docs/Marxan Good Practices Handbook v2 2010.pdf.

Arnstein, S. R. 1969. "A Ladder of Citizen Participation." *Journal of the American Institute of Planners* 35: 216–24.

Balmford, A., K. J. Gaston, S. Blyth, A. James, and V. Kapos. 2003. "Global Variation in Terrestrial Conservation Costs, Conservation Benefits, and Unmet Conservation Needs." *Proceedings of the National Academy of Sciences of the United States of America* 100: 1,046–50.

Balmford, A., P. Gravestock, N. Hockley, C. J. McLean, and C. M. Roberts. 2004. "The Worldwide Costs of Marine Protected Areas." *Proceedings of the National Academy of Sciences of the United States of America* 101: 9,694–97.

Ban, N. C., V. Adams, R. L. Pressey, and J. Hicks. 2011. "Promise and Problems for Estimating Management Costs of Marine Protected Areas." *Conservation Letters* 4(3): 241–52.

Ban, N. C., and C. J. Klein. 2009. "Spatial Socioeconomic Data as a Cost in Systematic Marine Conservation Planning." *Conservation Letters* 2: 206–15.

Ban, N. C., M. Mills, J. Tam, C. C. Hicks, S. Klain, N. Stoeckl, M. C. Bottrill, J. Levine, R. L. Pressey, T. Satterfield, and K. M. A. Chan. Forthcoming. "Towards a Social-Ecological Approach for Conservation Planning: Embedding Social Considerations." In *Frontiers in Ecology and the Environment*. doi: 10.1890/110205.

Ban, N. C., C. Picard, and A. C. J. Vincent. 2008. "Moving towards Spatial Solutions in Marine Conservation, with Indigenous Communities." *Ecology and Society* 13: 32.

Ban, N. C., C. R. Picard, and A. C. J. Vincent. 2009. "Comparing and Integrating Community-Based and Science-Based Approaches in Prioritizing Marine Areas for Protection." *Conservation Biology* 23: 899–910.

Ban, N. C., and A. C. J. Vincent. 2009. "Beyond Marine Reserves: Exploring The Approach of Selecting Areas where Fishing Is Permitted, Rather Than Prohibited." *PLoS ONE* 4: e6258.

Beierle, T. C., and D. M. Konisky. 2001. "What Are We Gaining from Stakeholder Involvement? Observations from Environmental Planning in the Great Lakes." *Environment and Planning C: Government and Policy* 19 (4): 515–27.

Berkes, F. 2003. "Alternatives to Conventional Management: Lessons from Small-Scale Fisheries." *Environments* 31: 5–19.

Bradshaw, C. J. A., N. S. Sodhi, K. S. H. Peh, and B. W. Brook. 2007. "Global Evidence that Deforestation Amplifies Flood Risk and Severity in the Developing World." *Global Change Biology* 13: 2,379–95.

Bruner, A. G., R. E. Gullison, and A. Balmford. 2004. "Financial Costs and Shortfalls of Managing and Expanding Protected-Area Systems in Developing Countries." *Bioscience* 54: 1,119–26.

Cacho, O., G. R. Marshall, and M. Milne. 2005. "Transaction and Abatement Costs of Carbon-Sink Projects in Developing Countries." *Environment and Development Economics* 10: 597–614.

Carwardine, J., K. A. Wilson, S. A. Hajkowicz, R. J. Smith, C. J. Klein, M. Watts, and H. P. Possingham. 2010. "Conservation Planning when Costs Are Uncertain." *Conservation Biology* 24: 1,529–37.

Chan, K. M. A., and G. C. Daily. 2008. "The Payoff of Conservation Investments in Tropical Countryside." *Proceedings of the National Academy of Sciences* 105: 19,342–47.

Chan, K. M. A., J. Goldstein, T. Satterfield, N. Hannahs, K. Kikiloi, R. Naidoo, N. Vadeboncoeur, and U. Woodside. 2011. "Cultural Services and Non-Use Values." In *Natural Capital: Theory & Practice of Mapping Ecosystem Services*, edited by P. Kareiva, H. Tallis, T. H. Ricketts, G. C. Daily, and S. Polasky, 206–228. Oxford; New York: Oxford University Press.

Chan, K. M. A., A. Guerry, P. Balvanera, S. Klain, T. Satterfield, X. Basurto, A. Bostrom, R. Chuenpagdee, R. Gould, B. S. Halpern, N. Hannahs, J. Levine, B. Norton, M. Ruckelshaus, R. Russell, J. Tam, and U. Woodside. Forthcoming. "Where Are 'Cultural' and 'Social' in Ecosystem Services: A Framework for Constructive Engagement." *Bioscience* 6 (8): 744-56.

Chan, K. M. A., L. Hoshizaki, and B. Klinkenberg. 2011. "Ecosystem Services in Conservation Planning: Targeted Benefits or Co-benefits/Costs?" *PLoS ONE* 6 (9): e24378. http://dx.plos.org/10.1371/journal.pone.0024378.

Chan, K. M. A., R. M. Pringle, J. Ranganathan, C. L. Boggs, Y. L. Chan, P. R. Ehrlich, P. K. Haff, N. E. Heller, K. Al-Khafaji, and D. P. Macmynowski. 2007. "When Agendas Collide: Human Welfare and Biological Conservation." *Conservation Biology* 21: 59–68.

Chan, K. M. A., and M. Ruckelshaus. 2010. "Characterizing Changes in Marine Ecosystem Services." *F1000 Biology Reports 2010* 2: 54, doi: 10.3410/B2-54.

Chan, K. M. A., and T. Satterfield. Forthcoming. "Justice, Equity, and Biodiversity." In *The Encyclopedia of Biodiversity*, edited by S. Levin, G. C. Daily, R. K. Colwell, J. Lubchenco, H. A. Mooney, E.-D. Schulze, and D. Tilman. Oxford: Elsevier.

Chan, K. M. A., T. Satterfield, and J. Goldstein. 2012. "Rethinking Ecosystem Services to Better Address and Navigate Cultural Values." *Ecological Economics* 74: 8–18.

Chan, K. M. A., M. R. Shaw, D. R. Cameron, E. C. Underwood, and G. C. Daily. 2006. "Conservation Planning for Ecosystem Services." *PLoS Biology* 4 (11): 2,138–2,152. http://biology.plosjournals.org/perlserv/?request=get-document&doi=10.1371/journal.pbio.0040379.

Chomitz, K. M., K. Alger, T. S. Thomas, H. Orlando, and P. Vila Nova. 2005. "Opportunity Costs of Conservation in a Biodiversity Hotspot: The Case of Southern Bahia." *Environment and Development Economics* 10: 293–312.

Claus, C. A., K. M. A. Chan, and T. Satterfield. 2010. The Roles of People in Conservation. In *Conservation Biology for All*, edited by N. Sodhi and P. R. Ehrlich, 262–283. Oxford: Oxford University Press.

Conservation Measures Partnership. 2007. *Open Standards for the Practice of Conservation, Version 2.0.* Washington DC: Conservation Measures Partnership. http://www.sustainablenorthwest.org/wss/wssagenda/wss-web-library/miradi-web-library/Conservation_Standards_Version_2.0.pdf.

Costanza, R., R. d'Arge, R. de Groot, S. Farber, M. Grasso, B. Hannon, K. Limburg, S. Naeem, R. V. O'Neill, J. Paruelo, R. G. Raskin, P. Sutton, and M. van den Belt. 1997. "The Value of the World's Ecosystem Services and Natural Capital." *Nature* 387: 253–260.

Costello, C., and S. Polasky. 2004. "Dynamic Reserve Site Selection." *Resource and Energy Economics* 26: 157–174.

Daily, G. C., ed. 1997. *Nature's Services: Societal Dependence on Natural Ecosystems.* Washington DC: Island Press.

Egoh, B. N., B. Reyers, J. Carwardine, M. Bode, P. J. O'Farrell, K. A. Wilson, H. P. Possingham, M. Rouget, W. De Lange, D. M. Richardson, and R. M. Cowling. 2010. "Safeguarding Biodiversity and Ecosystem Services in the Little Karoo, South Africa." *Conservation Biology* 24: 1,021–30.

Egoh, B., B. Reyers, M. Rouget, M. Bode, and D. M. Richardson. 2009. "Spatial Congruence between Biodiversity and Ecosystem Services in South Africa." *Biological Conservation* 142: 553–62.

Egoh, B., M. Rouget, B. Reyers, A. T. Knight, R. M. Cowling, A. S. van Jaarsveld, and A. Welz. 2007. "Integrating Ecosystem Services into Conservation Assessments: A Review." *Ecological Economics* 63: 714–21.

Ellis, E., K. Goldewijk, S. Siebert, D. Lightman, and N. Ramankutty. 2010. "Anthropogenic Transformation of the Biomes, 1700 to 2000." *Global Ecology and Biogeography* 19: 589–606.

Environmental Protection Agency (EPA) Science Advisory Board. 2009. *Valuing the Protection of Ecological Systems and Services: A Report of the EPA Science Advisory Board.* Washington, DC: EPA. http://yosemite.epa.gov/sab/sabproduct.nsf/WebBOARD/SAB-09-012/$File/SAB Advisory Report full web.pdf.

Espinosa-Romero, M. J., K. M. A. Chan, T. McDaniels, and D. Mullen-Dalmer. 2011. "Structuring Decision-Making for Ecosystem-Based Management." *Marine Policy* 35: 575–83.

Fairbank, Maslin, Maullin, Metz and Associates, and Public Opinion Strategies. 2010. *Key Findings from Recent National Opinion Research on "Ecosystem Services."* Missoula, MT: The Nature Conservancy. http://ec2-75-101-128-201.compute-1.amazonaws.com/attached-files/0/99/9929/Rally2010_C09.pdf.

Ferraro, P. J. 2003. "Assigning Priority to Environmental Policy Interventions in a Heterogeneous World." *Journal of Policy Analysis and Management* 22: 27–43.

Game, E. T., G. Lipsett-Moore, R. Hamilton, N. Peterson, J. Kereseka, W. Atu, M. Watts, and H. Possingham. 2011. Informed Opportunism for Conservation Planning in the Solomon Islands. *Conservation Letters* 4: 38–46.

Gong, Y., G. Bull, and K. Baylis. 2010. "Participation in the World's First Clean Development Mechanism Forest Project: The Role of Property Rights, Social Capital, and Contractual Rules." *Ecological Economics* 69: 1,292–302.

Goulder, L. H., and R. N. Stavins. 2002. "Discounting: An Eye on the Future." *Nature* 419: 673–74.

Grimble, R., and K. Wellard. 1997. "Stakeholder Methodologies in Natural Resource Management: A Review of Principles, Contexts, Experiences and Opportunities." *Agricultural Systems* 55: 173–93.

Guerry, A., M. Ruckelshaus, K. Arkema, J. Bernhardt, G. Guannel, C. Kim, M. Marsik, M. Papenfus, M. Pinsky, J. Toft, G. Verutes, S. Wood, M. Beck, F. Chan, K. Chan, G. Gelfenbaum, B. Gold, B. Halpern, B. Labiosa, S. Lester, P. Levin, M. McField, M. Plummer, S. Polasky, P. Ruggiero, D. Sutherland, and H. Tallis. 2012. "Modelling Benefits from Nature: Using Ecosystem Services to Inform Coastal and Marine Spatial Planning." *International Journal of Biodiversity Science, Ecosystem Services & Management* 2012: 1–15.

Guo, Z. W., Y. M. Li, X. M. Xiao, L. Zhang, and Y. L. Gan. 2007. "Hydroelectricity Production and Forest Conservation in Watersheds." *Ecological Applications* 17: 1,557–62.

Guo, Z. W., X. M. Xiao, and D. M. Li. 2000. "An Assessment of Ecosystem Services: Water Flow Regulation and Hydroelectric Power Production." *Ecological Applications* 10: 925–36.

Heal, G. 1998. *Valuing the Future.* New York: Columbia University Press.

Hoag, H. 2010. "Banking on Biodiversity." *Momentum* 8–15. http://www.environment.umn.edu/momentum/issue/3.1f10/cover_biodiversity.html.

Iceland, C., C. Hanson, and C. Lewis. 2008. *Identifying Important Ecosystem Goods & Services in Puget Sound.* Washington, DC: World Resources Institute.

Jacobs, M. 1997. "Environmental Valuation, Deliberative Democracy and Public Decision-Making Institutions." In *Valuing Nature? Ethics, Economics and the Environment*, edited by J. M. Foster, 211–31. New York: Routledge.

Kareiva, P., and M. Marvier. 2011. *Conservation Science: Balancing the Needs of People and Nature.* Greenwood Village, CO: Roberts & Co.

Kareiva, P., H. Tallis, T. H. Ricketts, G. C. Daily, and S. Polasky. 2011. *Natural Capital: Theory & Practice of Mapping Ecosystem Services.* Oxford: Oxford University Press.

Kessler, B. L. 2004. *Stakeholder Participation: A Synthesis of Current Literature.* Silver Spring, Maryland: NOAA.

Klain, S., and K. M. A. Chan. "Participatory Mapping of Ecosystem Services for Spatial Planning." Unpublished manuscript.

Klain, S., K. M. A. Chan, and T. Satterfield. "Bundles of Marine Cultural Ecosystem Services and Values." Unpublished manuscript.

Klein, C. J., N. C. Ban, B. S. Halpern, M. Beger, E. T. Game, H. S. Grantham, A. Green, T. J. Klein, S. Kininmonth, E. Treml, K. Wilson, and H. P. Possingham. 2009. "Prioritizing Land and Sea Conservation Investments to Protect Coral Reefs." *PLoS ONE* 5: e12431.

Klein, C. J., C. Steinbeck, A. J. Scholz, and H. P. Possingham. 2008. "Effectiveness of Marine Reserve Networks in Representing Biodiversity and Minimizing Impact to Fishermen: A Comparison of Two Approaches Used in California." *Conservation Letters* 1: 44–51.

Knight, A. T., R. M. Cowling, and B. M. Campbell. 2006. "An Operational Model for Implementing Conservation Action." *Conservation Biology* 20: 739–50.

Knight, A., R. Cowling, M. Rouget, A. Balmford, A. Lombard, and B. Campbell. 2008. "Knowing but Not Doing: Selecting Priority Conservation Areas and the Research–Implementation Gap." *Conservation Biology* 22: 610–17.

Kremen, C., N. M. Williams, R. L. Bugg, J. P. Fay, and R. W. Thorp. 2004. "The Area Requirements of an Ecosystem Service: Crop Pollination by Native Bee Communities in California." *Ecology Letters* 7: 1,109–19.

Kurz, W. A., S. J. Beukema, and M. J. Apps. 1997. "Carbon Budget Implications of the Transition from Natural to Managed Disturbance Regimes in Forest Landscapes." *Mitigation and Adaptation Strategies for Global Change* 2: 405–21.

Kurz, W. A., C. C. Dymond, G. Stinson, G. J. Rampley, E. T. Neilson, A. L. Carroll, T. Ebata, and L. Safranyik. 2008. "Mountain Pine Beetle and Forest Carbon Feedback to Climate Change." *Nature* 452: 987–90.

Levin, S. A., and W. C. Clark. 2010. "Toward a Science of Sustainability." Report presented at the US National Science Foundation Conference, Toward a Science of Sustainability, Airlie Center, Warrenton, VA. http://www.nsf.gov/mps/dms/documents/SustainabilityWorkshopReport.pdf.

Levine, J. S., K. M. A. Chan, and T. Satterfield. "The Meanings of Conservation Biology: Using Mental Models to Demystify Normative Thinking in Conservation." Unpublished manuscript.

Lombard, A., R. Cowling, J. Vlok, and C. Fabricius. 2010. "Designing Conservation Corridors in Production Landscapes: Assessment Methods, Implementation Issues, and Lessons Learned." *Ecology and Society* 15: 7.

Lubchenco, J., and N. Sutley. 2010. "Proposed US policy for Ocean, Coast, and Great Lakes Stewardship." *Science* 328: 1,485–86.

Luck, G. W., K. M. A. Chan, and J. P. Fay. 2009. "Protecting Ecosystem Services and Biodiversity in the World's Watersheds." *Conservation Letters* 2: 179–88.

Margules, C. R., and R. L. Pressey. 2000. "Systematic Conservation Planning." *Nature* 405: 243–53. http://courses.washington.edu/esrm200/Margules_Pressey_Conservation_Planning_Nature_2000.pdf.

McCauley, D. J. 2006. "Selling Out on Nature." *Nature* 443: 27–28.

McCrea-Strub, A., D. Zeller, U. Rashid Sumaila, J. Nelson, A. Balmford, and D. Pauly. 2011. "Understanding the Cost of Establishing Marine Protected Areas." *Marine Policy* 35: 1–9.

McLeod, K., and H. Leslie, ed. 2009. *Ecosystem-Based Management for the Oceans*. Washington, DC: Island Press.

McShane, T. O., P. D. Hirsch, T. C. Trung, A. N. Songorwa, A. Kinzig, B. Monteferri, D. Mutekanga, H. V. Thang, J. L. Dammert, M. Pulgar-Vidal, M. Welch-Devine, J. Peter Brosius, P. Coppolillo, and S. O'Connor. 2011. "Hard Choices: Making Trade-Offs between Biodiversity Conservation and Human Well-Being." *Biological Conservation* 144: 966–72.

Millennium Ecosystem Assessment. 2005. *Ecosystems and Human Well-Being: Synthesis*. Washington, DC: Island Press.

Moore, J., A. Balmford, T. Allnutt, and N. Burgess. 2004. "Integrating Costs into Conservation Planning across Africa." *Biological Conservation* 117: 343–50.

Murdoch, W. W., S. Polasky, K. A. Wilson, H. P. Possingham, P. Kareiva, and M. R. Shaw. 2007. "Maximizing Return on Investment in Conservation." *Biological Conservation* 139: 375–88.

Mushove, P., and C. Vogel. 2005. "Heads or Tails? Stakeholder Analysis as a Tool for Conservation Area Management." *Global Environmental Change Part A* 15: 184–98.

Naidoo, R., and W. L. Adamowicz. 2006. "Modeling Opportunity Costs of Conservation in Transitional Landscapes." *Conservation Biology* 20: 490–500.

Naidoo, R., A. Balmford, R. Costanza, B. Fisher, R. E. Green, B. Lehner, T. R. Malcolm, and T. H. Ricketts. 2008. "Global Mapping of Ecosystem Services and Conservation Priorities." *Proceedings of the National Academy of Sciences* 105: 9,495–500.

Naidoo, R., A. Balmford, P. J. Ferraro, S. Polasky, T. H. Ricketts, and M. Rouget. 2006. "Integrating Economic Costs into Conservation Planning." *Trends in Ecology & Evolution* 21: 681–87.

Naidoo, R., and T. Iwamura. 2007. "Global-Scale Mapping of Economic Benefits from Agricultural Lands: Implications for Conservation Priorities." *Biological Conservation* 140: 40–49.

Naidoo, R., and T. H. Ricketts. 2006. "Mapping the Economic Costs and Benefits of Conservation." *PLoS Biology* 4: 2,153–64.

National Research Council (US). Committee on Assessing and Valuing the Services of Aquatic and Related Terrestrial Ecosystems. 2005. *Valuing Ecosystem Services: Toward Better Environmental Decision-Making*. Washington, DC: The National Academies Press.

Naughton-Treves, L., R. Grossberg, and A. Treves. 2003. "Paying for Tolerance: Rural Citizens' Attitudes toward Wolf Depredation and Compensation." *Conservation Biology* 17: 1,500–1,511.

Nelson, E. J., and G. C. Daily. 2010. "Modelling Ecosystem Services in Terrestrial Systems." *F1000 Biology Reports 2010* 2: 53, doi: 10.3410/B2-53.

Nelson, E., G. Mendoza, J. Regetz, S. Polasky, H. Tallis, D. R. Cameron, K. M. A. Chan, G. C. Daily, J. Goldstein, P. M. Kareiva, E. Lonsdorf, R. Naidoo, T. H. Ricketts, and M. R. Shaw. 2009. "Modeling Multiple Ecosystem Services, Biodiversity Conservation, Commodity Production, and Tradeoffs at Landscape Scales." *Frontiers in Ecology and the Environment* 7: 4–11.

Nordman, E., and J. E. Wagner. 2010. "Incorporating Acquisition Costs in Forestland Open Space Programs: Lessons from Conservation Biology and Applications." *Journal of Forestry* 108: 282–87.

Norgaard, R. B. 2010. "Ecosystem Services: From Eye-Opening Metaphor to Complexity Blinder." *Ecological Economics* 69: 1,219–27.

Noss, R. F., C. Carroll, K. Vance-Borland, and G. Wuerthner. 2002. "A Multicriteria Assessment of the Irreplaceability and Vulnerability of Sites in the Greater Yellowstone Ecosystem." *Conservation Biology* 16: 895.

Osmond, M., S. Airame, M. Caldwell, and J. Day. 2010. "Lessons for Marine Conservation Planning: A Comparison of Three Marine Protected Area Planning Processes." *Ocean & Coastal Management* 53: 41–51.

Ostrom, E. 2007. A Diagnostic Approach for Going Beyond Panaceas. Proceedings of the National Academy of Sciences of the United States of America 104: 15,181–87.

———. 2009. "A General Framework for Analyzing Sustainability of Social-Ecological Systems." *Science* 325: 419–22.

Palmer, M. A., and S. Filoso. 2009. "Restoration of Ecosystem Services for Environmental Markets." *Science* 325: 575–76.

Pasquini, L., C. Twyman, and J. Wainwright. 2010. "Toward a Conceptual Framework for Blending Social and Biophysical Attributes in Conservation Planning: A Case-Study of Privately-Conserved Lands." *Environmental Management* 5: 659–70.

Polasky, S. 2008. "Why Conservation Planning Needs Socioeconomic Data." *Proceedings of the National Academy of Sciences* 105: 6,505–6.

Polasky, S., J. D. Camm, and B. Garber-Yonts. 2001. "Selecting Biological Reserves Cost-Effectively: An Application to Terrestrial Vertebrate Conservation in Oregon." *Land Economics* 77: 68–78.

Portney, P. R., and J. P. Weyant, ed. 1999. *Discounting and Intergenerational Equity*. Washington, DC: Resources for the Future Press.

Pressey, R. L., and M. C. Bottrill. 2009. "Approaches to Landscape- and Seascape-Scale Conservation Planning: Convergence, Contrasts and Challenges." *Oryx* 43: 464–75.

Reed, M. S. 2008. "Stakeholder Participation for Environmental Management: A Literature Review." *Biological Conservation* 141: 2,417–31.

Richardson, E. A., M. J. Kaiser, G. Edwards-Jones, and H. P. Possingham. 2006. "Sensitivity of Marine-Reserve Design to the Spatial Resolution of Socioeconomic Data." *Conservation Biology* 20: 1,191–202.

Ricketts, T. H., J. Regetz, I. Steffan-Dewenter, S. A. Cunningham, C. Kremen, A. Bogdanski, B. Gemmill-Herren, S. S. Greenleaf, A. M. Klein, M. M. Mayfield, L. A. Morandin, A. Ochieng, and B. F. Viana. 2008. "Landscape Effects on Crop Pollination Services: Are There General Patterns?" *Ecology Letters* 11: 499–515.

Satterfield, T., S. Klain, R. Gregory, and K. Chan. 2011. "Intangibles, Measurement and Tradeoffs across Cultures of Environmental Management." Paper presented at the Environmental Norms, Institutions, and Policy Workshop, Stanford Humanities Center, Stanford University.

Schlosberg, D. 2007. *Defining Environmental Justice: Theories, Movements, and Nature.* Oxford: Oxford University Press.

Schreckenberg, K., I. Camargo, K. Withnall, C. Corrigan, P. Franks, D. Roe, L. Scherl, and V. Richardson. 2010. *Social Assessment of Conservation Initiatives: A Review of Rapid Methodologies*. London: International Institute for Environment and Development.

Serret, Y., and N. Johnstone. 2006. *The Distributional Effects of Environmental Policy.* Cornwall UK: Edward Elgar Publishing.

Sinden, J. A. 2004. "Estimating the Opportunity Costs of Biodiversity Protection in the Brigalow Belt, New South Wales." *Journal of Environmental Management* 70: 351–62.

Sodhi, N. S., and P. R. Ehrlich, eds. 2010. *Conservation Biology for All.* Oxford: Oxford University Press.

Steffan-Dewenter, I., M. Kessler, J. Barkmann, M. M. Bos, D. Buchori, S. Erasmi, H. Faust, G. Gerold, K. Glenk, S. R. Gradstein, E. Guhardja, M. Harteveld, D. Hertel, P. Höhn, M. Kappas, S. Köhler, C. Leuschner, M. Maertens, R. Marggraf, S. Migge-Kleian, J. Mogea, R. Pitopang, M. Schaefer, S. Schwarze, S. G. Sporn, A. Steingrebe, S. S. Tjitrosoedirdjo, S. Tjitrosoemito, A. Twele, R. Weber, L. Woltmann, M. Zeller, and T. Tscharntke. 2007. "Tradeoffs between Income, Biodiversity, and Ecosystem Functioning during Tropical Rainforest Conversion and Agroforestry Intensification." *Proceedings of the National Academy of Sciences* 104: 4,973–78.

Stewart, R., and H. Possingham. 2005. "Efficiency, Costs and Trade-Offs in Marine Reserve System Design." *Environmental Modeling and Assessment* 10: 203–13.

Sumaila, U. R., and C. Walters. 2005. "Intergenerational Discounting: A New Intuitive Approach." *Ecological Economics* 52: 135–42.

TEEB—The Economics of Ecosystems and Biodiversity for National and International Policy Makers—Summary: Responding to the Value of Nature. 2009. United Nations Environment Programme. http://www.teebweb.org/Portals/25/Documents/TEEB_D1 summary FINAL DOC.pdf.

The Economist. 2005. "Are You Being Served?" *The Economist,* April 21, 76–78.

Turner, W. R., K. Brandon, T. M. Brooks, R. Costanza, G. A. B. da Fonseca, and R. Portela. 2007. "Global Conservation of Biodiversity and Ecosystem Services." *Bioscience* 57: 868–73.

Underwood, E. C., M. R. Shaw, K. A. Wilson, P. Kareiva, K. R. Klausmeyer, M. F. McBride, M. Bode, S. A. Morrison, J. M. Hoekstra, and H. P. Possingham. 2008. "Protecting Biodiversity when Money Matters: Maximizing Return on Investment." *PLoS ONE* 3: e1515.

Venter, O., W. F. Laurance, T. Iwamura, K. A. Wilson, R. A. Fuller, and H. P. Possingham. 2009. "Harnessing Carbon Payments to Protect Biodiversity." *Science* 326: 1,368.

Venter, O., E. Meijaard, H. Possingham, R. Dennis, D. Sheil, S. Wich, L. Hovani, and K. Wilson. 2009. "Carbon Payments as a Safeguard for Threatened Tropical Mammals." *Conservation Letters* 2: 123–29.

Vitousek, P. M., H. A. Mooney, J. Lubchenco, and J. M. Melillo. 1997. "Human Domination of Earth's Ecosystems." *Science* 277: 494–99.

Wainger, L. A., and J. W. Boyd. 2009. "Valuing Ecosystem Services." In *Ecosystem-Based Management for the Oceans,* edited by K. McLeod and H. Leslie. Washington, DC: Island Press.

Watts, M. E., I. R. Ball, R. S. Stewart, C. J. Klein, K. Wilson, C. Steinback, R. Lourival, L. Kircher, and H. P. Possingham. 2009. "Marxan with Zones: Software for Optimal Conservation Based Land- and Sea-Use Zoning." *Environmental Modelling & Software* 24: 1,513–21.

Wilson, M. A., and R. B. Howarth. 2002. "Discourse-Based Valuation of Ecosystem Services: Establishing Fair Outcomes through Group Deliberation." *Ecological Economics* 41: 431–43.

Wilson, K. A., M. F. McBride, M. Bode, and H. P. Possingham. 2006. "Prioritizing Global Conservation Efforts." *Nature* 440: 337–40.

World Resources Institute (WRI), World Business Council on Sustainable Development (WBCSD), and Meridian Institute. 2008. *The Corporate Ecosystem Services Review: Guidelines for Identifying Business Risks and Opportunities Arising from Ecosystem Change.* United Nations Environment Programme.

3

Scale and conservation planning

Robert S. Unnasch, PhD; and Jason W. Karl, PhD

Conservation planning has been widely embraced as a way to efficiently allocate limited resources for biodiversity conservation to aspects most in need of protection or sustainable management. As conservation planners focus on a specific conservation questions, one of their first decisions is to determine the appropriate spatial analysis scales. Creation of successful long-term strategies for protection or management requires explicit consideration of biological and ecological patterns at different scales. Species differ in their size, mobility, physiology, and life history, and each responds to the same physical setting in different ways and at different scales (Addicott et al. 1987; Kotliar and Wiens 1990). Thus, the planning scale appropriate for one species is rarely fully appropriate for others.

Different species, communities, and ecological processes occur over different quantifiable spatial and temporal ranges. As a result, each can be observed over a set of characteristic scales. Additionally, ecosystem threats, and effective methods and actions for alleviating them, have characteristic scales at which they can and must be observed. Although specific threats to individual species, or small groups of species, can be recognized, ranking biodiversity elements or ecosystem threats to determine priorities for conservation is a challenging but necessary task.

A conservation planning method based explicitly in theories of conservation biology (e.g., redundancy, resilience, and complementarity) that results in actions being taken to preserve biodiversity has a better chance of long-term success and is less biased than a purely ad hoc or reactionary approach (Margules and Pressey 2000; Groves et al. 2000). However, the complex nature of ecosystems and their component biodiversity and integrative function (see O'Neill et al. 1986) makes conservation planning difficult at best. Reducing this complexity so that planning becomes manageable requires planners to use an analysis scale appropriate to the questions being addressed and the mechanisms available for implementation.

The obvious patterns of scale perceived by the conservation planner rarely coincide with the spatial structure and configuration driving the behavioral and ecosystem processes that are the subject of conservation efforts. For example, Poiani et al. (2000) proposed four levels of spatial scale to guide conservation planning efforts (figure 3.1): regional (millions of hectares), coarse (tens of thousands to millions of hectares), intermediate (hundreds to tens of thousands of hectares), and local (hectares to thousands of hectares).

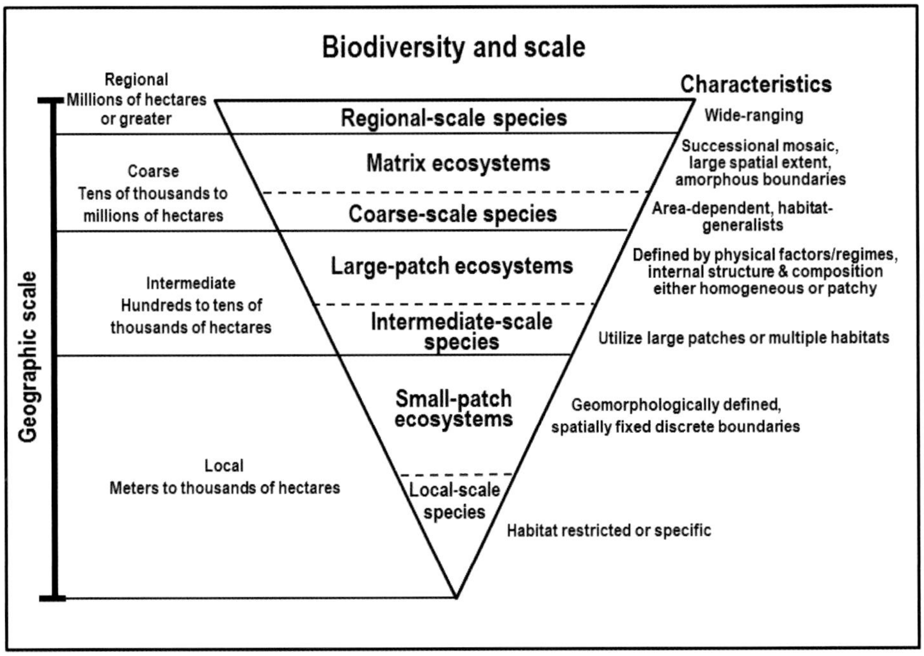

Figure 3.1. Biodiversity and scale interactions as envisioned by Poiani et al. (2000) produce four logical divisions for conservation planning. Figure from Poiani et al. (2000).

Sanderson et al. (2002) suggested that "landscape species," by merit of their extensive use of a landscape, capture the needs of many other species, and that planning for persistence of these species would benefit conservation of many other species as well. Such planning frameworks, however, focus largely on the spatial *extent* of use, while ignoring the *grain* and pattern within a planning area (see discussion below). Andelman and Fagan (2000) reported that such "umbrella" species provide for the conservation needs of other species no better than do randomly selected species. This is a manifestation of the problematic issues of spatial scale. Choosing appropriate target species from a habitat and ecosystem perspective is discussed in detail in Brock and Atkinson (chapter 6).

Planning for conservation at a given scale should include the following:

- Conservation targets—any discrete element of biodiversity significant to conservation objectives such as species, subspecies, unique populations, or communities
- Threats—anthropogenic or other stressors that negatively impact conservation targets
- Conservation actions appropriate to the given scale

Ideally, planning should be framed at a variety of scales. Using a hierarchical (nested) framework of scales ensures that a wide range of conservation targets and threats can be addressed. It also helps translate conservation priorities between scales, providing larger-scale context to specific actions and showing how fine-scale actions can contribute to coarse-scale conservation priorities. The conservation planner may be able to address only a single question at a particular scale within the hierarchy at any one time, but must remain cognizant of its relationship to planning at other scales. Modern geographic information systems (GIS) facilitates the use of different scales and analysis of their interrelationships, making the explicit consideration of scale in conservation planning even more imperative. Decisions about scale are made in the early stages of a systematic conservation planning process, as described by Ban et al. (chapter 8, table 8.1).

The choice of scale for a conservation plan necessarily is limited by the available data. The scale at which data is collected determines not only the detail at which features on the ground can be represented, but also the number of features that can be resolved. Both of these factors affect the inferences that can be drawn from spatial data (see discussion below). In most cases, the base data consists of land cover and infrastructure (see chapters 4 and 5) along with topographical data (elevation, slope, and aspect). Which datasets are used depends upon the target of interest and the scale at which it responds to the environment (e.g., how species perceive their habitat). For additional discussion of modeling assumptions related to scale, see Johnston et al. (chapter 14).

In this chapter, we review concepts of scale and provide examples of how scaling issues can affect conservation planning, including selection of conservation targets, identification and impacts of threats, delineation of conservation priority areas, and the influence and aggregation of local actions on realizing broader conservation goals. To reiterate, we propose that planning for conservation at a given scale should include conservation targets, threats, and possible actions appropriate to that scale, and be conducted within a framework recognizing a variety of scales. We provide two examples of nested planning hierarchies and discuss what conservation questions are relevant at different scales within each.

Working with scale

"Scale" has several definitions in spatial analysis and ecology, causing confusion as to the impact of scale on biodiversity conservation. Nowhere is this more apparent than in conservation planning, because at least three operative definitions of "scale" must be considered: cartographic scale, analysis scale, and ecological scale. To discuss the impacts of scale on conservation planning, it is necessary to first clarify these definitions.

Generally defined, "scale" is a characteristic of a set of observations that have a definable grain and extent (Turner et al. 1989). "Grain" is the finest level of observable detail, and "extent" is the maximum size of the area under consideration (Wheatley and Johnson 2009). Together, grain and extent set the limits on what can be observed with a given dataset. Elements or patterns that occur below the grain size cannot be detected. Conversely, patterns greater than the extent generally cannot be resolved, because the total of observations within the extent represent only a portion of the entire pattern (Wiens 1989; Burnett and Blaschke 2003). An appropriately defined scale can reveal patterns in data while filtering out unnecessary information or noise (Wu and David 2002).

Not all observations have spatial scale. Wheatley and Johnson (2009) differentiate between spatial observations and scalar observations (i.e., observations having scale). In the former, location is important for describing the attribute being measured, but because the level of spatial resolution (the grain) and maximum area of consideration (the extent) have not been specified, location-based observations, in themselves, do not have scale. A collection of point observations of the presence of a species of conservation concern could be considered spatial observations without scale. In the United States, each state maintains a Conservation Data Center or Natural Heritage Program that collects and manages such data.

For spatial observations, references to scale reflect the accuracy and precision of the observations in representing real-world features. All spatial data has a scale at which the data was originally acquired, whether by digitization of an aerial photograph, collection by a global positioning systems (GPS) unit, or collection by an Earth-orbiting satellite. That native scale dictates not only what features can be observed in the data, but also the precision with which they can be recorded (see discussion below). For vector data (points, lines, and polygons), grain generally is expressed in terms of the cartographic scale at which the data was collected. With raster data, the grain is the resolution (dimension of one side of a pixel) of the raster. Spatial data displayed on a map takes on a cartographic scale. Cartographic scale is the relationship between features on a map and the corresponding objects on the ground. A representative fraction (e.g., 1 inch equals 1 mile) is an example of cartographic scale.

Most important from the standpoint of spatial analysis for conservation planning is the analysis scale. The analysis scale defines the grain and extent relative to an analysis objective, rather than being intrinsic to the data. For example, if the analysis objective is to evaluate habitat connectivity, the extent will be larger, and the grain also may be larger, for a wide-ranging species like a wolverine (*Gulo gulo*) than for a narrow-ranging species like a boreal toad (*Bufo boreas*). For spatial analysis, project boundaries define the extent. Grain refers to the smallest units that are being considered and to which all other datasets are aggregated. In some cases, the grain units for conservation planning are of consistent shape and size (e.g., square cells or hexagons), but they can also be variable in shape and size (e.g., hydrologic units, counties, or ecological land units). Specifying a set of features to serve as the analysis grain for conservation planning allows spatial observations to be translated into scalar observations and used with different data layers.

Also important for conservation planning is the ecological definition of scale. Grain and extent can be defined with respect to how ecological or management processes create patterns across a landscape. Both natural and human-caused processes have characteristic frequencies and magnitudes that result in landscape patterns detectable only at certain scales. In this regard, "grain" refers to the smallest elements upon which a process depends, and "extent" refers to the coarsest set of patterns a process produces (Addicott et al. 1987; Kotlair and Wiens 1990). For example, the finest unit, or grain, of a timber harvest typically would be a forest *stand*, and the extent would be the total area of harvestable forest land. Thus, from an ecological or management perspective, scale is tied directly to a specific process and is not an inherent property of a system (Wiens and Milne 1989). In other words, there is no single ecological scale, even for an individual species, and a significant challenge in conservation planning is defining the correct scales for assessing the effects of specific actions and ecological processes.

Conservation targets are often referred to as being either "coarse filter" or "fine filter," and some confusion may exist between these terms and the definitions of "coarse scale" and "fine scale." Whether a species is a coarse- or fine-filter target refers to its ability to act as a surrogate for one or more species in a conservation plan (Hunter and Kohm 1991; Noss 1987, 1996; Noss et al. 1997).

In theory, conservation of a coarse-filter target (also called umbrella, focal, or keystone targets) provides protection to many other targets. Fine-filter targets, however, generally are single species or narrowly defined plant communities that require explicit consideration in a conservation plan. Either they are not expected to be adequately conserved via planning for coarse-filter targets or have special conditions, such as critically low population or conservation status (e.g., threatened or endangered), that require special consideration. It is important to note that fine- and coarse-filter targets can be included in conservation plans at any scale. However, it is common practice to identify as "coarse" those targets requiring the largest spatial extent. This approach to defining conservation targets and how it relates to a focal species approach is discussed further by Brock and Atkinson (chapter 6).

Why scale is important to conservation planning

In the context of conservation planning, scale has important implications for the following:

- Selecting processes, targets, and threats for consideration
- The relationships between targets, their key attributes (Parrish et al. 2003), and processes or threats
- The kinds of data used and the applicability of planning results

The first reason scale is important to conservation planning is that effects of natural and human-caused processes and events can only be observed or measured across specific scale ranges. Populations and distributions of conservation targets vary in space and time in response to processes and events (figure 3.2). In one sense, species distribution is a response variable dependent upon habitat covariates such as vegetation type, elevation, and so on, as discussed by Dickson et al. (chapter 7), which, in turn, are determined by underlying physical and biotic processes. Successful conservation planning starts with knowing what processes benefit or threaten the persistence of a conservation target (Parrish et al. 2003). Hence, the processes relevant to a conservation target will help determine the scales at which conservation planning should take place.

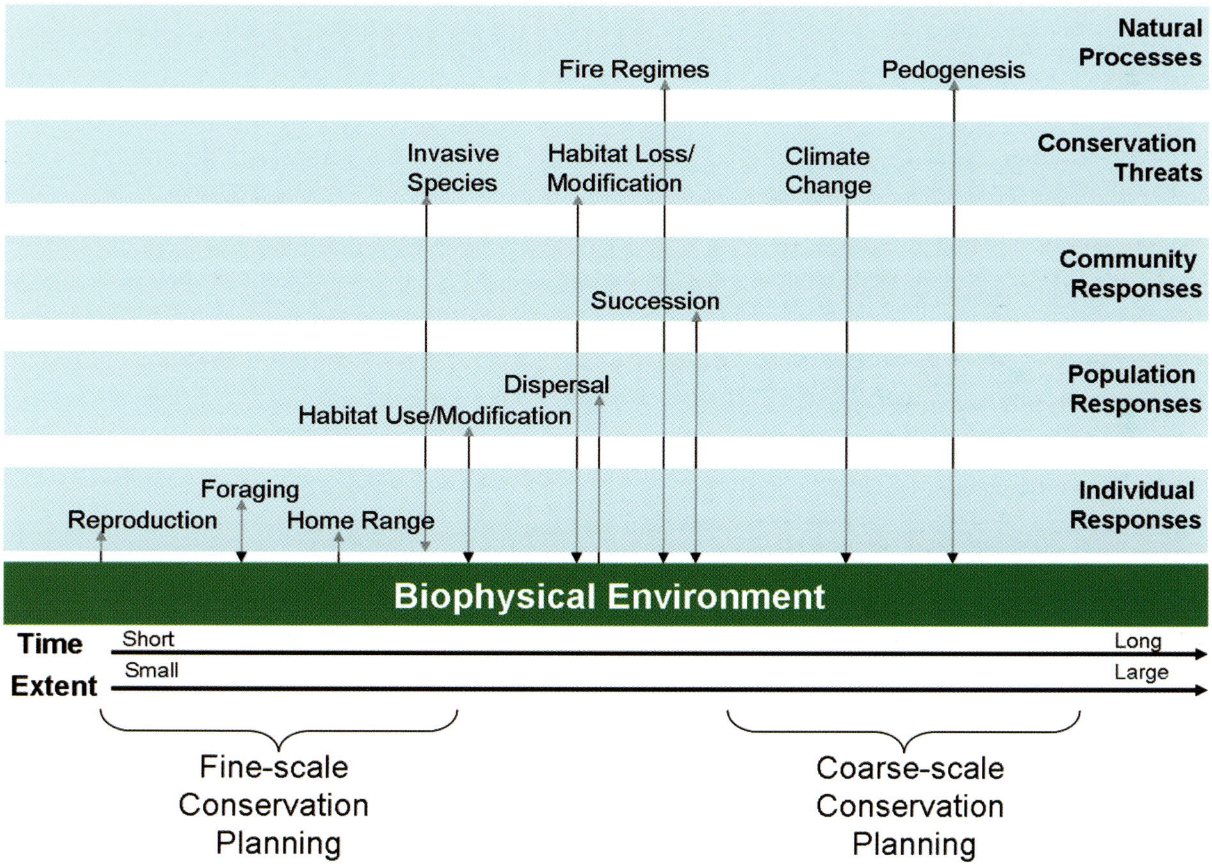

Figure 3.2. Biological, management, and biophysical processes over spatial and temporal scales. Processes at finer scales are constrained by those operating at longer or larger scales. Thus, conservation planning at fine scales must take these constraints into consideration. Figure from Jason W. Karl.

A second reason scale is important to conservation planning is that a species' associations with its environment can appear different when observed at different scales. Consider the association between sage grouse (*Centrocercus urophasianus*) and agriculture at three different scales. Historically, sage grouse were found in sagebrush habitats across much of the western United States. The loss of sagebrush habitat to agricultural development is one contributor to their widespread decline across their range. So, at a range-wide scale, sage grouse are negatively associated with agriculture. Sage grouse do use agricultural lands, however, because they are a rich source of the insects hens need for rearing their chicks. At an intermediate scale that encompasses multiple populations (e.g., the extent of one or several counties), good sage grouse habitat is considered to have a mix of high-quality sagebrush habitat and agricultural fields (Schroeder et al. 1999). So, at this scale, sage grouse are positively associated with agricultural lands. Agricultural lands, though, do not provide important resources, including cover, nesting habitat, or lekking areas that are important aspects of the sage grouse life cycle, and lack of these exposes sage grouse to increased risk of predation or disturbance. At the scale of an individual population, high-quality sagebrush habitat is the essential, and often is a limiting factor to sustaining sage grouse populations in the face of agricultural encroachment. Thus, at the finest scales, sage grouse are again negatively associated with agricultural lands. In a hierarchy of conservation plans, agriculture can be treated as a component of habitat or a threat, depending upon the scale of analysis.

A third reason scale is important to conservation planning is that the synthesis of different data layers into an analysis scale produces an output specific to that scale. Therefore, the results of a particular conservation plan cannot be scaled down because information has been lost in aggregating to the analysis scale, nor can results be predictably scaled up. A good example is the use of climate-change predictions from global climate models (see Bachelet, chapter 13). When results from these models are scaled down, their accuracy is limited to that of the original projection and thus do not account for local heterogeneity. Ideally, the choice of an analysis scale for conservation planning would be the result of consideration of the scales relevant to the conservation targets. In practice, however, this often is not done. Analysis scales are selected arbitrarily or by convenience, and, as a consequence, planning results may not be applicable to the scales at which actions will be taken.

Influence of cartographic scale on conservation planning data

Data issues

All data used in conservation planning has a specific, original scale that is defined by the processes of observing and recording real-world elements or events. Knowing the original scale of data used and selecting the right scales of data for conservation planning is crucial to generating valid results for the intended scale of application. Two issues are involved with the scale of data used in conservation planning.

First, the scale of a spatial dataset controls the precision with which real-world features can be represented and the accuracy with which they can be located. At very fine scales (e.g., 1:6,000 or greater), it is possible to represent most of the details of a feature or boundary, and to accurately define its coordinates. At a scale of

1:6,000, the width of a 0.5-mm line on a map represents a ground distance of approximately 3 m (10 ft), and it is possible to resolve very small features or small details of larger features from aerial imagery or using a differentially corrected GPS. As scale coarsens, however, the ability to resolve details decreases, and the error in placing features on a map increases. At a scale of 1:250,000, the same 0.5-mm line on the map is now 125 m (410 ft) wide in real-world distance. At this scale, it is no longer possible to resolve many of the fine-scale details of features, and misplacement of a feature point or corner by even a small amount translates into large errors. Accordingly, coarse-scale data may be suitable for coarse-scale regional conservation planning, but is not suitable for fine-scale, local planning efforts.

Second, the cartographic scale of a spatial dataset controls how completely a set of features can be represented. At very coarse cartographic scales, it may not be possible to resolve very small features. Consider the example of the roading in the Clearwater National Forest, Idaho (figure 3.3). At the 1:100,000 scale, many of the smaller logging roads are indiscernible—only the major roads can be seen and mapped accurately.

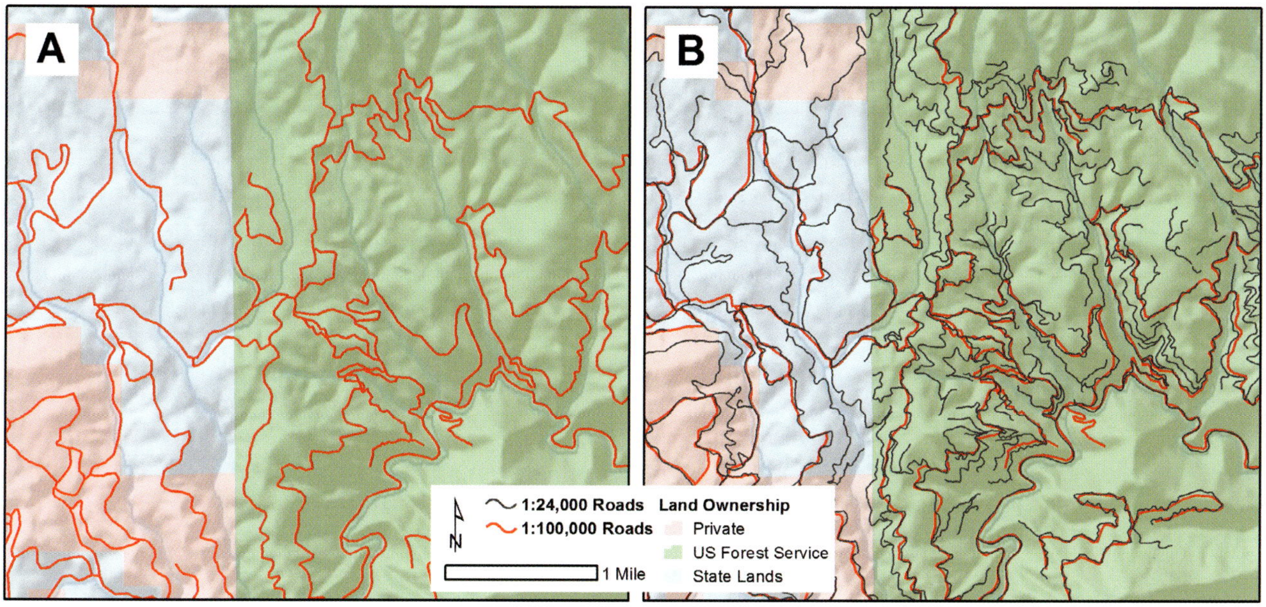

Figure 3.3. Road datasets for a forested landscape in northern Idaho. Using a roads layer with a native cartographic scale of 1:100,000 (A), there appears to be much roadless area. Using a roads layer (B) with a native cartographic scale of 1:24,000, however, the existence of many more roads becomes apparent. Also, the 1:24,000-scale layer better represents the location and detail of the roads shown in the coarser-scale layer. Land ownership from Clearwater National Forest; 1:24,000 roads from Clearwater National Forest; 1:100,000 roads from ESRI Data & Maps, 2005, courtesy of Esri.

The choice of cartographic scales used in conservation planning should be dictated, wherever possible, by the relationships between the conservation targets and the features captured by the datasets. The fact that small features can be missed when mapped at coarse cartographic scales may or may not be of consequence for coarse-scale conservation planning. If the smaller features missed are relatively unimportant to the overall conservation target (e.g., low-order or ephemeral streams for anadromous fish distribution), then their exclusion is not critical. However, if the small features are important (e.g., wetlands or vernal pools for amphibians) then a coarse cartographic-scale dataset may not suffice even for a coarse-scale planning effort.

In addition, factors of data availability and the practicality of working with datasets must be considered. Although it may be desirable to use the finest-scale cartographic data possible, such data may not be available across the entire region of interest. Also, the computer storage space and processing requirements for dealing with fine-scale cartographic datasets spanning very large regions may not be tractable, although this is becoming less of a problem with advances in computer technology.

Influence of scale on inferences from data

Scale has important implications for the type and quality of inferences that can be drawn from data. At a basic level, the grain and extent of a scale define the upper and lower bounds of an inference space. However, the effects of scale on a planner's ability to make inferences from data extend beyond grain and extent, impacting many factors relevant to conservation planning.

First, scale influences the kinds of questions that are important to ask: relevant questions for a conservation target are determined by the patterns and processes to which the target responds. Thus, a target's functional attributes must influence the conservation planning scales selected for that target. As discussed above, the effects of agricultural development on sage grouse vary across at least three distinct scales. Statewide planning for sage grouse seeks to determine important regions within sage grouse range for habitat conservation and restoration. Within smaller regions of a few million acres, the statewide prioritizations are refined based on indices of habitat quality to identify precisely where and how habitat conservation and restoration will occur locally. At the finest scale, the local priority areas are used as a context for designing and implementing individual projects. The objectives of planning at each scale are different, and the nested nature of the planning helps ensure that local actions contribute to meeting coarser-scale conservation goals.

Scales commonly used in conservation planning

Conservation planning generally is focused at three scales: global, regional, and local. At each of these scales, the objective of conservation planning is different. Global and regional conservation plans typically focus on *where* to conserve biodiversity. At a global scale, the objective is to identify rare, unique, or imperiled regions in need of conservation. At a regional scale, the objective is to identify combinations of the best-quality areas that, if protected, would ensure long-term persistence of conservation targets. Conservation planning at a local scale takes place at the level of management areas (e.g., preserves or wildlife areas), counties, or priority landscapes, and focuses more on *how* to conserve biodiversity. At this scale, site-specific factors (e.g., ownership in and around the site, disturbance and management history, and proximity and immediacy of threats) must be considered because they have considerable impact on not only the form that conservation action will take, but on its likelihood for success. For these reasons, local-scale conservation planning is sensitive to the extent of knowledge of the area's biodiversity (The Nature Conservancy et al. 2003). The Nature Conservancy's hierarchical approach provides a good illustration of planning across these scales (see below and the "How We Work" section of The Nature Conservancy website, http://www.nature.org). Similar approaches have been used by other organizations and agencies.

Global conservation planning

Several approaches to global-scale conservation have been developed. These have generally been the province of large, multinational conservation organizations. Conservation planning on a global scale began when the biosphere reserve concept was first proposed in 1974 by the United Nations Educational, Scientific, and Cultural Organization (UNESCO). Later, Conservation International's global assessment emphasized areas of high species richness and endemism. Under this global strategy, 25 biodiversity hotspots were identified that together support about 62% of the Earth's terrestrial vertebrates (Mittermeier et. al. 2000, 2005). The World Wildlife Fund has used a similar hot-spot approach that incorporates a measure of threat (Dinerstein and Wikramanayake 1993). In contrast, The Nature Conservancy has taken a standardized approach to global planning based on major habitat types. Global habitat assessments (Hoekstra et al. 2010) of the conservation status of ecoregions across the planet are used to identify conservation gaps and help establish the organization's priorities.

Regional conservation planning

Regional conservation planning is well established and has been conducted systematically in many parts of the world by many different organizations. In the United States, one of the first regional conservation planning efforts was the Gap Analysis Program (GAP; Scott et al. 1993). Originally conceived in Hawaii as a proactive way of identifying habitats to protect and to prevent endangerment or extinction of Hawaiian forest birds, a gap analysis was first formally prototyped in Idaho, and subsequently has been applied in almost every state. Land-cover data was classified from Landsat imagery and then used to develop predictive range maps for individual terrestrial vertebrate species on a state-by-state basis using habitat suitability models in a GIS. The goal of the program was to identify important plant communities and wildlife habitats not formally protected and thus under threat of being altered or destroyed.

The term "regional" can be applied to assessments over a wide range of extents using different data resolutions, or grain. Examples provided in this volume include wolverine habitat modeled over most of the state of Washington (Johnston et al., chapter 14) and bird habitat modeled over a 15,000-km^2 area in the Great Basin (Dickson et al., chapter 7), both using 30-m data, to focal species habitat modeled over a 194,824-km^2 area of interior cedar-hemlock forest (Williams, chapter 12) and the Canadian Rockies Ecoregional Assessment of more than 1.5 million hectares (discussed below), both using 1-km data.

The Wildlife Conservation Society's Landscape Species Approach (Sanderson et al. 2002) generally has been applied at a regional scale. The methodology originally involved defining five criteria for selecting landscape species—area requirements, heterogeneity, ecological function, vulnerability, and socioeconomic significance. This approach was designed as a transparent, replicable method for identifying species as "umbrellas" for site-based conservation (landscape species), identifying data gaps, ranking threats, and setting research priorities (Coppolillo 2004).

Conservation planning at a regional scale has been extremely important for marine environments and is discussed in detail by Ban et al. (chapter 8) with examples from the Australian Research Council Centre of Excellence (http://www.coralcoe.org.au/index.html).

The Nature Conservancy's ecoregional assessments

At regional scales, The Nature Conservancy uses ecoregions to identify priority areas for conservation. Ecoregions are not homogeneous areas, but are convenient for regional-scale planning because they are defined by natural boundaries of composition and function that often relate better to distribution patterns of biodiversity elements than do arbitrary geometric (e.g., square or hexagonal) or political (e.g., state or provincial) boundaries.

Ecoregional assessments center on ensuring the persistence of conservation targets, a subset of species, or ecological features within an ecoregion. Conservation goals focused on abundance and distribution are defined for each target based on available data and opinions of experts. A portfolio of conservation areas is crafted that seeks to satisfy the goals for each target as efficiently as possible. Threats to target persistence are evaluated, and the portfolio is prioritized to highlight those experiencing more severe and immediate threats. Thus, the primary outputs of an ecoregional assessment are the portfolio of conservation areas (i.e., a map of areas of biodiversity significance in the ecoregion) and a prioritization of which of those need attention first.

Originally, Groves et al. (2002) suggested seven major steps to an ecoregional assessment:

1. Identify conservation targets.
2. Collect information and identify information gaps.
3. Establish conservation goals for the targets.
4. Assess how well existing conservation areas protect targets.
5. Evaluate the ability of the targets to persist.
6. Design a portfolio of conservation areas that will help ensure target persistence.
7. Identify which conservation areas are priorities for action.

Subsequently, this framework has been supplemented with guidelines and "best practices" to not only help assure high-quality, credible ecoregional assessments, but also to increase the usefulness of the results (The Nature Conservancy and World Wildlife Fund 2006).

According to The Nature Conservancy et al. (2003), "The portfolio of areas of biodiversity significance should be regarded as the ultimate vision for the set of areas on-the-ground that require conservation attention." However, overreliance on conservation area maps may be counterproductive to conservation efforts in an ecoregion. Ecoregional assessments and their output are the product of the best data obtainable at a specific time and should be viewed as dynamic analyses that should be refreshed and improved regularly (Groves et al. 2000). Spatial distribution of biodiversity threats and costs of conservation can influence greatly the configuration of conservation areas, but these two parameters currently are not well understood. Additionally, the process of distilling "efficient" conservation areas from the input data (see discussion of Marxan by Williams, chapter 12) may result in viable target occurrences being excluded from the final conservation portfolio. These problems are compounded when the data available for an ecoregional assessment does not accurately depict the distribution of targets or ecological integrity. Thus, although a conservation area may be sufficient for ecoregion-wide conservation, it may not be necessary. Conversely, just because an area was not designated a conservation area does not mean it cannot contribute to ecoregional conservation.

The Nature Conservancy's ecoregional planning (see sidebar "The Nature Conservancy's ecoregional assessments") focusing on terrestrial, freshwater, and marine ecosystems has been documented by Groves et al. (2000, 2002), Groves (2003), The Nature Conservancy et al. (2003), and The Nature Conservancy and World Wildlife Fund (2006). The resolution for these ecoregional assessments are 6th-order hydrologic unit codes or hexagons ranging in area from 3–5 km^2. Predictive range maps for each targeted plant and animal species, all natural communities, and all ecosystems are developed (The Nature Conservancy 1997), and each unique analysis unit is characterized by the predicted presence or absence of each conservation target. The conservation planning software Marxan (Ball et al. 2009) is used to identify a portfolio of priority conservation areas. Ecoregional portfolios are then developed based on a three-tiered approach (Noss 1996): (1) portfolio sites strive to capture the spatial range of all targets, (2) all habitat needs for focal species are represented, and (3) all known occurrences of special elements (e.g., threatened and endangered species) are captured. Data layers for ecoregional planning in the United States have relied heavily upon GAP land cover at a grain of 30 m or 90 m.

Although a map of conservation areas generally has been considered the main outcome of an ecoregional assessment, the ancillary outputs and byproducts of an ecoregional assessment can be equally or even more valuable for achieving conservation. Data collected on target distributions and viability, in addition to supporting data gathered for the planning process, is useful for other planning exercises or spatial analyses. Examples include the contribution that alternative conservation areas may make to ecoregional conservation goals (see "Scaling down an ecoregional assessment," below) or the relative ecological value of parcels to be considered in a public-private land swap. Another important benefit of an ecoregional assessment is that it often unites stakeholders around a common conservation vision. Ecoregional assessments that include diverse, interdisciplinary teams from nongovernmental organizations, government agencies, and private partners produce robust results that can meet the needs of participating groups and galvanize support around specific areas and desired outcomes.

Local conservation planning

Major habitat and ecoregional assessments focus on setting goals and priorities for conservation, but local-level planning emphasizes developing and implementing strategies to address priorities, achieve goals, and measure outcomes. The Nature Conservancy develops local plans to identify priority landscapes and specific management areas.

Planning at local scales, a major focus of this book, is perhaps the most rapidly developing front in conservation planning. County governments, planning departments, and land-management agencies in the United States are becoming increasingly GIS-proficient and are embracing conservation goals in concert with traditional land-use and infrastructure planning. Local planning efforts often utilize GAP land-cover data with a grain of 30 m or color orthophotos at finer scales, and increasingly are using remote sensing products at even finer scales, such as QuickBird, Satellite Pour l'Observation de la Terre (SPOT), and lidar imagery.

Dickson et al. (chapter 7) provide an example of occupancy modeling at a local scale using land-cover data for a 135-km^2 focal study area derived from 10-m resolution SPOT 5 and 1-m resolution color infrared (CIR) imagery. Johnston et al. (chapter 14) provide examples of local analysis for spruce budworm outbreaks in a single drainage basin.

Conservation area (action) planning and landscape conservation plans

The Nature Conservancy has developed Conservation Action Planning (CAP) as an adaptive approach to defining conservation actions that ultimately result in effective on-the-ground protection of biodiversity. In CAP, teams of scientists and managers work iteratively through a series of steps to focus on key biodiversity elements and ecological attributes in need of conservation, identify threats to these elements, determine clear steps and timelines for addressing the threats, and define how to measure the success of actions taken (Poiani et al. 2006).

The current version of CAP has ten steps broken into five principal groups (table 3.1; The Nature Conservancy 2005). The first group deals with defining the scope of the plan in terms of who the stakeholders are, what the spatial and temporal extent of the plan is, and what the conservation targets are. The second group of steps analyzes the status and viability of the focal targets to identify threats, formulate specific strategies for ensuring long-term target persistence, and design programs for monitoring the effectiveness of conservation actions and the status of the targets. The final groups entail implementing the conservation strategies and measures identified previously, conducting monitoring, and using the data to adapt the process over time. Although CAP is typically presented in a linear fashion, in practice it is an iterative and nonlinear process by which information is updated as thinking evolves. To support its iterative nature, they created a customized Microsoft Excel spreadsheet (available from http://conserveonline.org/workspaces/cbdgateway/cap/resources/index_html/) to automate compilation of CAP data, facilitate information sharing and updating, and archive the process. The CAP process has also been integrated into the Miradi conservation-planning software (https:// miradi.org) developed by the Conservation Measures Partnership.

Table 3.1. The standards for the practice of conservation

Conceptualize the project
1. Identify people involved in the project. 2. Define project scale and focal conservation targets.
Plan strategies, actions, and measures
3. Assess viability of focal conservation targets. 4. Identify critical threats. 5. Develop conservation strategies. 6. Establish measures.
Implementing conservation strategies and measures
7. Develop work plans. 8. Implement.
Using results to adapt and improve
9. Analyze, reflect, and adapt. 10. Learn and share.
Capture and share learning
11. Document and share lessons learned. 12. Create a learning environment.
Source: Robert S. Unnasch and Jason W. Karl.

Conservation planning at other scales

Occasionally, conservation planning at one of the three typical scales is either not adequate or not appropriate. In these cases, supplemental planning between two scales can be implemented, or a separate planning effort can be performed at a specific scale. For instance, if a large disparity between regional and local planning exists in terms of spatial extent, targets and threats considered, or data availability, an intermediate-scale conservation plan can help refine regional priority areas and assist in translating local-scale actions to regional-scale conservation. As another example, wide-ranging carnivores and migratory birds may occupy areas larger than the typical local-scale plan, and a conservation plan targeted to a scale appropriate for their management may be needed. Below are two examples of conservation planning at scales intermediate between regional and local efforts.

Scaling down an ecoregional assessment

As part of the Forest Stewardship Council certification of more than 668,000 acres of private forest land in 2003, the Potlatch Corporation, a private timber company in northern Idaho, was required to identify forests with high conservation value (see Cauley et al. 2001). Staff from The Nature Conservancy's Idaho chapter were asked to conduct this assessment, using its newly completed Canadian Rocky Mountains ecoregional assessment (Rumsey et al. 2003); see figure 3.4. For different approaches to conservation priorities in this region, see figures 12.2 through 12.5 in chapter 12. These approaches did not include threats (vulnerability) except where they were reflected in land cover (see also chapter 5).

The original intent was to use the ecoregional priority areas to identify forests with high conservation value, but it was quickly discovered that the ecoregional assessment data was at the wrong resolution to be informative at this finer scale (the example shown in figure 3.4 was from this planning effort). A new conservation prioritization was conducted using a set of targets specific to northern Idaho and sets of more recent, higher-resolution data. The priority forest stands that resulted were selected so they would meet or exceed the contribution to ecoregional conservation goals identified in the original ecoregional assessment for the same region. Using a simple, richness-based method for determining conservation priorities, a new set of priority areas were identified that better reflected the distribution of targets and ecosystem quality.

This revised set of priority areas was more informative for actions on the ground, made roughly equivalent contributions to regional conservation goals, and placed the finer-scale areas in the context of the broader ecoregional conservation vision. The Canadian Rockies Ecoregional Assessment (Rumsey et al. 2003) delineated priority conservation areas totaling 52.6% of the study area (834,000 ha). Total area of the nested Potlatch conservation priority areas was similar, 51.5% (818,000 ha), but with a much different spatial configuration within the study area. Generally, these intermediate conservation priority areas were more efficient at capturing the distribution of the conservation targets and meeting conservation goals than were the Canadian Rockies Ecoregional Assessment areas.

The intermediate-scale planning also gave direction to local prioritization at the scale of ownership parcels or management units (e.g., stands). This effort illustrates a nested approach to conservation planning where

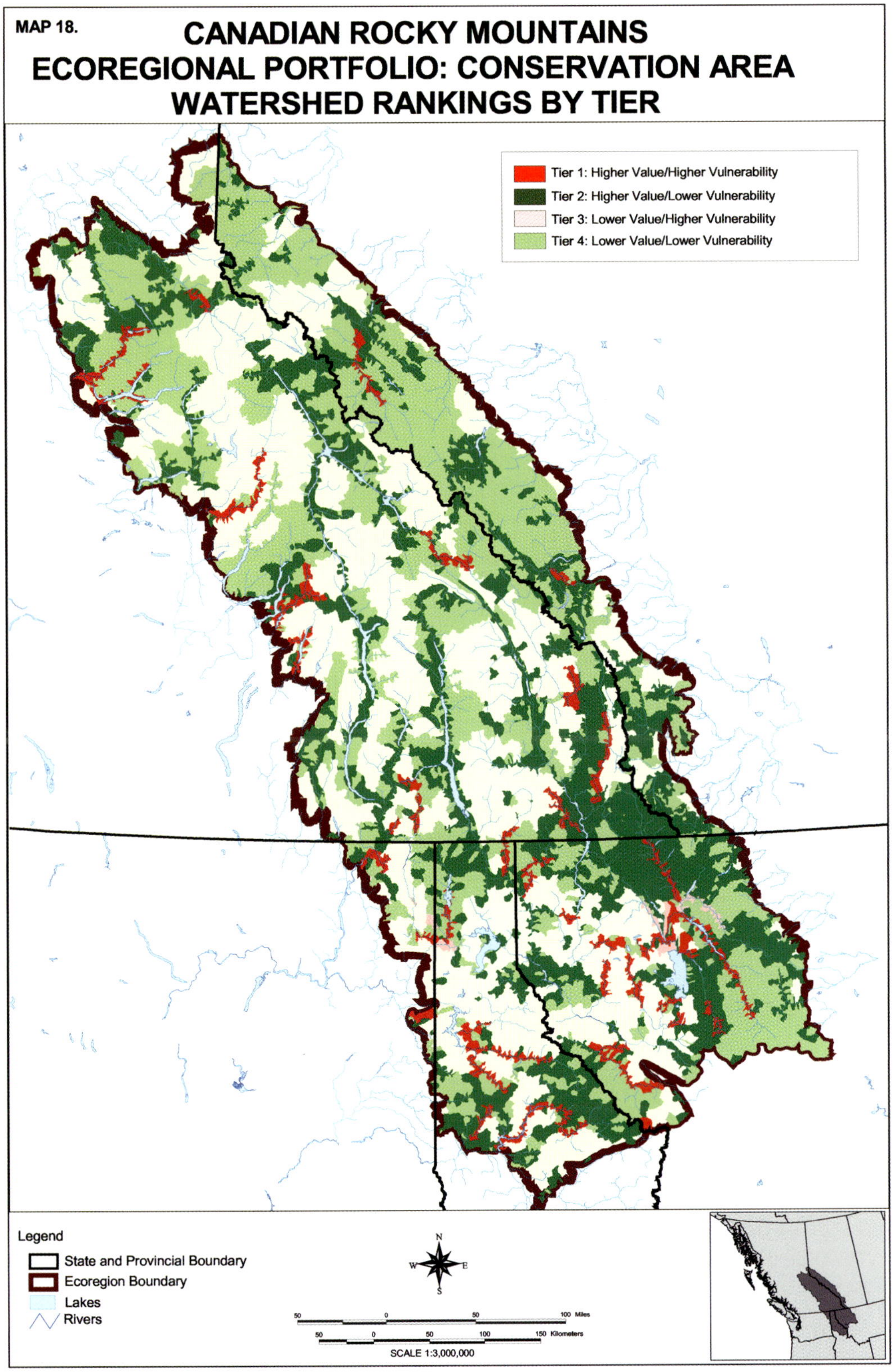

Figure 3.4. A conservation portfolio for the Canadian Rocky Mountains ecoregion. This portfolio was developed to meet specific abundance and distribution goals for 342 conservation targets. Figure from the Nature Conservancy of Canada.

it becomes clearer how specific areas or individual actions operating locally or over large areas contribute to regional conservation objectives.

Two unique features of this approach have relevance to conservation planning at multiple scales. First, finer-scale data provides results more useful at the scale of the local study area. Local priority areas provided similar contribution to ecoregional goals partly because the objectives (manifest by the selection of conservation targets) between the two levels of prioritization were similar. Second, dividing conservation targets into tiers by target type allowed use of a larger volume of information by recognizing differences in the quality and completeness of data representing the targets. The tier-one targets (ecological systems) were identified in a comprehensive coverage of the study area. This dataset had the coarsest spatial delineation and provided little information on the accuracy of the classification or the quality of the habitats within each ecological system. The tier-two targets were also identified across the entire study area using datasets with increased accuracy and detail, making it possible to determine attributes that might affect habitat quality. The tier-three targets were identified using less-than-complete datasets, but the information collected was assumed accurate and complete enough to represent generally the target distribution in the study area. Data used to identify tier-four targets was assumed incomplete and, except for a few well-known species, did not represent fully their distribution in the study area. Organizing targets into tiers by these definitions made it possible to begin prioritizing with the most complete and accurate data (tier two) and then incorporate less complete datasets in subsequent steps. Prioritizing based on all conservation targets lumped together, a common practice, could risk prioritizing areas based on personal knowledge rather than more objective criteria.

Dynamic modeling of ecosystem processes to prioritize restoration activities in the Castle Creek area, Idaho

When planning to implement practical conservation actions, it is important to consider the ecological and management processes affecting the conservation targets as carefully as the distribution of the target or the patterns of action across the landscape. This requires an understanding of the ecological communities in the landscape, and how they may respond at the scale of natural processes, including disturbance, and management actions, both related and unrelated to the conservation goal. One way to do this is through the use of state-and-transition (S/T) models, which are depictions of what is known about how an ecosystem is structured and how it functions (Westoby et al. 1989; Bestelmeyer et al. 2003).

Unlike simple successional models of ecological change, S/T models often incorporate nonlinear dynamics where disturbances, management events, or other processes cause a vegetation community to change to a new successional state or pathway (Westoby et al. 1989). A state in an S/T model is considered to be a plant community (or collection of plant communities that respond similarly to management or disturbance) that is relatively stable in relationship to natural and management processes and the environment. A transition is a shift from one state to another. In the strict sense of S/T modeling, a transition is not reversible without the addition of energy to the system, usually in the form of active management (Bestelmeyer et al. 2003). However, a community can cycle between states in the same successional pathway without undergoing a transition.

Knowledge of S/T models provides several advantages in local conservation planning. First, they explicitly model our understanding of how species and plant communities respond to processes and their environment.

This is crucial for formulating management actions that address threats and stresses to conservation targets and developing indicators to monitor the status and trend of targets. Second, the probability of transitions can be specified within the model. In developing these models, the limits of knowledge about an ecosystem become quickly apparent, and new research can be undertaken to expand the database. Third, with a quantified S/T model, software for dynamic simulation (i.e., change over time in response to deterministic and stochastic factors) can be used to explore what a landscape might look like in the future under different scenarios and the long-term effects of different management regimes.

As an example, a century of fire suppression in the western United States has caused ecological changes by disrupting the frequency of wildfire, a coarse-scale natural process. In southern Idaho, one effect of fire suppression has been the expansion of western juniper (*Juniperus occidentalis*) beyond its historic range. The natural wildfire process kept juniper largely restricted in its range in southern Idaho to places like rocky outcrops, cooler ridges and mountain tops, and isolated plateaus that typically did not burn. With fire suppression, stages such as sagebrush communities were replaced by juniper. In addition, juniper is a long-lived native species, and thick layers of litter that accumulated under larger trees suppressed the growth of other plants. As older, more mature juniper states replaced younger states, juniper expansion led to a decrease in the value of rangelands for livestock and as habitat for wildlife species such as sage grouse as the native shrubs, grasses, and forbs were excluded.

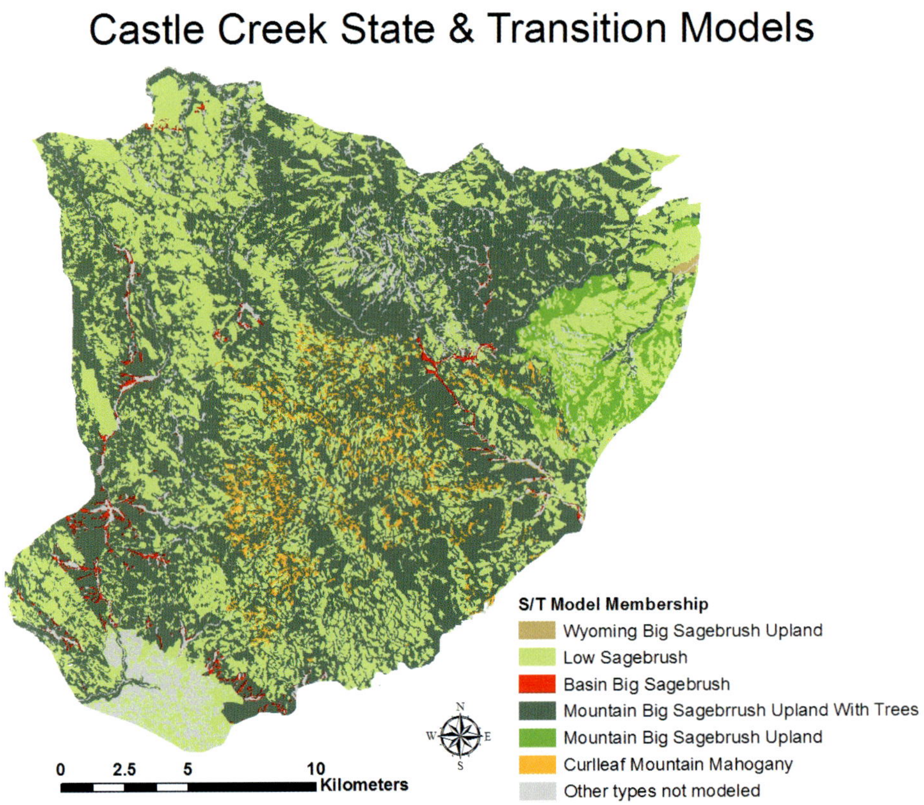

Figure 3.5. State-and-transition model states as mapped for the Castle Creek landscape. Six states, representing natural communities encompassing more than 95% of the area's extent were modeled. Multiple management scenarios were tested to examine the most cost-efficient suite of juniper control treatments. Simulations of multiple management scenarios were run, and projections of changes in the natural communities were made 50 years into the future. Figure from The Nature Conservancy of Idaho.

Planning at the scale of ecological and/or management actions is illustrated in the 117,900-acre Castle Creek landscape of Owyhee County, Idaho (figure 3.5), where the Bureau of Land Management (BLM) planned to stop the expansion of juniper and restore native sagebrush-steppe communities. Due to budget and time constraints, however, only a relatively small proportion of the entire landscape could be treated each year for juniper removal and restoration projects. The BLM had to re-evaluate. Which of many juniper management strategies, each consisting of many small treatment projects scattered across the landscape, would best meet the long-term goals for management of the entire Castle Creek landscape? To accomplish this, BLM partnered with The Nature Conservancy to conduct a prioritization based on dynamic simulation models of different management scenarios. These models needed to be developed at a local scale consistent with the scope of management actions, but able to be evaluated overall at a broader scale. A series of S/T models were created and simulations run for the Castle Creek landscape. Detailed reports of the methods and results are available at http://www.landscapetoolbox.org.

The 50-year simulations revealed only slight differences in the average conditions resulting from the different management scenarios, indicating that the current management efforts were insufficient to counter the expansion of juniper into the sage steppe. Not surprisingly, a management scenario in which resources allocated to juniper were doubled relative to current levels predicted the biggest decreases in juniper-encroached areas and the biggest gains in area of reference-state classes.

Conservation planning at scales for a specific target

The population structure and dynamics of many species require that conservation planning be done at multiple scales. The grizzly bear in the western United States, for example, is now found in increasingly isolated subpopulations. Effective conservation planning for this species requires a consideration of key ecological components and threats at regional, coarse, and local scales (see sidebar "Scales of planning for grizzly bear conservation" below). Planning at each scale addresses different variables, and it is important to state these explicitly, prior to initiating each planning effort. Local-scale planning, for example, would focus on the movement and home range needs of individual animals to ensure sufficient habitat patches to support a stable subpopulation. In contrast, regionwide planning would focus on landscape permeability to the target, potential movement barriers, and abating habitat fragmentation along key movement corridors.

Scales of planning for grizzly bear conservation

At a local scale, planning for grizzly bears (figure 3.6) as the "focal" species (see Brock and Atkinson, chapter 6) is thought to provide an umbrella effect. The term "umbrella species" has been applied to any species thought to confer conservation protection to other co-occurring species (Fleishman et al. 2000; Roberge and Angelstam 2004; Seddon and Leech 2008). In an analysis of the umbrella effect of

Figure 3.6. A grizzly bear. Photo by Lance Craighead.

63 candidate species in the Madison Valley of Montana, grizzly bears were found to have the highest aggregate rating, encompassing 29 of 40 habitats and 17 of 26 identified threats (Brock et. al. 2006).

Grizzly bears have evolved utilizing a metapopulation structure whereby distinct populations exchange genes over time with other populations. At a metapopulation level, dispersal between population centers is important for population persistence (Boyce et al. 2001; Levins 1970; Hanski and Gilpin 1997). Maintaining a distinct population segment (Rosen 2007; Rosen and Saundry 2008), or an ecosystem with grizzly bears, is thus a regional conservation problem that may, in fact, involve multiple ecoregions (Craighead and Vyse 1996). Planning for a metapopulation occurs at a scale that is broader than any single population (figure 3.7).

Over the past 150 years, grizzly bears in North America have experienced drastic population declines throughout their historical range. The historical grizzly bear metapopulation once found throughout the states of Montana, Wyoming, and Idaho now comprises distinct subpopulations in the Greater Yellowstone Ecosystem, the Northern Continental Divide Ecosystem, the Selkirk Ecosystem, and the Cabinet-Yaak-Purcells Ecosystem. A subpopulation that once inhabited the Selway-Bitterroot Ecosystem in central Idaho was extirpated. At the regional scale, planning for connectivity among subpopulations is critical and must be at a coarse scale that provides habitat features sufficient to maintaining a viable population in each ecosystem (see chapter 11).

Conservation of a subpopulation, such as that in the Greater Yellowstone Ecosystem, requires conservation planning across numerous land ownerships and multiple land uses. Habitat value, combined with human interactions, creates a source-sink dynamic across the landscape (Schwartz et al. 2006; also see Church, chapter 9). Core areas with habitat sufficient to support many individual bears are found almost exclusively on public lands. Many population sinks occur on private land. Habitat must be maintained across government agency jurisdictions with a variety of land-use strategies, and across private lands in between them, if the species is to be preserved.

To ensure that individual bears can find secure habitat patches and move to meet seasonal and/or lifetime needs, planning at a local scale is required. Bears use low-elevation grasses and wetlands in

Regional Scale Grizzly Bear Combined Analysis for the Central Rockies

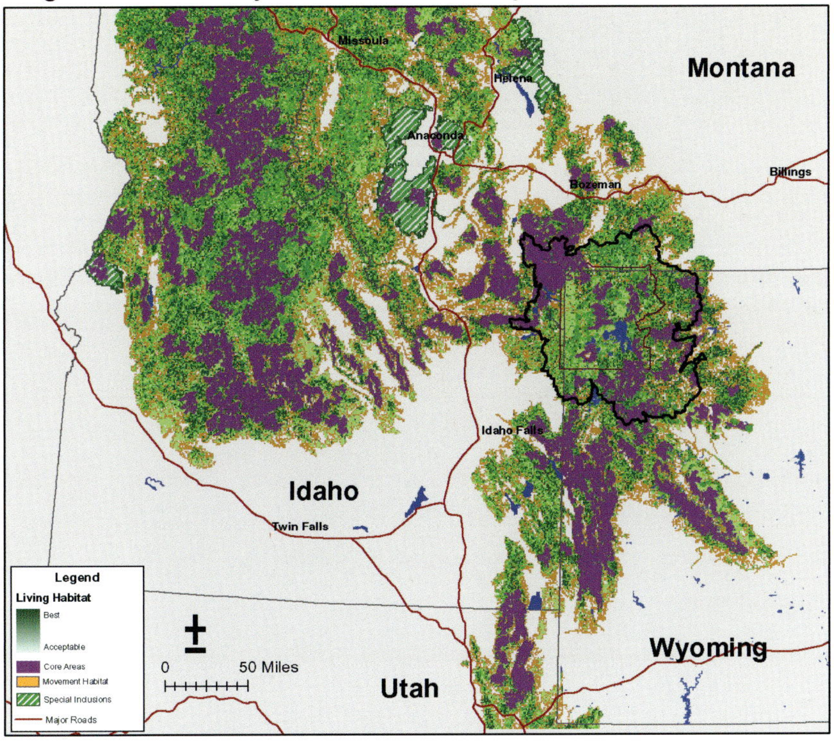

A

A Metapopulation Reserve to ensure long-term survival of the Grizzly Bear Population

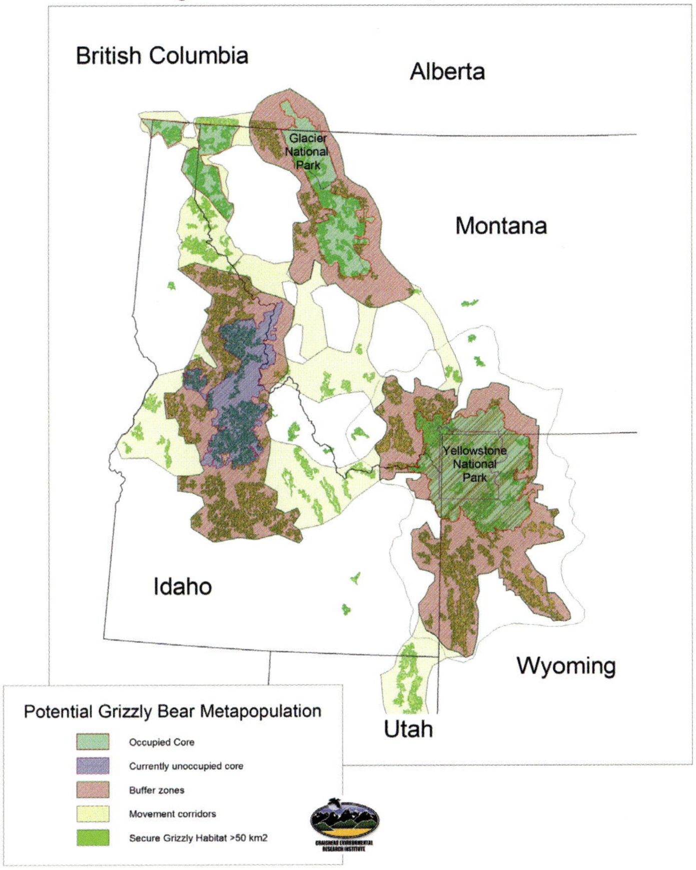

B

Figure 3.7. Grizzly bear habitat at two scales. Planning to maintain individual grizzlies must focus on the best habitat, shown in purple in (A), at a local or watershed level. To support a population requires large blocks of contiguous habitat (purple and green) at an ecoregional scale. For long-term persistence, planning must be at the metapopulation level (multiple ecoregions), with connectivity habitat as depicted in (B). State boundaries from ESRI Data & Maps, 2005, courtesy of ArcWorld Supplement; roads from ESRI Data & Maps, 2005, courtesy of TANA/GDT; streams/lakes from ESRI Data & Maps, 2005, courtesy of ArcWorld; grizzly habitat/grizzly connectivity from the Craighead Institute.

the spring, other plant resources at low and mid-elevations in summer, and foods at high elevations in fall. In a similar fashion, their large animal prey move to find forage at its most nutritious early stages. Grizzlies can perceive objects visually at distances up to several miles, and by scent, even further. Avoidance of human encroachment at low elevations has reduced grizzly bear habitat mostly to mountain forests and meadows. Data indicates that, on average, they avoid roads at around 500 m, dependent to some extent upon hearing. In some cases, microsites with consistent food resources can be identified at a local or site scale, and a key habitat such as an individual wetland can be protected. In many cases, however, the availability and nutritional value of food plants vary across the landscape from year to year, requiring a coarse-scale assessment. At the site scale, grizzly bears interact with the environment to locate food sources or den sites, and to avoid areas of high mortality risk. In terms of connectivity and linkage habitat, small areas at the site scale often are critical for allowing movement (see Singleton and MacRae, chapter 10).

Finally, all planning must result in actions that conform to the scales at which county policy and regulations apply and public lands are managed in order to be implemented effectively. Forest planning thus must be done at administrative scales, from the forest scale to the district scale to the site or stand scale, depending upon the objectives.

Scaling challenges and research needs

Much that has been learned about the different aspects of scaling since the earliest efforts at conservation planning has highlighted the need to consider scale throughout the process. However, although the importance of scale is now clearly recognized, and many techniques have been developed for identifying and describing scales, accommodating scale differences, and planning at different scales, significant gaps in our understanding of scale and its influences remain.

Foremost among these is the challenge of linking conservation planning between scales. We have discussed the need for conservation planning at multiple scales: at the coarse scale because the whole is greater than the sum of the parts, and at the fine scale because understanding the whole does not imply understanding of the parts. As we have indicated, however, translating conservation priorities between these scales and into effective action is not straightforward. We have attempted to illustrate how this can be done through intermediate-scale planning, but more research is needed here.

Importantly, most conservation planning efforts identify priorities at coarser scales, but effective conservation is most often accomplished through coordinated local actions. This kind of "thinking globally, but acting locally" situation poses the scaling challenge, because individual local actions often cannot be scaled up to meet larger-scale conservation objectives. This also raises a question as to what constitutes effective conservation at coarser scales, especially when the planning includes a diverse array of conservation targets.

Another significant issue is the quality and quantity of data and its impact on conservation planning at different scales. Although the availability of high-quality, fine-scale spatial data has increased dramatically over the past two decades and is expected to continue increasing, planners frequently will find it necessary to use data coarser than desired, or to attempt integrating data at different scales. This is especially problematic in data-poor regions like developing countries, where the need for conservation planning is greatest. Combining data of different scales is technically possible and has been well studied by geographers, but this can produce results that are either misleading or not meaningful at one or all scales. The implications of defining conservation priorities from combined data such as this are not well understood.

Considerable gaps remain in our understanding of scaling, independent of data quality issues. Scaling discontinuity (Wu and Li 2006) is one of these. An invasive species, for example, may appear rare and patchily distributed at the local scale, yet continuous at the regional scale. Using only local-scale data, one would conclude that this species is of little conservation importance, but the regional data would suggest it to be a huge pending threat.

Likewise, as the scale of a conservation plan changes, new questions can arise because of perceived change in relationships between targets and threats. As we discussed, an example of this is the sage grouse. At coarse scales, sage grouse are negatively affected by agricultural development due to habitat loss and fragmentation. At finer scales, however, agricultural fields play an important role as foraging habitat. Unfortunately, clear understanding of this kind about the multiscale relationships between conservation targets and threats is currently limited to only a few well-studied species.

The complex relationships between conservation targets and threats can, in part, be addressed by improving understanding of and focusing conservation planning on natural and human-caused processes. Most conservation planning, to date, has focused on patterns of target and threat distribution that result from processes occurring across landscapes. Although such efforts have filled a critical need and identified significant priorities, they are limited in their ability to plan for changing local-scale development pressures or global climate change. Increasingly, though, dynamic process-based models are being used to identify how current conservation priorities will be affected by specific threats over time. Such approaches will require improved conceptual and quantitative models of ecosystem functioning and site potential. Much has been done in developing these kinds of models recently (e.g., LANDFIRE and Ecological Site Description (ESD) models, Bestelmeyer et al. 2009; Forbis et al. 2006), but the application of these models to conservation planning is still in the early stage.

References

Addicott, J. F., J. M. Aho, M. F. Antolin, D. K. Padilla, J. S. Richardson, and D. A. Soluk. 1987. "Ecological Neighborhoods: Scaling Environmental Patterns." *Oikos* 49: 340–46.

Andelman, S. J., and W. F. Fagan. 2000. "Umbrellas and Flagships: Efficient Conservation Surrogates or Expensive Mistakes?" *Proceedings of the National Academy of Sciences* 97: 5,954–59.

Ball, I. R., H. P. Possingham, and M. Watts. 2009. "Marxan and Relatives: Software for Spatial Conservation Prioritisation." In *Spatial Conservation Prioritisation: Quantitative Methods and Computational Tools,* edited by A. Moilanen, K. A. Wilson, and H. P. Possingham. Oxford: Oxford University Press.

Bestelmeyer, B. T., J. R. Brown, K. M. Havstad, G. Chavez, R. Alexander, and J. E. Herrick. 2003. "Development and Use of State and Transition Models for Rangelands." *Journal of Range Management* 56: 114–26.

Bestelmeyer, B. T., A. J. Tugel, G. L. Peacock Jr., D. G. Robinett, P. L. Shaver, J. R. Brown, J. E. Herrick, H. Sanchez, and K. M. Havstad. 2009. "State-and-Transition Models for Heterogeneous Landscapes: A Strategy for Development and Application." *Rangeland Ecology and Management* 62: 1–15.

Boyce, M. S., B. M. Blanchard, R. R. Knight, and C. Servheen. 2001. *Population Viability for Grizzly Bears: A Critical Review.* International Association for Bear Research and Management, Monograph Series 4. http://www.cfc.umt.edu/grizzlybearrecovery/pdfs/Boyce et al. 2001.pdf.

Brock, B. L., E. C. Atkinson, C. Groves, A. Toivola, T. Olenicki, and L. Craighead. 2006. *A Wildlife Conservation Assessment of the Madison Valley, Montana.* Bozeman, MT: Wildlife Conservation Society, Greater Yellowstone Program. http://www.wcsnorthamerica.org/DesktopModules/Bring2mind/DMX/Download.aspx?EntryId=5581&PortalId=37&DownloadMethod=attachment.

Burnett, C., and T. Blaschke. 2003. "A Multi-Scale Segmentation/Object Relationship Modelling Methodology For Landscape Analysis." *Ecological Modeling* 168: 233–49.

Cauley, H. A., C. M. Peters, R. Z. Donovan, and J. M. O'Connor. 2001. "Forest Stewardship Council Forest Certification." *Conservation Biology* 15: 311–12.

Coppolillo, P., H. Gomez, F. Maisels, and R. Wallace. 2004. "Selection Criteria for Suites of Landscape Species as a Basis for Site-Based Conservation." *Biological Conservation* 115 (3): 419–30.

Craighead, L., and E. R. Vyse. 1996. "Brown/Grizzly Bear Metapopulations." In *Metapopulations and Wildlife Conservation,* edited by D. R. McCullough. Washington, DC: Island Press.

Dinerstein, E., and E. D. Wikramanayake. 1993. "Beyond Hotspots—How to Prioritize Investments to Conserve Biodiversity in the Indo-Pacific Region." *Conservation Biology* 7: 53–65.

Fleishman, E., D. D. Murphy, and P. F. Brussard. 2000. "A New Method for Selection of Umbrella Species for Conservation Planning." *Ecological Applications* 10: 569–79.

Forbis, T. A., L. Provencher, L. Frid, and G. Medlyn. 2006. "Great Basin Land Management Planning Using Ecological Modeling." *Environmental Management* 38: 62–83.

Groves, C. R. 2003. *Drafting a Conservation Blueprint: A Practitioner's Guide to Planning for Biodiversity.* Washington, DC: Island Press.

Groves, C. R., D. B. Jensen, L. L. Valutis, K. H. Redford, M. L. Shaffer, J. M. Scott, J. V. Baumgartner, J. V. Higgins, M. W. Beck, and M. G. Anderson. 2002. "Planning for Biodiversity Conservation: Putting Conservation Science into Practice." *BioScience* 52: 499–512.

Groves, C., L. L. Valutis, D. Vosick, B. Neely, K. Wheaton, J. Touval, and B. Runnels. 2000. *Designing a Geography of Hope: A Practitioners Handbook to Ecoregional Conservation Planning.* Arlington, VA: The Nature Conservancy.

Hanski, I. A., and M. E. Gilpin. 1997. *Metapopulation Biology: Ecology, Genetics, and Evolution.* San Diego, CA: Academic Press.

Hoekstra, J., J. L. Molnar, M. Jennings, C. Revenga, M. D. Spalding, T. M. Boucher, J. C. Robertson, T. J. Heibel, and K. Ellison. 2010. *The Atlas of Global Conservation: Changes, Challenges, and Opportunities to Make a Difference.* Berkeley: University of California Press.

Hunter, M. L. Jr., and K. A. Kohm. 1991. "Coping with Ignorance: The Coarse-Filter Strategy for Maintaining Biodiversity." In *Balancing on the Brink of Extinction: The Endangered Species Act and Lessons for the Future,* edited by K. A. Kohm, 266–81. Covelo, CA: Island Press.

Kotlair, N. B., and J. A. Wiens. 1990. "Multiple Scales of Patchiness and Patch Structure: A Hierarchical Framework for the Study of Heterogeneity." *Oikos* 59: 253–60.

Levins, R. 1970. "Extinction." In *Some Mathematical Problems in Biology,* edited by M. Gesternhaber. Providence, RI: American Mathematical Society.

Margules, C. R., and R. L. Pressey. 2000. "Systematic Conservation Planning." *Nature* 405: 243–53.

Mittermeier, R. A., P. R. Gil, M. Hoffman, and J. Pilgrim. 2005. *Hotspots Revisited: Earth's Biologically Richest and Most Endangered Terrestrial Ecoregions.* Amsterdam: Amsterdam University Press.

Mittermeier, R. A., N. Myers, and C. G. Mittermeier. 2000. *Hotspots: Earth's Biologically Richest and Most Endangered Terrestrial Ecoregions.* Conservation International. Amsterdam: Amsterdam University Press.

Noss, R. F. 1987. "From Plant Communities to Landscapes in Conservation Inventories—A Look at The Nature Conservancy (USA)." *Biological Conservation* 41: 11–37.

———. 1996. "Ecosystems as Conservation Targets." *Trends in Ecology & Evolution* 11: 351–51.

Noss, R. F., M. A. O'Connell, and D. D. Murphy. 1997. *The Science of Conservation Planning. Habitat Conservation under the Endangered Species Act.* Washington, DC: Island Press.

O'Neill, R. V., D. L. Deangelis, J. B. Waide, and T. F. H. Allen 1986. *A Hierarchical Concept of Ecosystems.* Princeton, NJ: Princeton University Press.

Parrish, J. D., D. P. Braun, and R. S. Unnasch. 2003. "Are We Conserving What We Say We Are? Measuring Ecological Integrity within Protected Areas." *BioScience* 53: 851–60.

Poiani, K. A., J. V. Baumgartner, S. C. Buttrick, S. L. Green, E. Hopkins, G. D. Ivey, K. P. Seaton, and R. D. Sutter. 2006. "A Scale-Independent, Site Conservation Planning Framework in The Nature Conservancy." *Landscape and Urban Planning* 43: 143–56.

Poiani, K. A., B. D. Richter, M. G. Anderson, and H. E. Richter. 2000. "Biodiversity Conservation at Multiple Scales: Functional Sites, Landscape, and Networks." *BioScience* 50: 133–46.

Roberge, J., and P. Angelstam. 2004. "Usefulness of the Umbrella Species Concept as a Conservation Tool." *Conservation Biology* 18: 76–85.

Rosen, T. 2007. "The Endangered Species Act and the Distinct Population Segment Policy." *Ursus* 18 (1): 109–16.

Rosen, T. (lead author), and P. Saundry (topic editor). 2008. "Distinct Population Segment Policy (1996) under the Endangered Species Act, United States." In *Encyclopedia of Earth,* edited by C. J. Cleveland. Washington, DC: Environmental Information Coalition, National Council for Science and the Environment. (First published in the *Encyclopedia of Earth* June 18, 2008; Last revised date: June 18, 2008.)

Rumsey, C., M. Wood, B. Butterfield, P. J. Comer, D. Hilary, M. Bryer, C. R. Carroll, G. Kittel, and K. J. Torgerson. 2003. *Canadian Rocky Mountains Ecoregional Assessment,* vol. 1: Report. Victoria, BC: The Nature Conservancy and the Nature Conservancy of Canada.

Sanderson, E. W., K. H. Redford, A. Vedder, P. B. Coppolillo, and S. E. Ward. 2002. "A Conceptual Model for Conservation Planning Based on Landscape Species Requirements." *Landscape and Urban Planning* 58: 41–56.

Schroeder, M. A., J. R. Young, and C. E. Braun. 1999. "Greater Sage-Grouse (*Centrocercus urophasianus*), The Birds of North America Online," edited by A. Poole. Ithaca: Cornell Lab of Ornithology; Retrieved from the Birds of North America Online: http://bna.birds.cornell.edu/bna/species/425.

Schwartz, C. C., R. B. Harris, and M. A. Haroldson. 2006. "Impacts of Spatial and Environmental Heterogeneity on Grizzly Bear Demographics in the Greater Yellowstone Ecosystem: A Source-Sink Dynamic with Management Consequences." In *Temporal, Spatial and Environmental Influences on the Demographics of Grizzly Bears in the Greater Yellowstone Ecosystem,* edited by C. C. Schwartz, M. A. Haroldson, G. C. White, R. B. Harris, S. Cherry, K. A. Keating, D. Moody, and C. Servheen. *Wildlife Monographs* 161: 57–68. http://www.cas.umt.edu/facultydatabase/FILES_Faculty/1152/Schwartz et al 2006 Source sink dynamics Wild Monog 161.pdf.

Scott, J. M., F. Davis, B. Csuti, R. F. Noss, B. Butterfield, C. Groves, H. Anderson, S. Caicco, F. D'Erchia, T. C. Edwards, Jr., J. Ulliman, and R. G. Wright. 1993. "Gap Analysis: A Geographical Approach to Protection of Biological Diversity." *Wildlife Monographs* 123: 1–41.

Seddon, P. J., and T. Leech. 2008. "Conservation Short Cut, or Long and Winding Road? A Critique of Umbrella Species Criteria." *Oryx* 42: 240–45.

The Nature Conservancy. 1997. *Designing a Geography of Hope: Guidelines for Ecoregional-Based Conservation in The Nature Conservancy.* Arlington, VA: The Nature Conservancy.

———. 2005. *Conservation Action Planning: Developing Strategies, Taking Action, and Measuring Success at Any Scale—Overview of Basic Practices.* Arlington, VA: The Nature Conservancy.

The Nature Conservancy and World Wildlife Fund. 2006. *Standards for Ecoregional Assessments and Biodiversity Visions.* Arlington, VA: The Nature Conservancy.

The Nature Conservancy, World Wildlife Fund, Wildlife Conservation Society, Conservation International, and BirdLife International. 2003. *A Resource Guide to Terrestrial Conservation Planning at the Regional Scale.* Arlington, VA: The Nature Conservancy.

Turner, M. G., R. V. O'Neill, R. H. Gardner, and B. T. Milne. 1989. "Effects of Changing Spatial Scale on the Analysis of Landscape Pattern." *Landscape Ecology* 3: 153–62.

Westoby, M., B. L. Walker, and I. Noy-Meir. 1989. "Opportunistic Management for Rangelands Not at Equilibrium." *Journal of Range Management* 42: 266–74.

Wheatley, M., and C. Johnson. 2009. "Factors Limiting Our Understanding of Ecological Scale." *Ecological Complexity* 6: 150–59.

Wiens, J. A. 1989. "Spatial Scaling in Ecology." *Functional Ecology* 3: 385–97.

Wiens, J. A., and B. T. Milne. 1989. "Scaling of 'Landscapes' in Landscape Ecology, or, Landscape Ecology from a Beetle's Perspective." *Landscape Ecology* 3: 87–96.

Wu, J., and J. L. David. 2002. "A Spatially Explicit Hierarchical Approach to Modelling Complex Ecological Systems: Theory and Applications." *Ecological Modeling* 153: 7–26.

Wu, J., and H. Li. 2006. Concepts of Scale and Scaling. In *Scaling and Uncertainty Analysis in Ecology: Methods and Applications,* edited by J. Wu, K. B. Jones, H. Li, and O. L. Loucks. Dordrecht, the Netherlands: Springer.

4

Land-cover data: The foundation for conservation planning

Thomas J. Olenicki

Land-cover data commonly forms the foundation for conservation planning. Used alone, it offers the potential to identify and conserve unique or rare plant communities but often is used as variables defining wildlife habitat across the landscape. Conservation biology has long relied on the idea that protecting habitat for surrogate wildlife species will protect habitat for other species with similar requirements, as in concepts of focal species (Lambeck 1997), umbrella species (Wilcox 1984), flagship species (Caro and O'Doherty 1999), and indicator species (Landres et al. 1988). Land-cover–derived habitat is often a primary component for defining core areas (Church, chapter 9) and connectivity habitat (Singleton and McRae, chapter 10) that subsequently forms the spatial basis for other conservation planning analyses, such as population viability (Stanton and Akçakaya, chapter 11), optimization (Williams, chapter 12), ecosystems services valuation (Chan et al., chapter 2), and predicted land-cover response to climate change (Bachelet, chapter 13). Selection of land-cover data is therefore a critical consideration affecting the entire conservation planning process.

Widespread land-cover data was not commonly available until the first remote sensing satellite dedicated to providing information about Earth's surface was launched in 1972, followed by development of global positioning systems (GPS) in the 1990s and the availability of advanced computers and software. The advent of these sophisticated tools, coupled with the needs of resource managers, environmental modelers, and policy makers, has led to increased development and use of remote sensing to produce land-cover data. Digital land-cover data is now among the most popular data used for resource applications (Thogmartin et al. 2004). Use of remote sensing for land-cover classification is often the only practical method for covering large areas, and both the techniques and applications in this field are steadily improving. However, remote sensing specialists and resource managers often have had difficulty reconciling the needs of resource management with the limitations and capabilities of remote sensing data (Hoffer 1994).

The progression in land-cover mapping has led to a divergence of approaches due to various needs, types of remote sensing equipment, and methods for processing and interpreting remote sensing data. One ideal classification of land use and land cover to meet all applications does not exist, and it is unlikely that one could ever be developed (Anderson et al. 1976; Franklin and Wulder 2002). Many land-cover mapping schemes tend to be distinct and incompatible with each other, and often are applicable only for the application or area of interest for which they were designed. Adams (1999) likened the different land-cover classifications to the mythical Tower of Babel, where everyone is working hard but speaking different languages, and stated, "The progress that has been made so far is despite the large number of schemes, not because of them." A myth persists that land-cover data is accurate and up-to-date (Estes and Mooneyhan 1994). This myth is perpetuated by practitioners using whatever land-cover data is readily available without fully considering the effects of data limitations on results. Conservation planning occurs at different spatial and temporal scales, across jurisdictional boundaries and extents differing in availability of land-cover data, and by different organizations. As the foundation for conservation planning, selection of land-cover data is critically important to minimize the potential accumulation of errors throughout the process and allow the integration of conservation efforts.

The intent of this chapter is to provide considerations about the selection and use of land-cover data for delineating habitat as part of the conservation planning process. The focus is on the vegetative aspect of land cover. Vegetation is a strong, although complex, indicator of the ecological function of natural systems (Grossman et al. 1998). Nearly all animal life is in some way dependent on vegetation for food and shelter, and the influence of vegetation on water cycles and albedo (sunlight reflectance and adsorption) ultimately affects our global climate. Our ability to accurately characterize, map, and monitor the vegetation around us is fundamental for understanding the world we live in, assessing environmental issues, conserving biodiversity, and, ultimately, for conservation planning.

Definition of land cover

A definition of "land cover" and the distinction between "land cover" and "land use" is fundamental to any discussion of the selection and use of land-cover data. In a strict sense, the use of "land cover" should be limited to the vegetation and human-made features on Earth's surface and consequently does not include exposed rock, bare soil, or bodies of water (Di Gregorio and Jansen 1998). Burley (1961) adheres to this definition by defining the term as "the vegetational and artificial constructions covering the land surface." In practice, the scientific community usually includes unvegetated categories, as in the expanded definition of Di Gregorio and Jansen (1998): land cover "is the observed (bio)physical cover on the Earth's surface." In contrast, the term "land use" reflects human activities and inputs to produce, change, or maintain land cover (Choudry and Jansen 1998). Many land-cover classifications include categories that reflect land use; classes such as cropland, rangeland, and urban/recreational grasses indicate the human activity that produces the vegetational aspect defining the land-cover class. The inclusion of information on land use in land-cover data may provide additional information on human activity to aid the conservation planning process. For example, human activity will modify values of living habitat, core area, and connectivity habitat in land designated as pasture/hay compared to a similar grassland where human activity is nonexistent. Methods to aggregate and incorporate data on land use into conservation plans are discussed by Theobald (chapter 5).

This chapter is geared toward conservation planners who use digital land-cover data for delineating wildlife habitat, but may also help remote sensing specialists better understand the needs and requirements of conservation planners and other resource managers. Remote sensing is not the topic of this chapter, although the reliance on remote sensing to classify land cover requires a basic understanding of the subject. Therefore, I briefly review pertinent remote sensing topics before discussing the need to link remote sensing data with relevant vegetation on the ground and match land-cover data with the specific needs and scale of wildlife species used in conservation planning. I then discuss some commonly available land-cover classifications and present a conceptual classification scheme tailored to the needs of defining habitat for conservation planning. The conceptual classification is intended to help readers understand the many options and processes available for classifying land cover and provide considerations for anyone planning on developing their own classification. Finally, I provide a few recommendations for future research and conservation applications.

Fundamental considerations of remote sensing data

Remote sensing is "the science and art of obtaining information about an object, area, or phenomenon through the analysis of data acquired by a device that is not in contact with the object, area, or phenomenon under investigation" (Lillesand et al. 2008). The earliest application of remote sensing is the use of aerial photography. By the 1850s, balloons and pigeons were carrying cameras for photos used in land surveys. A passenger of Wilbur Wright provided the first known photograph from an airplane in 1908 (Belward and Eva 2004). In the mid-1940s, Francis J. Marschner began mapping major land-use associations for the entire United States, using aerial photographs taken during the late 1930s and the early 1940s (Anderson et al. 1976). Aerial photographs still provide valuable information, but the typical method of manually interpreting air photos limits the areal extent over which their use is practical.

Today, satellite-based sensors supply the majority of remote sensing data. Multispectral data from passive sensors (measuring the amount of energy reflected and emitted within distinct wavelengths across the electromagnetic spectrum) provided the earliest form of satellite data and still are the most common basis for land-cover classification. Next-generation sensors will undoubtedly improve the abilities for classifying land cover when they become fully operational. Active sensors that emit their own energy source, such as light detection and ranging (lidar) and synthetic aperture radar (SAR), offer the ability to directly measure physical parameters of vegetation, such as height, that must be inferred from passive sensors. Hyperspectral imagery is a specialized type of passive sensor that slices the electromagnetic spectrum into many more discrete bands (often hundreds) that may even allow the detection of specific plant species. However, many sensors are still in the research phase and are not accessible for most researchers (Turner et al. 2003) or too expensive to be practical for use over large areas (Donoghue et al. 2004). For conservation planning at this time, multispectral imagery and their derived land-cover data provide the best options and are the focus of my discussions on remote sensing.

Many technical considerations must be accounted for when working with remote sensing data. Clouds and cloud shadows commonly limit acquisition of data. Water vapor and particulate matter in the atmosphere as well as the sun's angle influence the amount of energy that initially strikes Earth's surface and passes back through the atmosphere to sensors. Topography and vegetation create shadows that add further variability to the energy measured by sensors. Geometric correction of satellite data is required to ensure locations on the ground conform to locations in a satellite image. Additional information and methods to account for these and other technical issues can be found in the remote sensing literature and should be consulted whenever working with remote sensing data or the products it produces.

Classification accuracy and precision

Within the remote sensing field, the term "accuracy" defines correctness—the agreement between an assumed standard and the predicted class of data—and "precision" defines the degree of detail of land-cover classification (Campbell and Wynne 2011). The distinction and interaction between these terms is important. Accuracy of land-cover data is important to prevent confusion among classes, but the precision of land-cover classes determines the data's usefulness for specific applications. Precision of land-cover data relative to the scale, species, and landscape characteristics of a project is therefore an important consideration. Defining habitat for animals with strict requirements for plant species or vegetative structure obviously requires land cover of greater precision than for generalist herbivores. Unfortunately, most land-cover classifications are pushed to their limits, and attempts to increase classification precision for the same data and location usually decrease accuracy (figure 4.1, x-axis label 1).

Accuracy assessments of land-cover classifications typically are reported using an error matrix that compares locations of known land cover with predicted classes. Two types of errors—errors of omission and commission—can occur for each land-cover class. As an example, for a grassland class, an error of omission occurs where a location of known grassland is omitted from the grassland classification and assigned a different class. An error of commission occurs where the location is classified as grassland but it is actually another class; the classification process has committed an error by assigning "grassland" where it is not present. The distinction between errors is important because a classification could achieve 100% accuracy for known grassland locations by delineating the entire area as grassland if errors of commission are not considered.

In addition to identifying types of errors, error matrices provide two overall measures of accuracy for each class. Producer's accuracy indicates the percent of correctly classified locations relative to the number of known locations for each type; for example, 400 locations correctly classified as forest among 500 known forest locations results in a producer's accuracy of 80%. Consumer's accuracy indicates the reliability of the

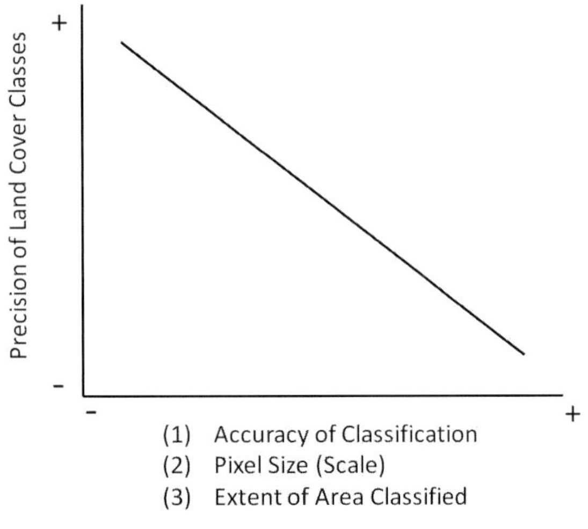

Figure 4.1. General relationships between precision of land-cover data and (1) accuracy of the classification scheme, (2) pixel size of satellite imagery (scale of data), and (3) extent of area classified. Figure from Thomas J. Olenicki.

data for predicting land cover by calculating the percent of correctly classified locations relative to the total number of locations the classification process predicts for that type. Therefore, the correct classification of 400 locations as forest among 800 locations classified as forest in the predictive map results in a consumer's accuracy of 50%.

Error matrices can be difficult to understand fully, but they help in the selection of land-cover data. In addition to reporting the accuracy of each class, they indicate where the greatest amount of confusion occurs as a way to assess the consequences of misclassification. For a species using mesic shrub habitat, the effects of misclassification will likely be reduced if a mixed mesic shrub class is misclassified as mixed mesic conifer than as an urban class.

Spatial resolution

Spatial resolution refers to the size of the area on the ground that contains information from a single pixel of imagery. In this respect, spatial resolution is analogous to the scale of observations (Woodcock and Strahler 1987). Spatial resolution of common imagery is ~ 1 km for advanced very high resolution

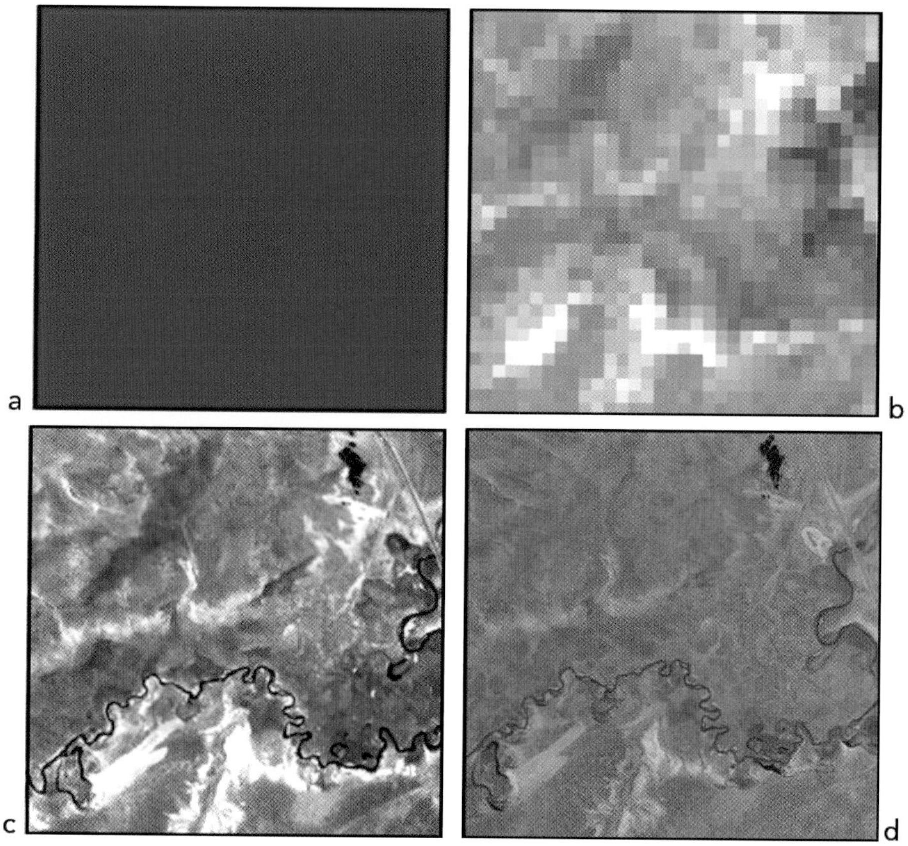

Figure 4.2. Remote sensing data at different spatial resolutions for the same 1-km-wide location. Increased detail becomes visible as spatial resolution increases from (a) 1-km AVHRR data to (b) 30-m Landsat imagery, (c) 4-m IKONOS imagery, and (d) a 1-m orthophoto. Data in (b) and (d) courtesy of US Geological Survey; satellite image (c) from GeoEye.

radiometer (AVHRR) data and ~ 30 m for most bands of Landsat data from 1982 onward. The multispectral bands of IKONOS satellite imagery have a spatial resolution of 3.2 m when looking straight down (at nadir). However, IKONOS and some other sensors often acquire data at oblique angles to increase the spatial coverage and temporal frequency, which can increase spatial resolution considerably, depending on acquisition angle. Figure 4.2 provides a comparison of the spatial resolution for several common multispectral sensors and what they "see" on the ground. Note that sensors actually sample a circular region rather than the square pixels portrayed in digital data and, therefore, the sensors detect reflectance from objects within the center rather than the entire ground surface area.

Spectral resolution

Spectral resolution refers to the specific wavelengths within the electromagnetic spectrum that a sensor measures. Multispectral sensors typically provide information in the blue, green, red, and infrared bands, but the number of bands and specific wavelengths they measure are sensor dependent. Landsat TM collects data in seven discrete bands and IKONOS measures four bands; AVHRR was originally designed for meteorological purposes and commonly uses only two bands for vegetation classification. Differences in reflection characteristics among objects form the basis for classifying land cover.

In addition to using information from individual bands, there are various ways of combining bands to concentrate information or target specific aspects of land cover. Vegetation indices such as the Normalized Difference Vegetation Index (NDVI; Rouse et al. 1973) was one of the earliest, and is probably the most common index that attempts to measure quantity or vigor of vegetation. Other approaches reduce the information from multiple bands to a limited number of derived variables that concentrate information using such methods as the tasseled cap transformation (Kauth and Thomas 1976) or principal component analysis (e.g., Singh and Harrison 1985).

Temporal resolution

Temporal resolution is the amount of time between repeat coverage for a sensor. The AVHRR sensor images most of Earth's surface daily, while Landsat takes 16 days between repeat coverage.

Trade-offs among data sources

Selection of remote sensing data and derived land cover is often a trade-off between spatial, temporal, and spectral resolution in addition to trade-offs in cost, data volume, and image footprint. The spatial resolution of IKONOS and QuickBird may be beneficial for defining fine-scale patterns in land cover over small areas, but the cost, amount of data, and difficulty obtaining timely data over large areas makes their use prohibitive in many situations. Similarly, the high temporal frequency of AVHRR data is ideal to track vegetation phenology, but its course spatial resolution and limited spectral resolution reduce the ability to characterize complex landscapes.

Linking remote sensing with vegetation on the ground

In the classic work on vegetation mapping, Kuchler (1967) defined vegetation as "the mosaic of plant communities in the landscape" and stated that "this definition implies that vegetation consists of more or less distinct mappable units." Defining land cover is essentially an effort to detect and map these patterns across the landscape. Unfortunately, there is a lack of coordination and standardization within the natural science community for defining plant communities on the ground. Botanists and field biologists have developed various systems for classifying vegetation that are frequently incompatible, may not be applicable to the needs of conservation planning, and that often cannot be extrapolated across large areas using remote sensing techniques. The difficulty and lack of consistency in classifying vegetation on the ground only becomes more pronounced in efforts to produce relevant land-cover data over large areas using satellite imagery.

Field classification of vegetation

Botanists and field biologists typically classify vegetation as either habitat types (Daubenmire 1952) delineating potential vegetative at climax conditions or as cover types defining existing vegetative conditions. Types are commonly named using one indicator or dominant species from the overstory (if present) and one from the understory, for example, the big sagebrush/Idaho fescue and Idaho fescue/bluebunch wheatgrass types described by Mueggler and Stewart (1980). Depending on the seral stage, habitat types often do not indicate the actual vegetation on the ground, and two areas of very different plant species composition can be classified as the same habitat type. Classification of habitat types using remote sensing data generally is not possible, and the use of field data based on habitat types is problematic in remote sensing applications.

Recent trends have centered on using cover types (existing vegetation) rather than habitat types. Several standards for classifying existing vegetation have been proposed but have not been adopted fully. The Federal Geographic Data Committee (FGDC) National Vegetation Classification Standards established an initial hierarchical classification with nine levels (FGDC 1997). The seven upper levels of the FGDC standards are based primarily on physiognomy; the two proposed lower levels, although not finalized, are based on floristic attributes. Floristic standards were drafted by the Ecological Society of America (ESA) Panel on Vegetation Classification (Jennings et al. 2004). Final adoption of classification standards will promote the consistent classification of existing vegetation in the field and facilitate descriptors of land cover. However, it is unlikely that multispectral data will be able to classify vegetation accurately at the floristic level of the proposed standards. These sensors cannot "see" through dense tree and shrub canopies to classify vegetation of the understory and have difficulty differentiating among homogenous grass types, much less between mixed types containing similar life-forms, such as Idaho fescue/bluebunch wheatgrass and Idaho fescue/tufted hairgrass.

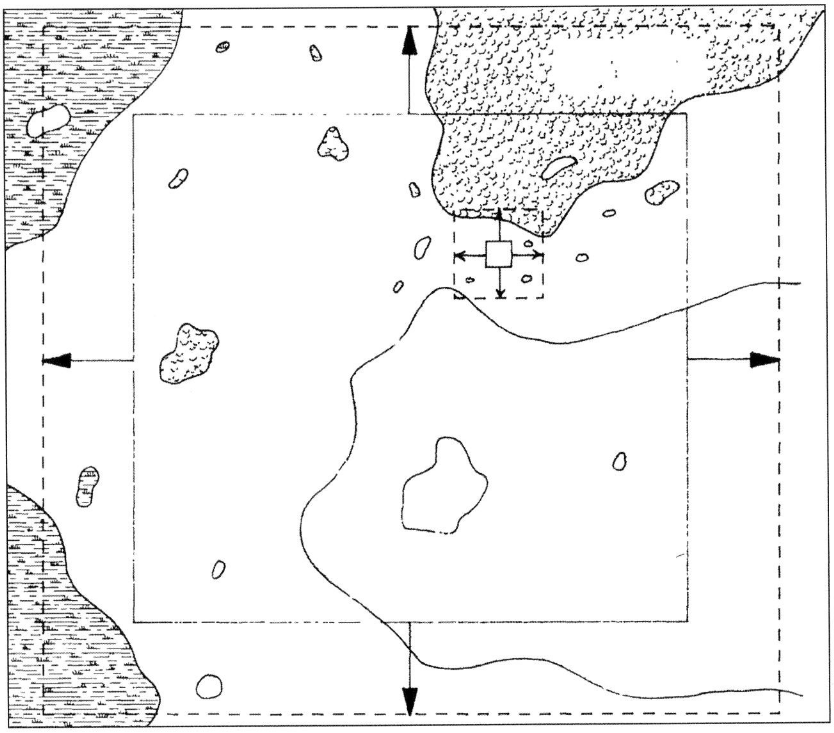

Figure 4.3. The relationship between grain and extent. Figure from Wiens (1989). Reprinted with permission from John Wiley and Sons Ltd.

In the ecological field, detecting landscape patterns is a function of the size of sample units (grain) and the extent of the area under investigation (see Unnasch and Karl, chapter 3). The interaction between grain and extent in heterogeneous vegetation (figure 4.3) is concisely described by Wiens (1989): "As the extent of the study is increased (large squares), landscape elements that were not present in the original study area are encountered. As the grain of samples is correspondingly increased (small squares), small patches that initially could be differentiated are now included within samples and the differences among them are averaged out."

Wiens (1989) further describes how spatial variance changes depending on grain and extent, and discusses the relationship between spatial and temporal scaling. Two important points elucidate the influence of extent and grain on field classification of vegetation:

1. As grain increases, the number of classes and distinction between classes decreases because more variability is encountered within each sample.
2. As the extent increases, the number of vegetation classes increases because more classes are encountered.

Remote sensing classification of vegetation

The concepts of grain and extent for field classification of vegetation apply to classification of vegetation using remote sensing data. "Grain" is analogous to pixel size in remote sensing, and "extent" corresponds to "area of interest." As pixel size increases, the variability in land cover sampled within each pixel also increases. The advent of finer-resolution satellite imagery has greatly increased the ability to classify small patches of vegetation that are typical of heterogeneous landscapes (Mehner et al. 2004). Therefore, all other factors being equal, greater precision in land-cover classes is possible with data of higher resolution (smaller pixels; figure 4.1, *x*-axis label 2).

Similar to the increase in extent from an ecological perspective, the number of vegetation types will typically increase as the area of interest increases. The greater number of vegetation types makes it more difficult to discriminate between classes, and vegetation types are often compressed into broader classes to maintain accuracy. Precision often declines as the extent increases (figure 4.1, *x*-axis label 3).

A well-known and common problem when classifying vegetation with remote sensing data is the occurrence of mixed pixels: pixels containing more than one vegetation type. These pixels can be misclassified because the overall reflectance from mixed pixels may be dissimilar to any of the vegetation types they contain, or the reflectance resembles vegetation types that are not present (Campbell and Wynne 2011). Additionally, reflectance from pure pixels containing a single land-cover class will vary due to differences in the amount and heterogeneity of plant species present. Smaller pixel size reduces the number of mixed pixels relative to the same number of classes using coarser imagery (Campbell and Wynne 2011), but higher-resolution imagery still suffers from problems of mixed pixels (Foody 2002).

Classifying and checking the accuracy of land-cover data require accurate matching between locations of data collected on the ground and corresponding pixels of satellite imagery. Current GPS produce very accurate locations, often within 1 m, depending on the receiver type and GPS satellite configuration. However, modern techniques are still not able to make remote sensing data conform perfectly to locations on Earth's surface. Loveland et al. (2000) identified image registration as an important area of future research. The best technique is to match known points on the imagery (referred to as ground control points [GCPs]) with GPS ground locations and "warp" the imagery to conform. Even with this process, a general method to deal with inaccuracies in geographical locations is to sample an area equivalent to 3 × 3 pixels (~ 90 m × 90 m for Landsat data) and apply field data to the center pixel of the corresponding imagery. Difficulty in matching locations on the ground with specific pixels further degrades the accuracy and precision of land-cover data.

Because patterns on the landscape may be scale-dependent, pixel size and the uniform sampling grid of satellite may affect the ability to classify vegetation patterns on the ground. Marceau et al. (1994) concluded that remote sensing data is not independent of the sampling grid used for its acquisition, that neglecting the scale and aggregation level can produce haphazard results with little correspondence to geographical entities, and that there is not a unique spatial resolution appropriate for all situations. Nijland et al. (2009) found they could improve classification accuracy for the variable of interest within their study area by resampling the original 5-m imagery to 7 m. Ideally, the perfect land-cover classification system would have the ability to adjust the spatial resolution of data to match the process or variable of interest on the landscape. Marceau and Hay (1999) provide an excellent review of the issue for conservation planners involved with remote sensing or land-cover classification derived from remote sensing.

Matching land cover data with habitat needs of wildlife

Identifying important areas for conservation planning is often determined, in part, by habitat suitability models based on the needs of wildlife. Wildlife select specific habitats, an idea David Lack (1933) may have been the first to propose. Habitat selection may be floristic, determined predominately by structural components, or be a combination of the two. For example, bison require graminoid habitats (Reynolds et al. 1982), but fisher generally avoid open areas and use a range of plant communities containing a large amount of vegetative structure (Jones and Garton 1994). Sage grouse are associated strongly with sagebrush habitat but require specific structural components of sagebrush and amounts of herbaceous cover within sagebrush habitat, depending on season (Connelly et al. 2000). Land-cover classes for habitat suitability models must match the type and scale of habitat requirements for the wildlife species of interest.

Every species observes the environment through its own unique suite of scales for space and time (Levin 1992). Landscape heterogeneity and resource requirements influence the decision-making processes. Hierarchical decisions can result in habitat use and selection at multiple scales (Owen 1972; Senft et al. 1987; Bailey et al. 1996; Kie et al. 2002; Boyce et al. 2003; Shipley 2007). Johnson (1980) suggested the following orders of selection:

1. First-order selection defines the physical or geographical range of a species.
2. Second-order selection determines home range.
3. Third-order selection pertains to usage of habitat within the home range.
4. Fourth-order selection relates to the procurement of food items at a site.

In the case of large herbivores, Senft et al. (1987) proposed regional, landscape, and plant community scales of habitat selection. Bailey et al. (1996) expanded spatial and temporal selection by large herbivores to include bite, feeding station, patch, feeding site, camp, and home range levels of selection. Selection decisions can vary among levels (Jiang and Hudson 1993; Searle et al. 2005), and it is important to match land-cover attributes not only with the species, but also with the order of selection under consideration.

As the order of selection increases from a regional or geographical level to a feeding-site level, the extent of the search area decreases (scale becomes larger) and the specificity of vegetation parameters increases. Obviously, the exact level and rate at which these variables change is dependent on the species in question. Grizzly bears are a wide-ranging species that vary their habitat use among seasons and across their range. Their home ranges are much larger and their habitat requirements are less specific than for many other species (e.g., boreal toads) and the rate of change from initial selection orders to the procurement of food items is overall much greater.

The level of precision required for land-cover data to address differences in habitat specificity, scale of selection, and corresponding differences among wildlife can be generalized as in figure 4.4. The required

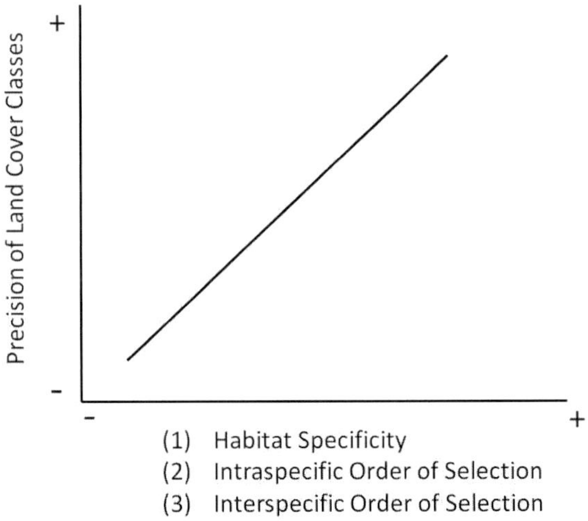

Figure 4.4. General relationships between the required precision of land-cover data to delineate habitat and (1) habitat specificity of the species of concern, (2) intraspecific order of selection for each species, and (3) interspecific order of selection for all selected species. Figure from Thomas J. Olenicki.

precision increases as the specificity of habitat requirements increases from habitat generalists to habitat-obligate species (figure 4.4, x-axis label 1). For an individual species, the required precision also increases as the order of selection increases to finer levels of selection (figure 4.4, x-axis label 2).

Land-cover precision to delineate habitat for the same order of selection varies among species; third-order selection for grizzly bears equates to a similar size of area and specificity of habitat components as second-order selection for narrower-ranging species that have more specific habitat requirements. Figure 4.4 (x-axis label 3) indicates the required difference in precision of land-cover data for different selection levels among species.

Selection and examples of land-cover data

Modern sensors and techniques combined with readily available remote sensing data have improved land-cover classification for a variety of applications. The overall availability of land-cover data often gives the impression that data is available for any application and location. Although land-cover data exists for virtually the entire Earth, in many situations, available land cover may not be appropriate for delineating habitat suitability and other conservation uses.

The relationships presented in figures 4.1 and 4.4 are not understood fully, but the concepts they represent should help match requirements of land-cover data with selected wildlife species for a specific project and indicate selection levels that can be delineated. Conversely, the figures may help select wildlife species based on available land-cover data. Conservation planning typically uses a suite of wildlife species that range over various extents and habitat types. The required precision of land-cover data for projects is determined by wildlife species with the most specific requirements and the finest levels of selection the planners wish to identify. Land-cover classes can always be grouped into broader categories (e.g., grouping classes of specific coniferous tree species into a general conifer class), but the ability to separate broad classes into finer levels is extremely limited (e.g., using aspect and elevation to make assumptions where specific tree species occur within a broad conifer class). The following examples of commonly available land-cover data represent differences in spatial resolution, extent, and precision of data that often limit its use in many applications.

Global land-cover databases

Global land-cover data treating all parts of the world equally and representing actual land cover of our planet has been available since the 1990s (Belward and Eva 2004). Table 4.1 (Strand et al. 2007) lists four sources of readily available global land-cover data. Theobald (chapter 5, table 5.2) provides others. The major advantages of these land-cover classifications are their ability to cover any location on Earth with a consistent classification scheme. One of these classifications, the AVHRR global land cover (Hansen et al. 2000), is available with a spatial resolution of 1 km, 8 km, and 1 degree, having either 13 or 14 land-cover classes depending on spatial resolution. Classes defined by this data are very broad and limited to habitat suitability modeling at a regional or geographical scale. However, this data offers the ability to identify general areas for closer examination, and the high temporal resolution (e.g., moderate resolution imaging spectroradiometer [MODIS]) data is beneficial for change detection.

Table 4.1. Sources of global land cover

Land cover name	Download site	Spatial resolution
Global Land Cover 2000	http://bioval.jrc.ec.europa.eu/products/glc2000/products.php	1 km
MODIS Products	https://lpdaac.usgs.gov/products/modis_products_table	500 m and 1 km
UMD Global Land Cover Classification	http://glcf.umiacs.umd.edu/data/landcover/	1 km, 8 km, 1 degree
Global Land Cover Characterization	http://edc2.usgs.gov/glcc/glcc.php	1 km
Source: Strand et al. 2007.		

National databases

A selection of national land-cover databases for North America is discussed below. Additional data is presented by Theobald (chapter 5, table 5.4) and discussed in relation to land use and threats.

National Land Cover Database

The National Land Cover Database (NLCD), available from the Multi-Resolution Land Characteristics Consortium (http://www.epa.gov/mrlc/), is the product of a long-term, multipartner project dedicated to land-cover products. Originally released in 1992, newer versions using revised methodology were released in 2001 and updated in 2006 to include percent tree canopy and percent urban imperviousness. Slight modifications were made in classes (table 4.2). Some classes describe land use rather than land cover; these classes are discussed by Theobald (chapter 5, table 5.4). Data is based on Landsat imagery and is available for the conterminous United States, Alaska, and Puerto Rico.

For delineating habitat, there are two main advantages of the NLCD data compared to the global land cover previously discussed:

1. Better representation of the spatial heterogeneity of cover classes that 30-m pixels of Landsat imagery provide compared to the 1-km pixels for AVHRR.
2. Estimates of canopy coverage for the 2001 version. Actual classes of NLCD data are still fairly broad and best suited for delineating lower orders of habitat selection, but estimates of canopy structure increase habitat modeling capabilities for species that rely on forest structure.

Classification accuracy was assessed by mapping region for the 1992 NLCD data but has not yet been conducted for the 2001 version. For Federal Region 8 (Montana, Wyoming, North Dakota, South Dakota, Utah, and Colorado), overall consumer's accuracy for single pixels is 60% (table 4.2). A closer look at accuracy for individual classes indicates that 8 of the classes have an estimated accuracy of 15% or less. Cunningham (2006) and Thogmartin et al. (2004) provide good discussions on the use of NLCD data for habitat studies as well as some of the reasons for classification errors.

Table 4.2. Comparison between 1992 and 2001 National Land Cover Data (NLCD) classes with consumer's and producer's accuracy for Federal Region 8 of 1992 classes

1992 NLCD			2001 NLCD
Classes	Consumer's accuracy[1]	Producer's accuracy[1]	Classes
Open Water	0.74	0.94	Open Water
Perennial Ice/Snow	0.08	0.02	Perennial Ice/Snow
Low Intensity Residential	0.68	0.38	Developed, Open Space
High Intensity Residential	0.42	0.61	Developed, Low Intensity
Commercial/Industrial/Transportation	0.3	0.09	Developed, Medium Intensity
			Developed, High Intensity
Bare Rock/Sand/Clay	0.38	0.43	
Quarries/Strip Mines/Gravel Pits	0.62	0.28	
Transitional Barren	0.01	0.15	Barren Land
Deciduous Forest	0.13	0.17	Deciduous Forest
Evergreen Forest	0.61	0.72	Evergreen Forest
Mixed Forest	0.15	0.02	Mixed Forest
Shrubland	0.77	0.62	Shrub/Scrub
Orchards/Vineyards/Other	n/a	n/a	
Grassland/Herbaceous	0.67	0.8	Grassland/Herbaceous
Pasture/Hay	0.36	0.47	Pasture/Hay
Row Crops	0.79	0.36	Cultivated Crops
Small Grain	0.38	0.59	
Fallow	0	0	
Urban/Recreational Grasses	0.09	0.55	
Woody Wetlands	0.03	0.1	Woody Wetlands
Emergent Herbaceous Wetlands	0.11	0.25	Emergent Herbaceous Wetlands
			Dwarf Shrub[2]
			Sedge/Herbaceous[2]
			Moss[2]

Source: Data courtesy of US Geological Survey.
Note: An accuracy assessment has not been completed for 2001 data.

[1] Agreement is defined as a match between the primary or alternate reference land-cover label and mapped land-cover class of the corresponding single pixel.
[2] Alaska only.

GAP analysis land cover

The intent of the Gap Analysis Program (GAP) is to identify and maintain nonthreatened animal species and plant communities that are not covered by other legislation and may not occur on existing conservation lands. Two goals of this project are to map land cover and to predict species distribution across the United States. GAP is probably the most comprehensive source of information pertaining to land-cover data and species distribution in the United States. Many research projects, publications, and land-cover classifications have been produced under this program. GAP land cover, which is based on Landsat imagery, is available for download at http://gapanalysis.usgs.gov/data/land-cover-data/.

The first GAP land-cover data was produced independently for each state, and the lack of standardized methods caused problems when attempting to conduct conservation planning in states with more generalized land-cover classes as well as for areas crossing state boundaries. New GAP land-cover released since this chapter was first written solves the following issue relating to GAP data, but the following example is still pertinent when a conservation area of interest encompasses land-cover data from multiple sources.

Figure 4.5 indicates a location along the Montana and Wyoming border within Yellowstone National Park where initial versions of GAP land-cover data for both states overlapped. The red polygon indicated in the upper left portion of figure 4.5 comprises ~8,028 acres classified by Wyoming GAP as 80% lodgepole pine and 20% subalpine meadow. Figure 4.6 indicates this same polygon classified under Montana GAP, where each color represents different land-cover classes. Compared to Wyoming data, the greater precision and improved spatial representation of vegetation types for Montana land-cover data increase the ability to define specific habitats and greater intraspecific and interspecific orders of selection. Delineating habitat in this and similar situations where multiple sources of land cover are used is typically constrained by the most generalized data.

Cross-walking of vegetation types has been attempted in these situations to produce consistent land cover, but it is often a best-guess exercise when field data is not used. Detailed information on floristic composition and canopy cover are required to correctly cross-walk between vegetation types (Brohman and Bryant 2005; FGDC 1997, 2008; Jennings et al. 2004). My own experience suggests that independently applying a habitat rating system to each classification and combining datasets also produces inconsistencies among data sources.

Classifications using IKONOS or QuickBird imagery

Launches of IKONOS, QuickBird, and the newest SPOT (Satellite Pour l'Observation de la Terre) sensors in recent years have greatly increased the spatial resolution of available imagery. IKONOS imagery has shown the ability to classify percent tree cover (Snyder et al. 2005) as well as individual trees (Read et al. 2003), and to differentiate between vegetation types in dry shrub/grassland (Depew 2004). At this point, no widespread classifications are available using these types of imagery. The extent of each scene is generally small, data is fairly costly, and the high spatial resolution results in very large datasets. The same could once be said for Landsat data, and these issues will be less of a problem in the future. Although these types of data offer increased capabilities, their current use for habitat modeling is limited to small extents and as an intermediate sampling tool between ground surveys and coarser-grained imagery.

Vegetation Resource Inventory of British Columbia

Contrary to many land-cover classifications that use satellite imagery and focus on differences in plant species, the Vegetation Resource Inventory (VRI) of British Columbia (RIC 2002) is based on the physiognomy of vegetation as determined by air photo interpretation. Figure 4.7 diagrams the sequential splitting process and subsequent classes within a polygon initially classified as "vegetated." The process used and the classes defined likely produce one of the most useful and accurate classification systems for modeling wildlife

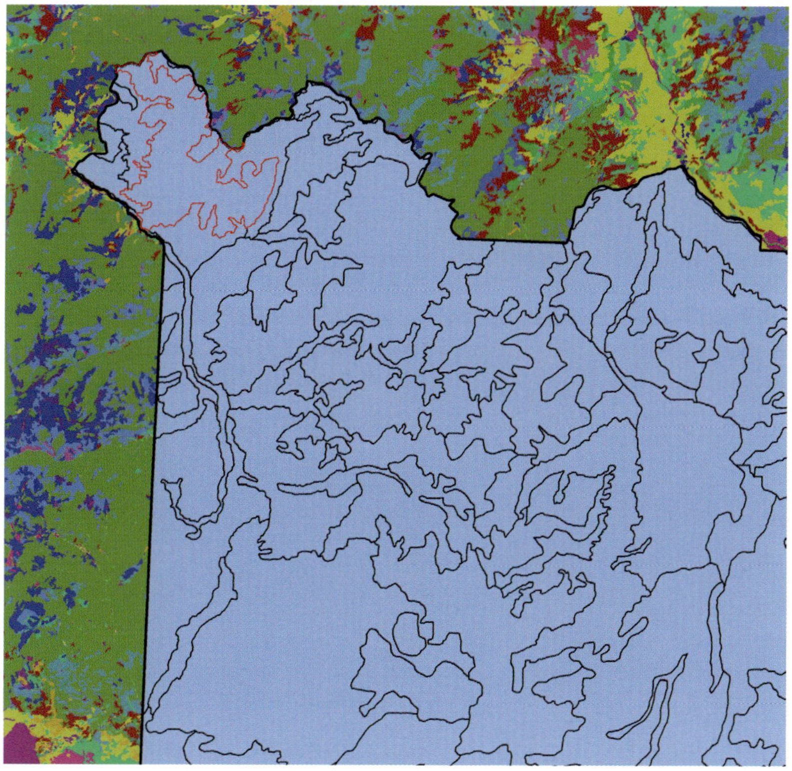

Figure 4.5. First-generation GAP land cover at the border between Montana and Wyoming in Yellowstone National Park, indicating the differences in spatial representation of land cover due to changes in classification methods between states. Data courtesy of US Geological Survey.

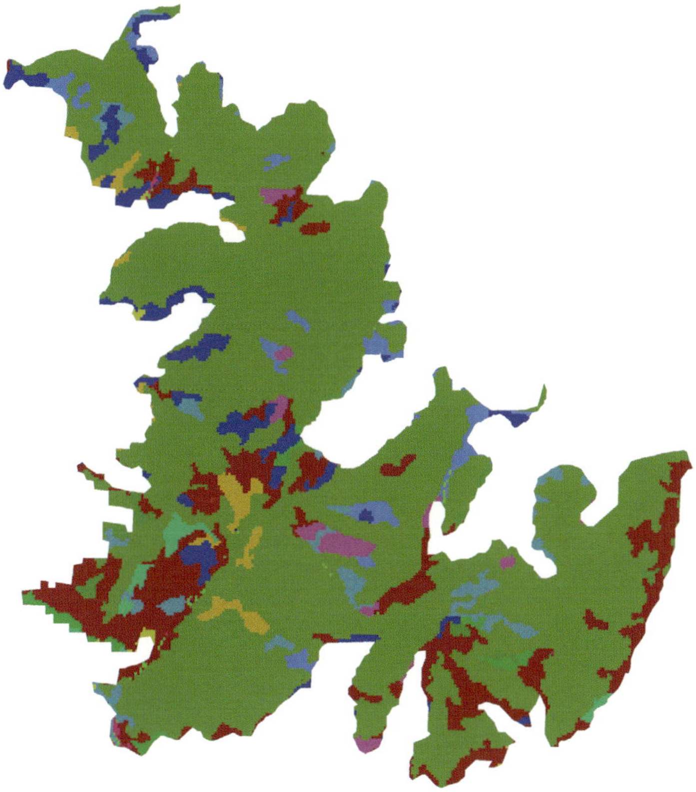

Figure 4.6. Land-cover classes for the red polygon in figure 4.5. Wyoming GAP data classified this entire 8,028-acre polygon as 80% lodgepole pine and 20% subalpine meadow, while the Montana system classified it into eight classes as represented by each color. Data courtesy of US Geological Survey.

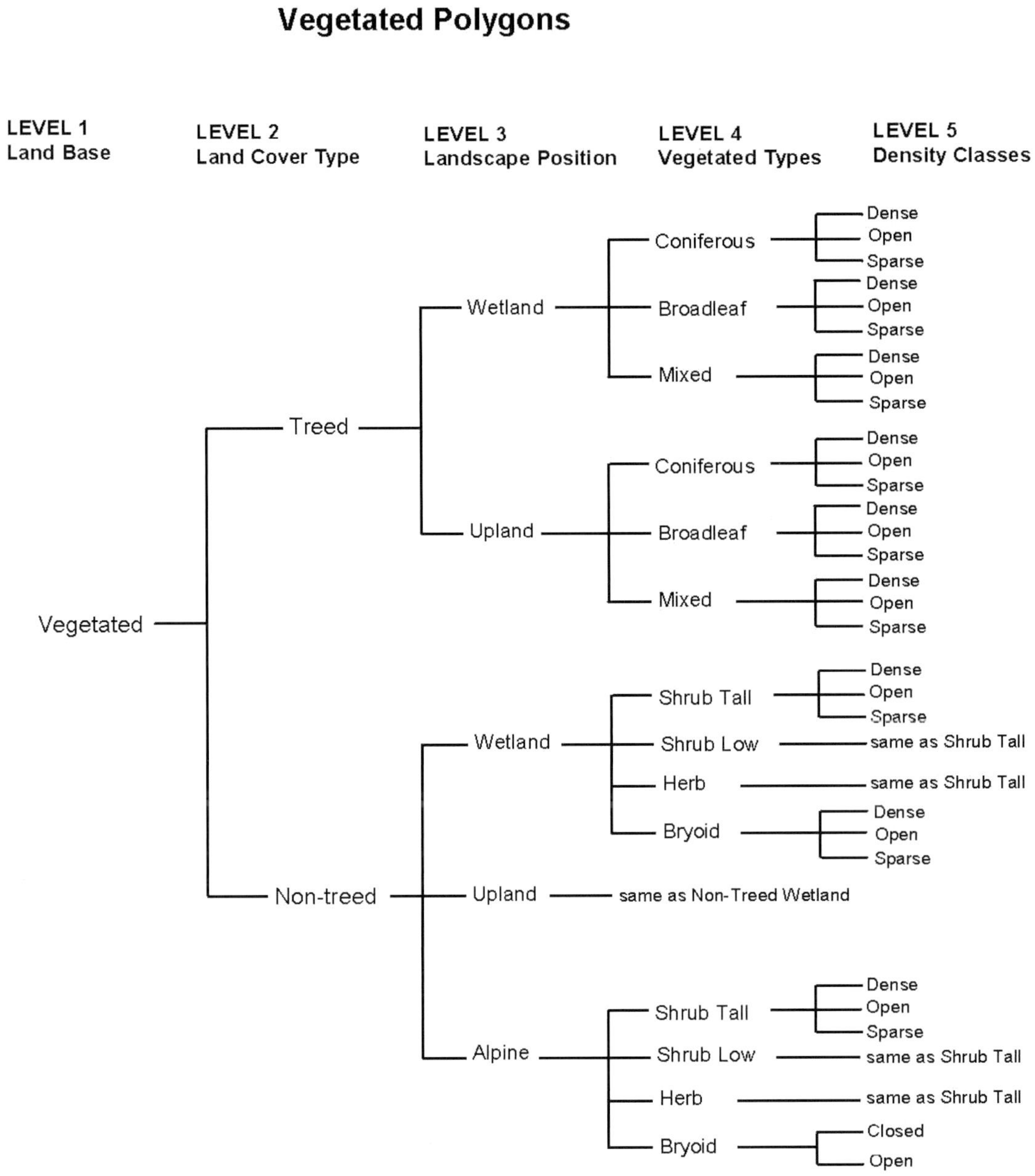

Figure 4.7. Hierarchical classification of vegetated polygons for the VRI classification. Figure derived from Integrated Land Management Bureau.

habitat. Although accuracy of air photo interpretation varies (Strand et al. 2007), some of the problems associated with using satellite imagery are also reduced, such as issues with mixed pixels.

By itself, vegetative structure defined by VRI is useful for modeling habitat for many wildlife species. Additionally, physiognomic classes provided by VRI combined with ancillary data allow easier prediction of plant species occurrence than trying to estimate vegetative structure using the floristic classes of other land-cover data. I used landscape position, vegetation type, and density identified by VRI, combined with elevation, aspect, and slope, to predict plant species as a component of seasonal feeding and living habitat models for seven focal species across the 16.2 million hectare Muskwa-Kechika Management Area in northern

British Columbia (see Heinemeyer et al. 2004). Models proved reliable when validated with telemetry and aerial survey data. However, drawbacks to any classification using air photo interpretation include the intensive labor and associated costs required to classify large areas.

Classification using digital orthophotos

Interpretation of air photos has long been an accurate method for land-cover classification, but even with the availability of digital orthophotos that allow on-screen classification, the process remains time-consuming and impractical over large areas. The ability of computers to discern the same objects that photo interpreters see would increase the use of this readily available data. Miller et al. (2004) were able to use image-processing software and digital air photos to identify tree canopies. Akbari et al. (2003) had limited success with machine processing in an urban area.

Color air photos are similar to satellite imagery in that each pixel is composed of distinct values for red, green, and blue that produce the displayed color. The key difference is that color air photos cover the full range of these colors rather than specific wavelengths that most sensors target. Nonetheless, color air photos can be decomposed into their individual bands and treated like satellite imagery in a hybrid cross between remote sensing classification and air photo interpretation. I used this approach in a situation where Montana GAP data was too coarse for the small area over which I was developing habitat suitability models.

In some situations, the more detailed the feature becomes, the greater the variation detected within a class rather than between classes (Grenzdorffer and Bill 1994). I encountered this problem in my hybrid approach: more variation was detected within tree shadows than among classes of coniferous trees. The solution was to aggregate each of the three bands from a 1-m to a 5-m pixel size, which smoothed values within classes. I then ran an unsupervised classification and used photo interpretation of the original air photo to assign Montana GAP descriptors to output classes from the unsupervised classification. The end result was land-cover data that was more accurate and at a finer scale (figure 4.8), but maintained the Montana GAP classes used in my habitat models. A limitation to this method is that it may work only within the extent of individual air photos. Variation in sun angle from air photos taken at different times of the day, and differences in sun angle and plant

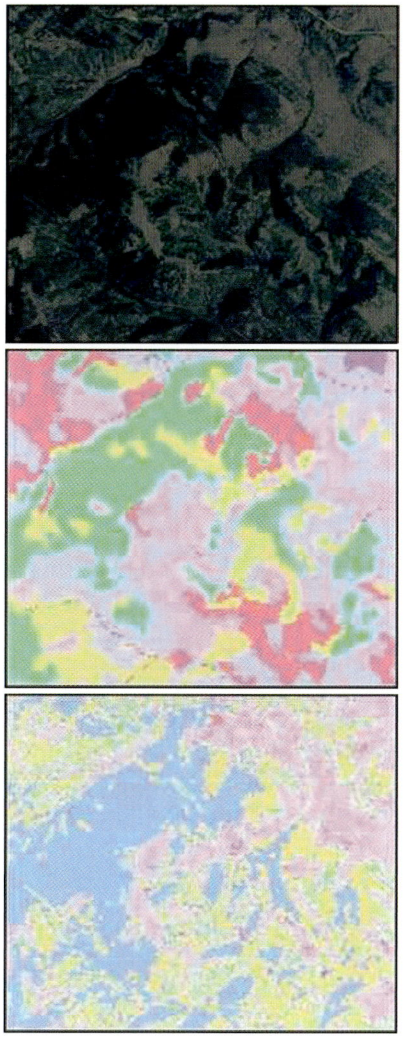

Figure 4.8. Original orthophoto (top), Montana GAP classification of the area (middle), and hybrid classification approach using a combination of remote sensing software and photo interpretation (bottom). Courtesy of the Montana Natural Resource Information System.

phenology of air photos taken in different months, made it difficult to standardize band values in photo mosaics to allow classification over large areas.

Stand-alone software and modules or extensions for many remote sensing packages (e.g., ENVI Feature Extraction Module) are now available to identify shapes within air photos and other high-resolution data. These products base their classification on the combination of shape and reflectance of spectrally similar units rather than on spectral differences alone. Walker and Blaschke (2008) used this process to classify vegetation types, buildings, and impervious surfaces within an urban area. These products offer the potential to classify vegetation over large areas using air photos alone or in conjunction with satellite imagery and more common techniques.

Development of project-specific land-cover classifications

In some conservation planning processes, existing land cover may not have the appropriate resolution or does not capture ecological features needed to model wildlife habitat or identify special elements. Developing land-cover data for a project may be the only alternative. Remotely sensed imagery likely exists at various scales for any area on the planet or can be specifically ordered. Image analysis software is also readily available (e.g., ENVI, or Erdas Imagine). Although standardization of methods is the ultimate goal, specialized land-cover classification will continue to occur until technology provides the ability to determine all desired land-cover variables. Unfortunately, no one land-cover classification system currently exists that meets all conservation planning needs.

A habitat-specific conceptualized classification

With the goal of working toward a standard approach focused on wildlife habitat, the following conceptualized land-cover classification provides ideas for anyone needing to develop land-cover data for a specific project, enhance existing land-cover data, and to help readers understand the processes involved in producing such data. It is a hierarchical classification in which the ecological variable of interest at each level determines the methodology, helps ensure equal accuracy among classes and applicability over large areas, allows aggregation of categories, and provides comparability with future data. The scheme considers the following 10 criteria (listed verbatim from Anderson et al. [1976]) for land-cover and land-use classification:

1. The minimum level of interpretation accuracy in the identification of land use and land-cover categories from remote sensor data should be at least 85%.
2. The accuracy of interpretation for the several categories should be about equal.
3. Repeatable or repetitive results should be obtainable from one interpreter to another and from one time of sensing to another.
4. The classification system should be applicable over extensive areas.
5. The categorization should permit vegetation and other types of land cover to be used as surrogates for activity.
6. The classification system should be suitable for use with remote sensor data obtained at different times of the year.

7. Effective use of subcategories that can be obtained from ground surveys or from the use of larger-scale or enhanced remote sensor data should be possible.
8. Aggregation of categories must be possible.
9. Comparison with future land-use data should be possible.
10. Multiple uses of land should be recognized when possible.

The classification is geared toward large areas using Landsat imagery, but it should be applicable to higher-resolution imagery such as QuickBird or IKONOS when the area of interest covers smaller areas and specific habitat requirements. The combination of field data with higher-resolution imagery and digital air photos can also be used to create large training sites for Landsat imagery and aid in accuracy assessment of the resulting classification.

This conceptual workflow (figure 4.9) is currently untested and is intended to be a starting point for classification and further discussion. An array of habitat variables important to various wildlife species is produced throughout the process. It should provide options for other conservation needs and offer points of discussion between resource managers and remote sensing specialists for accurately linking vegetation on the ground with digital land-cover data. I have incorporated key ecological variables and processes throughout the classification.

Step 1 (figure 4.9) brings inputs from Landsat or higher-resolution imagery into the process, and subsequent boxes indicate the resulting classification at each level. Table 4.3 provides brief class descriptions at each level. Suggested methods for each level are indicated along the arrows connecting the boxes and red ovals indicate the use of ancillary data at each step.

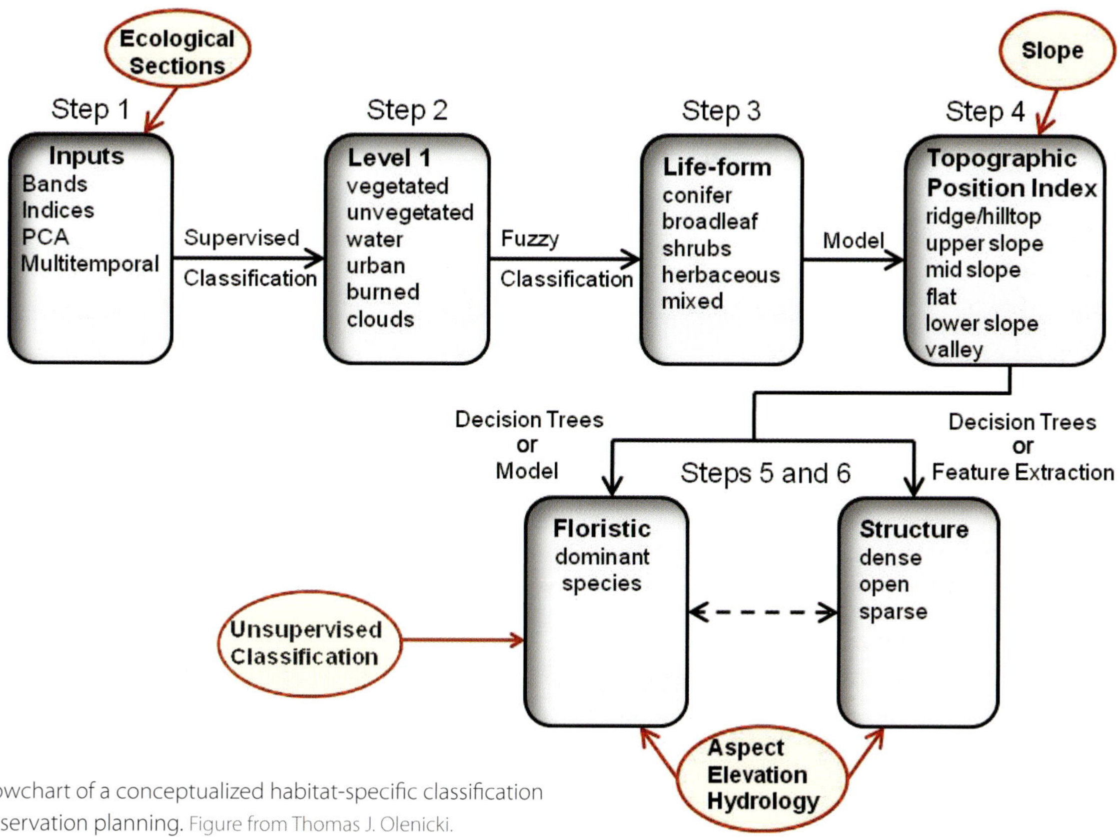

Figure 4.9. Flowchart of a conceptualized habitat-specific classification system for conservation planning. Figure from Thomas J. Olenicki.

Table 4.3. Levels of classification and and class attributes of habitat-specific conceptualized classification for conservation planning

Level 1	Level 2: Life-form	Level 3: Landscape position	Levels 4 and 5: Structure classes (for each class in level 2) and floristics
Vegetated (> 10% total vegetation cover)			
	Conifers (> 75% conifers)		
		Ridge/hilltop	
		Upper slope	
		Mid slope	
		Flat	
		Lower slope	
		Valley	
			Seedling and sapling sized trees, closed (> 50% crown closure)
			Seedling and sapling sized trees, open (10–49% crown closure)
			Mature trees (pole size and greater) dense (> 60% crown closure)
			Mature trees (pole size and greater) open (26–60% crown closure)
			Mature trees (pole size and greater) sparse (10–25% crown closure)
	Broadleaf (> 75% broadleaf)		
		Ridge/hilltop	
		Upper slope	
		Mid slope	
		Flat	
		Lower slope	
		Valley	
			Seedling and sapling sized trees, closed (> 50% crown closure)
			Seedling and sapling sized trees, open (10–49% crown closure)
			Mature trees (pole size and greater) dense (> 60% crown closure)
			Mature trees (pole size and greater) open (26–60% crown closure)
			Mature trees (pole size and greater) sparse (10–25% crown closure)
	Mixed Conifer/Broadleaf (< 75% of conifer or broadleaf)		
		Ridge/hilltop	
		Upper slope	
		Mid slope	
		Flat	
		Lower slope	
		Valley	
			Seedling and sapling sized trees, closed (> 50% crown closure)
			Seedling and sapling sized trees, open (10–49% crown closure)
			Mature trees (pole size and greater) dense (> 60% crown closure)
			Mature trees (pole size and greater) open (26–60% crown closure)
			Mature trees (pole size and greater) sparse (10–25% crown closure)

Level 1	Level 2: Life-form	Level 3: Landscape position	Levels 4 and 5: Structure classes (for each class in level 2) and floristics
	Shrubs (> 75% of the vegetated cover is shrubs)		
		Ridge/hilltop	
		Upper slope	
		Mid slope	
		Flat	
		Lower slope	
		Valley	
			Dense shrub (> 60% canopy cover)
			Open shrub (26–60% canopy cover)
			Sparse shrub (10–25% canopy cover)
	Herbaceous (> 75% of the vegetated cover is herbaceous)		
		Ridge/hilltop	
		Upper slope	
		Mid slope	
		Flat	
		Lower slope	
		Valley	
			Dense herbaceous (> 60% canopy cover)
			Open herbaceous (26–60% canopy cover)
			Sparse herbaceous (10–25% canopy cover)
	Mixed shrub/herb (< 75% of the vegetated cover is either shrub or herbaceous)		
		Ridge/hilltop	
		Upper slope	
		Mid slope	
		Flat	
		Lower slope	
		Valley	
			Dense mixed (> 60% canopy cover)
			Open mixed (26–60% canopy cover)
			Sparse mixed (10–25% canopy cover)
Barren Land (< 10% cover)			
Water			
Urban			
Standing Burned Forest (< 3 years prior to satelite acquisition date)			
Clouds			
Source: Thomas J. Olenicki.			

Step 1: Inputs

Preprocessing techniques for satellite imagery include geometric, atmospheric, and topographic corrections, calculations of vegetation indices, and methods for compressing information from multiple bands into fewer bands. A review of their benefits and applications can be found in most remote sensing books. Landsat imagery, with a greater number of bands, offers more options for preprocessing and enhancement. QuickBird or IKONOS data often are complicated by increased spatial resolution and off-nadir angles of acquisition. Wu et al. (2008) provides a discussion and methods for topographic correction of QuickBird data that may be relevant for some applications.

Inputs to the proposed process are individual bands of satellite imagery, images derived from principal components analysis, and a vegetation index such as NDVI. Images from two dates, early and late in the growing season, are desirable to identify phenological changes. Ecological sections (McNab et al. 2007) for the area of interest should be used to stratify the classification process by relatively homogeneous physical and biological components. Classifying ecological sections individually should improve accuracy and limit misclassification to classes within each ecological section rather than classes covering the entire area of interest.

Step 2: Classification to level 1

A supervised classification using one of the hard classifier algorithms assigns membership into the six broad classes of level 1 (table 4.3). Confusion among classes at this level may have the greatest consequences on accurate habitat delineation and the corresponding conservation planning process. This initial classification is essentially a dichotomous selection separating potential habitat for most terrestrial species (vegetated) from nonhabitat (other classes) using a hard classifier to make this distinction. Consumer's accuracy greater than 90% for each class is desired to reduce the perpetuation of errors in subsequent levels. An accuracy assessment should be made and methods adjusted if necessary to increase accuracy of all classes. In particular, errors of commission for the vegetated class that incorrectly classify locations as vegetated should be reduced. The burned forest class was included to identify this habitat for species dependent on it, for the potential uniqueness in reflectance characteristics, and for the ability to reclassify it as early successional forest in subsequent years. A class of burned grasslands and shrub/grasslands may be included for the same reasons. Areas classified as barren, water, urban, and clouds have reached their endpoint in the classification, but those identified as vegetated continue in the classification process.

Step 3: Classification to life-form

This step makes a physiognomic distinction of vegetated pixels as conifers, broadleaf trees, shrubs, herbaceous, or mixed life-forms (when individual classes cannot be defined). The use of imagery from different dates and a vegetation index such as NDVI may be especially helpful in this step due to differences in phenology among classes. Fuzzy classification was chosen to provide a value for membership in each class rather than an either-or classification. Values listed for percent composition (table 4.3) are intended to be starting points. Because reflectance and shadow characteristics may differ between stands of young and old trees, the potential to define two age classes of trees should be examined at this step.

Step 4: Calculation of topographical position index

At a finer scale than ecological section, vegetation at a site often depends on its topographic position. Step 4 is a modeling exercise to calculate the six topographic positions (Weiss 2001) indicated in table 4.3 as ancillary data for subsequent classification. A topographic position index (TPI) is calculated (e.g., with the ArcGIS extension developed by Jenness Enterprises in 2006: http://www.jennessent.com/arcview/tpi.htm). Calculation of TPI is scale sensitive, and the accuracy of matching topographic classes with fine-scale changes in vegetation is likely influenced by local topography of the classification area. The same ground-truth data that typically is used for accuracy assessment of classifications might be used to adjust TPI calculations to match vegetation changes in the area of interest. Additionally, the use of TPI singly or in combination with aspect calculations may be beneficial in correcting some of the topographical effects of QuickBird imagery discussed by Wu et al. (2008).

Steps 5 and 6: Structural classes and dominant floristics

The final steps classify vegetation into structural classes and identify the dominant plant species to the extent possible. Separate classifications are conducted for the five life-forms identified in step 3. Assignment of final floristic and structural classes may require a feedback loop to assist classification and is the reason they are shown in figure 4.9 as simultaneous steps connected by a dashed line.

Structural classes

Structural classes (canopy coverage of the upper stratum for tree and shrub classes that also contain an understory) for the life-forms determined in step 3 are presented in table 4.3. Canopy closure for two age classes of trees are proposed in the event that either step 3 or this step can differentiate between them.

Decreases in photosynthetic activity as plants senesce during the growing season, as indicated by changes in individual bands of imagery and decreases in most vegetation indices between early and late overpass dates, may prove helpful for determining canopy coverage. However, a number of factors will influence the amount of change detected between dates as an indication of canopy coverage. Weather, location, and topographic location will influence plant senescence and the ability to detect change. Sagebrush actively grows and flowers in the fall, potentially resulting in higher NDVI values during that time. Green needles of most conifers minimize seasonal changes due to senescence, although larches (genus *Larix*) have deciduous leaves that will result in greater changes where they are present. Exceptions such as these can be factored into a classification scheme, but they underscore the fact that ecological process must always be considered during the classification process. Separate classification of ecological sections will help in this respect.

Two methods are proposed for determining structural classes. Decision trees have been used to classify canopy coverage (Herold et al. 2003), in addition to classifying vegetation at finer spatial resolution than traditional methods (Joy et al. 2003). Although decision trees require a large sample size (Pal and Mather 2003), they are nonparametric with the advantages of using different types of response variables; provide invariance to transformation of explanatory variables; and have the capacity for interactive exploration, description, and prediction of explanatory variables that is useful in ecological studies (De'ath and Fabricius 2000). The software tool Random Forests (Breiman 2001), essentially an extension of the decision tree process, has improved the efficiency and accuracy of this process (Gislason et al. 2006).

The use of feature extraction software, such as the ENVI Feature Extraction Module or Feature Analyst by Visual Learning Systems, offers a new approach to the remote sensing field when used in conjunction with high-resolution imagery (e.g., Walker and Blaschke 2008; San Souci and Doyle 2006). These programs essentially attempt to provide the machine-learned equivalent of hand digitizing visibly homogenous polygons. Tree and shrub canopies for an entire area of interest might be classified using digital orthophotos or at specific locations as training data for more conventional classification methods using Landsat imagery (e.g., Herold et al. 2003).

Floristic classification

Decision trees or the modeling of plant species occurrence using variables identified in previous steps are the proposed methods for floristic classification. Separate unsupervised classifications of each life-form identified in step 3 can provide additional input to help differentiate species. The number of unsupervised classes is likely variable among life-forms, locations of interest, and types of imagery. For instance, in decision tree or model output based on known ecological principles, conifers occurring on ridges and upper slopes above 8,000 feet in certain ecological sections and with specific reflectance characteristics are likely to be whitebark pine. Hydrologic layers may also improve the ability for floristic classification. Depending on the scale for calculation of topographic position, valleys containing streams or rivers may define riparian corridors and corresponding types of riparian vegetation. The ability to determine vegetative structure may help with the floristic classification of some life-forms and vice versa. Determining the floristic composition of classes containing sparse amounts of trees and shrubs may be especially difficult. A possibility is to use an overstory/understory nomenclature in which species in only one layer are specified (e.g., scattered shrubs/Idaho fescue).

Recommendations for conservation applications and future research

Success in conservation planning relies on acceptance and implementation of our work. Because land-cover data often forms the foundation of conservation planning, it is critical that limitations in data accuracy and precision are understood prior to its use and it is known how these limitations affect results. The following recommendations may increase the credibility of results that rely on land-cover data.

1. Land-cover data must match the question of interest; habitat modeling or any other work should be done within the limitations of the land-cover data being used. For example, delineating habitat for a grassland species requires high accuracy of grassland classes. Most land-cover data is more applicable for wide-ranging habitat generalists than for habitat obligates and for coarser rather than finer levels of selection.

2. Even if an accuracy assessment is available for the data used, accuracy can be variable across the classification area. It is a good idea to conduct an accuracy assessment for the specific project area using field data or photo interpretation. Even the best model will not produce accurate results if the input data has errors.
3. No land-cover classification is completely accurate. A brief discussion of the confusion among classes (from an accuracy assessment) and of the limitations in the data may explain reasons for discrepancies between habitat predictions and locations of known use or avoidance. These explanations may allow better acceptance of discrepancies and acceptance of conservation plans.
4. FGDC standards for collection of vegetation data should be followed to the extent possible for collection of all field data.

The ability to better match land-cover data with the species, selection order, and area of interest would increase accuracy of individual projects and consistency among projects. Although determination of the true relationships of the conceptual graphs in figures 4.1 and 4.4 is complex, a rating system could be developed for the use of these figures. Even the relative locations for a range of species and selection orders in figure 4.4 (e.g., brood-rearing habitat for sage grouse and home range for grizzly bears), and their corresponding locations in figure 4.1, could serve as a guideline for matching land-cover data with a specific application.

New technologies are continually advancing land-cover classification. The full potential of high-resolution IKONOS and QuickBird imagery is only beginning to be realized. The ability to estimate structural components of vegetation using lidar (Lefsky et al. 2002) or the fusion of hyperspectral and lidar data (Mundt et al. 2006) has been demonstrated. In the meantime, we cannot wait for new technologies to advance; we must strive to make improvements with the tools we have.

References

Adams, J. M. 1999. "A Suggestion for an Improved Vegetation Scheme for Local and Global Mapping and Monitoring." *Environmental Management* 23: 1–13.

Akbari, H., L. S. Rose, and H. Taha. 2003. "Analyzing the Land Cover of an Urban Environment Using High-Resolution Orthophotos." *Landscape and Urban Planning* 63: 1–14.

Anderson, J. R., E. E. Hardy, J. T. Roach, and R. E. Witmer. 1976. *A Land Use and Cover Classification System for Use with Remote Sensor Data.* US Geological Survey Professional Paper 964. http://landcover.usgs.gov/pdf/anderson.pdf.

Bailey, D. W., J. E. Gross, E. A. Laca, L. R. Rittenhouse, M. B. Coughenour, D. M. Swift, and P. L. Sims. 1996. "Mechanisms that Result in Large Herbivore Grazing Distribution Patterns." *Journal of Range Management* 49: 386–400.

Belward, A., and H. Eva. 2004. "Global Land Cover Mapping Using Data from Earth Observing Satellites; Current Status and Future Perspectives." In *Remote Sensing in Land Cover/Land Use Change Detection Workshop.* The Macaulay Institute, Craigiebuckler, Aberdeen, UK.

Boyce, M. S., J. S. Mao, E. H. Merrill, D. Fortin, M. G. Turner, and J. M. Fryxell. 2003. "Scale and Heterogeneity in Habitat Selection by Elk in Yellowstone National Park." *Ecoscience* 10: 421–31.

Breiman, L. 2001. "Random Forests." *Machine Learning* 45: 5–32.

Brohman, R., and L. Bryant. 2005. "Existing Vegetation Classification and Mapping Technical Guide." USDA Forest Service General Technical Report WO-67. Washington, DC: Ecosystem Management Coordination Staff, Washington Office. http://www.fs.fed.us/r5/rsl/publications/rsmapping/ExistingVegClassMappingGuide2005.pdf.

Burley, T. M. 1961. "Land Use of Land Utilization." *The Professional Geographer* 13: 18–20.

Campbell, J. B., and R. H.Wynne. 2011. *Introduction to Remote Sensing*, 5th ed. New York: The Guilford Press.

Caro, T. M., and G. O'Doherty. 1999. "On the Use of Surrogate Species in Conservation Biology." *Conservation Biology* 13: 805–14.

Choudhury, K., and J. M. Jansen, eds. 1998. *Terminology for Integrated Resource Planning and Management.* Rome: Food and Agriculture Organization of the United Nations (FAO).

Connelly, J. W., M. A. Schroeder, A. R. Sands, and C. E. Braun. 2000. "Guidelines to Manage Sage Grouse Populations and their Habitats." *Wildlife Society Bulletin* 28: 967–85.

Cunningham, M. A. 2006. "Accuracy Assessment of Digitized and Classified Land Cover Data for Wildlife Habitat." *Landscape and Urban Planning* 78: 217–28.

Daubenmire, R. 1952. "Forest Vegetation of Northern Idaho and Adjacent Washington, and Its Bearing on Concepts of Vegetation Classification." *Ecological Monographs* 22: 301–30.

De'ath G., and K. E. Fabricius. 2000. "Classification and Regression Trees: A Powerful yet Simple Technique for Ecological Data Analysis." *Ecology* 81: 3,178–92.

Di Gregorio, A., and L. J. M. Jansen. 1998. *Land Cover Classification System (LCCS): Classification Concepts and User Manual.* Environment and Natural Resources Service, GCP/RAF/287/ITA Africover—East Africa Project and Soil Resources, Management and Conservation Service. Rome: Food and Agriculture Organization of the United Nations (FAO).

Depew, J. J. 2004. "Habitat Selection and Movement Patterns of Cattle and White-Tailed Deer in a Temperate Savanna." MS thesis, Texas A&M University, College Station, Texas. http://txspace.di.tamu.edu/bitstream/handle/1969.1/2616/etd-tamu-2004B-2-RLEM-Depew.pdf?sequence=4.

Donoghue, D. N. M., P. J. Watt, and K. B. McManus. 2004. "Remote Sensing of Growth Dynamics of Sitka Spruce Plantation Forests in Upland Britain." Paper presented at the Remote Sensing and Photogrammetry Society Annual Conference, Macaulay Institute, Craigiebuckler, Aberdeen, UK. http://www.macaulay.ac.uk/workshop/remotesensing2004/DD_Full_paper.pdf.

Estes, J. E., and W. Mooneyhan. 1994. "Of Maps and Myths." *Photogrammetric Engineering & Remote Sensing* 60: 517–24.

FGDC (Federal Geographic Data Committee). 1997. *Vegetation Classification Standard.* FGDC-STD-005, Vegetation Subcommittee, Federal Geographic Data Committee, FGDC Secretariat. Reston, VA: US Geological Survey.

———. 2008. *National Vegetation Classification Standard, Version 2.* FGDC-STD-005-2008 (Version 2). Vegetation Subcommittee, Federal Geographic Data Committee, FGDC Secretariat. Reston, VA: US Geological Survey. http://www.fgdc.gov/standards/projects/FGDC-standards-projects/vegetation/NVCS_V2_FINAL_2008-02.pdf.

Foody, G. M. 2002. "Status of Land Cover Classification Accuracy Assessment." *Remote Sensing of Environment* 80: 185–201.

Franklin, S. E., and M. A. Wulder. 2002. "Remote Sensing Methods in Medium Spatial Resolution Satellite Data Land Cover Classification of Large Areas." *Progress in Physical Geography* 26: 173–205.

Gislason, P. O., J. A. Benediktsson, and J. R. Sveinsson. 2006. "Random Forests for Land Cover Classification." *Pattern Recognition Letters* 27: 294–300.

Grenzdorffer, G., and R. Bill. 1994. "Digital Orthophotos for Mapping and Interpretation in Hybrid GIS Environments." In *The Proceedings of the 4th European Conference on Geographical Information Systems,* 1,845–56. Paris: EGIS.

Grossman, D., D. Faber-Langendoen, A. S. Weakley, M. Anderson, P. Bourgeron, R. Crawford, K. Goodin, S. Landaal, K. Metzler, K. D. Patterson, M. Pyne, M. Reid, and L. Sneddon. 1998. "International Classification of Ecological Communities: Terrestrial Vegetation of the United States." In *The National Vegetation Classification System: Development, Status, and Applications,* vol. 1. Arlington, VA: The Nature Conservancy. http://www.csu.edu/cerc/researchreports/documents/TerrestrialVegetationUnitedStatesVolumeI.pdf.

Hansen, M. C., R. S. DeFries, J. R. G. Townshend, and R. Sohlberg. 2000. "Global Land Cover Classification at 1 km Spatial Resolution Using a Classification Tree Approach." *International Journal of Remote Sensing* 21: 1,331–64.

Heinemeyer, K., R. Tingey, K. Ciruna, T. Lind, J. Pollock, B. Butterfield, J. Griggs, P. Iachetti, C. Bode, T. Olenicki, E. Parkinson, D. Rumsey, and D. Sizemore. 2004. *Conservation Area Design for the Muskwa-Kechika Management Area (MKMA),* Nature Conservancy of Canada, Round River Conservation Studies, Dovetail Consulting. http://science.natureconservancy.ca/resources/docs/MK_CAD_V1.pdf.

Herold, N. D., G. Koeln, and D. Cunnigham. 2003. "Mapping Impervious Surfaces and Forest Canopy Using Classification and Regression Tree (Cart) Analysis." *ASPRS 2003 Annual Conference Proceedings*. Anchorage, Alaska. http://nemo.uconn.edu/tools/impervious_surfaces/pdfs/Herold_etal_2003.pdf.

Hoffer, R. M. 1994. "Challenges in Developing and Applying Remote Sensing to Ecosystem Management." In *Remote Sensing and GIS in Ecosystem Management,* edited by V. A. Sample, 25–40. Washington, DC: Island Press.

Jennings, M., D. Faber-Langendoen, R. Peet, O. Loucks, D. Glenn-Lewin, A. Damman, M. Barbour, R. Pfister, D. Grossman, D. Roberts, D. Tart, M. Walker, S. Talbot, J. Walker, G. Hartshorn, G. Waggoner, M. Abrams, A. Hill, and M. Rejmanek. 2004. *Guidelines for Describing Associations and Alliances of the U.S. National Vegetation Classification* vol. 46. The Ecological Society of America Vegetation Classification Panel, Ecological Society of America.

Jiang, Z., and R. J. Hudson. 1993. "Optimal Grazing of Wapiti (*Cervus elaphus*) on Grassland: Patch and Feeding Station Departure Rules." *Evolutionary Ecology* 7: 488–98.

Johnson, D. H. 1980. "The Comparison of Usage and Availability Measurements for Evaluating Resource Preference." *Ecology* 61: 65–71.

Jones, J. L., and E. O. Garton. 1994. "Selection of Successional Stages by Fishers in Northcentral Idaho." In *Martens, Sables, And Fishers: Biology and Conservation,* edited by S. W. Buskirk, A. Harestad, and M. Ralph, 377–87. Ithaca, NY: Cornell University Press.

Joy, S. M., R. M. Reich, and R. T. Reynolds. 2003. "A Non-parametric, Supervised Classification of Vegetation Types on the Kaibab National Forest Using Decision Trees." *International Journal of Remote Sensing* 24: 1,835–52.

Kauth, R. J., and G. S. Thomas. 1976. "The Tasseled Cap—A Graphic Description of the Spectral-Temporal Development of Agricultural Crops as Seen by Landsat." 4B41–4B51. Symposium on Machine Processing of Remotely Sensed Data, Laboratory for Applications of Remote Sensing, Purdue University, West Lafayette, IN.

Kie, J. G., R. T. Bowyer, M. C. Nicholson, B. B. Boroski, and E. R. Loft. 2002. "Landscape Heterogeneity at Different Scales: Effects on Spatial Distribution of Mule Deer." *Ecology* 83: 530–44.

Kuchler, A. W. 1967. *Vegetation Mapping.* New York: Ronald Press.

Lack, D. 1933. "Habitat Selection in Birds. With Special Reference to the Effects of Afforestation on the Breckland Avifauna." *The Journal of Animal Ecology* 2: 239–62.

Lambeck, R. J. 1997. "Focal Species: A Multi-species Umbrella for Nature Conservation." *Conservation Biology* 11: 849–56.

Landres, P. B., J. Verner, and J. W. Thomas. 1988. "Ecological Use of Vertebrate Indicator Species: A Critique." *Conservation Biology* 2: 316–28.

Lefsky, M. A., W. B. Cohen, G. G. Parker, and D. J. Harding. 2002. "Lidar Remote Sensing for Ecosystem Studies." *BioScience* 52: 19–30.

Levin, S. A. 1992. "The Problem of Pattern and Scale in Ecology: The Robert H. MacArthur Award Lecture." *Ecology* 73: 1943–1967.

Lillesand, T. M., R. W. Kiefer, and J. W. Chipman. 2008. *Remote Sensing and Image Interpretation.* 6th ed. New York: John Wiley and Sons.

Loveland, T. R., B. Reed, J. F. Brown, D. O. Ohlen, Z. Zhu, L. Yang, and J. W. Merchant. 2000. "Development of a Global Land Cover Characteristics Database and IGBP DISCover from 1 km AVHRR Data." *International Journal of Remote Sensing* 21: 1,301–30.

Marceau, D. J., and G. J. Hay. 1999. "Remote Sensing Contributions to the Scale Issue." *Canadian Journal of Remote Sensing* 25: 357–66.

Marceau, D. J., P. J. Howarth, and D. J. Gratton. 1994. "Remote Sensing and the Measurement of Geographical Entities in a Forested Environment. 1. The Scale and Spatial Aggregation Problem." *Remote Sensing of Environment* 49: 93–104.

McNab, W. H., D. Cleland, J. A. Freeouf, J. J. E. Keys, G. J. Nowacki, and C. A. Carpenter 2007. *Description of Ecological Subregions: Sections of the Conterminous United States.* US Department of Agriculture Forest Service General Technical Report WO-76B, Washington, DC. http://www.edc.uri.edu/atmt-dss/ReportForecast/SectionDescriptions.pdf.

Mehner, H., M. Cutler, D. Fairbairn, and G. Thompson. 2004. "Remote Sensing of Upland Vegetation: The Potential of High Spatial Resolution Satellite Sensors." *Global Ecology and Biogeography* 13: 359–69.

Miller, D. R., A. D. Cameron, and G. Zagalikis. 2004. *The Monitoring of Tree and Stand Characteristics Using Digital Photogrammetry and Image Analysis Techniques.* Macaulay Institute, Craigiebuckler, Aberdeen, UK. http://www.macaulay.ac.uk/workshop/remotesensing2004/DM_Full_paper.pdf.

Mueggler, W. F., and W. L. Stewart. 1980. *Grassland and Shrubland Habitat Types of Western Montana.* US Department of Agriculture Forest Service General Technical Report INT-66. Ogden, UT. http://www.fs.fed.us/rm/pubs_int/int_gtr066.pdf.

Mundt, J. T., D. R. Streutker, and N. F. Glenn. 2006. "Mapping Sagebrush Distribution Using Fusion of Hyperspectral and Lidar Classifications." *Photogrammetric Engineering & Remote Sensing* 72: 47–54.

Nijland, W., E. A. Addink, S. M. De Jong, and F. D. Van der Meer. 2009. "Optimizing Spatial Image Support for Quantitative Mapping of Natural Vegetation." *Remote Sensing of Environment* 113: 771–80.

Owen, M. 1972. "Some Factors Affecting Food Intake and Selection in White-Fronted Geese." *The Journal of Animal Ecology* 41: 79–92.

Pal, M., and P. M. Mather. 2003. "An Assessment of the Effectiveness of Decision Tree Methods for Land Cover Classification." *Remote Sensing of Environment* 86: 554–65.

Read, J. M., D. B. Clark, E. M. Venticinque, and M. P. Moreiras. 2003. "Application of Merged 1-m and 4-m Resolution Satellite Data to Research and Management in Tropical Forests." *Journal of Applied Ecology* 40: 592–600.

Reynolds, H. W., R. D. Gladholt, and W. L. Hawley. 1982. "Bison." In *Wild Mammals of North America: Biology, Management, and Economics,* edited by J. A. Chapman and G. A. Feldhammer, 972–1,007. Baltimore, MD: Johns Hopkins University Press.

RIC (Resources Inventory Committee). 2002. *Vegetation Resources Inventory: The B. C. Land Cover Classification Scheme.* Ministry of Sustainable Resource Management Terrestrial Information Branch for the Terrestrial Ecosystems Task Force—Vegetation Resources Inventory Committee, the Province of British Columbia.

Rouse, J. W., R. H. Haas, J. A. Schell, and D. W. Deering. 1973. "Monitoring Vegetation Systems in the Great Plains with ERTS." In *Third Earth Resources Technology Satellite-1 Symposium, Volume 1: Technical Presentations,* NASA SP-351, edited by Stanley C. Freden, Enrico P. Mercanti, and Margaret A. Becker, 309–17. Washington, DC: NASA.

San Souci, J. W., and J. T. Doyle 2006. "Tamarisk Mapping and Monitoring Using High-Resolution Satellite Imagery." In *Monitoring Science and Technology Symposium: Unifying Knowledge for Sustainability in the Western Hemisphere,* edited by C. Aguirre-Bravo, P. J. Pelicane, D. P. Burns, and S. Draggan. Fort Collins, CO: US Department of Agriculture, Forest Service, Rocky Mountain Research Station.

Searle, K. R., N. T. Hobbs, and L. A. Shipley. 2005. "Should I Stay or Should I Go? Patch Departure Decisions by Herbivores at Multiple Scales." *Oikos* 111: 417–24.

Senft, R. L., M. B. Coughenour, D. W. Bailey, L. R. Rittenhouse, O. E. Sala, and D. M. Swift. 1987. "Large Herbivore Foraging and Ecological Hierarchies." *BioScience* 37: 789–99.

Shipley, L. A. 2007. "The Influence of Bite Size on Foraging at Larger Spatial and Temporal Scales by Mammalian Herbivores." *Oikos* 116: 1,964–74.

Singh, A., and A. Harrison. 1985. Standardized Principal Components. *International Journal of Remote Sensing* 6: 883–96.

Snyder, M. N., S. J. Goetz, and R. K. Wright. 2005. "Stream Health Rankings Predicted by Satellite Derived Land Cover Metrics." *Journal of the American Water Resources Association* 41: 659–77.

Strand, H., R. Hoft, J. Strittholt, L. Miles, N. Horning, E. Fosnight, and W. Turner, eds. 2007. *Sourcebook on Remote Sensing and Biodiversity Indicators.* Technical Series no. 32. Montreal: Secretariat of the Convention on Biological Diversity.

Thogmartin, W. E., A. L. Gallant, M. G. Knutson, T. J. Fox, and M. J. Suarez. 2004. "A Cautionary Tale Regarding Use of the National Land Cover Dataset 1992." *Wildlife Society Bulletin* 32: 970–78.

Turner, W., S. Spector, N. Gardiner, M. Fladeland, E. Sterling, and M. Steininger. 2003. "Remote Sensing for Biodiversity Science and Conservation." *Trends in Ecology and Evolution* 18: 306–14.

Walker, J., and T. Blaschke. 2008. "Object-Based Landcover Classification for the Phoenix Metropolitan Area: Optimization vs. Transportability." *International Journal of Remote Sensing* 29: 2,021–40.

Weiss, A. 2001. "Topographic Position and Landform Analysis." Poster Presentation, Esri User Conference, San Diego, CA.

Wiens, J. A. 1989. "Spatial Scaling in Ecology." *Functional Ecology* 3: 385–97.

Wilcox, B. A. 1984. "In situ Conservation of Genetic Resources: Determinants of Minimum Area Requirements." In *National Parks, Conservation and Development,* Proceedings of the World Congress on National Parks, edited by J. A. McNeely and K. R. Miller, 18–30. Washington, DC: Smithsonian Institution Press.

Woodcock, C. E., and A. H. Strahler. 1987. "The Factor of Scale in Remote Sensing." *Remote Sensing of Environment* 21: 311–32.

Wu, J., M. E. Bauer, D. Wang, and S. M. Manson. 2008. "A Comparison of Illumination Geometry-Based Methods for Topographic Correction of QuickBird Images of an Undulant Area." *Journal of Photogrammetry & Remote Sensing* 63: 223–36.

5

Integrating land use and landscape change with conservation planning

David M. Theobald, PhD

Increasingly, scientists are concerned about the influence of land-cover changes and human activities on the long-term conservation of biodiversity (Foley et al. 2005). In this chapter, I briefly review how land use and landscape change typically are incorporated into conservation planning and describe how human modifications associated with particular land uses can be better integrated within conservation plans. This is important because: "[p]eople and their activities and uses of the landscape must be included in the vision and actions of conservation. . . . What this means is that conservation needs more than information on land cover alone[,] and should be linked with data on land use . . . to confer different conservation values on places" (Wiens et al. 2008, 163). That is, threats that are represented in land-cover datasets account for a small number of important types of human activities that might put biodiversity at risk. It follows that conservation planners need to do a better job of mapping various land uses and to find rigorous ways of incorporating them into conservation plans.

A common way to integrate human modifications with a conservation plan is to identify direct threats associated with human modification that affect, or stress, biodiversity targets (Salafsky et al. 2008; figure 5.1). Data on land use is used to help understand which conservation elements should be targeted and where they currently are located, what the threats are from human land-use changes, and how different planning actions might protect elements or mitigate these threats or lead to restoration.

Numerous challenges and methodological issues are related to quantifying and modeling land-use/land-cover dynamics, and many people have argued for the need to develop a science of land change (e.g., Rindfuss et al. 2004; Turner et al. 2007). A key to these discussions is the distinction between land cover and land use. Land cover is "the biophysical state of the earth's surface and immediate subsurface" (Turner et al. 1995, 20; also see Olenicki, chapter 4). Land use is the purpose for which the land is used, which "involves both the manner in which the biophysical attributes are manipulated and the intent underlying that manipulation"

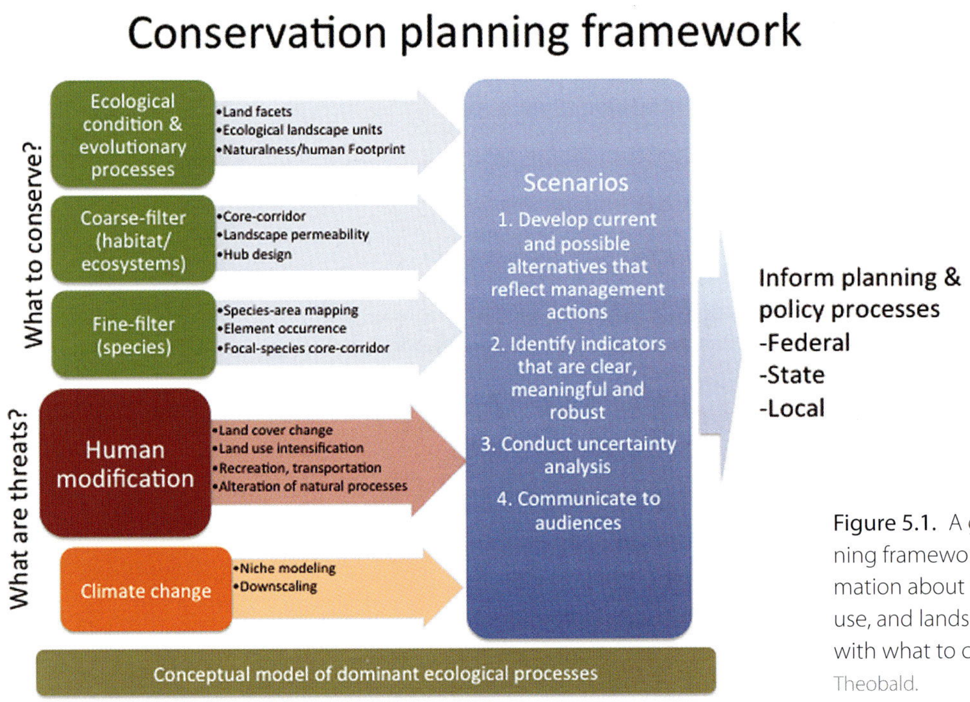

Figure 5.1. A general conservation planning framework that places data and information about human modification, land use, and landscape change into context with what to conserve. Figure from David M. Theobald.

(Turner et al. 1995, 20). For example, logging, like recreation, is a land use that often occurs on forested land-cover types; grazing is a land use occurring on grassland cover; agriculture is a land use, but cropland is a land cover, and wildlife refuge is a land use on wetland cover types. Often, the intensity of a land use is sufficient to cause a conversion of a land cover from a natural state to a human-dominated land-cover type (e.g., grassland to cropland). But although some land uses may not cause a visible conversion of land cover, they may provide sufficient alteration of natural composition, structure, or function to impact biodiversity (e.g., intense grazing on a grassland). For example, Schlesinger et al. (2008) found that human activity was a more important stressor than habitat loss from development in explaining bird species richness. Thus, land use and land cover provide different types of information useful to understand effects on ecosystem structure and processes. Typically, major human activities that result in conversion from natural to human-modified cover are captured adequately in a land-cover map, but there can be additional, often subtle, yet important human activities that usually have substantial consequences for biodiversity and affect both ecosystem structure and function.

Land use and conservation planning

Although these concepts and the emerging science of land change are foundational to conservation planning, both the variety and urgency of threats have spurred conservation practitioners to work cooperatively and look beyond an exclusive focus on land cover. Table 5.1 provides a slightly generalized listing of the human activities identified by the Conservation Measures Partnership (Salafsky et al. 2008) to provide a general sense of the range and variety of human activities that are considered to be threats to conservation of biodiversity. These include residential and commercial land uses (often referred to as "urban development"), extractive resource use (for example, agriculture, energy production, and mining), transportation infrastructure, invasive species (which are often facilitated indirectly by disturbing native species; directly through purposeful introduction; or by unknowingly transporting plants, pests, or seeds), hunting and collecting, other human intrusions, and general modification of ecological processes. Note that to date, most efforts to depict human land uses have focused on urban development and transportation. This is partially because these activities result in conversion of land cover, and spatial data on them is more readily available, yet a variety of other activities may impact biodiversity.

Table 5.1. A comprehensive list of human activities that may threaten biodiversity

Group	Type
1. Residential and commercial development	Housing and urban areas Commercial and industrial areas Tourism and recreation areas
2. Agriculture and aquaculture	Annual and perennial nontimber crops Wood and pulp plantations Livestock farming and ranching Marine and freshwater aquaculture
3. Energy production and mining	Oil and gas drilling Mining and quarrying Renewable energy (solar, wind, etc.)
4. Transportation and service corridors	Roads and railroads Utility and service lines Shipping lanes Flight paths
5. Biological resource use	Hunting and collecting terrestrial animals Gathering terrestrial plants Logging and wood harvesting Fishing and harvesting aquatic resources
6. Human intrusions and disturbance (nonconsumptive)	Recreational activities War, civil unrest, and military exercises Work and other activities
7. Natural system modifications	Fire and fire suppression Dams and water management use Other ecosystem modifications
8. Invasive/problematic species and genes	Invasive nonnative/alien species Problematic native species Introduced genetic material
9. Pollution	Household sewage and urban wastewater Industrial and military effluents Agricultural and forestry effluents Garbage and solid waste Airborne pollutants Excess energy (heat, sound, light)

Source: Salafsky et al. 2008.
Note: From the Conservation Measures Partnership.

The conservation planning framework provides a general context for both identifying and organizing how spatial data on human activities can be used to examine the ways land use and human activities can modify ecological systems. Building on this framework provides a sturdy foundation to convert this data into an index for each factor (or threat) as well as to combine the indicators into an overall score. Schultz (2001) provides a useful cautionary review of how ad hoc methods have been challenged in the past. There are formal ways to convert data into an index and combine indicators into an overall score, based in multi-attribute utility theory (Hajkowicz 2006), which uses five general steps:

1. Identify relevant threats to include, striving for comprehensiveness and avoiding redundancy when identifying each indicator.
2. Obtain spatial datasets that best represent each threat (indicator), being explicit about what is being measured, what the units are, and what scale (grain) is.
3. Convert raw values for each threat (indicator) into commensurable units, typically standardizing or normalizing them from 0 to 1.0.
4. Specify through a set of weights or equations the relative importance of each indicator.
5. Conduct a sensitivity analysis to examine the robustness of the solution for the specified weighting structure.

In subsequent sections, I emphasize steps 2 and 4 by describing the commonly used, readily available spatial datasets on human modification, at global and national/regional scales. Then I describe some common ways that these datasets are compiled to generate composite datasets through some spatial modeling process. Finally, I conclude this chapter with a discussion of some general recommendations, caveats, and future opportunities.

Commonly available land-use and land-cover datasets

A variety of data is used for understanding the patterns and trends of human land use and modification. Rather than attempting to provide an exhaustive list of datasets, I focus on those that are commonly used and readily available. Multiple global land-use and land-cover datasets are available (table 5.2). Not surprisingly, these provide different estimates of the extent of human-modified land-cover types because of the different purposes, sensors, and scales used to generate the datasets (figure 5.2). Yet the range of global estimates is substantial: from 0.20% to 0.45% (a difference of a factor of 2.25) for urban/built-up areas and from 14.16% to 22.69% (a difference of a factor of 1.6) for cropland agriculture (table 5.3). Comparing these global datasets to more detailed national datasets for the United States (~100–1000 times the resolution, with approximately the same number of cover types) also reveals large differences. The global datasets estimate 10–30% of the extent of urban areas compared to the national datasets, although they are much more similar for agricultural areas (85–140%). It is difficult to draw a strong conclusion from these comparisons, except to underscore the fact that datasets are highly variable, and fine-scale information is often lost when aggregated to coarser scales.

Although land-use and land-cover data is central to understanding potential impacts of human activities on biodiversity, the variety of additional threats outlined in table 5.1 should be considered more fully in a comprehensive conservation plan. Table 5.4 provides a listing of recommended, commonly available datasets

Table 5.2. Commonly used and readily available datasets on global land use/land cover

Name and description	Source
Gridded Population of the World v3—Depicts the "nighttime" distribution of human population across the globe at 30 ft. (~ 1 km) Global Rural-Urban Mapping Project—Human population density	Center for International Earth Science Information Network http://sedac.ciesin.columbia.edu/gpw/
Nighttime lights —Defense Meteorological Satellite Program-Operational Linescan System—records low levels of visible near-infrared radiance at night at ~ 2.7 km (Sutton et al. 2009)	National Geophysical Data Center http://www.ngdc.noaa.gov/dmsp/downloadV4composites.html
Global Land Cover 2000—European Commission, Joint Research Centre, Global Environment Monitoring Unit at 1 km	European Commission, Joint Research Centre http://bioval.jrc.ec.europa.eu/products/glc2000/products.php
GlobCover—European Space Agency, Ionia GlobCover	European Space Agency http://www.esa.int
MODIS—Land Cover (MOD12Q1) at 500 m	Goddard Space Flight Center http://modis-land.gsfc.nasa.gov/landcover.html
Global Human Footprint v2 (Last of the Wild)—A composite measure of the Human Influence Index (HII) normalized by biome and realm, created from population density, human land use and infrastructure, and human access at 1 km (Sanderson et al. 2002)	Center for International Earth Science Information Network http://sedac.ciesin.org/data/set/wildareas-v2-human-footprint-geographic
Global map of accessibility—each location represents the time it would take to travel (via road, rail, boat, etc.) to the nearest city of at least 50,000 people in 2000.	European Commission, Joint Research Centre http://bioval.jrc.ec.europa.eu/products/gam/index.htm

Source: David M. Theobald.

Table 5.3. The percent of agricultural and urban cover types for different general-purpose land-cover/land-use datasets (ocean areas excluded)

	GLC	GlobCover	MODIS	NLCD 2001	NLCD 2006
Urban/built-up	0.20%	0.23%	0.45%	-	-
Agricultural	15.74%	22.69%	14.16%	-	-
US Urban	1.00%	0.40%	1.54%	5.32%	5.46%*
US Ag	20.30%	18.81%	32.29%	22.33%	22.24%

Source: David M. Theobald.
*This reduces to 2.89% after removing urban cover types that occur as linear features along roads (nonhighway).

for the United States. Many efforts to date emphasize a single aspect of human modification, for example, roads (Riitters and Wickham 2003; Watts et al. 2007; Sutton et al. 2009) or land cover (Hoekstra et al. 2005). Increasingly, land-use analysis has provided a more comprehensive perspective by combining population with land use and land cover (e.g., Ellis and Ramankutty 2008) or by generating a composite "human footprint" approach that includes urban and agricultural cover and transportation infrastructure (e.g., Sanderson et al. 2002). Most recently, efforts are expanding to include additional threats such as extractive resource uses (e.g., Leu et al. 2008; Theobald 2010; Theobald et al. 2012), traffic volume (Theobald 2010; Theobald et al. 2012), and impacts to freshwater resources (Esselman et al. 2011). Even so, significant data gaps remain and it seems that more progress could be made by considering a yet broader variety of potential threats in conservation plans (table 5.4).

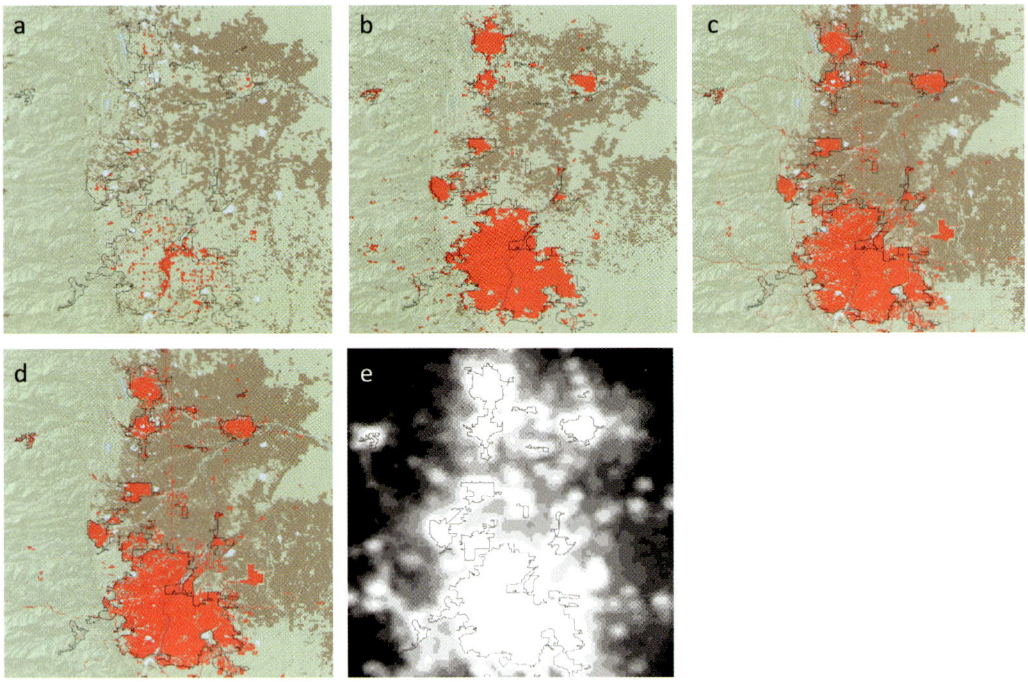

Figure 5.2. A comparison of land-use and land-cover datasets: (a) Global Land Cover, (b) MODIS, (c) National Land Cover Data (NLCD) 2006, (d) NLCD 2006 filtered, and (e) Defense Meteorological Satellite Program (DMSP) "nightlights." The geographic area is the northern Front Range of Colorado, including the cities of Fort Collins, Estes Park, Loveland, Greeley, Longmont, Boulder, and Denver (top to bottom); urban areas in red, cropland in brown, and "natural" in green. Global land cover from US Geological Survey; MODIS from NASA; NLCD 2006 from US Geological Survey; NLCD 2006 filtered from US Geological Survey; DMSP "nightlights" from NOAA.

Table 5.4. Datasets and data gaps related to biodiversity/ecosystem threats for the United States

Threat type	Dataset	Strengths	Weaknesses	Data gaps
1. Residential and commercial development				
Housing and urban areas	National Land Cover Dataset v2 (2001, 2006); NLCD retrofit (1992, 2001), NLCD v1 (1992) (http://www.mrlc.gov) US Geological Survey Gap Land Cover v1 (http://gapanalysis.usgs.gov/data/land-cover-data/) LANDFIRE (http://www.landfire.gov) Housing density from census blocks (e.g., ICLUS/SERGoM)	Standard product and multidate allows change detection Distinguishes semicultural classes (e.g., plantations) Can compare to "potential natural" vegetation (biophysical settings) Quantifies patterns of disturbance beyond urban fringe	Urban cover misses residential <1 dwelling unit per acre Coarser and variable sized analytical unit	Age of development Direct measures of human activity such as noise, light, pets
Commercial and industrial areas	National Land Cover Dataset v2	Medium and high intensity development classes	Cannot differentiate from high-density residential or specific uses (e.g., retail, wholesale, service, etc.)	Urban land uses
Tourism and recreation areas	Protected Areas Database–US (CBI v1.1, GAP v1.2) (http://www.protectedlands.net/padus/)	Infer tourism and recreation areas from management types	Does not identify spatial distribution of use within protected area or seasonality	Number of visitors, mode of use (motorized/non), pets

Threat type	Dataset	Strengths	Weaknesses	Data gaps
2. Agriculture and aquaculture				
Annual and perennial nontimber crops	NLCD v2 (classes 81, 82) Cropland Data Layer (http://www.nass.usda.gov)	Basic cropland and pasture types Compiled state by state, nonstandardized	Does not distinguish irrigated land and has "striping" from tillage practices	Need for irrigated, tilled, "native" grasslands, vineyards
Wood and pulp plantations	US Gap Land Cover v1 (http://gapanalysis.usgs.gov/data/land-cover-data/)	Timber harvest and plantations	Difficult to distinguish through spatial pattern	Commercial plantations
Livestock farming and ranching	NLCD v2 (classes 52, 71) Census of Agriculture (http://www.agcensus.usda.gov)	Infer grazing from land-cover types Number of livestock animals by type	Most data is available only at county level	Grazing allotments and intensity (AUM), type Fence lines, predation control
Water diversions, transbasin pipelines, irrigation canals	USGS National Hydrography Dataset (http://nhd.usgs.gov/)	Consistent, standardized and strong stewardship for improving data Location of structures	Little information on volume or flow	Irrigated areas, perennial/intermittent, actual annual/monthly flow volume
Water storage (reservoirs)	USGS National Hydrography Dataset	Location of ponds, lakes, reservoirs	Natural or human-made is not consistently attributed	Reservoir volume
Field drain tiles	Not known			No centralized, standardized dataset
Marine and freshwater aquaculture	Census of Aquaculture (Goldburg et al. 2001)	General indication but reported at state level	Reported at state level	No standardized dataset of stocking of lakes, sport fish
3. Energy production and mining				
Oil and gas drilling	USGS National Oil and Gas Assessment 1995 (http://energy.cr.usgs.gov/) State oil and gas commissions	Private industry have updates More up-to-date, standardization and compilation issues	Locations are often imprecise, not a publicly available dataset	Well locations at surface Includes drill pad, road, other infrastructure
Mining and quarrying	USGS Mineral Resources Data System (http://tin.er.usgs.gov/mrds/) USGS Significant Topographic Change (1974–99) (http://topochange.cr.usgs.gov/)	Standardized database, point locations	Surface area/extent of disturbance not well documented	Aggregate (e.g., gravel) resources as a centralized database
Renewable energy (solar, wind, etc.)	NREL Energy (http://www.nrel.gov/gis/data.html) Western Renewable Energy Zones (http://www.westgov.org/rtep/224-wrez-zone-identification-a-technical-analysis) Wind turbine locations (http://www.fort.usgs.gov/WindEnergy/)	General maps of resource locations	Need standardized, funded data-collection efforts	

Threat type	Dataset	Strengths	Weaknesses	Data gaps
4. Transportation and service corridors				
Roads and railroads	Census Bureau TIGER 2010 (http://www.census.gov) National Transportation Atlas 2011 National Bridge Inventory	Comprehensive, consistent database Provides road use (traffic volume) Locations and characteristics of bridges	Poorer quality in rural public lands (i.e., BLM, USFS)	Structures specifically designed to enhance movement (overpasses, underpasses) Deer- and elk-proof fencing
Utility and service lines	Power lines and utilities (http://sagemap.wr.usgs.gov/)	Significant power lines used to be available Private industry datasets are available	Limited and out-of-date public datasets	Pipelines and other utilities not in the public domain
Shipping lanes	Shipping lanes (http://www.nceas.ucsb.edu/globalmarine/impacts)			Barge traffic on inland waterways
Flight paths	Federal Aviation Administration, Bureau of Transportation Statistics			Spatiotemporal, 3D data
5. Biological resource use				
Hunting and collecting terrestrial animals	State Fish and Game hunter tags and success rates		Not standardized, difficult to compile, coarse analytical unit (game management unit)	
Gathering terrestrial plants				No known source
Logging and wood harvesting	US Forest Service	General data on harvest	Detailed consistent spatial data is limited and not centralized	
Groundwater pumping	State water engineer offices	Locations of wells	Older wells are imprecise, data on year and volume limited	
Fishing and harvesting aquatic resources	State Fish and Game fishing locations		Not standardized, not readily available	
6. Human intrusions and disturbance (nonconsumptive)				
War, civil unrest, and military exercises				No known source
Work and other activities	Habitat modification (e.g., chaining, fertilizing) Illegal crossing of international borders		Only partially mapped in land cover datasets	No known source No known source
7. Natural system modifications				
Fire and fire suppression	GeoMAC (http://www.geomac.gov)	Fire perimeters		
Dams and water management use	National Inventory of Dams (Army Corp of Engineers (http://nid.usace.army.mil) USGS Water Use (http://water.usgs.gov/watuse/)	Locations of dams, structural information	Storage area incomplete, some locational errors	Small reservoirs
Other ecosystem modifications				

Threat type	Dataset	Strengths	Weaknesses	Data gaps
8. Invasive/problematic species and genes				
Invasive nonnative/alien species	National Institute for Invasive Species Science (http://www.niiss.org)			
Problematic native species				No known source
Introduced genetic material				No known source
Disease and pests	USDA APHIS			No known source
9. Pollution				
Household sewage, industrial effluents, and urban wastewater	US EPA Permit Compliance System; National Pollutant Discharge Elimination System; Toxic Release Inventory (http://www.epa.gov/enviro/geo_data.html)	Detailed, consistent dataset	Spatial linkage to NHD dataset	
Military effluents				No known source
Agricultural and forestry effluents	Census of Agriculture		Fertilizer use available only at county scale	Confined agricultural feeding operations
Garbage and solid waste	US EPA Resource Conservation and Recovery Act waste sites			
Airborne pollutants	US EPA Air Facility System	Detailed, consistent dataset		
Excess energy (heat, sound, light)	DMSP Night lights (for light)			No known source for sound, heat
Accidental spills				No known source

Source: David M. Theobald.

Combining multiple indicators into a composite index of degree of human modification

In addition to selecting what threats to include and which datasets best represent a threat (figure 5.3), it is likely that the multiple layers will need to be simplified or combined into a single, composite (or cumulative) data layer. Of course, this is one of the primary strengths of GIS technology—the ability to conduct overlay analyses and/or generate weighted combinations of multiple layers. However, some important guidelines have emerged (e.g., Schultz 2001; Hajkowicz 2006) to generate rigorous indicators that include clear, easy-to-explain, simple models that allow for unambiguous interpretation and treat ordinal and ratio values appropriately (i.e., conform to the additivity axiom; Hajkowicz 2006). I illustrate these issues by describing a composite index of the degree of human modification, building on the approach described in Theobald et al. 2012, which incorporates threats from land-cover conversion, housing density, road presence, road use (traffic), and extractive resources (cropland, mining, and oil and gas wells).

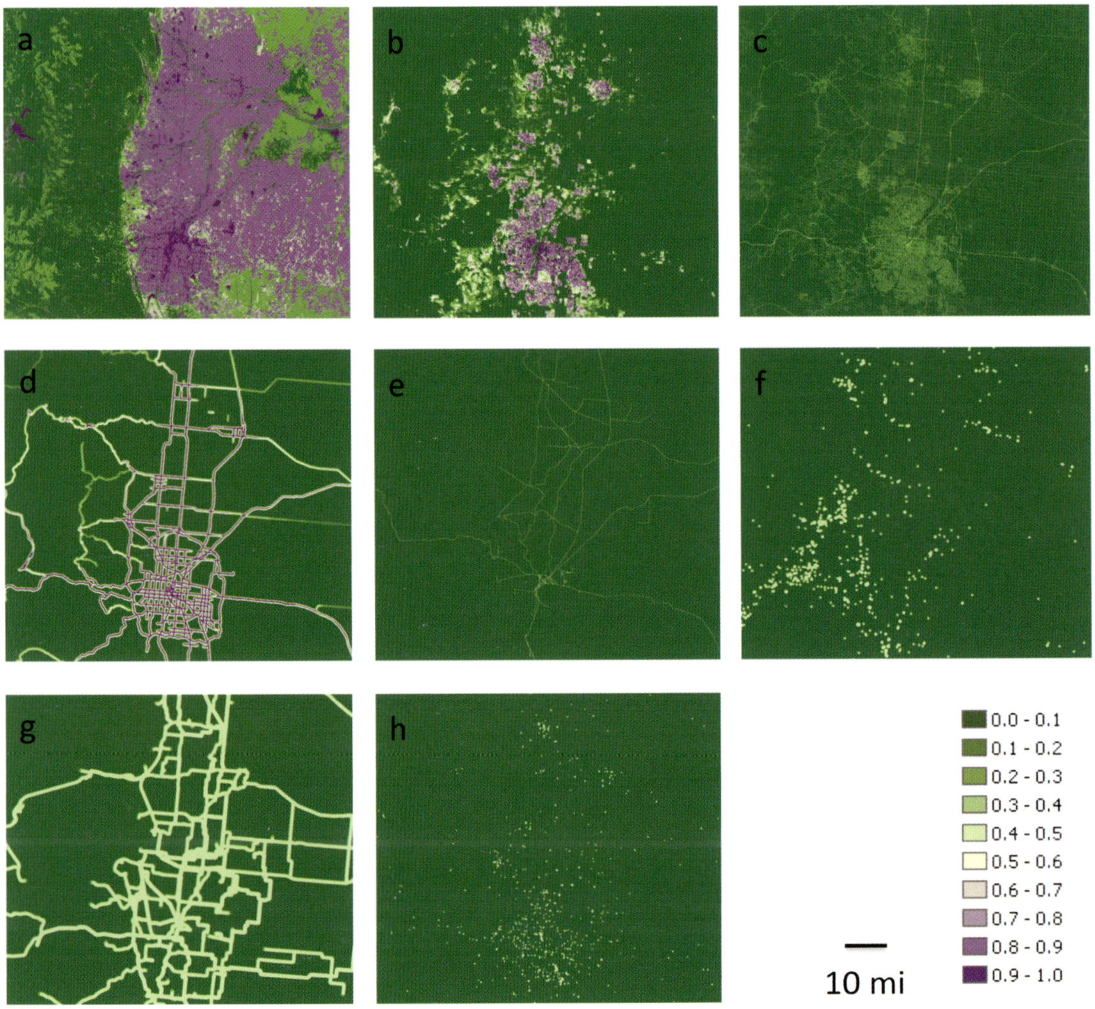

Figure 5.3. Illustration on the Front Range of Colorado, USA, showing eight threats: (a) land cover, (b) housing density, (c) road footprint, (d) highway traffic, (e) railways, (f) mines, (g) power lines, and (h) communication and cell towers. The percentage of area within each cell modified by human activities or structures increases from dark green to yellow to purple, with darkest purple cells being 90% or more modified. Global land cover from United States Geological Survey; MODIS from NASA; NLCD 2006 from US Geological Survey; NLCD 2006 filtered from US Geological Survey; DMSP "nightlights" from National Oceanic Atmospheric Administration; housing density from the ICLUS/SERGoM; road footprint from US Census Bureau; highway traffic from the National Transportation Atlas; mines and power lines from US Geological Survey; communication and cell towers from the Federal Communications Commission.

The degree of human modification is measured as the proportion of a given area (typically a cell, e.g., 30 m × 30 m) that has been considerably modified by human activities or structures. This provides a clear, physical basis to understand and explain what is being measured. Moreover, the metric is well bounded, ranging from a low of 0.0 to a maximum value of 1.0. To estimate the degree of human modification, I developed an empirical estimate of the proportion of land cover modified. These estimates of human modification were generated by interpreting land use from high-resolution color aerial photography (ca. 2006) based upon 6,000 randomly located samples—creating detailed polygons that represent human modification. To transfer this data over as a complete surface, I calculated the average degree of human modification as a function of each of 13 major land-cover types from the US Geological Survey Gap Land Cover v1 dataset. I calculated that urban/commercial cells had an average degree of human modification of 0.85; cropland, 0.68; plantation/other disturbed lands, 0.24; grasslands, 0.17; forests, 0.07; and shrublands, 0.05.

Because land-cover maps such as NLCD do not adequately capture residential areas of density lower than about one housing unit per acre, I also included a measure to capture likely human modification associated with lower-density housing and residential land use. That is, the degree of human modification was based on residential housing density, h, based on the equation developed by Leinwand et al. (2010), where $h = 0.1627 * ln(x) + 0.773$ ($R^2 = 0.827$), and x is housing units per ha.

For the physical footprint of various types of roads, I calculated r, the proportion of a 30-m cell occupied by a road as represented by Census 2010 data. Based on an ad hoc analysis from high-resolution aerial photography, I estimated that highways occupied 100% of a 30-m cell (or a 1.0 proportion). Secondary roads occupied 50%; local roads, 30%; and dirt/4WD roads, 10%.

To account for road use, I converted the annual average daily traffic, which is reported as the number of vehicles per day. To account for assumed impacts to areas adjacent to these highways, such as noise, lights, pollution, and collision, I applied a quadratic kernel density that assumes the impact declines with distance to 1 km away from a road (Forman et al. 2003; Theobald 2010).

To incorporate datasets on oil and gas, I converted the surface location of active wells into a well density surface by converting locations of active wells using a kernel density function (1-km radius). I assigned a human-modification factor for wells of 0.5 for $d > 2.0$ per km^2 and 0.25 for 0.1–2.0 per km^2 (Copeland et al. 2009).

Lands that have mining operations (US Geological Survey Mining Resources Data System) and substantial topographic changes associated with mining activities (US Geological Survey Topographic Change) were assigned a value of 1.0.

I calculated an index using Defense Meteorological Satellite Program (DMSP) "night lights" values for 2009 (Sutton et al. 2009) using the natural log and then normalized. I also estimate the degree of human modification by power lines by assigning a value of 0.5 and then calculating a kernel density function and for cell and radio towers (using a 0.5-km radius).

The next step was to combine the values for each index (for all layers) at a given location (cell). Generally a weighted-linear combination is used (e.g., Weighted Sum tool or Weighted Overlay tool), where the values at each layer are multiplied by a weight, and then summed (or sometimes multiplied). However, typically there are three challenges to using an additive or multiplicative operation to generate the degree of human modification from multiple threat layers. First, many of the threats are not independent of one another (e.g., urban land cover tends to have higher road density). Most linear combination methods assume independence of factors, and if factors are correlated—which is more and more likely with an increased number of factors (all other things equal)—unknown weighting of different factors can occur, making interpretation of the results problematic.

Second, some locations might be completely impacted by a single threat, yet when combined with data from multiple attributes, the effect of the single layer will be weakened (averaged away). For example, a cell occupied by an open-pit mine might reasonably be assumed to be completely modified/impacted by mining, but likely would have a score of 0.0 for other factors such as housing density, road effect, or perhaps even land cover (because of misclassification errors). Consequently, the effect of that major impact gets reduced simply as an artifact of having additional indicators or data layers in the model.

Third, for threats that are reasonably independent of one another, it makes sense that multiple threats are cumulative (e.g., a cropland with high density of oil and gas wells and power lines crossing overhead). This is represented reasonably well in a linear combination but can become perverse, because using a model with many threat layers can lead to locations with a cumulative index value that exceeds 1.0. Normalizing the cumulative index by the maximum value solves this problem but can lead to underestimation of a location completely affected by a single threat.

To illustrate the effects of different mathematical combinations of values that represent the value at a single location from multiple factors, imagine we are combining three threat layers, A, B, and C (e.g., land cover, housing density, and road density). Their values can potentially range from 0.0 to 1.0, but to make the example simpler, they will have only values of 0.1, 0.5, and 0.9 (table 5.5).

A few common techniques are to calculate the mean (arithmetic or geometric), sum, or product. The mean value in the example above ranges from 0.1 to 0.9. The geometric mean is computed by multiplying threat values together, then calculating the *n*th root (where *n* is the number of threats). Because values of 0.0 for at least one threat is common, such values cause mathematical difficulties (it is not possible to take the root of 0.0). The cumulative sum of the threats ranges from 0.3 to 2.7, which seems illogical in that it results in values exceeding 1.0, meaning greater than 100% modification. Calculating the product results in very conservative (low), nonlinear distribution of values, and requires extremely high values for all threats for the final index value to be high.

Another set of combination methods comes from fuzzy logic, which allows values and threats to belong to multiple sets of membership—that is, it does not assume complete independence among factors (or threats). "Fuzzy And" is conservative in that it takes the minimum value of the inputs; "Fuzzy Or" takes the maximum value of the inputs (the maximum value was used by Theobald et al. 2012). "Fuzzy Sum" allows some correlation between factors but also assumes some accumulation of effects (and is very similar to conditional probabilities among factors as well). For many situations related to mapping human modification using multiple attributes, the Fuzzy Sum approach seems most appropriate (table 5.5 and figure 5.4).

Table 5.5. Illustration of combination formulas at a single location (or cell) for multi-attribute (multilayer) analysis for three threats with values of 0.1, 0.5, and 0.9

	Possible combinations of values, iteration									
Threats/Combos	1	2	3	4	5	6	7	8	9	10
A	0.1	0.1	0.1	0.1	0.5	0.1	0.1	0.5	0.5	0.9
B	0.1	0.1	0.1	0.5	0.5	0.5	0.9	0.5	0.9	0.9
C	0.1	0.5	0.9	0.5	0.5	0.9	0.9	0.9	0.9	0.9
Sum	0.300	0.700	1.100	1.100	1.500	1.500	1.900	1.900	2.300	2.700
Average	0.100	0.233	0.367	0.367	0.500	0.500	0.633	0.633	0.767	0.900
Product	0.001	0.005	0.009	0.025	0.125	0.045	0.081	0.225	0.405	0.729
Fuzzy And(min)	0.100	0.100	0.100	0.100	0.500	0.100	0.100	0.500	0.500	0.900
Fuzzy Or (max)	0.100	0.500	0.900	0.500	0.500	0.900	0.900	0.900	0.900	0.900
Fuzzy Sum	0.271	0.595	0.919	0.775	0.875	0.955	0.991	0.975	0.995	0.999

Source: David M. Theobald.

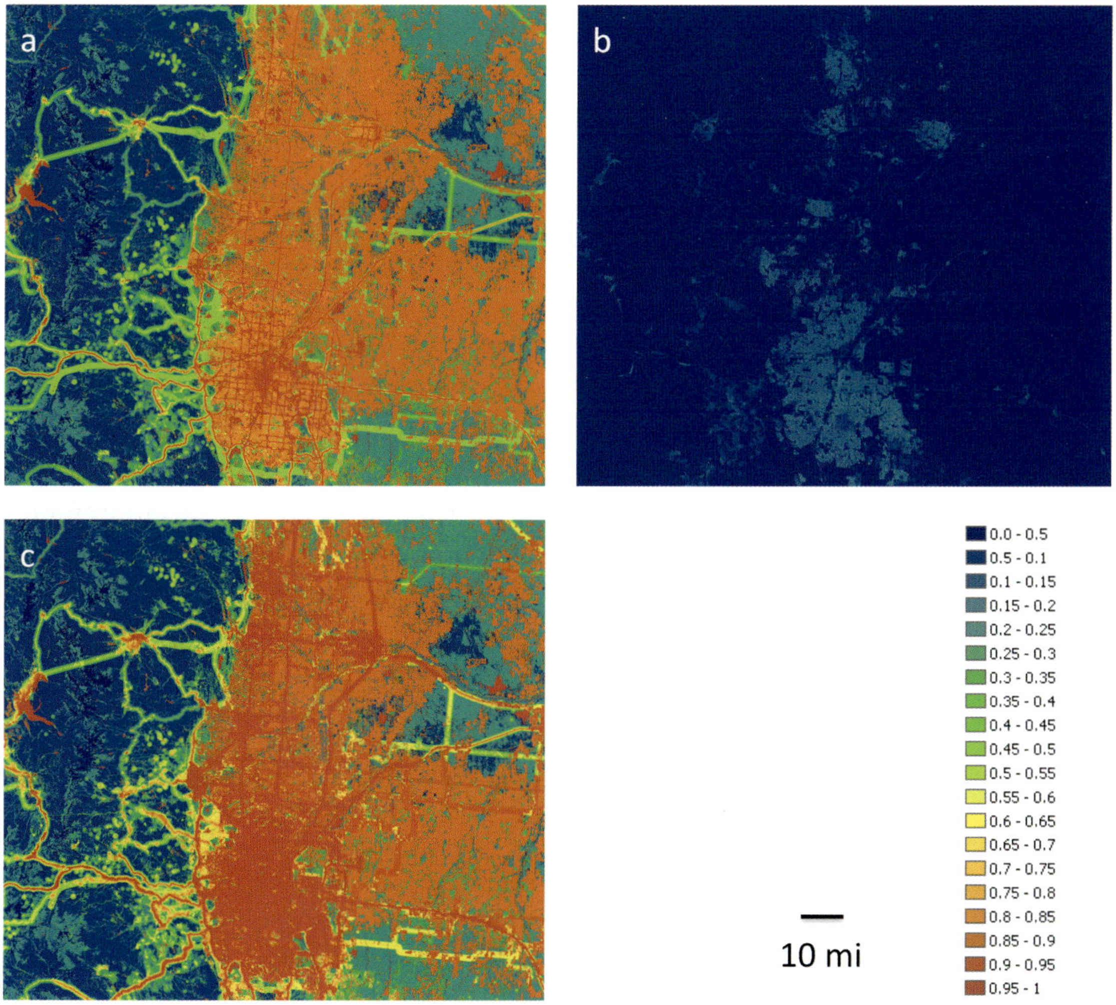

Figure 5.4. Maps illustrating different mathematical combinations of eight threats for the Front Range of Colorado, USA: (a) maximum value, (b) mean value, and (c) Fuzzy Sum. The maximum and Fuzzy Sum provide similar results with regard to showing strong differences between highly urbanized areas (e.g., Denver, bottom center) and remote protected areas (the dark blue area in the northwest quadrant is Rocky Mountain National Park), though the Fuzzy Sum provides stronger differentiation because it includes partial effects of additional (possibly compensatory) threats. The percentage of area within each cell modified by a combination of multiple threat factors increases from dark blue to green to yellow to red, with the darkest red cells being 95% or greater modified. Note that the mean value is quite low (~10%) because of the high proportion of 0 values in threats, and that values for large water bodies were arbitrarily set to 1.0, assuming a terrestrial focus. Global land cover from US Geological Survey; MODIS from NASA; NLCD 2006 from US Geological Survey; NLCD 2006 filtered US Geological Survey; DMSP "nightlights" from NOAA; housing density from the ICLUS/SERGoM; road footprint from US Census Bureau; highway traffic from the National Transportation Atlas; mines and power lines from US Geological Survey; communication and cell towers from FCC; Rocky Mountain National Park boundary from Conservation Biology Institute; water bodies from US Census Bureau.

Forecasting threats

In addition to examining stressors and current threats, increasingly, conservation planning is incorporating forecasted threats in an attempt to anticipate probable future land-use changes. Many models of varying complexity and data requirements have been developed to forecast land-use change, and Sohl et al. (2010) provide a useful review of some of the important issues in developing such models.

In concert with a variety of collaborators, I have constructed a series of land-use-change models to develop a regional growth model for the United States to forecast patterns of housing density based on population projections. The Spatially Explicit Regional Growth Model (SERGoM) relates historical growth patterns and accessibility to urban and protected lands to develop a national, fine-grained (1 ha) dataset on historical, current, and forecasted housing density (Theobald 2005). This data easily can be used to extend common methods used in conservation planning (such as a gap analysis), to incorporate threats, and to target conservation action in a more direct fashion by incorporating land-use changes to differentiate levels, sources, and locations of human land-use change (Theobald 2003). The data is available on a number of conservation planning websites, such as Landscope (http://www.landscope.org/explore/threats/sprawl/). SERGoM was run to generate five scenarios of possible growth patterns consistent with the embedded assumptions in the Special Report on Emission Scenarios from the International Panel on Climate Change Report (see Bierwagen et al. 2010). These land-use-change scenarios are particularly useful when used to examine a reasonable range of growth rates and patterns for potential consequences to land-cover change and impacts on biodiversity (e.g., Robles et al. 2008).

Summary

A central theme of this chapter is to show how conservation planners can better incorporate data on important types of human activities that might threaten biodiversity. Although land-cover datasets are a key aspect of representing human activities, many additional threats should be considered for which data is readily available. The recently established threats framework (Salafsky et al. 2008) provides an excellent basis on which to ensure a comprehensive and transparent approach. Part of the utility of a framework is that it also can expose important shortcomings—in this case, gaps in readily available data. The most notable among these data and knowledge gaps are (1) an oversimplified representation of urban areas as "built-up" that neglects the wide variety of effects likely to occur between different industries, activities, treatments, and so on, as well as being unable to detect any possible improvements through green design, low-impact development, and so on; (2) accounting for the spatial pattern of human use in and around protected areas (i.e., ecotourism, recreation); (3) limited spatial data of grazing allotments and their use; (4) uncritical examination of the extent of ecosystem manipulation by management activities focused on traditional hunting, fishing, and predator control; and (5) better fine-grained data on ecological condition that is often calculated through field-level data collection (e.g., presence of invasive species). Some of these gaps occur because the human activity varies greatly both within and across multiple agencies and owners. A prime example of this is the challenge of compiling land ownership and management. Recently, great strides have been made to improve the standardization and coordination of this data (e.g., Protected Areas Database–US), and this process provides a useful example to learn from.

I also argue that the composite approach to calculate the degree of human modification described here advances the incorporation of threats into conservation planning because it provides a clear conceptual basis (figure 5.1) for establishing the context of how the data and information will be used, builds on the threat framework that helps to provide comprehensiveness while avoiding redundancy, provides a clear basis that is easy to interpret and adheres to the principle of parsimony, and uses an appropriate formula (Fuzzy Sum) that adheres to guidelines developed through multi-attribute utility theory. Of course, this should be continually tested and revised as new data becomes available, conceptual frameworks emerge (such as land change science), and practical guidance is developed by conservation practitioners.

Land-use data can be combined with land-cover data (chapter 4) to develop suitability models of wildlife habitat (as discussed by Dickson et al., chapter 7; and Johnston et al., chapter 14) or aquatic and marine habitat (Ban et al., chapter 8). Land-use considerations are extremely important to determine what is functional habitat and what is not, in delineating secure core habitats (Church, chapter 9), and in mapping connectivity (Singleton and McRae, chapter 10).

Acknowledgments

Thanks to Sarah Reed for helpful comments and suggestions on an earlier draft.

References

Bierwagen, B., D. M. Theobald, C. R. Pyke, A. Choate, P. Groth, J. V. Thomas, and P. Morefield. 2010. "National Housing and Impervious Surface Scenarios for Integrated Climate Impact Assessments." *Proceedings of the National Academy of Sciences* 107 (49): 20,887–92.

Copeland, H. E., K. E. Doherty, D. E. Naugle, A. Pocewicz, and J. M. Kiesecker. 2009. "Mapping Oil and Gas Development Potential in the US Intermountain West and Estimating Impacts to Species." *PLoS ONE* 4 (1): e7400.

Ellis, E. C., and N. Ramankutty. 2008. "Putting People in the Map: Anthropogenic Biomes of the World." *Frontiers in Ecology and Environment* 6: 439–47.

Esselman, P. C., D. M. Infante, L. Wang, D. Wu, A. R. Cooper, and W. W. Taylor. 2011. "An Index of Cumulative Disturbance to River Fish Habitats of the Conterminous United States from Landscape Anthropogenic Activities." *Ecological Restoration* 29 (1): 133–51.

Foley, J. A., R. DeFries, G. P. Asner, C. Barford, G. Bonan, and S. R. Carpenter. 2005. "Global Consequences of Land Use." *Science* 309: 570–74.

Forman R. T. T., D. Sperling, J. A. Bissonette, A. P. Clevenger, C. D. Cutshall, V. H. Dale, L. Fahrig, R. France, C. R. Goldman, K. Heanue, J. A. Jones, F. J. Swanson, T. Turrentine, and T. C. Winter. 2003. *Road Ecology: Science and Solutions.* Washington, DC: Island Press.

Goldburg, R. J., M. S. Elliott, and R. L. Naylor. 2001. *Marine Aquaculture in the United States: Environmental Impacts and Policy Options.* Arlington, VA: Pew Oceans Commission.

Hajkowicz, S. 2006. "Multi-attributed Environmental Index Construction." *Ecological Economics.* 57: 122–39.

Hoekstra, J. M., T. M. Boucher, T. H. Ricketts, and C. Roberts. 2005. "Confronting a Biome Crisis: Global Disparities of Habitat Loss and Protection." *Ecology Letters* 8: 23–29.

Leinwand, I. I. F., D. M. Theobald, J. Mitchell, and R. L. Knight. 2010. "Landscape Dynamics at the Public-Private Interface: A Case Study in Colorado." *Landscape and Urban Planning* 97 (3): 182–93.

Leu, M., S. E. Hanser, and S. T. Knick. 2008. "The Human Footprint in the West: A Large-Scale Analysis of Anthropogenic Impacts." *Ecological Applications* 18 (5): 1,119–39.

Rindfuss, R. R., S. J. Walsh, B. L. Turner II, J. Fox, and V. Mishra. 2004. "Developing a Science of Land Change: Challenges and Methodological Issues." *Proceedings of the National Academy of Sciences* 101 (39): 13,976–81.

Riitters, K. H., and J. D. Wickham. 2003. "How Far to the Nearest Road?" *Frontiers of Ecology and Environment* 1 (3): 125–29.

Robles, M. D., C. H. Flather, S. M. Stein, M. D. Nelson, and A. Cutko. 2008. "The Geography of Private Forests that Support At-Risk Species in the Coterminous United States." *Frontiers in Ecology and the Environment* 6 (6): 301–7.

Salafsky, N., D. Salzer, A. J. Stattersfield, C. Hilton-Taylor, R. Neugarten, S. H. M. Butchart, B. Collen, N. Cox, L. L. Master, S. O'Connor, and D. Wilkie. 2008. "A Standard Lexicon for Biodiversity Conservation: Unified Classifications of Threats and Actions." *Conservation Biology* 22 (4): 897–911.

Sanderson, E. W., M. Jaiteh, M. A. Levy, K. H. Redford, A. V. Wannebo, and G. Woolmer. 2002. "The Human Footprint and the Last of the Wild." *BioScience* 52 (10): 891–904.

Schlesinger, M. D., P. N. Manley, and M. Holyoak. 2008. "Distinguishing Stressors Acting on Land Bird Communities in an Urbanizing Environment." *Ecology* 89 (8): 2,302–14.

Schultz, M. T. 2001. "A Critique of EPA's Index of Watershed Indicators." *Journal of Environmental Management* 62: 429–42.

Sohl, T. L., T. R. Loveland, B. M. Sleeter, K. L. Sayler, and C. A. Barnes. 2010. Addressing Foundational Elements of Regional Land-Use Change Forecasting. *Landscape Ecology* 25 (2): 233–47.

Sutton, P. C., S. A. Anderson, C. D. Elvidge, B. T. Tuttle, and T. Ghosh. 2009. "Paving the Planet: Impervious Surface as Proxy Measure of the Human Ecological Footprint." *Progress in Physical Geography* 33: 510–27.

Theobald, D. M. 2003. "Targeting Conservation Action through Assessment of Protection and Exurban Threats." *Conservation Biology* 17 (6): 1,624–37.

———. 2005. "Landscape Patterns of Exurban Growth in the USA from 1980 to 2020." *Ecology and Society* 10 (1): 32. http://www.ecologyandsociety.org/vol10/iss1/art32/.

———. 2010. "Estimating Changes in Natural Landscapes from 1992 to 2030 for the Conterminous United States." *Landscape Ecology* 25 (7): 999–1,011.

Theobald, D. M., S. E. Reed, K. Fields, and M. Soule. 2012. "Connecting Natural Landscapes Using a Landscape Permeability Model to Prioritize Conservation Activities in the United States." *Conservation Letters* 5 (2): 123–33.

Turner, B. L. II, E. F. Lambin, and A. Reenberg. 2007. "The Emergence of Land Change Science for Global Environmental Change and Sustainability." *Proceedings of the National Academy of Sciences* 104 (52): 20,666–71.

Turner, B, L. II, D. Skole, S. Sanderson, G. Fischer, L. Fresco, and R. Leemans. 1995. *Land-Use and Land-Cover Change Science/Research Plan.* International Geosphere-Biosphere Programme Report No. 35/Human Dimensions of Global Environmental Change Program. http://www.ihdp-cnc.cn/publish/Science Plans and report series/ www.ihdp.uni-bonn.de/html/publications/reports/report07/luccsp.htm.

Watts, R. D., R. W. Compton, J. H. McCammon, C. L. Rich, S. M. Wright, T. Owens, and D. S. Ouren. 2007. "Roadless Space of the Conterminous United States." *Science* 316: 736–40.

Wiens, J. A., M. G. Anderson, and T. Boucher. 2008. "Land Cover and Conservation: From Protected Areas to Landscapes." In *North American Land Cover Summit,* edited by J. C. Campbell, K. B. Jones, J. H. Smith, and M. T. Koeppe. Washington, DC: Association of American Geographers. http://www.aag.org/galleries/nalcs/CH10.pdf.

6

Selecting species as targets for conservation planning

Brent L. Brock and Eric C. Atkinson

"Ultimately, questions such as what type of pattern is required in a landscape, or at what rate a given process should proceed, cannot be answered without reference to the needs of the species in that landscape. Therefore, we cannot ignore the requirements of species if we wish to define the characteristics of a landscape that will ensure their retention" (Lambeck 1997).

Humans have been setting aside areas for conservation for millennia (Margules and Pressey 2000). For reasons ranging from spiritual and aesthetic ("intrinsic values" of Callicott [2006]) to purely economic and functional factors (utilitarian values), we, as a species, continue to design, implement, and manage areas for wildlife conservation. Today, we face the worldwide dilemma of how to best provide for the long-term viability of native species, communities, and ecological processes in an era of rapid loss of natural areas. Furthermore, much of the conservation efforts practiced in the past have not and do not guarantee the long-term survival of much of the world's biodiversity (Andelman and Fagan 2000; Redford and Richter 1999; Robinson 2006; Soule and Sanjayan 1998).

The ultimate goal of conservation planning is to slow, stop, or even reverse degradation and loss of habitat that results in loss of biodiversity and ecosystem services provided by natural areas. The rapid pace and global extent of habitat loss requires conservation practitioners to quickly and efficiently develop and implement conservation plans, within existing political constraints and with limited funds, that have a high probability of achieving desired goals and objectives. Given these constraints, conservation practitioners need practical and effective shortcuts to maximize the return on investment of extremely limited conservation resources.

This chapter discusses use of surrogate species as shortcuts for developing conservation plans intended to ensure long-term survival of native species and functioning ecosystems within a planning area. We provide a brief overview of the use of surrogate species in conservation planning and include some general suggestions for practitioners seeking to implement this approach. Finally, we propose a new framework for selecting species as targets for site-based conservation planning.

The surrogate species concept

To decide how and what to conserve, conservation practitioners must look for avenues to more readily identify and implement ways to practice on-the-ground conservation that are ecologically relevant, politically feasible, ethical, and able to be economically implemented (Robinson 2006). Surrogate species have been widely used (and abused) to provide one such avenue. Surrogate species are "species that are used to represent other species or aspects of the environment to attain a conservation objective" (Caro 2010). In theory, focusing conservation activities on a small set of species will provide adequate protection for a much larger set of species or aspects of the environment. In practice, this is not always the case.

The surrogate species concept suffers from a confusing array of terms and buzzwords with overlapping, and often multiple, definitions designed to meet an equally confusing array of objectives (Caro 2010; Regan et al. 2002). "Umbrella," "focal," "indicator," "keystone," "flagship," and "landscape" species are just a sample of the terms (briefly described below) used to describe different applications of the surrogate species concept. Given the multitude of criteria by which surrogate species are selected, it is not surprising that the concept has received a great deal of criticism. Andelman and Fagan (2000) examined various surrogate species schemes used to populate surrogate species suites drawn from conservation databases containing threatened, endangered, and at-risk species. These ad hoc schemes performed poorly in protecting areas important for regional biota, leading to questions about the utility of using such species as surrogates for protecting regional biodiversity. In his review of the surrogate species approach, Simberloff (1998) provided a detailed critique of the shortcomings of the most popular selection criteria and declared single-species management as "passé." Other reviews have been only slightly less scathing. Caro and O'Doherty (1999) reviewed surrogate species approaches and argued their efficacy has been impeded by the haphazard use of terminology and methods (see also Caro 2010). They recommended that surrogate species should be used with greater care and that species should be chosen according to explicitly stated criteria designed to meet previously defined conservation goals. Seddon and Leech (2008) proposed revised criteria for selecting umbrella species but questioned whether the rigorous assessment needed to select appropriate umbrella species provided much of a conservation shortcut.

However, several studies analyzing existing data in areas set aside to protect target species have found those areas to contain higher biodiversity of nontarget species than nearby areas or areas chosen at random (Betrus et al. 2005; Caro 2003; Caro 2010; Suter et al. 2002; Rondinini and Boitani 2006). It seems the shortcomings of the surrogate species approach has more to do with the haphazard ways it has been implemented than the concept itself. This conclusion has support from a recent meta-analysis of the umbrella species concept (Branton and Richardson 2011), which found consistently higher species richness or abundance at sites containing umbrella species but little or equivocal support for a given criterion for selecting umbrella species. Regardless of the debate over surrogate species, they remain popular in conservation assessment and planning. "Surrogate species are a necessary shortcut to pursuing conservation objectives, given the shortage of conservation funding and time, coupled with the complexities of species distributions and the complicated way that different species respond to environmental perturbation" (Caro 2010).

Choosing surrogate species

The best approach for choosing surrogate species depends on the conservation objectives the practitioner wishes to attain. Caro (2010) identifies three major categories with little overlap in objectives: "those that help in locating areas of conservation significance, those that help in documenting the effects of environmental change on biological systems . . . and those employed in public relations efforts." One traditional use of surrogate species has been in protected area reserve design (Caro 2010; Margules and Pressey 2000; Vane-Wright et al. 1991, but see also Groves 2003 and Unnasch and Karl, chapter 3, for other multiscale approaches). In this application, the locations, sizes, shapes, and configurations of proposed protected areas are designed to meet the requirements of one or several species. But surrogate species, particularly focal species (see below), have been used to guide management of whole communities or ecosystems based on a few species.

Carefully chosen surrogate species can inform practitioners about where conservation efforts should be focused, and what should be done to achieve conservation objectives. Prior to engaging in any species selection process, practitioners should think broadly about the various roles species-based conservation may play in achieving conservation objectives. For example, a group of species may help identify locations of areas to be protected for conservation through designation, purchase, or easement. But they may also be used to guide land use and management on areas both inside and outside protected area boundaries. Private lands may contain crucial seasonal habitats for species that also rely on nearby protected areas for survival. In addition, they may provide important movement corridors that allow species to disperse between discontinuous protected areas, allowing for genetic exchange and long-term population viability. In these cases, mismanagement of nearby private lands can severely impact the ability of protected areas to support biodiversity (e.g., Hansen et al. 2005) or natural phenomena such as long-distance migration (Berger 2004). In other cases, conservation of wildlife on private lands may be a goal regardless of nearby protected areas. Conservation on private lands requires working with landowners, land-use planners, and other stakeholders to implement measures to attain conservation objectives. Species are useful for providing parameters for developing conservation measures that have a high likelihood of success and that stakeholders can understand. By basing measures on carefully selected surrogate species, practitioners also gain an understanding of which other species are likely to benefit from conservation actions.

As previously mentioned, a plethora of terms and definitions have been developed to apply the concept of surrogate species to different conservation objectives and scales. Later in this chapter, we describe some of the common applications, each of which has strengths and weaknesses. A surrogate species can also provide multiple functions. For example, elk (*Cervus canadensis*) might function as an umbrella species by identifying areas of winter range for protection, which also protects habitat for grassland songbirds. In other portions of the planning area, they might function as focal species by guiding land-use decisions that maintain movement corridors for a variety of species. And in all areas, they may serve as a flagship species to garner public support for proposed conservation action. For most conservation planning purposes, we believe surrogates based on multiple species (e.g., Lambeck 1997; Sanderson et al. 2002), and multiple criteria (e.g., Fleishman et al. 2000; Sanderson et al. 2002; Seddon and Leech 2008) provide the most robust species-based foundation for conservation planning. Further, successful conservation often relies on multiple strategies to achieve favorable outcomes, such as identifying areas for protection as well as improving management and stewardship both inside and outside of protected areas. Furthermore, no conservation plan is likely to succeed without public and political support. Therefore, surrogate species selection schemes that integrate multiple approaches may provide the best basis for informing conservation action.

Selected surrogate species approaches

Brief descriptions of some of the common approaches to selecting surrogate species are presented below. This is not an exhaustive list. For a thorough review of all surrogate species approaches and more extensive descriptions, we refer readers to Caro (2010). The approaches listed here are designed to meet different conservation objectives. They are included because they represent important concepts that have been combined in the integrative approaches described later in this chapter.

Vulnerable species. Vulnerable species are species that are likely to suffer significant population declines and possible extinction across all, or significant portions, of their natural ranges without conservation intervention on their behalf. In an effort to stem the loss of global biodiversity, conservationists historically have concentrated on the conservation of rare species; those identified as "threatened," "endangered," or "imperiled." Other monikers, such as "at risk" and "species of concern," have been used, each with its own criteria for designation. The status of globally threatened species can be found at the International Union for Conservation of Nature Red List of Threatened Species, http://www.iucnredlist.org, and NatureServe's databases at http://www.natureserve.org/explorer. Vulnerable species have been used to identify conservation priority areas or "hot spots" (most commonly at the scale of a continent or country) that contain high concentrations of species in need of conservation. At smaller scales, rare or vulnerable species probably serve best as indicators of ecosystem health or a full complement of biodiversity, provided their vulnerability is linked to management or conservation practices (see below).

Endemic species. Endemic species are species that are found within an area, irrespective of its size, and nowhere else. By this definition, all species are endemic at some scale. Therefore, conservation practitioners often define endemics as species with small geographical ranges (Terborgh and Winter 1983). These narrow endemics vary in their rarity and sometimes exhibit habitat specialization. Groves (2003) urges conservation planners to include these species in their conservation plans because such local efforts may exemplify the few opportunities for their conservation. However, endemism is often a poor predictor of species richness or the number of threatened species (Orme et al. 2005). Perhaps more importantly, conservation of endemic species may not serve to conserve other species because the geographic range of an endemic species often is limited.

Umbrella species. Umbrella species are species whose conservation is hoped to provide for the conservation of other community members and their habitats. A variety of criteria have been used to define them (Seddon and Leech 2008). Caro (2003) provided evidence that conservation reserves in East Africa originally designed to protect populations of wide-ranging mammals (generally considered umbrella species) now appear to support higher abundances of nontarget mammals than persist in areas surrounding the parks. Therefore, it appears these parks were at least successful in protecting habitat for a wide range of species beyond those for which the reserves were originally intended. Although ideally umbrella species selection should be prospective, in practice species selected are often charismatic vertebrates with legal protection (Betrus et al. 2005) and their value as umbrellas seems more of an afterthought. Fleishman et al. (2000) developed an a priori method of selecting umbrella species based on rarity, sensitivity to human land use, and mean percentage of co-occurring species. Using these criteria to select butterflies as umbrella species indicated that, in theory, the umbrella effect could transfer across taxa (Betrus et al. 2005). Seddon and Leech (2008) reviewed 17 published criteria for umbrella species selection and recommended seven criteria be used for selection:

1. Well-known natural history and ecology
2. Large home range size
3. High probability of population persistence
4. Co-occurrence with other species
5. Management needs benefit other species
6. Moderate sensitivity to human disturbance
7. Easily sampled or observed

However, they noted that the resources required to evaluate these criteria may render this approach less of a shortcut than managers would hope. A recent meta-analysis by Branton and Richardson (2011) of the umbrella species concept found that areas containing putative umbrella species contained higher species richness than control areas, suggesting there is merit to the umbrella species concept, but they found little support for the criteria used to identify umbrella species.

Focal species. Lambeck (1997) advanced a method for addressing perceived limitations in using single species, and especially umbrella species, as a basis for defining conservation requirements. His mechanism, assembling a suite of "focal" species drawn from the total species inhabiting an area, had the goal of serving the complete assemblage of species through conservation efforts directed toward this suite. Lambeck presented a hierarchical decision-making process that started out by assessing whether each species in the landscape was vulnerable. Subsequently, the at-risk species were characterized by the type of threat(s) they face. Species were linked to limiting factors in landscape pattern (habitat, habitat matrix, and so on) and limiting processes (i.e., predation, fire, and so on). What Lambeck hoped to produce was a method to link species-based conservation methods with those based upon the conservation of representative landscape patterns and processes (Franklin 1993; Hansen et al. 1993; Orians 1993).

Indicator species. An indicator species is defined as "an organism whose characteristics (e.g., presence or absence, population density, dispersion, reproductive success) are used as an index of attributes too difficult, inconvenient, or expensive to measure for other species or environmental conditions of interest" (Landres et al. 1988). Caro (2010) devoted four chapters to discussing indicator species, reflecting the myriad uses, subcategorizations, and definitions that have been applied to indicator species as surrogates for conservation. Indicator species have been used to identify areas of high biodiversity, to monitor the level or effects of pollutants, and as indicators of habitat alteration or fragmentation, to name just a few applications. In the United States, agencies such as the US Forest Service were mandated to address management goals for forested lands based upon management indicator species (MIS). MIS are vertebrate or invertebrate species selected for monitoring habitat conditions in the forest because their population changes are believed to indicate the effects of management activities (1982 National Forest Management Act planning regulations [36 CFR 219.19(a)(1)]). This approach has come under intense scrutiny and critique (Landres et al. 1988; Niemi et al. 1997; Simberloff 1998) because MIS have not always been strongly tied to habitats or management actions, nor do they often serve as useful indicators for other species or ecosystem health.

Keystone/Strong interactor species. A keystone species is one whose effect is large, and disproportionately large relative to its abundance (Power et al. 1996). Since the inception of the term "keystone," its use and application in ecology and conservation biology has been problematic (Paine 1966, 1969, 1995; Power et al. 1996). Nevertheless, functional and pragmatic definitions have been advanced that force practitioners to think of ecological processes through time when selecting species and defining areas important to conserve. For the purpose of selecting focal species as surrogates to promote broad conservation outcomes, the keystone concept should include what MacArthur (1972) called "strong interactors," which he defined as a species whose loss was likely to cause considerable change in community structure.

Flagship species. Flagship species are used to raise public awareness of conservation issues (Western 1987) and have proved to be rallying calls for species-specific conservation (Caro and O'Doherty 1999). They are often large-bodied mammals, typically carnivores, and their ecological function may be secondary, in terms of engaging public support, to their charisma. Shortcomings to basing conservation priorities on these species include the fact that they may not be representative of other species in the local assemblage in habitat association, requirements, and threats. In addition, if public support for conservation is focused on a single charismatic species, support for conserving an area may be lost if the conservation of flagship species fails (Simberloff 1998).

An integrated surrogate species approach

Landscape species. In 2002, Sanderson et al. (2002) advanced a model for conservation that integrates several of the previously described approaches for using surrogate species in conservation planning. This methodology relies upon a stepwise selection process, producing a suite of focal species based on the requirements of "landscape species" that "use large, ecologically diverse areas and often have significant impacts on the structure and function of natural ecosystems" (Redford et al. 2003). The authors argued that because landscape species require large areas containing different habitat and land-use types, they encompass the needs of many other species. Further, because landscape species require areas that are maintained at a landscape scale, they require landscape-level processes (e.g., disturbance regimes such as fire or windthrow) to maintain functional landscapes that can support their needs, and therefore, presumably many other species' needs. Therefore, landscape species not only provide umbrellas for sympatric species, but also encompass the structure and ecological function of the landscape on which many other species depend. Coppolillo et al. (2004) developed a method for selecting suites of focal species that inform practitioners about *where* to focus conservation activities and *how* to perform them, based on five criteria:

1. Area requirements
2. Heterogeneity of habitat
3. Vulnerability to environmental stressors
4. Ecological function
5. Social and economic significance

Thus, the landscape species approach (LSA) has the goal of incorporating the strengths of several surrogate species approaches (umbrella, vulnerable, flagship, indicator, and keystone species) within a single species selection framework.

We applied the LSA to develop a conservation assessment of the Madison Valley, Montana (Brock et al. 2006). Our experience with the LSA has been very favorable, and the resulting assessment continues to form the basis for ongoing conservation efforts in the valley. However, we did discover challenges in applying the LSA. As Caro (2010) astutely observed, "its data requirements are extensive." For example, applying the LSA to the Madison Valley required cursory examination of 410 native vertebrate species, of which 63 formed the final candidate pool to select from. Compiling information and assigning category scores for each of these species was a considerable undertaking, placing the time and expense of species selection beyond most project budgets. In addition, cross-correlation among several categories leads to triple-weighting of species that utilize large areas. For example, species with large home ranges often utilize a variety of habitat types and are affected by a number of environmental stressors, even if they aren't particularly sensitive to or threatened by them. Species like black bear (*Ursus americanus*) tend to rank high as landscape species even though they have proven adaptable to living in severely altered and fragmented landscapes.

Despite these shortcomings, we find the transparent and integrated approach compelling, and the benefits of using the LSA are considerable. With modification to address the problem of triple-weighting species with larger area requirements, we would recommend use of the LSA if budgets permit. However, the data requirements are likely to make the LSA impractical for all but large, well-funded projects. Instead, we offer an alternative approach in the following section that attempts to address some of the limitations of the LSA while borrowing heavily from its concepts.

Refinements and improvements in best practices

The highest success in conserving complete assemblages of naturally occurring species is obtained through maintaining or restoring intact natural ecosystems (Noss 1996; Soule et al. 2003; Sanderson et al. 2002). Intact ecosystems can be defined as areas that contain complete communities of native species with viable populations within the natural range of variability, such that complex interactions among individuals and species are maintained. To select species that can serve as surrogates for intact ecosystems, we propose a process that combines, within a single framework, the utility of the umbrella species concept with an ecosystem approach. We refer species selected using this framework as "focal species" because the purpose of the collective species suite is to identify limiting processes that need to be maintained or restored, but species within the suite may serve surrogate roles other than as focal species as defined by Lambeck (1997). The comprehensive "ecosystem umbrella" that is developed by combining multiple functional umbrellas (figure 6.1) attempts to ensure that conservation planning maintains sufficiently large areas to support viable populations, contains adequate distribution of the right habitat types to sustain most species in the native community, provides protection from threats to assure population persistence, and maintains natural processes to sustain complex interactions. In addition to sustaining natural communities and ecosystem function, this ecosystem umbrella can incorporate human values. For additional methodologies to involve stakeholders and their values, see Chan et al. (chapter 2).

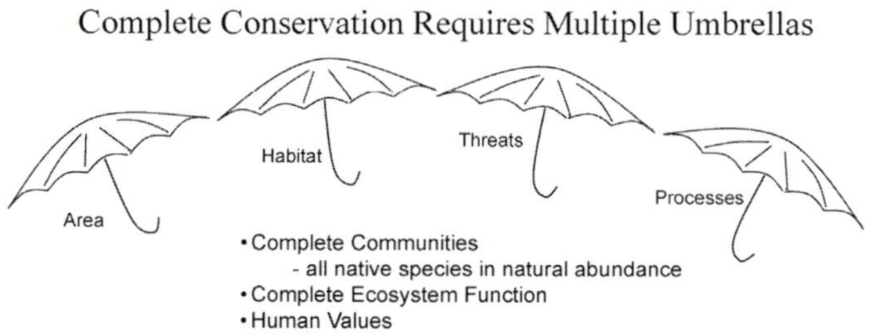

Figure 6.1. Component umbrellas needed for ecosystem-level conservation. Figure from Brent L. Brock.

A framework for selecting focal species for ecosystem conservation

Borrowing from Maslow's theory of human needs (Maslow 1943), we developed a hierarchy of conservation needs (figure 6.2) as the basis for prioritizing species selection. At the base of the pyramid, we place "area" as the first consideration because conserving large geographic blocks is likely to encompass a variety of habitat types in sufficient quantity to meet the requirements of many native species. Additionally, widely ranging species are more likely to suffer from habitat fragmentation and conflict with human values than species with small area requirements (Sanderson et al. 2002), although there are exceptions where ecologically rich habitat types are disproportionately used by humans (e.g., riverine habitats and shorelines). Next in the hierarchy is "habitat" because area alone will not conserve a complete assemblage of species unless a full complement of habitat types is available to support all species. Therefore, focal species should be selected to ensure that all major vegetation types required to provide habitats are covered under the combined umbrella. Even if sufficiently large areas of habitat are available, additional threats (e.g., competition with nonnative species, unregulated harvest, or introduced disease) may prevent the persistence or recovery of species in the native community. Therefore, "security from threats" is placed third in the hierarchy of needs. Because habitats are not static through time, ecosystem-level planning must provide for the continuation of natural processes necessary for maintaining a dynamic equilibrium of habitat patches required by the native species community. Placing "ecological processes" fourth in the hierarchy of needs acknowledges the importance of maintaining habitat dynamics and species interactions over long, including evolutionary, time frames. Finally, habitat protection, mitigation of threats, and management of many important ecological processes (e.g., fire and flood), require social acceptance before the political will and adequate resources can be allocated to address these challenges. Therefore, "social acceptance" rounds out our hierarchy of conservation needs.

The hierarchy of needs provides a prioritization scheme implemented by identifying focal species attributes associated with each need, and using selection criteria that prioritize the attributes of the greatest needs (table 6.1). The species selection process is designed to find the fewest possible species that satisfy all identified needs, using a mathematical optimization process known as the "set covering problem" (Vazirani 2001). The prioritization scheme is central to this process because species that cover the greatest amount

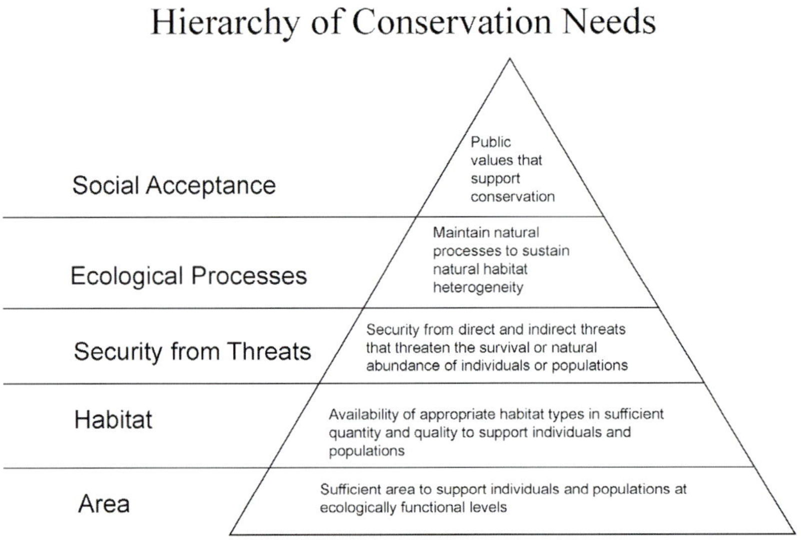

Figure 6.2. Hierarchy of conservation needs. Figure from Brent L. Brock and Eric C. Atkinson.

Table 6.1. Focal species attributes and selection criteria associated with conservation needs

Conservation need	Focal species attribute	Selection criteria
Area	Landscape species	Select large-area generalists to protect sufficient area and diversity of habitat types.
Habitat	Habitat types	Make sure suite of focal species covers all major habitat types in the area.
Security	Threats	Make sure suite of focal species covers all major threats that impact wildlife populations in the area.
Ecological processes	Ecological processes and key species	Include key species needed to maintain natural community, and species that depend on ecological processes to sustain ecologically functional populations.
Social acceptance	Socioeconomic values	Identify values that complement or conflict with conservation objectives.

Source: Brent L. Brock.

of needs at the bottom of the pyramid are likely to cover the greatest amount of subordinate needs with respect to ecosystem conservation. In other words, species selected first in the process tend to have the most demanding conservation requirements and therefore provide an umbrella for the greatest number of species within the natural community. A complete ecosystem umbrella is achieved when all identified needs are covered by the combined focal species suite. Therefore, successful conservation of the species suite should require management or mitigation of all elements affecting the needs of the total species community and provide ecosystem-level protection.

To make focal species selection more time- and cost-effective without sacrificing conservation utility, we developed a selection process that is similar to the landscape species approach (Sanderson et al. 2002) but with reduced requirements for data acquisition and analyses. This approach requires as little as a half to a whole day of preparation by one or two staff knowledgeable about wildlife in the area of interest, followed by a one-day workshop attended by area wildlife experts and stakeholders. This approach attempts to yield a focal species suite providing multiple umbrellas covering all major habitat types, all major conservation threats, complete ecosystem function, and major socioeconomic factors. Like the LSA, the selection method begins with species having large area requirements and facing multiple threats, and builds upon the list of focal species until all conservation targets are covered under the species suite umbrella. Our method differs by relying on local expertise rather than assigning somewhat subjective scores by gleaning information from the literature to calculate indices as is required under the LSA. Although the quantitative aspect of the LSA is powerful and can provide information that is useful beyond the species selection process, it can be prohibitively expensive and time-consuming for areas that contain species represented by a rich body of literature. Because local experts are already abreast of the literature in their areas of expertise and are familiar with the area of interest, they are a highly cost-effective information source for species selection. This expert-driven process also allows the attributes of each candidate species to be judged in combination to account for relevant factors that may be masked using a strictly quantitative approach.

Our selection methods are based on the assumption that the goal is to select a species suite that provides a complete umbrella for protection at the ecosystem level. This means that successful protection and conservation of all species contained within the suite would result in the following:

- Conserving sufficient quantity and quality of all major habitat types supporting ecologically functional populations (Conner 1988) of all native species present in the planning area

- Implementing mitigation strategies to reduce the impact of activities appreciably threatening the persistence of any native species to levels that ensure a high likelihood of persistence for the next 100 years
- The continuation of all natural ecosystem processes (either through natural occurrence or simulation by prescription) that are necessary for the long-term persistence of all native species within their natural range of variability of abundance
- Respecting the importance of wildlife to local economies and culture, and implementing strategies maximizing positive (and minimizing negative) effects of wildlife on these economies while preserving cultural values

In reality, no suite of focal species will produce a perfect umbrella, and gaps will occur. In addition, implementing a conservation plan that perfectly protects and manages all focal species is unlikely. But a focal species suite that comes as close as possible to providing the theoretical umbrella can help stakeholders understand trade-offs that inevitably will be required as decisions are made about priorities and which actions to implement.

To test this approach, a focal species suite was developed for the Madison Valley in southwest Montana, USA. The Madison Valley lies within the Greater Yellowstone Ecosystem; the Madison River is one of the major river drainages flowing from Yellowstone National Park. This area supports a rich assemblage of native species, including large ungulates and carnivores such as elk (*Cervus canadensis*), moose (*Alces alces*), grizzly bear (*Ursus arctos*), and wolf (*Canis lupus*), in addition to iconic wilderness species such as wolverine (*Gulo gulo*) and lynx (*Lynx canadensis*). A wildlife assessment was completed in 2006 by the Wildlife Conservation Society, the Craighead Institute, and the Madison Valley Ranchlands Group, using the LSA (Brock et al. 2006). During this assessment we were able to compare the focal species suite selected using our abbreviated method with the suite selected using the quantitative LSA. The results of this example are referred to in the methods that follow for illustrative purposes.

Methods

A facilitator should convene a workshop to systematically select a suite of focal species, based on expert knowledge of well-rounded and evenly represented participants. Before doing this, the facilitator should prepare a focal species list, as described in the next section.

Preworkshop preparation

The first step in the focal species selection process is to define clearly the geographic boundary of the planning area (see also Unnasch and Karl, chapter 3). It is impossible to make specific recommendations regarding the appropriate size of a planning area, but some general guidelines can be suggested:

1. The planning area extent should be sufficiently large to allow conservation at scales appropriate for meeting conservation targets, but not necessarily large enough to meet all conservation targets as a self-contained unit. For example, the Madison Valley is not large enough to contain sustainable populations of species such as grizzly bear or gray wolf, but these species are important targets for maintaining wildlife habitat, and the valley provides sufficient wildlife habitat that it is critical regionally to the overall function of the Greater Yellowstone Ecosystem.

2. Consider the intended stakeholders expected to implement the plan (see also chapter 2). The scale of conservation planning should match the scale of management it is intended to initiate. For example, conservation planning at a broad scale (coarse spatial resolution over a large extent) may be appropriate if recommendations resulting from the planning are to be implemented by federal agencies that operate at equally broad scales. But conservation plans at this scale may be useless to decision makers at the county level who need higher-resolution planning for decisions about land use over a few square miles or less. Therefore, the size of the planning area should be selected to allow for analyses and planning at resolutions high enough to provide useful information to guide the intended decision makers. As a guideline, resolutions of 1–3 m work well for use at the parcel level, 30–90 m for county or small reserve planning, and 0.9–10 km for broad-scale planning at the state, province, or federal level.

For delineating planning unit boundaries, we recommend using hydrologic drainage boundaries rather than political boundaries. Drainage divides often follow natural ecological boundaries that are biologically meaningful to wildlife species, and the fractal nature of basins makes it easy to scale the planning area to match conservation objectives simply by increasing or decreasing the hydrologic unit code size of drainages used (see the Natural Resources Conservation Service Watershed Boundary Dataset at http://www.nrcs.usda.gov/wps/portal/nrcs/main/national/water/watersheds/dataset). In our example, the Madison Valley is an example of a fifth-level hydrologic unit, or watershed.

The second step is to create a preliminary list of major habitat types within the planning unit. The goal is to retain habitats that, if lost, would result in a loss of native species richness or significantly alter community composition within the project boundary. This list is necessarily somewhat subjective and may be modified later, but creating an initial list saves time during the workshop. Major habitat types should be defined rather broadly to minimize the number included, but comprehensive enough to include all habitats needed to support all native inhabitants. Differences in species community composition provide a useful guideline for defining major habitats. For example, in the Madison Valley, montane forest Douglas fir and lodgepole pine forests support very similar vertebrate communities, so they were lumped into montane conifer habitat but remained separate from the subalpine conifer classification because of the different species assemblages found in those habitats. Existing land-cover classifications (such as from the Gap Analysis Program) can provide a useful starting point for creating a major habitat list because the classes can be grouped quickly into appropriate, and biologically meaningful, categories for the planning area.

The final step for workshop preparation is to generate a list of native species ranked by their area requirements. This list should be exhaustive, including all native extant and locally extinct species, to encourage discussion about potential restoration efforts that may result from the planning process. Species lists can be generated quickly using several tools available on the World Wide Web. In particular, the WildFinder tool by the World Wildlife Fund (http://www.worldwildlife.org/science/wildfinder/intro.html) provides an interactive map that allows the user to query by ecoregions of the world and place-names to export a vertebrate species list to an electronic spreadsheet. Workshop planners must edit the resulting list to remove species not native to the particular area of interest and add species that were omitted from the WildFinder list. For the Madison Valley, we found information about geographic distribution in the NatureServe online database; state comprehensive wildlife plans can be very useful for verifying and editing the initial species list. Although this list may need later modification, creating an initial list saves time during the workshop and expedites the species selection process.

Once the species list is created, species are sorted according to their area requirement ranks. Several hours should be allotted to complete this task, but is essential for ensuring that species selection can be accomplished during a single-day workshop. To aid this process, a list of functional groups has been developed (table 6.2) that shows species' area requirements and ranked species within these categories. All species assigned to a category are considered to have area requirements larger than any species assigned to categories below, regardless of rank. But within a category, species are assigned a rank of 1–3, where rank 1 is for species with the largest area requirements within the category, and 3 is for species with the smallest. Ranks were not assigned to species within "restricted," "habitat specialist," or "microhabitat specialist" categories because species within these categories have such small area requirements that they would not be selected as focal species based on area criteria. However, those species may be selected as focal species based on other criteria, as will be described later in this chapter.

Table 6.2. Functional groups based on species' area requirements

Large landscape	Classic landscape species that range widely and typically occupy multiple habitat types during a season. Large-landscape species depend upon large areas of contiguous or functionally well-connected habitat for survival. Individuals may have seasonal home ranges extending beyond the boundaries of conservation planning area.
Mesoscale	Species that require moderately sized areas, typically less than 500 hectares. Mesoscale species may not have inherently large area requirements to complete their life histories but tend to move over relatively large areas to meet these requirements. These species may have poorly defined territories or be nonterritorial, or may even defend multiple-use territories.
Habitat selectors	Habitat-selector species may travel moderate to large distances between habitats, but foraging, breeding, or other activities are concentrated in relatively few habitat types compared to the number of habitats crossed while moving from one habitat patch to the next. They are searching and targeting specific habitat types. The tendency is for these species to forage across a wide variety of habitat types but become temporarily sedentary when they encounter preferred or highly desirable habitat patches. Sedentary periods vary, but typically occur on the scale of days to months and less than the period of seasonal use in species that exhibit seasonal migration or movements. In some cases (e.g., flighted birds), habitat selectors make daily movements between nesting/roosting and foraging habitats but may not always utilize the same patches from day to day. The species may "trapline" on various ecological and temporal scales, with their preferred or targeted habitats appearing as beads on a string or nodes on a spider's web. Individuals may traverse geography in a directional walk.
Restricted or sedentary	These are species with relatively small home ranges, on a scale of a few hectares or less.
Habitat specialist	Species in which the majority of seasonal use is tied to a small number of habitat types; those habitat types are often restricted in patch size. Species range may be relatively widespread.
Microhabitat specialists	Species that are specialized for, and/or restricted to, very specific habitat types that occur as very small habitat patches (e.g., warm springs). These species spend an entire season, or often their entire lives, within a single restricted habitat patch and should not be confused with habitat-selector species that make periodic movements in search of specialized microhabitats. These species may have their best fit within the Special Elements category.
Source: Brent L. Brock.	

While categorizing land area requirements, consider the importance of habitat connectivity for the species in question. For example, an American white pelican may range over a wider area in a season than even a large-landscape species such as a grizzly bear. However, the ability of a pelican to fly between discrete habitat patches is dictated mainly by the distance between patches, not by habitat quality or disturbances on the ground between patches. In contrast, the ability of a grizzly bear to access a sufficient amount of habitat to meet its needs is strongly dependent upon habitat connectivity between patches. Therefore, a grizzly bear is likely to require a considerably larger amount of contiguous habitat than a pelican, despite having a smaller total home range size.

Once all species in the list are assigned to a category and ranked, the list can be sorted according to area requirements. Note that the order of species with the same category and rank combination is not important and will not affect the outcome of the selection process.

Convening the workshop

Armed with the local species list, the facilitator organizes a workshop of participants with expertise in the important taxa to be represented and knowledge of the area and its human economies, if possible.

Step 1: Select a panel of experts. The participant list will depend upon the needs of the particular planning area and the availability of experts in the region. Following are some useful criteria for assembling an expert panel:

- Biologists well-versed in the distribution, natural history, and conservation status of a group of taxa locally. In the United States, Natural Heritage Program personnel should be included and may be able to assist with identifying other participants for the workshop.
- Participants representing each class included in the species occurrence list. For example, a mammologist, an ornithologist, and a herpetologist would be included.
- Biologists representing or associated with wildlife management agencies and land-management agencies with wildlife experience.
- Community ecologists that may be especially aware of ecosystem processes and threats (i.e., fire ecology, global climate change).
- Consider relevant invertebrate or plant ecology experts if there is sufficient concern that species or associations would be endangered by their nonparticipation. This may be best implemented through a postmeeting distribution of the habitat list and be targeted to experts requesting input regarding the umbrella-coverage value of these habitats for species within their area of expertise. In general, we believe that invertebrates and plants serve as indicators of ecosystem health rather than as umbrellas protecting habitats or species at the scale of this approach. Therefore, their inclusion in the initial focal species selection process may not be warranted, but their importance in conserving complete natural assemblages should not be overlooked.
- Identify any socioeconomic values within the planning area that must be considered or addressed before effective conservation planning can proceed. Such values may include (1) extractive industries that may compete with wildlife conservation values, such as mining or timber harvest, (2) certain types of recreation and values that may complement (or diminish) conservation, such as wildlife tourism, hunting, or motorized recreation, and (3) cultural or economic values, such as cattle ranching, crop farming, or traditional uses by indigenous people. In some cases, it may be beneficial to add focal species as special elements to represent important socioeconomic values, but in many cases, it is sufficient to just document the issues those values bring to the planning process so they can be addressed during subsequent phases of conservation planning.

Step 2: Review the species list. Begin with review and finalization of the native species list developed prior to the workshop. It may be most efficient to have experts review those species within their area of expertise. Species with no historical evidence of occurrence in the planning area, or species present only as extralimital occurrences, should be removed from the list unless there are compelling conservation reasons to retain them (e.g., endangered species needing special protection). These species may be also addressed through other means, such as special elements. Threatened or endangered species, or species with significant conservation concerns, should also be noted.

Finally, in some cases, there may be species that simply must be included as focal species regardless of the selection outcome. A species may warrant such a priori selection, for example, if it holds such an important biological or socioeconomic value in the area that its inclusion in any conservation assessment would be mandatory.

Table 6.3. Candidate species and major habitat types identified in the Madison Valley

Candidate species			Major habitat types
American badger	Grizzly bear	Rough-legged hawk	Agriculture
American beaver	Hispid pocket mouse	Ruffed grouse	Alpine meadow
American bison	Hoary bat	Sage grouse	Aspen
American black bear	Least chipmunk	Sagebrush vole	Foothill shrub/xeric woodland
American kestrel	Little brown bat	Sharp-shinned hawk	Lentic water
American marten	Long-billed curlew	Short-eared owl	Lotic water
American mink	Long-eared myotis	Silver-haired bat	Mesic shrub
American pika	Long-eared owl	Snowshoe hare	Montane conifer
Big brown bat	Long-legged myotis	Southern red-backed vole	Native grassland
Bighorn sheep	Long tailed vole	Spotted bat	Nonnative grass
Black-tailed jackrabbit	Long-tailed weasel	Spruce grouse	Recently burned forest
Blue grouse	Meadow vole	Striped skunk	Riparian forest
Bobcat	Merriam's shrew	Swainson's hawk	Riparian shrub
Boreal chorus frog	Montane shrew	Tiger salamander	Rock/cliff/talus
Boreal owl	Montane vole	Townsend's big-eared bat	Shrub-steppe
Bull snake	Moose	Turkey vulture	Subalpine forest
Bushy-tailed woodrat	Mountain cottontail	Uinta chipmunk	Wetlands (marsh)
California myotis	Mule deer	Uinta ground squirrel	
Canada lynx	Muskrat	Vagrant shrew	
Cinereus shrew	North American porcupine	Water shrew	
Columbian ground squirrel	Northern flying squirrel	Water vole	
Common raven	Northern goshawk	Western harvest mouse	
Cooper's hawk	Northern grasshopper mouse	Western heather vole	
Cougar	Northern harrier	Western jumping mouse	
Coyote	Northern pocket gopher	Western rattlesnake	
Deer mouse	Northern pygmy-owl	Western screech-owl	
Desert cottontail	Northern raccoon	Western toad	
Dwarf shrew	Northern river otter	Western small-footed myotis	
Elk	Northern saw-whet owl	White-footed mouse	
Ermine	Northern shrike, Ord's kangaroo rat	White-tailed deer	
Ferruginous hawk	Prairie vole, Preble's shrew	White-tailed jackrabbit	
Fisher	Pronghorn	Wolverine	
Golden eagle	Red fox	Wyoming ground squirrel	
Golden-mantled ground squirrel	Red squirrel	Wyoming pocket gopher	
Gray wolf	Red-tailed chipmunk	Yellow-bellied marmot	
Great Basin pocket mouse	Red-tailed hawk	Yellow-pine chipmunk	
Great gray owl	Richardson's ground squirrel	Yuma myotis	
Great horned owl			

Source: Brent L. Brock and Eric C. Atkinson.

Step 3: Review major habitats. Make sure that all important habitat types are included (see table 6.3), and omit habitats that do not meet the criteria for "major habitat type" as described in the preworkshop planning steps.

Step 4: List threats. With the final species and habitat lists complete, a list of major threats to those species or habitats should be compiled. In general, "major threats" are threats that are likely to appreciably impact species community composition or ecosystem function.

	Major threats	Key species	Important ecological processes	Important socioeconomic values
	Dewatering Exotic disease Farming Fencing Fire Fire suppression Fishing Grazing Harvest (hunting) Homesite development Invasive aliens Loss of ecosystem integrity Management/predator control Migratory sensitivity Mineral mining Motorized recreation Nonmotorized recreation Nonnative introductions Oil/gas pollution Power lines Road chemicals/sedimentation Roading Timber harvest Vehicular traffic Weed/pest control	American beaver American bison Elk? Gray wolf	Periodic fire Grazing Riparian flooding	Sport hunting and fishing Cattle ranching and rural lifestyle Open space, wild areas, and wildlife viewing opportunities

Table 6.4. Focal species selection proceeding by cohorts for the Madison Valley.

Species pool of mesoscale and higher area requirements	Land area requirement category	Cohort 1	Cohort 2	Cohort 3	
Wolverine	LL-1	Wolverine			
Sage grouse	LL-1	Sage grouse			
Grizzly bear	LL-2	**Grizzly bear**			
Ferruginous hawk	LL-2	Ferruginous hawk			
Gray wolf	LL-3	Gray wolf	Gray wolf		
Golden eagle	LL-3	Golden eagle	Golden eagle		
American bison	LL-4		American bison		
Elk	LL-4		**Elk**		
Long-billed curlew	LL-4		Long-billed curlew		
Cougar	LL-5			Cougar	
American black bear	LL-6			American black bear	
Canada lynx	Meso-1			Canada lynx	
Western toad	Meso-1			**Western toad**	
Turkey vulture	Meso-1			Turkey vulture	
Boreal owl	Meso-2				
Pronghorn	Meso-2				
Bighorn sheep	Meso-2				
Western rattlesnake	Meso-2				
Common raven	Meso-3				
Moose	Meso-3				
Tiger salamander	Meso-3				
Northern goshawk	Meso-4				
Coyote	Meso-4				
Fisher	Meso-4				
Mule deer	Meso-4				
Boreal chorus frog	Meso-4				
Bull snake	Meso-4				
Blue grouse	Meso-5				
Bobcat	Meso-5				
Rough-legged hawk	Meso-6				
White-tailed deer	Meso-6				
Great horned owl	Meso-7				
Red-tailed hawk	Meso-7				
Swainson's hawk	Meso-7				
Northern river otter	Meso-7				
American marten	Meso-8				

Source: Brent L. Brock and Eric C. Atkinson.
Note: Candidate species are selected systematically by cohort according to area requirements. One species from each cohort is selected for the final focal specie
LL = Large landscape: in this example, requiring hundreds of square miles per population.
Meso = mesoscale: in this example, requiring less area than LL species.

	Cohort 4	Cohort 5	Cohort 6	Special elements added to complete umbrella	Focal species suite
					Grizzly bear
				Sage grouse	**Elk**
					Western toad
					Bighorn sheep
					Northern goshawk
					Northern river otter—replaced by beaver as special element
					Special elements: **Sage grouse** **Pronghorn** **Westslope cutthroat** **American beaver** **Black-backed woodpecker**
	Boreal owl				
	Pronghorn			Pronghorn	
	Bighorn sheep				
	Western rattlesnake				
	Common raven				
	Moose	Moose			
	Tiger salamander	Tiger salamander			
		Northern goshawk			
		Coyote			
		Fisher			
		Mule deer			
		Boreal chorus frog			
		Bull snake			
			Blue grouse		
			Bobcat		
			Rough-legged hawk		
			White-tailed deer		
			Great-horned owl		
			Red-tailed hawk		
			Swainson's hawk	Westslope cutthroat	
			Northern river otter	American beaver—key species (replaces otter)	
				Black-backed woodpecker	
			Processing stopped because no additional habitat types would be covered under remaining candidate species' umbrella.		

The following three criteria are useful to help consistently decide what to include in the list of major threats. Threats that meet all three criteria should be included in the list:

1. Is the threat occurring within the project area boundary?
2. Is the threat significant enough to jeopardize the persistence of any native species within the project boundary, or does the threat appreciably impact population size of multiple species or a recognized keystone species?
3. Can the threat be mitigated? If the threat cannot be mitigated, changes to policy, regulation, or law should be considered.

In addition to local threats, experts also should consider threats occurring outside the planning area boundaries, but that have significant consequences to wildlife populations within the planning area.

For each threat identified in the previous steps, list the species likely to be impacted by the threat. This list will be referred to later to ensure that the suite of focal species selected includes representation of all major threats.

Step 5: Make initial species selection. Selecting the suite of focal species begins once separate lists are available of species categorized by (1) land area requirements, (2) threats, and (3) major habitat types. The outcome of the selection process is a suite of focal species, the umbrella of threats that will be mitigated, and the umbrella of habitat types that will be preserved by conserving those species. If any a priori focal species, including those that represent important socioeconomic values (from the species review process), were chosen by the panel, those species form the beginning of the focal species suite, and the threats and habitat types covered by those species are added to the appropriate umbrella lists.

Selection begins by choosing a cohort of species with the highest area requirement categories. The number of species to include in a cohort is somewhat arbitrary and should be decided by the panel before the selection process begins. In the Madison example (table 6.4), we used a minimum cohort size of five. In areas with high species richness, it may be practical to use larger minimum cohort sizes, and vice versa for areas of low species richness.

For the final focal species suite, select the species from the first cohort that is the most sensitive to threats (i.e., the species whose viability is most influenced by local mitigation of threats). (Note: If a priori species selection was made, then initial species selection is complete. Habitats and threats under a priori species umbrellas should be listed, and the first cohort of candidates should be reviewed according to the process described in step 6 below). For the species selected, compile another list with the habitat types that fall under that species' umbrella (table 6.5). A species provides a complete habitat umbrella if conserving the species requires a sufficient amount of a given habitat type to also accommodate the needs of all other native species that depend on that habitat. A good test for including a habitat under a species' umbrella is to ask, "Would the species likely persist if this habitat type were lost?" If the answer is no, then the species provides a complete umbrella for that habitat. If the answer is yes, then either the species does not provide an umbrella for the habitat, or it provides only partial coverage for a habitat type ("P" notation in table 6.5). Finally, in this new list, add the threats that would be addressed by successful conservation of the chosen species.

Step 6: Build the species suite. After the initial species has been selected, choose the next cohort of candidate species (table 6.4). Starting with the previous cohort of candidates, discard all species with land area requirements equal to, or larger than, those of the species chosen in the previous step. Add species from the next highest area requirement categories until the minimum cohort size is reached. As in the initial selection cohort, species within an area requirement category are added or discarded as a block.

Table 6.5. Selected focal species for the Madison Valley, with associated habitats and threats covered by species and cumulative umbrellas

Focal species	Habitats covered under umbrella	Habitats outside cumulative umbrella	Threats covered under umbrella	Threats outside cumulative umbrella	Comments
Grizzly bear	Alpine meadow Aspen (P) Mesic shrub Montane conifer Riparian shrub (P) Shrub-steppe (P) Subalpine forest	Agriculture Aspen (P) Foothill shrub/xeric woodland Lentic water Lotic water Native grassland Nonnative grassland Recently burned forest Riparian forest riparian shrub (P) Rock/cliff/talus Shrub-steppe (P) Wetlands/marsh	Exotic disease (P) Migratory sensitivity (P) Management/predator control Mineral mining Motorized recreation Nonmotorized Oil/gas (P) Recreation Residential development (P) Road Timber harvest Vehicular traffic	Dewatering Exotic disease (P) Farming Fencing Fire Fire suppression Fishing Grazing Harvest (hunting) Invasive aliens Loss of ecosystem integrity Migratory sensitivity (P) Nonnative introductions Oil/gas (P) Pollution Power lines Residential development (P) Road chemicals/ sedimentation Weed/pest control	
Elk	Agriculture Aspen (P) Foothill shrub/xeric woodland Native grassland Nonnative grassland Recently burned forest (P) Riparian forest (P) Riparian shrub (P) Shrub-steppe (P)	Aspen (P) Lentic water Lotic water Recently burned forest Riparian forest (P) rock/ cliff/talus Shrub-steppe (P) Wetlands/marsh	Grazing Harvest (hunting) Invasive aliens (P) Migratory sensitivity (P) Residential development (P)	Dewatering Exotic disease (P) Farming Fencing Fire Fire suppression Fishing Invasive aliens (P) Loss of ecosystem integrity Migratory sensitivity (P) Nonnative introductions Oil/gas (P) Pollution Power lines Road chemicals/ sedimentation Weed/pest control	
Western toad	Aspen (P) Lentic water Lotic water (P) Recently burned forest (P) Riparian forest (P) Shrub-steppe (P) Wetlands/marsh	Aspen (P) Lotic water (P) Recently burned forest (P) Riparian forest (P) rock/ cliff/talus Shrub-steppe (P)	Dewatering (P) Exotic disease (P) Farming (P) Fire suppression (P) Invasive aliens (P) Loss of ecosystem integrity Nonnative introductions (P) Road chemicals/ sedimentation (P) Pollution Weed/pest control	Dewatering (P) Exotic disease (P) Farming (P) Fencing Fire Fire suppression (P) Fishing Invasive aliens (P) Migratory sensitivity (P) Nonnative introductions (P) Oil/gas (P) Power lines Road chemicals/ sedimentation (P)	

Focal species	Habitats covered under umbrella	Habitats outside cumulative umbrella	Threats covered under umbrella	Threats outside cumulative umbrella	Comments
Bighorn sheep	Rock/cliff/talus Shrub-steppe (P)	Aspen (P) Lotic water (P) Recently burned forest (P) Riparian forest (P) Shrub-steppe (P)	Exotic disease (P) Farming (P) Invasive aliens (P) Migratory sensitivity (P) Nonnative introductions (P) Road chemicals/ sedimentation (P)	Dewatering (P) Exotic disease (P) Farming (P) Fencing Fire Fire suppression (P) Fishing Migratory sensitivity (P) Oil/gas (P) Power lines	
Northern goshawk	Aspen (P) Recently burned forest (P) Riparian forest (P) Shrub-steppe (P)	Lotic water (P) Recently burned forest (P) Shrub-steppe (P)	Fire	Dewatering (P) Exotic disease (P) Farming (P) Fencing Fire suppression (P) Fishing Migratory sensitivity (P) Oil/gas (P) Power lines	
Northern river otter (replaced by beaver)	Lotic water	Recently burned forest (P) Shrub-steppe (P)	Dewatering (P)	Dewatering (P) Exotic disease (P) Farming (P) Fencing Fire suppression (P) Fishing Invasive aliens (P) Migratory sensitivity (P) Oil/gas (P) Power lines	Replaced by American beaver because beaver covers lotic water but also is an engineer species important for ecosystem function
Special elements					
Sage grouse	Shrub-steppe	Recently burned forest (P)	Farming (P) Oil/gas (P) Power lines	Dewatering (P) Exotic disease (P) Fencing Fire suppression (P) Fishing Migratory sensitivity (P)	Included to cover shrub-steppe
Pronghorn		Recently burned forest (P)	Fencing Migratory sensitivity (P)	Dewatering (P) Exotic disease (P) Fire suppression (P) Fishing	Included to cover threat of fencing and long distance migration
Westslope cutthroat		Recently burned forest (P)	Dewatering (P) Exotic disease (P) Fishing	Fire suppression (P)	Included to cover dewatering and indirect effects of fishing (introduced exotics)
American beaver	Lotic water	Recently burned forest (P)		Fire suppression (P)	Replaced otter because of engineer-species role of beaver
Black-backed woodpecker	Recently burned		Fire suppression (P)		Included to cover fire-dependent species

Source: Brent L. Brock and Eric C. Atkinson.
(P) = partial umbrella for this habitat type or threat

Repeat step 6 for additional cohorts until one of the following stopping conditions is met:

1. All habitat types are provided full coverage under the selected species suite.
2. No additional habitat types are covered by any species in a candidate of cohorts. (Completion of a partial umbrella left open by previous selections counts as an additional habitat type.)
3. There are no more species with land area requirement classes of mesoscale or higher to consider.

Step 7: Complete umbrellas for habitat and threats. Choose species to represent any remaining habitat types or threats not included under the selected species suite. Begin with habitat types by starting at the top of the full species list (sorted in descending order of area requirements) and work down until the panel can agree on a suitable species to provide an umbrella for one or more of the habitat types. In some cases, the panel may have to decide between choosing a single species to represent two or more habitat types, versus two or more species with higher area requirements to represent one of each remaining type. In most cases, choosing a single species to cover multiple habitat types is preferred because it reduces the total number of species (and thus subsequent effort for analysis) in the completed species suite. However, there may be mitigating circumstances that warrant selecting multiple species. If any threats remain uncovered after all habitat umbrellas have been completed, they should be reviewed to make sure the panel still agrees the threats are sufficiently severe within the study area to warrant inclusion in the focal species umbrella. If so, then repeat the above procedure to complete all necessary threat umbrellas.

Step 8: Evaluate gaps and include special elements as needed. Special elements are species added to fill specific gaps in the conservation umbrella but are otherwise treated the same as other focal species in the suite for subsequent analysis and planning. Once all major habitat types and threats are covered by the selected focal species suite, the final step is to select special elements to ensure the focal species suite will protect ecological integrity and address important socioeconomic values within the study area. Ecological integrity is addressed by verifying that all key species and ecological processes are covered under the existing focal species umbrella. Socioeconomic concerns are addressed more indirectly during this workshop and may require a separate stakeholder workshop as well as a social assessment, a stakeholder analysis, and the inclusion of spatial human-use data (see Chan et al., chapter 2).

Results

Species selection is complete when all major habitat types, threats, ecological processes, and socioeconomic values are linked to one or more species within the selected suite and therefore covered under the ecosystem umbrella (figure 6.1). The result should be a suite of species providing robust protection of the ecosystem as a whole if the needs of all species are accommodated through conservation action.

Discussion

The ecosystem umbrella

The structured consensus building process outlined above is designed to identify a focal species suite using a minimal amount of time and resources. The approach is flexible and provides greater opportunity to capitalize on available expertise than does a strictly quantitative method (and some quantitative methods are also similarly subjective). Because of the nature of the approach, the resulting focal species suite is likely to vary for a given area depending upon the participants involved, with varying results increasingly likely in areas with a large amount of functional overlap between species. However, the actual make-up of the focal species suite is less important than the conservation umbrella the suite provides. This approach is designed to consistently create a focal species suite providing multiple umbrellas that combine to form a complete conservation umbrella. Hence, regardless of the actual species suite, the resulting suite should provide a conservation umbrella that will do the following if populations of those species are maintained:

- Conserves all major habitat types in sufficient amounts and spatial configuration to support all native wildlife species
- Provides a scientific basis for developing conservation plans to mitigate human activities that threaten the long-term viability of a complete native species assemblage
- Ensures the continued influence of all important natural processes needed to sustain a complete assemblage of native species
- Maintains or restores natural community composition of native species by ensuring that all key species needed to maintain natural biodiversity are conserved
- Identifies important socioeconomic factors in the planning area that could be capitalized upon to implement conservation strategies, or must be mitigated to accomplish conservation goals

Our method relies on similar selection criteria as the LSA, so we would expect overlap in the species selected (table 6.6). We selected 10 species using our methods, versus 15 species chosen using the LSA. Eight (47%) of the species in the combined list were chosen by both methods. Our methods selected two species (American beaver and northern goshawk) that were not included in the LSA, and the LSA included seven species (arctic grayling, Columbia spotted frog, moose, red-naped sapsucker, warbling vireo, wolverine, and yellow warbler) that were not included in the ecosystem umbrella suite. However, two species in the LSA suite (moose and wolverine) were selected a priori to take advantage of ongoing research in the area, rather than quantitatively, so the overlap between methods is actually higher.

Guidelines for application of approach

Although the ecosystem umbrella provided by a carefully chosen focal species suite is comprehensive, it is important to recognize its limitations. Foremost is the fact that conservation planning based on a focal species suite does not necessarily guarantee the persistence of all species within the planning area. The area required to maintain a viable population is scale-dependent and varies widely among species. In many cases, the planning area may be too small to support even a minimum viable population, let alone an ecologically functional population. Therefore, subsequent steps in the planning process should carefully examine the habitat and area requirements needed to sustain a population of each species in the focal suite, particularly those chosen in the early rounds of the selection process that have the highest area requirements. If population viability is a desired goal, it should be explicitly addressed later in the planning process (see Stanton and Akçakaya, chapter 11). It is also important that planners remain flexible on the use of chosen focal species in subsequent phases of the planning process. Some focal species may not be suitable representatives in later stages of planning. For example, data or expert knowledge to develop reliable habitat models may be lacking

Table 6.6. Species selected for the Madison Valley using the landscape species and ecosystem umbrella approaches

Species	Landscape species approach	Ecosystem umbrella approach
American beaver (*Castor canadensis*)		X
Arctic grayling (*Thymallus arcticus*)	X	
Bighorn sheep (*Ovis canadensis*)	X	X
Black-backed woodpecker (*Picoides arcticus*)	X	X
Columbia spotted frog (*Rana luteiventris*)	X	
Elk (*Cervus canadensis*)	X	X
Greater sage-grouse (*Centrocercus urophasianus*)	X	X
Grizzly bear (*Ursus arctos*)	X	X
Moose (*Alces alces*)	X	
Northern goshawk (*Accipiter gentilis*)		X
Pronghorn (*Antilocapra americana*)	X	X
Red-naped sapsucker (*Sphyrapicus nuchalis*)	X	
Warbling vireo (*Vireo gilvus*)	X	
Western toad (*Anaxyrus boreas*)	X	X
Westslope cutthroat trout (*Oncorhynchus clarki lewisi*)	X	X
Wolverine (*Gulo gulo*)	X	
Yellow warbler (*Dendroica petechia*)	X	

Source: Brent L. Brock and Eric C. Atkinson.

for a species that otherwise provides an excellent umbrella. In such cases, planners must adapt to circumvent obstacles without weakening the conservation umbrella. The selection process provides a framework for choosing substitute species or finding other solutions because important species-specific elements have been recorded during the workshop.

Regardless of the limitations, this approach offers an expedited method for selecting a focal species suite to provide a comprehensive and robust conservation umbrella. Although focal species selection is only one of many steps in the conservation planning process, it provides the foundation upon which the ultimate conservation plan is built. A key point to remember is that a broad-scale plan, such as a statewide one, provides a big picture in terms of regional connectivity and core habitat, but actual on-the-ground conservation must be based on planning performed at a finer resolution through hierarchies of scale (Unnasch and Karl, chapter 3). In our experience, on-the-ground conservation activities are most effective when they are based upon bottom-up decisions: local stakeholders need to be empowered to make biologically appropriate conservation decisions. A large-scale plan can provide guidance, but it must be pieced together by devising a multitude of smaller-scale plans using the same overall approach. Using focal species and their habitats as conservation targets can provide a means of transferring plans across scales.

References

Andelman, S. J., and W. F. Fagan. 2000. "Umbrellas and Flagships: Efficient Conservation Surrogates or Expensive Mistakes?" *Proceedings of the National Academy of Sciences of the United States of America* 97: 5,954–59.

Berger, J. 2004. "The Last Mile: How to Sustain Long-Distance Migration in Mammals." *Conservation Biology* 18: 320–31.

Betrus, C. J., E. Fleishman, and R. B. Blair. 2005. "Cross-Taxonomic Potential and Spatial Transferability of an Umbrella Species Index." *Journal of Environmental Management* 74: 79–87.

Branton, M., and J. S. Richardson. 2011. "Assessing the Value of the Umbrella-Species Concept for Conservation Planning with Meta-Analysis." *Conservation Biology* 25: 9–20.

Brock, B. L., E. C. Atkinson, C. Groves, A. Toivola, T. Olenicki, and L. Craighead. 2006. *A Wildlife Conservation Assessment of the Madison Valley, Montana.* Wildlife Conservation Society, Greater Yellowstone Program, Bozeman, MT. http://www.wcsnorthamerica.org/DesktopModules/Bring2mind/DMX/Download.aspx?EntryId=5581&PortalId=37&DownloadMethod=attachment.

Callicott, J. B. 2006. "Conservation Values and Ethics." In *Principles of Conservation Biology,* 3rd ed., edited by M. J. Groom, G. K. Meffe, and C. R. Carroll, 111–135. Sunderland, MA: Sinauer Associates.

Caro, T. M. 2003. "Umbrella Species: Critique and Lessons from East Africa." *Animal Conservation* 6: 171–81.

———. 2010. *Conservation by Proxy: Indicator, Umbrella, Keystone, Flagship, and Other Surrogate Species.* Washington, DC: Island Press.

Caro, T. M., and G. O'Doherty. 1999. "On the Use of Surrogate Species in Conservation Biology." *Conservation Biology* 13: 805–14.

Conner, R. N. 1988. "Wildlife Populations: Minimally Viable or Ecologically Functional?" *Wildlife Society Bulletin* 16: 80–84.

Coppolillo, P., H. Gomez, F. Maisels, and R. Wallace. 2004. "Selection Criteria for Suites of Landscape Species as a Basis for Site-Based Conservation." *Biological Conservation* 115: 419–30.

Fleishman, E., D. D. Murphy, and P. F. Brussard. 2000. "A New Method for Selection of Umbrella Species for Conservation Planning." *Ecological Applications* 10: 569–79.

Franklin, J. F. 1993. "Preserving Biodiversity—Species, Ecosystems, or Landscapes." *Ecological Applications* 3: 202–5.

Groves, C. R. 2003. *Drafting a Conservation Blueprint: A Practitioner's Guide to Planning for Biodiversity.* Washington, DC: Island Press.

Hansen, A. J., S. L. Garman, B. Marks, and D. L. Urban. 1993. "An Approach for Managing Vertebrate Diversity across Multiple-Use Landscapes." *Ecological Applications* 3: 481–96.

Hansen, A. J., R. L. Knight, J. M. Marzluff, S. Powell, K. Brown, P. H. Gude, and K. Jones. 2005. "Effects of Exurban Development on Biodiversity: Patterns, Mechanisms, and Research Needs." *Ecological Applications* 15(6): 1,893–1,905.

Lambeck, R. J. 1997. "Focal Species: A Multi-Species Umbrella for Nature Conservation." *Conservation Biology* 11: 849–56.

Landres, P. B., J. Verner, and J. W. Thomas. 1988. "Ecological Uses of Vertebrate Indicator Species—A Critique." *Conservation Biology* 2: 316–28.

MacArthur, R. 1972. "Strong, or Weak, Interactions?" *Transactions of Connecticut Academy of Arts and Sciences* 44: 177–88.

Margules, C. R., and R. L. Pressey. 2000. "Systematic Conservation Planning." *Nature* 405: 243–53.

Maslow, A. H. 1943. "A Theory of Human Motivation." *Psychological Review* 50: 370–96.

Niemi, G. J., J. M. Hanowski, A. R. Lima, T. Nicholls, and N. Weiland. 1997. "A Critical Analysis on the Use of Indicator Species in Management." *Journal of Wildlife Management* 61: 1,240–52.

Noss, R. F. 1996. "Ecosystems as Conservation Targets." *Trends in Ecology & Evolution* 11: 351–351.

Orians, G. H. 1993. "Endangered at What Level? *Ecological Applications* 3: 206–8.

Orme, C. D. L., R. G. Davies, M. Burgess, F. Eigenbrod, N. Pickup, V. A. Olson, A. J. Webster, T. S. Ding, P. C. Rasmussen, R. S. Ridgely, A. J. Stattersfield, P. M. Bennett, T. M. Blackburn, K. J. Gaston, and I. P. F. Owens. 2005. "Global Hotspots of Species Richness Are Not Congruent with Endemism or Threat." *Nature* 436: 1,016–19.

Paine, R. T. 1966. "Food Web Complexity and Species Diversity." *The American Naturalist* 100: 65–75.

———. 1969. "A Note on Trophic Complexity and Community Stability." *The American Naturalist* 103: 91–93.

———. 1995. "A Conversation on Refining the Concept of Keystone Species." *Conservation Biology* 9: 962–64.

Power, M. E., D. Tilman, J. A. Estes, B. A. Menge, W. J. Bond, L. S. Mills, G. Daily, J. C. Castilla, J. Lubchenco, and R. T. Paine. 1996. "Challenges in the Quest for Keystones." *BioScience* 46: 609–20.

Redford, K. H., P. Coppolillo, E. W. Sanderson, G. A. B. DaFonseca, E. Dinerstein, C. Groves, G. Mace, S. Maginnis, R. A. Mittermeier, R. F. Noss, D. Olson, J. G. Robinson, A. Vedder, and M. Wright. 2003. "Mapping the Conservation Landscape." *Conservation Biology* 17: 647–647.

Redford, K. H., and B. D. Richter. 1999. "Conservation of Biodiversity in a World of Use." *Conservation Biology* 13: 1,246–56.

Regan, H. M., M. Colyvan, and M. A. Burgman. 2002. "A Taxonomy and Treatment of Uncertainty for Ecology and Conservation Biology." *Ecological Applications* 12: 618–28.

Robinson, J. G. 2006. "Conservation Biology and Real-World Conservation." *Conservation Biology* 20: 658–69.

Rondinini, C., and L. Boitani. 2006. "Differences in the Umbrella Effects of African Amphibians and Mammals Based on Two Estimators of the Area of Occupancy." *Conservation Biology* 20: 170–79.

Sanderson, E. W., K. H. Redford, A. Vedder, P. B. Coppolillo, and S. E. Ward. 2002. "A Conceptual Model for Conservation Planning Based on Landscape Species Requirements." *Landscape and Urban Planning* 58: 41–56.

Seddon, P. J., and T. Leech. 2008. "Conservation Short Cut, or Long and Winding Road? A Critique of Umbrella Species Criteria." *Oryx* 42: 240–45.

Simberloff, D. 1998. "Flagships, Umbrellas, and Keystones: Is Single-Species Management Passé in the Landscape Era?" *Biological Conservation* 83: 247–57.

Soule, M. E., J. A. Estes, J. Berger, and C. M. Del Rio. 2003. "Ecological Effectiveness: Conservation Goals for Interactive Species." *Conservation Biology* 17: 1,238–50.

Soule, M. E., and M. A. Sanjayan. 1998. "Ecology—Conservation Targets: Do They Help?" *Science* 279: 2,060–61.

Suter, W., R. F. Graf, and R. Hess. 2002. "Capercaillie (*Tetrao urogallus*) and Avian Biodiversity: Testing the Umbrella-Species Concept." *Conservation Biology* 16: 778–88.

Terborgh, J., and B. Winter. 1983. "A Method for Siting Parks and Reserves with Special Reference to Columbia and Ecuador." *Biological Conservation* 27: 45–58.

Vane-Wright, R. I., C. J. Humphries, and P. H. Williams. 1991. "What to Protect-Systematics and the Agony of Choice." *Biological Conservation* 55: 235–54.

Vazirani, V. V. 2001. *Approximation Algorithms.* Berlin: Springer-Verlag.

Western, D. 1987. "Africa's Elephants and Rhinos: Flagships in Crisis." *Trends in Ecology and Evolution* 2: 343–46.

7

Identification of habitat and assessment of habitat quality for conservation of terrestrial animals

Brett G. Dickson, PhD; Steven E. Sesnie, PhD; Erica Fleishman, PhD; and David S. Dobkin, PhD

Modification and loss of the terrestrial habitats of many animal species is occurring globally at unprecedented rates and extents, but the magnitude of these changes can be challenging to document. The persistence of most animal populations varies in part as a function of the amount and configuration of their habitats. Accurate models and maps of land cover, topography, and the location, quantity, and quality of habitats are critical for effective guidance of management of these species.

Methods for classifying land cover and topography and for quantifying relationships between animals and their environment are advancing rapidly. The recent development of spatially and temporally extensive data on biotic and abiotic features (e.g., vegetation type and structure, topography, road networks) is improving the ability to accurately delineate and evaluate habitats for animals in many regions around the world. Many of these datasets are available at no cost from government institutions and websites. Access to geographic information systems (GIS) software and statistical and spatial modeling packages (including freeware, such as the R environment; http://www.r-project.org/) that are used to develop habitat models also has improved.

Habitat is a concept with many biotic and abiotic dimensions and about which much has been written and debated. Convenient delineations of habitat may not encompass the spatial and temporal dynamics of an area or species (Morrison and Hall 2002; Mitchell 2005; Johnson 2007), which are essential for modeling and mapping. In this chapter, we adopt the definition of habitat presented by Hall et al. (1997): "the resources and conditions present in an area that produce occupancy—including survival and reproduction—by a given organism." Hall et al. (1997) also emphasized that habitat "implies more than vegetation or vegetation structure." We further define habitat as spatially and temporally explicit, with multiple attributes that can be modeled, mapped, and related to patterns of species occupancy across large extents. This operational definition of habitat was developed from earlier concepts, including that of Fretwell (1972), who defined the habitat of a species as "any portion of the surface of the Earth where the species is able to colonize and live at some density greater than zero." In reference to birds that nest and primarily forage in terrestrial systems, Hutto (1985) more narrowly defined habitat as "a spatially contiguous vegetation type that appears more or less homogeneous throughout and is physiognomically distinctive from other such types."

By definition, habitat is suitable for occupancy, reproduction, and survival. Thus, suitable habitat is redundant and unsuitable habitat an oxymoron. Habitat quantity, configuration, and quality vary in space and time. However, habitat can be occupied or unoccupied. Species-environment interactions at a given point in time generally depend on the resources needed by the species at that time for movement, feeding, growth, reproduction, and survival. Thus, when and why a location is occupied may depend on the life-history stage or ecology of the species or population, or on demographic or environmental stochasticity. If a location is known, or reasonably can be assumed, to support occupancy, reproduction, and survival, it likely is habitat. When a previously occupied location no longer supports occupancy, reproduction, and survival by a species, it is no longer habitat.

The concept of habitat may sometimes be extended to multiple species occupying a given location. Because few species exhibit the same response to abiotic and biotic features and processes, habitat is often described as species-specific (e.g., Hall et al. 1997). Planning for the conservation of multiple species requires evaluation of each species' habitat and attributes of habitat shared among species. In some cases, management of environmental features that are components of habitat for multiple species may benefit many species simultaneously (Hunter 2005; Noon et al. 2009), reducing the number of species that require individually tailored management strategies. The effectiveness of this approach, sometimes called "coarse filter," relies on identification of environmental features that affect occupancy, reproduction, and survival of multiple species, and on accurately predicting the response of a high proportion of those species to perturbations of those features.

Habitat quality is an extension of the concept of habitat. Usually, higher quality equates with higher frequency of use (or greater probability of occupancy) by an animal species. Accurately identifying terrestrial habitat and evaluating habitat quality for terrestrial animals is crucial for conservation planning. In this chapter, we first provide an overview of contemporary data and methods for identifying habitat and evaluating habitat quality. There is a substantial literature on modeling the location and quality of habitat. Johnston et al. (chapter 14) provide further information on modeling approaches and alternatives.

In the later sections of this chapter, we focus in detail on occupancy-based statistical modeling as a current best method to determine which environmental characteristics (or covariates) are most strongly associated with where a species is found and are most likely to characterize habitat accurately. We illustrate this method with a case study of habitat and occupancy modeling that combines multiple years of data on occurrence of two species of breeding birds in the central Great Basin (Nevada, USA). We base the models on ground-based and remotely sensed environmental data. Site-level estimates of occupancy for each species are extended to models of habitat quality that only include GIS-based habitat variables (hereafter, we refer to these models as "spatially explicit"). This method is practical and can be used for many species and taxonomic groups if the data is available.

Building and interpreting occupancy-based statistical models involves several steps after selecting the species to be modeled (see Brock and Atkinson, chapter 6). To provide reliable output, occupancy modeling requires a statistically robust measure of species-level occupancy determined from field surveys. The models also require high-quality spatial data on land cover and topography, which is freely available in the United States. Data with a resolution of 10 m or less can be obtained for some land-cover types that are not accurately classified at a 30-m resolution. Next, environmental covariates strongly associated with occupancy are determined from the occupancy, vegetation, and physiographic data. These covariates are then used to map potential habitat and evaluate habitat quality. Once a species' habitat has been accurately estimated or identified, information on the location and quality of habitat can be used to establish conservation priorities and strategies. If data of sufficient quality is not available for the area and species on which a management plan is focused, and if there are substantial time or funding constraints, it may be necessary to utilize an expert-based (e.g., Martin et al. 2012) or more simple statistical approach to model habitat, some of which are described below or elsewhere in this book.

A brief history of efforts to model and map habitat for terrestrial animals

Modern habitat mapping may have begun with Ian McHarg, who used the technology available to him—paper maps and Mylar overlays—to explore alternative patterns of human land use relative to environmental variables (McHarg 1969). With the advent of computer modeling and other digital technologies, the use and integration of spatially referenced information on the location of animals and their habitat became more commonplace and sophisticated. Some of the first models of relationships between animals and their habitat were developed in the late 1970s and early 1980s (e.g., Patton 1978; Thomas 1979; Salwasser et al. 1980; Schamberger et al. 1982). These models offered a means to identify or predict the distributions of species and their habitats that were otherwise difficult to observe. Such modeling was rapidly applied by scientists in government agencies, such as the US Forest Service and US Fish and Wildlife Service, whose mandates included managing animal species or their habitats. An international symposium held in California in October 1984, "Modeling Habitat Relationships of Terrestrial Vertebrates," produced the first book on modern techniques for habitat modeling (Verner et al. 1986).

In the late 1980s and early 1990s, wildlife habitat models were one of the first applications to benefit from the development of GIS and increasingly accurate and spatially extensive data on land cover (e.g., Lancia et al. 1986; Donovan et al. 1987; Pereira and Itami 1991). Widespread use of habitat models essentially began in 1988 with the initiation of the Gap Analysis Program (GAP; Scott et al. 1993, 2002). The foundation of GAP was a map of land cover derived from Landsat imagery. This data was used to project the ranges of individual species of terrestrial vertebrates on a state-by-state basis in the United States. The goal of the program was to identify vegetation alliances and habitats of wild animals that spanned all classes of land protection. The intent was that unprotected habitats or vegetation associations then would become a focus for conservation efforts. Several large environmental organizations began conservation planning in the 1990s and further developed computer applications for modeling the relationship between animals and the environment. For example, The Nature Conservancy projected geographic ranges of plants and animals and selected conservation targets on the basis of biogeographically defined units of analysis and land-cover data (The Nature Conservancy 1997). Each unit was characterized according to the predicted presence or absence of species. So-called ecoregional planning developed into a three-tiered approach (Noss 1996) of representation of vegetation types (coarse filter), high-priority or focal species (fine filter), and special elements (e.g., threatened or rare plant communities; see Brock and Atkinson, chapter 6). Habitat for focal species was identified on the basis of the predicted presence or absence of species. Perhaps because of inherent inaccuracies in identification of habitat due to the use of simple observed or projected occurrence data, habitat modeling for focal species was less emphasized than coarse-filter and special-element approaches.

Current best practices for modeling and mapping quality of habitat for terrestrial animals

More recently, many wildlife habitat models have been developed on the basis of personal knowledge (expert-based), often augmented by published literature, or on the basis of statistical models and empirical data on animal locations (e.g., logistic regression, resource selection functions [RSFs]). Habitat models rely on accurate knowledge of habitat use by animals that are often difficult to observe. The history of their development reveals increasing accuracy in determining, for example, habitat use or occupancy.

Expert-based models of habitat

Although they are more difficult to validate than empirical models, expert-based models of habitat quality may be the only option in areas where empirical data is sparse or nonexistent. In most cases, one or more experts rank the relative habitat quality of all land-cover classes for the species in question. Often a lookup table or weighting matrix is created in a GIS. Ranks (say, from 0 to 10) are assigned to each land-cover class on the basis of its assumed habitat quality, and these values are visualized in the digital landscape. The accuracy of this approach depends on the accuracy of the land-cover classification and other model inputs, and the relationship between actual animal associations with land-cover classes and the estimates of the experts. Because the subjective rankings are not a true scale of measurement (i.e., a value of 4 does not necessarily correspond to twice the quality of a value of 2), they are not a true representation of habitat quality or animal associations (Wolman 2006). Examples of recent expert opinion models are presented by Johnston et al. (chapter 14).

Statistical and empirically based models of habitat

Statistical models are the fundamental inferential tools used to estimate quantitative relationships between animals and their habitat. All model types have assumptions or limitations that affect which model might be most useful for a given application (see Johnston et al., chapter 14). Generalized linear models (GLMs; see

Neter et al. 1996) are among the most common types used to model habitat quality. For example, a logistic regression relates binary data on the presence or absence of a species to site-specific covariates and can be used to estimate the relative probability of occupancy at a given site (on the basis of a logit link function; see Hosmer and Lemeshow 2000). Models based on presence-only data also can be used to evaluate species occupancy (or distribution) and habitat. They are becoming increasingly popular for inference across large areas (e.g., maximum entropy [Phillips et al. 2006], genetic algorithm for rule-set prediction [Stockwell and Peters 1999], and ecological niche factor analysis [Hirzel et al. 2002], but see Brotons et al. 2004; Jones et al. 2010). Where the response variable in a statistical model is continuous (e.g., abundance, density, replicated estimates of abundance), many other models are available for estimating relationships between animals and their habitat, including GLMs, generalized additive models (Hastie and Tibshirani 1990), and various hierarchical model forms (e.g., generalized linear mixed models based on maximum likelihood or Bayesian procedures; Royle and Dorazio 2008). A substantial literature exists on the comparative implementation and efficacy of these different methods for statistical modeling of animal habitat (e.g., Guisan and Zimmermann 2000; Scott et al. 2002; Elith et al. 2006; Fitzgerald et al. 2009).

Building on the empirical data inputs and fundamental structure of a logistic regression model, RSF (Manly et al. 2002), and occupancy (*sensu* MacKenzie et al. 2006) models frequently are used to infer habitat quality. Typically, RSF models relate animal location data (presence and absence, or presence only, usually determined by telemetry or direct observation) to a range of environmental characteristics (or covariates) associated with animal locations, such as temperature, elevation, and vegetation structure or composition, to estimate a function that is proportional to the probability of habitat (or resource) use (Boyce et al. 2002). When implemented within a GIS and coupled with model selection techniques, RSF models are powerful tools for evaluating and informing management of animal habitat over extensive areas (Boyce et al. 2002; Carroll et al. 2006). Occupancy models, which are described and implemented below, share and extend many of these desirable properties.

Data collection and use considerations

Habitat models are only as accurate as the data on which they are based. Data on animal location or movement can be overlaid with data on land cover and topography to quantify attributes of areas that are used for reproduction, foraging, and dispersal (see Millspaugh and Marzluff 2001). Compared to radio tracking on the basis of ground-based very high frequency (VHF) devices, which require manual collection of telemetry data in the field, satellite tracking is much less labor-intensive and can provide substantially more location data. Global positioning system (GPS) collars and tags, for example, have been used to more extensively and efficiently collect location data at regular time intervals for terrestrial and aquatic animals (e.g., Marcinek et al. 2001; Johnson et al. 2002). Data on dispersal may be particularly useful for modeling habitat quality and connectivity for species that occur as patchy populations or metapopulations (see discussion of populations in Stanton and Akçakaya, chapter 11). These technologies and techniques have limitations, however. Collars and tags have a finite, battery-determined lifetime, and accuracy may vary as a function of topography or land-cover type (e.g., D'Eon et al. 2002; Lewis et al. 2007). Moreover, body size, life history, or behavior of an animal can constrain what devices are practical and ethical. Research questions, ideal sample size, and inference limits also should be clearly determined prior to placing tracking devices on animals (Hebblewhite and Haydon 2010).

Where models or maps of habitat or habitat quality do not exist, it may be necessary to develop new information for conservation planning. The quality of this information will in part reflect the quality of field sampling and ground data. For example, sufficient space among animal survey locations ideally should be maintained such that a given individual is unlikely to move among those locations during a given sampling period in response to the observer. This separation of survey locations also minimizes potential spatial or temporal autocorrelation (i.e., lack of independence) among locations (Gutzwiller and Riffell 2007; Dickson, Noon et al. 2009). When survey locations are close together in space or time, statistical estimates (e.g., regression coefficients) of habitat variables will reflect measured values that are similar among locations, and can overestimate habitat quality and the extent to which a variable is associated with, for example, occupancy of a species (see Legendre 1993; Haining 2003). The space among survey locations and time between survey periods should always be guided by the biology of the focal animals. A preliminary field effort and basic spatial statistics can help determine adequate spacing. Especially within a GIS environment, there is potential for compounding uncertainty in predictions of habitat quality and species occurrence (Stoms et al. 1992; Henebry and Merchant 2002).

Many spatial representations of habitat or habitat quality are possible. Binary representations classify locations as either habitat or not-habitat, and have limited utility for conservation planning. Categorical or continuous representations assume different features represent different dimensions of habitat or habitat quality. Multidimensional representations can facilitate evaluation of whether, at a given time, a specific location may represent unoccupied habitat. Results from statistical models of habitat quality can be implemented within a GIS and used to generate spatially explicit products (maps; see below). The spatial and temporal extent and resolution at which organisms interact with the environment varies among species and life-history stages (Unnasch and Karl, chapter 3; and Johnston et al., chapter 14). Thus, the appropriate extent and resolution of data inputs and outputs for models of habitat quality will vary among species.

The availability of digital and remotely sensed data at multiple extents and resolutions has facilitated environmental characterization and predictions of habitat quality in locations that historically were inaccessible due to field or social and political conditions. The geographic extent over which it has become possible to conduct such work also has increased. Environmental data and species data should reflect the same period (e.g., month or season) of time unless there is an empirical basis for assuming that values of some environmental variables are unlikely to change substantially over the period for which explanatory or predictive models of occupancy are being developed. Moreover, a model that explains considerable information or variation in the data that was used for its creation may not make effective predictions (Fleishman et al. 2003). Predictions that are based on any algorithm should be regarded as testable hypotheses to be confronted with newly collected test data (i.e., external validation). Internal validation (e.g., cross-validation, bootstrapping) is preferable to no validation, but ultimately is insufficient to rigorously evaluate the accuracy of predictive models. Additionally, one cannot assume that values of response variables will be predicted accurately on the basis of values of covariates, including locations and points in time, that are outside the range used to build the model (Thomson et al. 2006).

Improving and integrating best practices for identification and assessment of habitat quality

Characterization of habitat quality usually is based on a statistical model that relates empirical data on the distribution or population dynamics of a species to different abiotic and biotic features, including topography, land cover, and land use. For example, recently developed empirical, statistical, remote sensing, and GIS-based methods can be used to model and map probability of occupancy, which often is positively correlated with—or assumed to serve as a surrogate measure of—habitat quality, for many species of animals. These methods can be used to inform management for species and their habitats across multiple spatial and temporal extents and resolutions and in rapidly changing landscapes.

Modeling topography

A variety of ArcGIS (Esri) algorithms and extensions are available to model topographic features and ecosystem processes (see Pelletier 2008 and Horning et al. 2010). Topographic metrics derived from a digital elevation model (DEM), such as slope, aspect, and elevation, are commonly incorporated into habitat models as surrogates for vegetation and other attributes of habitat (e.g., Osborne et al. 2001; Clevenger et al. 2002; Shriner et al. 2002).

Topographic attributes that may influence habitat quality, such as ruggedness, solar radiation, and soil moisture, now can be extracted from free digital elevation datasets, such as the National Elevation Dataset (NED), Shuttle Radar Topography Mission (SRTM) data, and Global Digital Elevation Map (GDEM). Topography Tools (http://arcscripts.esri.com/details.asp?dbid=15996) and DEM Surface Tools (http://www.jennessent.com/arcgis/surface_area.htm) are free extensions to ArcGIS that each offer a set of models for deriving topographic indices linked to vegetation. For example, a topographic position index (TPI) measures whether the elevation of a pixel is higher or lower than that of surrounding pixels. This information can be transformed into categories of slope position and landform that, in turn, can be used to model animal-environment relationships (e.g., Dickson and Beier 2007). Topography also is correlated with land use. Measures of human activity often can be poorly represented in land-cover maps, underestimating effects on habitat quality and extent.

Three-dimensional data acquired from light detection and ranging (lidar) remote sensing affords many new avenues for modeling habitat attributes and quality (Vierling et al. 2008). ArcGIS extensions such as LIDAR Analyst (http://www.vls-inc.com/lidar_analyst.htm) enhance opportunities to process and use lidar data for developing accurate models of land surface features at spatial resolutions appropriate for many habitat modeling applications (e.g., < 1 to 10s of meters). Models of bare-earth elevation, topography, and above-ground measures of vegetation structure and successional stage derived from lidar data increasingly is used to model animal habitat (Bradbury et al. 2005; Martinuzzi et al. 2009).

Modeling land cover

Land cover encompasses natural features, such as dominant vegetation, rock, and water (see Olenicki, chapter 4), as well as agricultural fields, urban areas, and other features produced by human activity (see Theobald, chapter 5, and section on land use below). Land-cover data is a primary input in models of species occurrence and habitat quality.

The availability of remotely sensed images at a variety of spatial, temporal, and spectral resolutions has increased capacity to develop, enhance, and update land-cover and land-use data. National mapping efforts, such as the GAP, Landscape Fire and Resource Management Planning Tools (LANDFIRE) project, and National Land Cover Dataset (NLCD), provide free digital maps of land cover for most of the United States and territories that were derived from Landsat satellite images at a 30-m spatial resolution. Alliance-level vegetation categories used by GAP were developed with the aim of mapping animal habitat and species occurrence. The NLCD data has generalized land-cover classes to accurately map and monitor changes in cover types such as forest and agricultural lands. A large body of literature exists on the topics of modeling and classifying land cover important to habitat quality models (see Olenicki, chapter 4).

Environmental variables other than topography and vegetation may be components of habitat or affect habitat quality. For example, mutualists or parasites require the presence of other species; data on those species' occurrence or distribution can be an important covariate. The presence of pathogens can reduce habitat quality for some species while improving habitat quality for others. Human caused disturbance, land use, or climate change can facilitate or enhance pathogen presence (e.g., Sturrock et al. 2011).

Modeling land use and the human footprint

Land use emphasizes the link between social and economic activity and the environment (see Theobald, chapter 5). Common land-use classes include urban development, livestock grazing, and agriculture. In essence, land use describes how humans use land cover, although additional information often is needed to differentiate land cover from land use (Malczewski 2004). Classifying land use can be especially difficult when individual land-cover classes are associated with multiple land uses, such as agriculture, livestock grazing, and timber harvest. However, ancillary spatial data on, for example, road networks, rail and power lines, and human population density are easily obtained from US Census Bureau TIGER/Line files (http://www.census.gov). Similarly, spatial data on reservoirs, water developments for livestock, and canals can be obtained from local, regional, or national hydrography databases. Within a GIS, impervious surfaces (e.g., parking lots, roofs) and other human-made features can be delineated and digitized with automated feature extraction algorithms and existing aerial imagery, such as digital aerial photographs. Some or all of these datasets can be combined to produce a reasonable measure of human land use.

The sum extent and influence of human activities on Earth's natural features has been described as the human footprint (Sanderson et al. 2002). Modeling and mapping the current and predicted human footprint is a key element in assessing how land use may affect habitat quality in the future (Leu et al. 2008; Theobald, chapter 5). Often it is possible to generate alternative development scenarios and compare their potential effects on a species' habitat. For example, it is possible to analyze the effects of various wind farm sites, transmission line sites, and wind tower placement on greater sage-grouse (*Centrocercus urophasianus*) and its habitats (Unnasch and Karl, chapter 3).

Modeling probabilities of detection and habitat occupancy

Characterizing a species' habitat clearly is essential for modeling that habitat. However, whether a species is recorded as present or absent in a given location is affected by the probability of detecting the species. If a species is not detected, it does not mean that it is absent. Assuming a species always will be detected when present can result in unreliable inference (MacKenzie et al. 2005). Behavior and abundance of the species, land cover, ambient noise, weather, and observer ability all affect detection probability. Detection probability (p) can be defined as the probability of detecting a given species at a given site if it is present during a site visit. Occupancy (ψ) can be defined as the expected probability that a given site is occupied by the species (MacKenzie et al. 2006). The free software programs PRESENCE (Hines 2006) and MARK (White and Burnham 1999; http://warnercnr.colostate.edu/~gwhite/mark/mark.htm#Introduction) estimate the probability of occupancy of a given species at a given location on the basis of detection histories at that location (i.e., observed presence or absence at the location on each of multiple visits). These estimates can be improved by measuring different abiotic and biotic variables at the site and then evaluating the relationship between the variables and occupancy with information-theoretic and model selection techniques (see below and sidebar, "Information theory and model selection techniques for estimating habitat quality"). Estimates of occupancy that account for detection probability are almost always higher than naïve estimates (estimates based on the proportion of locations in which the species was detected; MacKenzie et al. 2006).

Detection probability also affects estimates of animal abundance, density, and survival. Many approaches and free software packages are available for generating these estimates of habitat quality, including programs MARK and DISTANCE (Thomas et al. 2010), which rely on repeated counts of individuals or animals, detection histories, or capture-recapture data. In addition, Bayesian methods are available for estimating some of these parameters (e.g., Royle and Dorazio 2008), and the free software program WinBUGS (Lunn et al. 2000; http://www.mrc-bsu.cam.ac.uk/bugs/) has become popular for these analyses. These estimation methods can become computationally intensive when population or environmental processes are highly dynamic or when survey designs are complex (Williams et al. 2002).

As data collection methods, technology, and computing power have advanced, it has become more feasible to use output from models of topography, land cover, and land use as inputs to models of detection and occupancy of animal species. In turn, outputs from these types of models can serve as inputs to statistical models of current and future habitat quality and connectivity (Singleton and McRae, chapter 10) across extensive areas.

Information theory and model selection techniques for estimating habitat quality

An information-theoretic approach to model selection and Akaike's Information Criterion (AIC) increasingly are used to objectively evaluate the relative importance of alternative, or candidate, models—de facto hypotheses—of relationships between demographic parameters (e.g., occupancy, abundance) and environmental attributes. These models may be informed by known empirical relationships or generated by specialists on the species or system. AIC is a maximum likelihood-based estimate of the relative distance between a given model (i.e., hypothesis) and the unknown true mechanism that generated the observed data (Burnham and Anderson 2002). Other information criteria are appropriate and easy to derive when, for example, sample sizes are small or data is not independently distributed. Given a dataset, the model in the candidate set with the lowest AIC value is the model that best approximates truth, and is set to zero. Within a candidate set of models, the difference among AIC values (ΔAIC) and AIC weights (model probabilities) are used to compare the relative strength of evidence in favor of a given model and to evaluate uncertainty in model selection (Anderson 2008). Burnham and Anderson (2002) and Anderson (2008) provide excellent introductions to information theory and model selection for inference in ecology, including model-averaging techniques. In addition, MacKenzie et al. (2006) provide a comprehensive description of methods for estimating occupancy and the use of model selection.

Often, an evaluation of the relation between demographic parameters and individual habitat variables may be of more interest than determining the best model of habitat quality. For example, AIC weights can be computed from AIC values and used to rank the strength of association between each variable and occupancy (Burnham and Anderson 2002). Similarly, the relative effect of each variable in a model on occupancy can be estimated with a Z statistic, which is easy to calculate and interpret, and does not require arbitrary statistical cutoffs (e.g., alpha levels or P values; Zar 1996).

Illustration of best practices

Modeling bird habitat and occupancy in the Great Basin

We applied empirical, statistical, remote sensing, and GIS-based techniques to model habitat quality for two species of songbirds with different a priori strengths of association with riparian areas. Our methods are derived from Dickson, Fleishman et al. (2009), free software for occupancy estimation, and common tools in ArcGIS software. Data for different variables was obtained or derived at different spatial and temporal extents and resolutions. These methods are practical and highly transferable among species and taxonomic groups. Using steps similar to those outlined below, a person can collect field data on species' locations (or use existing data), determine probability of occupancy, and model potential habitat with greater accuracy than with some other methods. Such models can then potentially be used to determine core areas, model connectivity, prioritize areas for conservation, and develop conservation plans (see subsequent chapters).

Data sources and acquisition

Our 15,000-km^2 study area (~ 8,000 km^2 montane) in the Great Basin encompasses four adjacent mountain ranges (figure 7.1). Canyons, many of which have perennial or ephemeral streams, drain the east and west slopes of the ranges. More than 95% of sampling locations are within the Humboldt–Toiyabe National Forest.

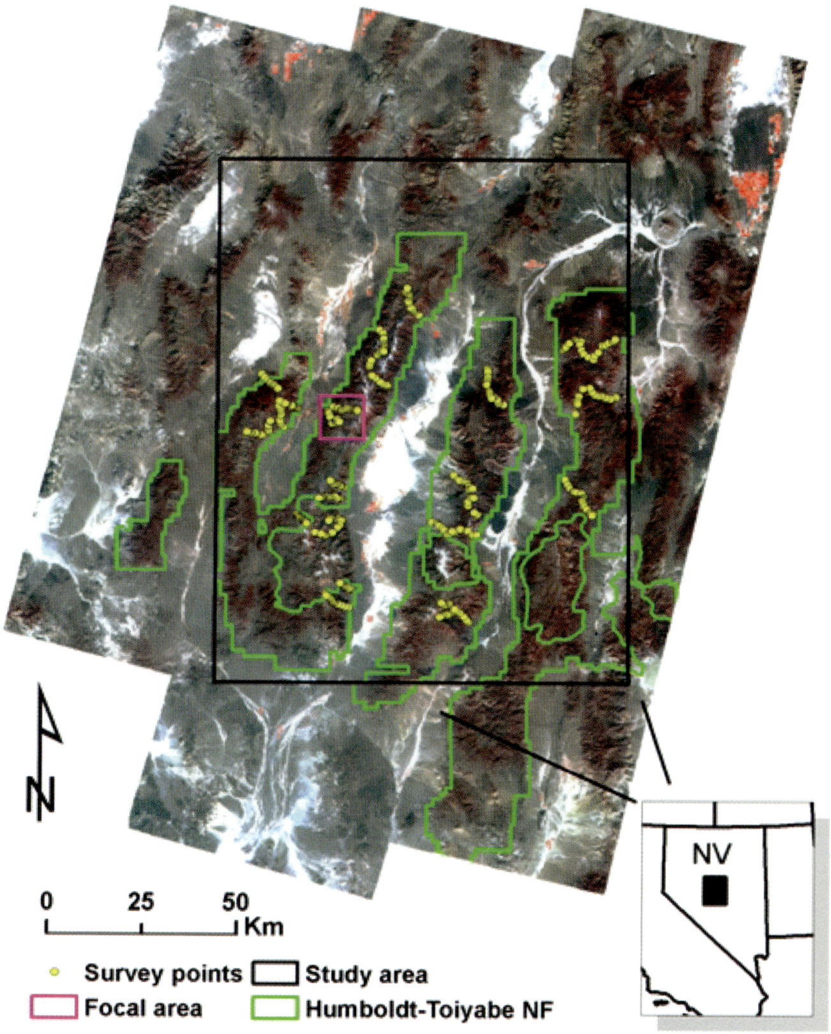

Figure 7.1. The 15,000-km^2 study area (including 135-km^2 focal area) in the central Great Basin (Nevada, USA) encompasses 301 survey points in four adjacent mountain ranges in Lander, Eureka, and Nye Counties. From west to east, these ranges are the Shoshone Mountains, Toiyabe Range, Toquima Range, and Monitor Range. SPOT 5 satellite imagery from University of California, Santa Barbara (UCSB) Resource Center for SPOT Imagery; USDA National Forest boundaries from the National Atlas of the United States; state boundaries from ESRI Data & Maps, 2009, courtesy of ArcUSA, US Census, Esri (Pop2008 field).

We surveyed birds during the breeding seasons (late May through June) of 2005–2009 with 75-m fixed-radius point counts. Most point centers were > 350 m apart, which minimized the probability that observer activity caused individuals to move among points. Points were located along the full elevational gradient of each canyon we sampled, typically with two or three points per 100-m vertical elevation change. Points were stratified and positioned to sample, in approximate proportion to areal extent, the dominant land-cover types throughout the canyons.

Geographic coordinates were recorded at each point with a GPS. During each visit, we recorded by sound or sight all birds using terrestrial habitat within the point. Each point was surveyed three times per year for five minutes per visit (Buckland et al. 2001; Siegel et al. 2001). All but two points were surveyed in two or more years. We surveyed a total of 301 points: 49 in the Shoshone Mountains, 124 in the Toiyabe Range, 59 in the Toquima Range, and 69 in the Monitor Range. The elevation of our survey points ranged from 1,915 to 3,210 m. We considered a species to be present at a point in a given year if it was detected during one or more of the three visits.

For this analysis, we focused on MacGillivray's warbler (*Oporornis tolmiei*), a riparian obligate, and spotted towhee (*Pipilo maculatus*), which breeds in both riparian and nonriparian vegetation. For both species, we developed one simple set of a priori models of habitat quality expressed as alternative, expert-generated hypotheses about the response of occupancy to multiple topographic and vegetation variables (henceforth

Table 7.1. Results of occupancy (ψ) model selection for MacGillivray's warbler and spotted towhee in the central Great Basin (Nevada, USA)

Model	K^1	AIC^2	ΔAIC^3	w^4
MacGillivray's warbler				
$\psi\ (FS + R^2 + C^2 + E^2 + PS)$	17	2085.0	0.00	0.906
$\psi\ (R^2 + C^2 + E^2 + PS)^5$	16	2090.6	5.60	0.055
$\psi\ (FS + C^2 + E^2 + PS)$	15	2091.6	6.57	0.034
$\psi\ (FS + R^2 + E^2 + PS)$	15	2095.6	10.53	0.005
$\psi\ (FS + R^2 + C^2 + E^2)$	16	2121.7	36.63	0.000
$\psi\ (.)$	9	2199.8	114.79	0.000
Spotted towhee				
$\psi\ (FS + R^2 + C^2 + E)$	17	2462.1	0.00	0.660
$\psi\ (FS + R^2 + C^2 + E + PS)$	18	2463.9	1.84	0.263
$\psi\ (FS + R^2 + E + PS)$	16	2467.8	5.72	0.038
$\psi\ (FS + C^2 + E + PS)$	16	2467.9	5.78	0.037
$\psi\ (R^2 + C^2 + E + PS)^5$	17	2473.3	11.20	0.002
$\psi\ (.)$	11	2502.6	40.50	0.000

Source: Brett G. Dickson.
Note: Modeled effects (i.e., habitat variables) include frequency of shrubs (*FS*), topographic roughness (*R*), percent woodland canopy cover (*C*), elevation (*E*), and percent cover of deciduous shrubs (*PS*). Frequency of shrubs was measured in the field at each survey point, whereas all other variables were derived from digital data. Squared term indicates quadratic form of variable. Null model denoted by (.). All models are conditioned on a single best model of detection probability (*p*), where *p* for MacGillivray's warbler and spotted towhee is a function of (*FS + R + PS*) and (*FS + R + C + E + PS*), respectively.

[1] Total number of model parameters, including those used to estimate *p*
[2] Akaike's Information Criterion
[3] AIC difference value
[4] AIC model weight
[5] Used to derive spatially explicit model in GIS

called habitat variables; table 7.1). We chose to use the same set to compare patterns of occupancy among species that frequently co-occur. We estimated occupancy as a function of variables that were measured empirically within an approximately 100-m radius of the survey point (i.e., ground data) or derived from remotely sensed data on topography and land cover.

To characterize vegetation composition and structure in the field, we established plots centered on the bird survey points (see Dickson, Fleishman et al. 2009). Vegetation at each plot was measured once between 2001 and 2009, and most woody vegetation (as measured here) changed little during our study period. We obtained 30-m resolution data for elevation, slope, and woodland canopy cover within our study area. We obtained land cover and topography data from the LANDFIRE project website. Raster (or grid) data for existing vegetation and terrain is available at no cost for the entire United States, and is accessed through a user-friendly interface. We downloaded estimates of elevation (in meters) and slope (degrees) that were created by the LANDFIRE project based on the NED (Gesch et al. 2002). Although the LANDFIRE project estimated canopy cover with the primary aim of modeling behavior and probability of fire (Rollins 2009), this data has proven useful in many other ecological applications. Percent woodland canopy cover was classified by the LANDFIRE project on the basis of Landsat TM imagery acquired from 1999 to 2001, and recently updated to account for extensive (> 400 ha) environmental disturbances (e.g., fire, insect outbreaks) through 2008. To maintain temporal consistency with our point counts, data for all of the habitat variables reflected conditions from 2005 through 2009.

Riparian areas (defined here by dominant deciduous vegetation that is typically associated with the presence of permanent or ephemeral streams) represent less than 2% of the land area of the Great Basin that occurs within Nevada. Riparian areas often are linear and only a few meters wide, with vegetation composition and structure that differs from that of upland communities (dominated by coniferous woodlands). Thus, riparian areas are often difficult to classify on the basis of 30-m resolution data and are poorly represented in land-cover maps.

Accordingly, we required high-resolution data to distinguish and characterize areas dominated by riparian vegetation. We obtained SPOT 5 multispectral satellite images (four spectral bands, 10-m pixels) from the archives of the University of California, Santa Barbara's Resource Center for SPOT Imagery. We acquired two sets of 21, 60 × 60-km SPOT scenes that covered the study area. The first set was captured from July through August 2005, when deciduous vegetation was photosynthetically active. The second set was captured during October and November 2007, when deciduous vegetation was without live leaves. The difference in the two image sets allowed us to better discriminate deciduous and evergreen vegetation. We acquired free, orthorectified color infrared digital aerial photographs (three spectral bands, 1-m pixels; 2006) that covered the study area from the Keck image data library at the University of Nevada, Reno (http://keck.library.unr.edu/). A 10-m resolution DEM that covered the study area was acquired from the Natural Resources Conservation Service Data Gateway (http://datagateway.nrcs.usda.gov/).

Analytical and GIS methods

Information on topography, such as variation in slope, sometimes can be used as a surrogate for vegetation heterogeneity at fine resolution. Using the Spatial Analyst extension to ArcGIS and a nearest neighbor function, we estimated topographic roughness as the standard deviation of slope (LANDFIRE) for all cells in a

3 × 3 cell (8,100-m^2) neighborhood. This moving-window function assigns the topographic roughness value to the center cell in each neighborhood.

Our sources of data for modeling and mapping land cover and riparian vegetation were DEMs, multispectral images, and ground data (see figure 7.2).

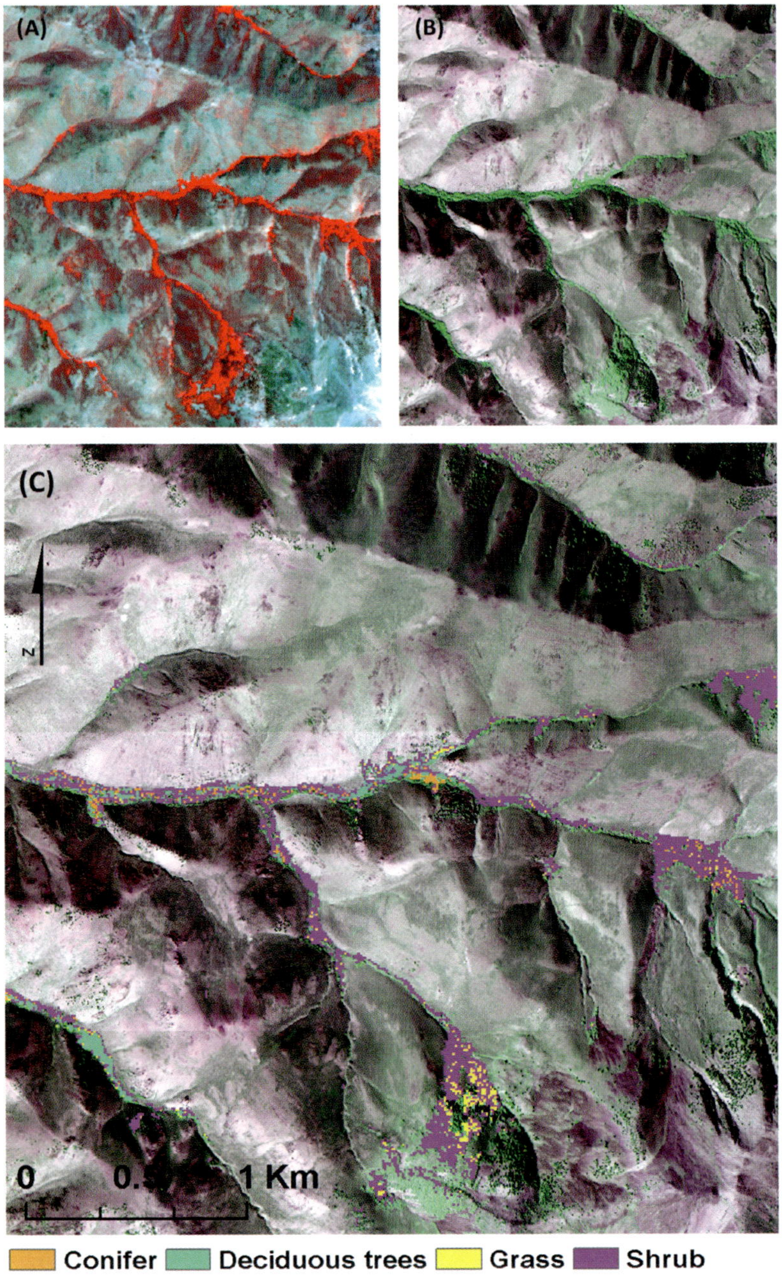

Figure 7.2. Custom land-cover data derived from (A) 10-m resolution SPOT 5 and (B) 1-m resolution color infrared (CIR) imagery was used to classify (C) four dominant vegetation classes within riparian zones in the study area. SPOT 5 satellite imagery from University of California, Santa Barbara (UCSB) Resource Center for SPOT Imagery; CIR digital aerial photography from Keck image data library, University of Nevada, Reno; DEM from US Geological Survey.

We used ArcGIS software to apply a basic image segmentation technique (Baatz and Schäpe 2000) to discriminate riparian areas on the basis of spatial and spectral data as opposed to applying a standard pixel-by-pixel classification to a single image. With the 30-m NED, we first used TPI and hydrological functions within ArcGIS to define canyon bottoms. We used principal drainage areas to define image segments. We orthorectified the SPOT 5 images with the 10-m resolution DEM and Erdas Imagine 9.2 image-processing software. Within canyon bottoms, we applied the Rulequest See5 (C.5) classification tree algorithm (Quinlan 1996; http://www.rulequest.com/) to the SPOT data to discriminate riparian vegetation from nonriparian vegetation classes. Within riparian image segments, or zones, we modeled vegetation class for each pixel on the basis of ground data and the higher spatial resolution color infrared imagery (1-m pixels, as opposed to 10-m for SPOT). We again applied Rulequest See5 to discriminate and map eight dominant vegetation classes, including coniferous trees, grasses, and deciduous trees and shrubs (figure 7.2). We used focal statistics to calculate the percent cover of each class within a moving window of 10 × 10 pixels, resampled to a 10-m pixel to match the spatial resolution of the SPOT data. On the basis of 10-fold cross-validation and error matrices (see Congalton and Green 1999), classification accuracies were generally high (> 90%) with the exception of coniferous trees (75% to 78%), which were sparsely distributed within riparian areas.

We used a moving-window function to estimate the mean elevation, roughness, and woodland canopy cover from values for all cells in a 75-m radius from the center of the survey point. The percentage of cover contributed by deciduous shrubs was estimated as the percentage of 10-m pixels dominated by deciduous cover within a circular moving window of 75-m radius. This procedure, also referred as data smoothing or filtering, can yield a more accurate estimate of habitat quality when the precise location of survey points or cell values is uncertain (see Hunsaker et al. 2001).

Prior to implementing occupancy models, we standardized and rescaled values for all habitat variables to a mean of zero and unit variance (also referred to as a z-score; Neter et al. 1996). Within ArcGIS, we calculated the mean and standard deviation of each habitat variable on the basis of values for all 30-m cells within the entire study area and then used the mean, standard deviation, and the Raster Calculator in the ArcGIS Spatial Analyst for Desktop extension to derive the rescaled value and create a new raster layer for each variable. The rescaling allowed us to evaluate our regression coefficients on the basis of common units. No pair of variables had a Pearson's correlation coefficient > 0.40, and therefore none was excluded from our analyses.

We created a data matrix that related the location of bird detections (i.e., survey points) to the standardized and rescaled value of each habitat variable at that location. We first created a point feature shapefile in the GIS on the basis of the universal transverse Mercator coordinates for each point. Next, we used the Sample tool in ArcGIS Spatial Analyst to intersect the shapefile with the raster data for each habitat variable. This procedure generated a table (.txt file) of values of habitat variables from all raster layers that were spatially related to each survey point. To this table, we also appended values of shrub frequency measured in the field. This new matrix was converted to a comma-separated values (.csv) file, the required input format for the occupancy modeling software.

On the basis of hypothesized relationships with the habitat variables, we estimated probabilities of detection (p) and occupancy (ψ) for each species. To generate a single estimate of p and ψ for 2005–2009, we formulated our candidate models (see sidebar, "Information theory and model selection techniques for estimating habitat quality") of p and ψ with the single-season analysis module in PRESENCE (v2.2), and we conditioned our estimates on the habitat variables. We assumed p varied among seasons (i.e., years), but was constant among the three visits within a year. We also assumed that ψ was constant among visits and among seasons. We focused our inference and spatial modeling on candidate models of ψ that were conditioned on a single best model of p. The occupancy estimation framework we used can accommodate missing observations within seasons (MacKenzie et al. 2002). For comparison, we report naïve estimates of occupancy for each species. We included in our candidate set of models of occupancy a null model that assumed ψ was constant (denoted by [.]) and not influenced by habitat variables. We considered candidate models with ΔAIC values < 4.0 as those that best approximated the data (see Anderson 2008), and Z values > 2.00 to be indicative of a relatively strong association between the response variable and covariate (see Dickson, Fleishman et al. 2009). To simplify this illustration, we do not use model-averaging techniques (but see MacKenzie et al. 2006 and Dickson, Fleishman et al. 2009).

For our statistical models of occupancy, we also considered whether predictor variables had quadratic relationships with occupancy (Zar 1996). We included a squared form of each variable in a given model and explored whether the relation with probability of occupancy was convex (hump-shaped; highest at high intermediate values of the variable) or concave (U-shaped; highest at low intermediate values of the variable) quadratic. We retained these variables only when they improved model fit (ΔAIC > 4.0). For each species, we used the regression coefficients estimated by PRESENCE and the Raster Calculator in ArcGIS Spatial Analyst to build a response surface (henceforth called a spatially explicit model) predicting the relative probability of occupancy at a site during the period of our analysis, which we assumed was positively correlated with habitat quality, at 30-m resolution. We used this model to project habitat quality across a subset of our study area (i.e., a focal area; see figure 7.1) on the basis of the best statistical model of occupancy that did not include ground data. Because the digital data layers describing land cover and topography were spatially continuous, we did not include ground measures of frequency of shrubs in our spatial models of occupancy. However, we retained this variable in our estimates of detection probability, on which the estimates of occupancy also were conditioned.

Modeling results

From 2005 through 2009, we recorded 441 MacGillivray's warbler and 520 spotted towhee detections. On average, we detected 88 (±1SD = 61) MacGillivray's warbler and 104 (46) spotted towhee per year. We estimated probabilities of p and ψ for MacGillivray's warbler as 0.28 (SE = 0.01) and 0.58 (0.04), respectively. The naïve estimate of occupancy was 0.49. Estimates of p and ψ for spotted towhee were 0.27 (0.01) and 0.73 (0.04), respectively. The naïve estimate of occupancy was 0.56.

Our candidate models and analyses of occupancy included a maximum of all five habitat variables (table 7.1). The convex quadratic forms of topographic roughness and percent woodland canopy cover always were more closely associated with occupancy than the linear forms. The form that was more strongly associated with occupancy of either species was consistent for all variables except elevation. The best model of occupancy of MacGillivray's warbler (ΔAIC = 0.0) included all five habitat variables and accounted for > 90% of the model weight. The best model of occupancy of spotted towhee included all variables except percent of deciduous shrubs (essentially a measure of riparian shrub cover), although this model was not clearly better than the next best model (ΔAIC = 1.84) that included all five variables (table 7.1). Overall, candidate model selection uncertainty was higher for spotted towhee. Models in candidate sets for both species that included habitat variables had much greater support than the null model, which assumed occupancy was constant.

Occupancy of MacGillivray's warbler had a strong positive association with the percent cover of deciduous shrubs, followed by high intermediate amounts of woodland canopy cover (convex quadratic form) and frequency of shrubs (table 7.2). Intermediate values of topographic roughness and elevation were more weakly associated with occupancy. Occupancy of spotted towhee was most strongly associated with frequency of shrubs (positive) and elevation (linear form, negative). High intermediate levels of canopy cover and topographic roughness also were relatively strongly (positively) associated with occupancy.

Table 7.2. Regression coefficients ($\hat{\beta}$), standard errors (SE), and Z-statistics (Z) for habitat variables included in the single best model (ΔAIC = 0) (table 7.1) of occupancy for MacGillivray's warbler and spotted towhee in the central Great Basin (Nevada, USA)

	MacGillivray's warbler			Spotted towhee		
Habitat variable	β	SE	Z^1	β	SE	Z^1
Frequency of shrubs	0.482	0.179	2.693	0.668	0.186	3.591
Topographic roughness	0.729	0.285	2.558	0.734	0.265	2.770
Topographic roughness^2	-0.084	0.052	-1.615	-0.105	0.049	-2.143
Percent woodland canopy cover	1.628	0.462	3.524	1.233	0.408	3.022
Percent woodland canopy cover^2	-0.726	0.218	-3.330	-0.392	0.160	-2.450
Elevation	0.862	0.563	1.531	-0.977	0.354	-2.760
Elevation^2	-0.417	0.286	-1.458	--	--	--
Percent cover of deciduous shrubs	2.147	0.594	3.614	--	--	--
Intercept	-1.347	0.425	--	0.562	0.351	--

Source: Brett G. Dickson.
Note: Estimates based on standardization and rescaling of all variables prior to analysis, and conditioned on a single best model of detection probability (table 7.1). For these models, a positive value of a given variable, along with a negative value for the squared form of that variable, indicates the relation with occupancy is convex (hump-shaped).
[1] Computed as β/SE
[2] Squared term indicates quadratic form of variable

Because they did not include ground data, our spatially explicit models of habitat quality (implemented in the GIS across a subset of our study area) did not reflect the best occupancy model in each candidate set, although they included all other habitat variables (table 7.1). Regardless, spatially explicit models had much greater support than the null models. The positive association of percent of deciduous shrubs with occupancy was evident in the model of occupancy of MacGillivray's warbler, and relative probabilities of occupancy were highest within or immediately adjacent to canyon bottoms (figure 7.3). Similarly, spotted towhee occupancy was highest in canyon bottoms, but occupancy was more closely associated with woodland canopy cover than with percent cover of deciduous shrubs (figure 7.3). Compared to MacGillivray's warbler, relative probability of occupancy of spotted towhee also was high in lower-elevation uplands with moderate levels of topographic roughness.

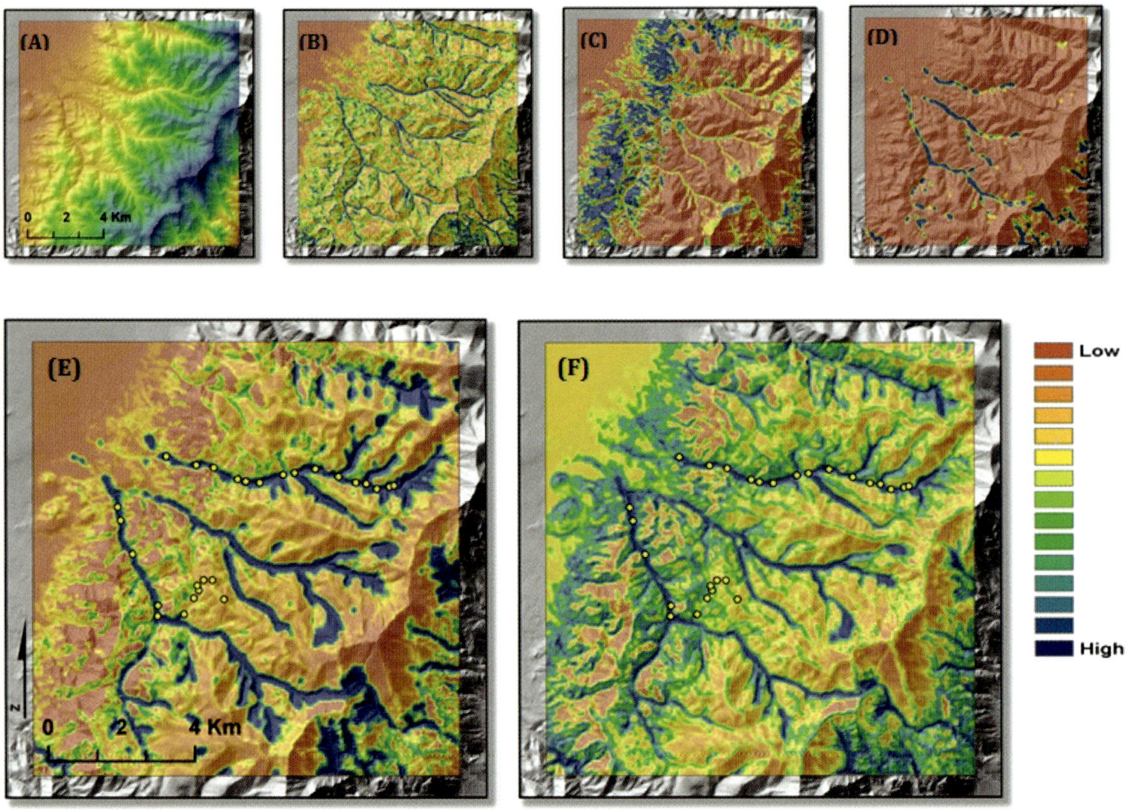

Figure 7.3. Habitat variables derived from digital sources of data: (A) elevation, (B) topographic roughness, (C) percent woodland canopy cover, and (D) percent cover of deciduous shrubs included in models of habitat quality (i.e., probability of occupancy) for (E) MacGillivray's warbler and (F) spotted towhee across the Great Basin focal area. Habitat variables and model outputs were derived at 30-m resolution. Yellow circles indicate bird survey points. SPOT 5 satellite imagery from University of California, Santa Barbara (UCSB) Resource Center for SPOT Imagery; CIR digital aerial photography from Keck image data library, University of Nevada, Reno; DEM and derived hillshade from US Geological Survey.

Discussion

Conservation planning occurs at a variety of spatial and temporal scales. The case study presented here encompassed four mountain ranges and five years. However, our methods are applicable to planning at smaller spatial scales and shorter time periods when objectives include rigorous evaluations of habitat quality for species that are the target of conservation or eradication.

Our methods for more precisely estimating probabilities of detection and occupancy for breeding birds can accommodate small sample sizes. Our estimates of detection probability always suggested detectability was imperfect (i.e., < 1.0). Accounting for detection probability (which was similar for both species), our estimates of occupancy were 18% and 30% higher than the naïve estimates for MacGillivray's warbler and spotted towhee, respectively. Although we chose to generate a single estimate of p and ψ for each species, single-season estimates can be of limited utility when habitat quality changes among years in response to ongoing changes in land use, weather, and climate. Dickson, Fleishman et al. (2009) present dynamic (i.e., multiyear) models of the seasonal occupancy patterns of birds in this system, including estimates of probabilities of colonization and local extinction, and we recommend this approach when sufficient data from multiple years is available.

The model selection approach revealed substantial differences in the relative support for models within and between our candidate set for each species. Only one model for MacGillivray's warbler and two models for spotted towhee had strong support. The better supported models of occupancy of MacGillivray's warbler always included percent cover of deciduous shrubs (derived from digital data), whereas the better supported models of occupancy of spotted towhee always included frequency of shrubs (ground data). Indeed, for MacGillivray's warbler, the habitat-based model of occupancy with the least support was the same as the model of spotted towhee occupancy with the most support (although the elevation variables were of different functional forms). These results demonstrate that two superficially similar species (insectivorous, shrub-nesting, passerine birds) occurring in the same area (e.g., at survey points or canyons) have different habitat associations.

The best models of occupancy always included a habitat variable measured in the field and some variables derived from digital data. Aside from estimates of p, we could not directly incorporate ground data on vegetation into our spatially explicit predictions of habitat quality. Differences in extent of measurements can be problematic for other efforts that must rely on variables measured at all locations across the study area. When a species is more closely associated with ground-based (i.e., fine-resolution) habitat characteristics, digital surrogates should be derived at that resolution. Using ΔAIC and AIC weights, we were able to estimate how much model selection uncertainty was associated with each spatially explicit model of occupancy in the candidate set. For example, the spatially explicit model of occupancy of MacGillivray's warbler was better than the spatially explicit model of occupancy of spotted towhee, but neither model was supported strongly ($< 4\Delta$AIC) relative to the best model for each species.

Potential application to future research

We believe that the thoughtful development of multiple alternative (i.e., competing) hypotheses about animal-habitat relationships is likely to be the most difficult part of any modeling exercise—perhaps as it should be. Although we purposely simplified our candidate set of occupancy models, it is important to strive for parsimony in model building, and consider only those attributes of species or habitat that can be measured with reasonable accuracy and estimated with reasonable precision. These models and methods should account for individuals that are detected imperfectly, which is the case in studies of most animal species.

In light of our model selection results, spatially explicit models of habitat quality are useful for exploring alternative future scenarios. Based on the estimates of individual habitat variables (i.e., regression coefficients) we generated for each occupancy model, we could manipulate one or more digital variables (e.g., reducing woodland canopy cover due to future fire, vegetation treatment, or climate change effects) and re-implement the model in a GIS. For a given area, pre- and postscenario model results can be compared and quantified on the basis of mapped outputs and common GIS tools, such as zonal statistics or histograms. Moreover, model-based probabilistic outcomes can be more intuitive than simplistic habitat quality indices, especially in situations where stakeholders or decision makers interact with spatial data. Probabilities can be easily reclassified in a GIS to produce map-based outputs that reflect multiple classes of habitat quality.

For example, a probability distribution can be presented with a quantile or equal area classification scheme and divided into classes that are defined by meaningful (and interpretable) breaks in the data, rather than by more arbitrary thresholds. Summarized model outputs in map form can highlight high-quality habitat and locations where management and riparian restoration activities may enhance habitat quality or connectivity for spatially overlapping species.

Variables measured in the field typically are not included in spatially explicit models of habitat quality like ours. Although our spatially explicit models had much greater support than the null model, models of habitat quality may not adequately represent animal-habitat relationships at fine spatial resolutions when only remotely sensed or digital data layers are used to build them. Moreover, these relationships can change through time, and not always in ways that are easily measured. Thus, the spatial and temporal scales of animal-habitat relationships should be considered prior to data collection, model building, and inference (Roloff et al. 2009). Because animals often have different habitat requirements during different seasons, submodels of habitat quality for different seasons also should be considered (see Johnston et al., chapter 14). For example, the British Columbia Resources Inventory Committee (Resources Inventory Committee 1999, 2002) developed models of food and cover for as many as four seasons depending on the focal species, whereas life events such as reproduction, migration, and hibernation are associated with both a specific habitat and time of year. This approach was used to develop a conservation area design for the Muskwa-Kechika region of British Columbia (Heinemeyer et al. 2004).

Many types of data on land cover and land use have not been validated extensively in the field or at resolutions that correspond to those at which models for individual species are developed (Corsi et al. 2000). Rigorous, independent validation of statistical models of animal-habitat relationships also is often not emphasized in the scientific literature or in practice. Without such validation, it may be assumed erroneously that models correctly project, for example, occupancy in new locations or time periods. Model projections that are inaccurate or without estimates of uncertainty cannot reliably inform conservation planning activities or management decisions.

Potential application to conservation planning and practice

Models and maps of animal-environment relationships, including habitat quality, are essential tools for modern conservation planning. Although these tools are evolving rapidly, we believe that the approaches and methods presented here are reliable for most applications. There is no substitute for good data, and one should endeavor to identify and incorporate the best available information into models and maps. Robust estimates of habitat quality will better facilitate the provision and conservation of resources necessary to support populations of species as climate changes (see Bachelet, chapter 13) and the human population continues to grow (see Theobald, chapter 5).

References

Anderson, D. R. 2008. *Model Based Inference in the Life Sciences: A Primer on Evidence.* New York: Springer.

Baatz, M., and A. Schäpe. 2000. "Multiresolution Segmentation: An Optimization Approach for High Quality Multi-Scale Image Segmentation." *Journal of Photogrammetry and Remote Sensing* 58: 12–23.

Boyce, M. S., P. R. Vernier, S. E. Nielsen, and F. K. A. Schmiegelow. 2002. "Evaluating Resource Selection Functions." *Ecological Modelling* 157: 281–300.

Bradbury, R. B., R. A. Hill, D. C. Mason, S. A. Hinsley, J. D. Wilson, H. Balzter, G. Q. A. Anderson, M. J. Whittingham, I. J. Davenport, and P. E. Bellamy. 2005. "Modelling Relationships between Birds and Vegetation Structure Using Airborne LiDAR Data: A Review with Case Studies from Agricultural and Woodland Environments." *Ibis* 147: 443–52.

Brotons, L., W. Thuiller, M. B. Araujo, and A. H. Hirzel. 2004. "Presence-Absence versus Presence-Only Modelling Methods for Predicting Bird Habitat Suitability." *Ecography* 27(4): 37–448.

Buckland, S. T., D. R. Anderson, K. P. Burnham, J. L. Laake, D. L. Borchers, and L. Thomas. 2001. *Introduction to Distance Sampling: Estimating Abundance of Biological Populations.* New York: Oxford University Press.

Burnham, K. P., and D. R. Anderson. 2002. *Model Selection and Multimodel Inference: A Practical Information-Theoretic Approach.* 2nd ed. New York: Springer-Verlag.

Carroll, C., R. L. Rodriguez, C. McCarthy, and K. M. Paulin. 2006. "Resource Selection Function Models as Tools for Regional Conservation Planning for Northern Goshawk in Utah." *Studies in Avian Biology* 31: 288–98.

Clevenger, A. P., J. Wierzchowski, B. Chruszcz, and K. Gunson. 2002. "GIS-Generated, Expert-Based Models for Identifying Wildlife Habitat Linkages and Planning Mitigation Passages." *Conservation Biology* 16: 503–14.

Congalton, R. G., and K. Green. 1999. *Assessing the Accuracy of Remotely Sensed Data: Principles and Practices.* New York: Lewis Publishers.

Corsi, F., I. De Leeuw, and A. Skidmore. 2000. "Species Distribution Modelling with GIS." In *Research Techniques in Animal Ecology,* edited by L. Boitani and T. K. Fuller, 389–434. New York: Columbia University Press.

D'Eon, R. G., R. Serrouya, G. Smith, and C. O. Kochanny. 2002. "GPS Radiotelemetry Error and Bias in Mountainous Terrain." *Wildlife Society Bulletin* 30: 430–39.

Dickson, B. G., and P. Beier. 2007. "Quantifying the Influence of Topographic Position on Cougar (*Puma concolor*) Movement in Southern California, USA." *Journal of Zoology* 271: 270–77.

Dickson, B. G., E. Fleishman, D. S. Dobkin, and S. R. Hurteau. 2009. "Relationship between Avifaunal Occupancy and Riparian Vegetation in the Central Great Basin (Nevada, U.S.A.)." *Restoration Ecology* 17: 722–30.

Dickson, B. G., B. R. Noon, C. H. Flather, S. Jentsch, and W. M. Block. 2009. "Quantifying the Multi-Scale Response of Avifauna to Prescribed Fire Experiments in the Southwest United States." *Ecological Applications* 19: 608–21.

Donovan, M. L., D. L. Rabe, and C. E. Olson. 1987. "Use of Geographic Information Systems to Develop Habitat Suitability Models." *Wildlife Society Bulletin* 15: 574–79.

Elith, J., C. H. Graham, R. P. Anderson, M. Dudik, S. Ferrier, A. Guisan, R. J. Hijmans, F. Huettmann, J. R. Leathwick, A. Lehmann, J. Li, L. G. Lohmann, B. A. Loiselle, G. Manion, C. Moritz, M. Nakamura, Y. Nakazawa, J. M. Overton, T. A. Peterson, S. J. Phillips, K. Richardson, R. Scachetti-Pereira, R. E. Schapire, J. Soberon, S. Williams, M. S. Wisz, and N. E. Zimmermann. 2006. "Novel Methods Improve Prediction of Species Distributions from Occurrence Data." *Ecography* 29: 129–51.

Fitzgerald, J. A., W. E. Thogmartin, R. Dettmers, T. Jones, C. Rustay, J. M. Ruth, F. R. Thompson, III, and T. Will. 2009. "Application of Models to Conservation Planning for Terrestrial Birds in North America." In *Models for Planning Wildlife Conservation in Large Landscapes*, edited by J. J. Millspaugh and F. R. Thompson III, 593–624. London: Elsevier.

Fleishman, E., R. Mac Nally, and J. P. Fay. 2003. "Validation Tests of Predictive Models of Butterfly Occurrence Based on Environmental Variables." *Conservation Biology* 17: 806–17.

Fretwell, S. D. 1972. *Populations in a Seasonal Environment.* Princeton, NJ: Princeton University Press.

Gesch, D. B., M. Oimoen, S. Greenlee, C. Nelson, M. Steuck, and D. Tyler. 2002. "The National Elevation Dataset." *Photogrammetric Engineering and Remote Sensing* 68: 5–11.

Guisan, A., and N. E. Zimmermann. 2000. "Predictive Habitat Distribution Models in Ecology." Ecological Modelling 135: 147–86.

Gutzwiller, K. J., and S. K. Riffell. 2007. "Using Statistical Models to Study Temporal Dynamics of Animal-Landscape Relations." In *Temporal Dimensions of Landscape Ecology: Wildlife Responses to Variable Resources,* edited by J. A. Bissonette and I. Storch, 93–118. New York: Springer-Verlag.

Haining, R. P. 2003. *Spatial Data Analysis: Theory and Practice.* Cambridge: Cambridge University Press.

Hall, L. S., P. R. Krausman, and M. L. Morrison. 1997. "The Habitat Concept and a Plea for Standard Terminology." *Wildlife Society Bulletin* 25: 173–82.

Hastie, T. J., and R. J. Tibshirani. 1990. *Generalized Additive Models.* New York: Chapman and Hall.

Hebblewhite, M., and D. T. Haydon. 2010. "Distinguishing Technology from Ecology: A Critical Review of the Use and Abuse of GPS Collar Data in Ecology." *Philosophical Transactions of the Royal Society B-Biological Sciences* 365: 2,303–12.

Heinemeyer K., R. Tingey, K. Ciruna, T. Lind, J. Pollock, B. Butterfield, J. Griggs, P. Iachetti, C. Bode, T. Olenicki, E. Parkinson, D. Rumsey, and D. Sizemore. 2004. *Conservation Area Design for the Muskwa-Kechika Management Area (MKMA).* Nature Conservancy of Canada, Round River Conservation Studies, Dovetail Consulting. http://science.natureconservancy.ca/resources/docs/MK_CAD_V1.pdf.

Henebry, G. M., and J. W. Merchant. 2002. "Geospatial Data in Time: Limits and Prospects for Predicting Species Occurrences." In *Predicting Species Occurrences: Issues of Accuracy and Scale,* edited by J. M. Scott, P. J. Heglund, M. L. Morrison, J. B. Haufler, M. G. Raphael, W. A. Wall, and F. B. Samson, 291–309. Covelo, CA: Island Press.

Hines, J. E. 2006. "PRESENCE2–Software to Estimate Patch Occupancy and Related Parameters." USGS-PWRC, accessed January 2008, http://www.mbr-pwrc.usgs.gov/software/presence.html).

Hirzel, A. H., J. Hausser, D. Chessel, and N. Perrin. 2002. "Ecological-Niche Factor Analysis: How to Compute Habitat-Suitability Map without Absence Data." *Ecology* 83: 2,027–36.

Horning, N., J. A. Robinson, E. J. Sterling, W. Turner, and S. Spector. 2010. *Remote Sensing for Ecology and Conservation: Handbook of Techniques.* New York: Oxford University Press.

Hosmer, D. W., and S. Lemeshow. 2000. *Applied Logistic Regression.* 2nd ed. New York: Wiley.

Hunsaker, C. T., M. F. Goodchild, M. A. Friedl, and T. J. Case, eds. 2001. *Spatial Uncertainty in Ecology: Implications for Remote Sensing and GIS Applications.* New York: Springer-Verlag.

Hunter, M. L. 2005. A Mesofilter "Conservation Strategy to Complement Fine and Coarse Filters." *Conservation Biology* 19: 1,025–29.

Hutto, R. L. 1985. "Habitat Selection by Nonbreeding, Migratory Land Birds." In *Habitat Selection in Birds,* edited by M. L. Cody, 455–76. Orlando, FL: Academic Press.

Johnson, C. J., K. L. Parker, D. C. Heard, and M. P. Gillingham. 2002. "Movement Parameters of Ungulates and Scale-Specific Responses to the Environment." *Journal of Animal Ecology* 71: 225–35.

Johnson, M. D. 2007. "Measuring Habitat Quality: A Review." *Condor* 109: 489–504.

Jones, C. C., S. A. Acker, and C. B. Halpern. 2010. "Combining Local- and Large-Scale Models to Predict the Distributions of Invasive Plant Species." *Ecological Applications* 20: 311–26.

Legendre, P. 1993. "Spatial Autocorrelation: Trouble or New Paradigm?" *Ecology* 74: 1,659–73.

Lancia, R. A., D. A. Adams, and E. M. Lunk. 1986. "Temporal and Spatial Aspects of Species-Habitat Models." In *Wildlife 2000,* edited by J. Verner, M. L. Morrison, and C. J. Ralph, 177–82. Madison: The University of Wisconsin Press.

Leu, M., S. E. Hanser, and S. T. Knick. 2008. "The Human Footprint in the West: A Large-Scale Analysis of Anthropogenic Impacts." *Ecological Applications* 18: 1,119–39.

Lewis, J. S., J. L. Rachlow, E. O. Garton, and L. A. Vierling. 2007. "Effects of Habitat on GPS Collar Performance: Using Data Screening to Reduce Location Error." *Journal of Applied Ecology* 44: 663–71.

Lunn, D. J., A. Thomas, N. Best, and D. Spiegelhalter. 2000. "WinBUGS—A Bayesian Modelling Framework: Concepts, Structure, and Extensibility." *Statistics and Computing* 10: 325–37.

MacKenzie, D. I., J. D. Nichols, G. B. Lachman, S. Droege, J. A. Royle, and C. A. Langtimm. 2002. "Estimating Site Occupancy Rates When Detection Probabilities Are Less Than One." *Ecology* 83: 2,248–55.

MacKenzie, D. I., J. D. Nichols, J. A. Royle, K. P. Pollock, L. L. Bailey, and J. E. Hines. 2006. *Occupancy Estimation and Modeling: Inferring Patterns and Dynamics of Species Occurrence.* San Diego, CA: Academic Press.

MacKenzie, D. I., J. D. Nichols, N. Sutton, K. Kawanishi, and L. L. Bailey. 2005. "Improving Inferences in Population Studies of Rare Species that Are Detected Imperfectly." *Ecology* 86: 1,101–13.

Malczewski, J. 2004. "GIS-Based Land-Use Suitability Analysis: A Critical Overview." *Progress in Planning* 62: 3–65.

Manly, B. F. J., L. L. McDonald, D. L. Thomas, T. L. MacDonald, and W. P. Erickson. 2002. *Resource Selection by Animals: Statistical Design and Analysis for Field Studies.* 2nd ed. London: Kluwer Academic Publishers.

Marcinek, D. J., S. B. Blackwell, H. Dewar, E. Freund, C. Farwell, D. Dau, A. Seitz, and B. A. Block. 2001. "Depth Movements and Muscle Temperatures of Pacific Bluefin Tuna Measured with Ultrasonic Telemetry." *Marine Biology* 138: 869–81.

Martin, T. G., M. A. Burgman, F. Fidler, P. M. Kuhnert, S. Low-Choy, M. McBride, and K. Mengersen. 2012. "Eliciting Expert Knowledge in Conservation Science." *Conservation Biology* 26: 29–38.

Martinuzzi, S., L. A. Vierling, W. A. Gould, M. J. Falkowski, J. S. Evans, A. T. Hudak, and K. T. Vierling. 2009. "Mapping Snags and Understory Shrubs for a LiDAR-Based Assessment of Wildlife Habitat Suitability." *Remote Sensing of Environment* 113: 2,533–46.

McHarg, I. L. 1969. *Design with Nature.* Garden City, NY: The Natural History Press.

Millspaugh, J., and J. M. Marzluff, eds. 2001. *Radio Tracking and Animal Populations.* San Diego, CA: Academic Press.

Mitchell S. C. 2005. "How Useful is the Concept of Habitat? A Critique." *Oikos* 110: 634–38.

Morrison, M. L., and L. S. Hall. 2002. "Standard Terminology: Toward a Common Language to Advance Ecological Understanding and Application." In *Predicting Species Occurrences: Issues of Accuracy and Scale,* edited by J. M. Scott, P. J. Heglund, M. L. Morrison, J. B. Haufler, M. G. Raphael, W. A. Wall, and F. B. Samson, 43–52. Covelo, CA: Island Press.

Neter, J., M. H. Kutner, C. J. Nachtsheim, and W. Wasserman. 1996. *Applied Linear Statistical Models.* 4th ed. Chicago, IL: WCB/McGraw-Hill.

Noon, B. R., K. S. McKelvey, and B. G. Dickson. 2009. "Multispecies Conservation Planning on US Federal Lands." In *Models for Planning Wildlife Conservation in Large Landscapes,* edited by J. J. Millspaugh and F. R. Thompson III, 51–84. London: Elsevier. http://www.rmrs.nau.edu/publications/noon_et_al/noon_et_al.pdf.

Noss, R. 1996. "Protected Areas: How Much Is Enough?" In *National Parks and Protected Areas,* edited by R. G. Wright, 91–120. Cambridge, MA: Blackwell Science.

Osborne, P. E., J. C. Alonso, and R. G. Bryant. 2001. "Modelling Landscape-Scale Habitat Use Using GIS and Remote Sensing: A Case Study with Great Bustards." *Journal of Applied Ecology* 38: 458–71.

Patton, D. R. 1978. *RUN WILD: A Storage and Retrieval System for Wildlife Habitat Information.* USDA Forest Service, General Technical Report RM-51. Fort Collins, CO: Rocky Mountain Forest and Range Experiment Station.

Pelletier, J. 2008. *Quantitative Modeling of Earth Surface Processes.* Cambridge: Cambridge University Press.

Pereira, J. M. C., and R. M. Itami. 1991. "GIS-Based Habitat Modeling Using Logistic Multiple-Regression—A Study of the Mt. Graham Red Squirrel." *Photogrammetric Engineering and Remote Sensing* 57: 1,475–86.

Phillips, S. J., R. P. Anderson, and R. E. Schapire. 2006. "Maximum Entropy Modeling of Species Geographic Distributions." *Ecological Modelling* 190: 231–59.

Resources Inventory Committee (RIC). 1999. *British Columbia Wildlife Habitat Rating Standards. Version 2.0.* Ministry of Environment, Lands and Parks. Resources Inventory Branch for the Terrestrial Ecosystems Task Force. Resources Inventory Committee, Province of British Columbia.

———. 2002. *Vegetation Resources Inventory: The B. C. Land Cover Classification Scheme. Version 1.3.* Ministry of Sustainable Resource Management. Terrestrial Information Branch for the Terrestrial Ecosystems Task Force—Vegetation Resources Inventory Committee. Resources Inventory Committee. Province of British Columbia.

Quinlan, J. R. 1996. "Improved Use of Continuous Attributes in c4.5." *Journal of Artificial Intelligence Research* 4: 77–90.

Rollins, M. G. 2009. "LANDFIRE: A Nationally Consistent Vegetation, Wildland Fire, and Fuel Assessment." *International Journal of Wildland Fire* 18: 235–49.

Roloff, G. J., M. L. Donovan, D. W. Linden, and M. L. Strong. 2009. "Lessons Learned from Using GIS to Model Landscape-Level Wildlife Habitat." In *Models for Planning Wildlife Conservation in Large Landscapes,* edited by J. J. Millspaugh and F. R. Thompson III, 287–320. London: Elsevier.

Royle, J. A., and R. M. Dorazio. 2008. *Hierarchical Modeling and Inference in Ecology: The Analysis of Data from Populations, Metapopulations and Communities.* San Diego, CA: Academic Press.

Salwasser, H., J. C. Capp, H. Black Jr., and F. F. Hurley. 1980. "The California Wildlife Habitat Relationships Program: An Overview." In *Management of Western Forests and Grasslands for Nongame Birds: Proceedings of the Workshop,* R. M. DeGraaf (tech. coord.), 369–78. USDA Forest Service, General Technical Report INT-86. Ogden, UT: Intermountain Forest and Range Experiment Station.

Sanderson, E. W., M. Jaiteh, M. A. Levy, K. H. Redford, A. V. Wannebo, and G. Woolmer. 2002. "The Human Footprint and the Last Wild." *BioScience* 52: 891–904.

Schamberger, M., A. H. Farmer, and J. W. Terrell. 1982. *Habitat Suitability Index Models: Introduction.* FWS-OBS-82-10. Washington, DC: USDI Fish and Wildlife Service.

Scott, J. M., F. Davis, B. Csuti, R. Noss, B. Butterfield, C. Groves, H. Anderson, S. Caicco, F. D'Erchia, T. C. Edwards Jr., J. Ulliman, and R. G. Wright. 1993. "Gap Analysis: A Geographical Approach to Protection of Biological Diversity." *Wildlife Monographs* 123: 1–41.

Scott, J. M., P. J. Heglund, M. L. Morrison, J. B. Haufler, M. G. Raphael, W. A. Wall, and F. B. Samson, eds. 2002. *Predicting Species Occurrences: Issues of Accuracy and Scale.* Covelo, CA: Island Press.

Shriner, S. A, T. R. Simons, and G. L. Farnsworth. 2002. "A GIS-Based Habitat Model for Wood Thrush, *Hylocichla mustelina,* in Great Smoky Mountain National Park." In *Predicting Species Occurrences: Issues of Accuracy and Scale,* edited by J. M. Scott, P. J. Heglund, M. L. Morrison, J. B. Haufler, M. G. Raphael, W. A. Wall, and F. B. Samson, 529–35. Covelo, CA: Island Press.

Siegel, R. B., D. F. Desante, and M. P. Nott. 2001. "Using Point Counts to Establish Conservation Priorities: How Many Visits Are Optimal?" *Journal of Field Ornithology* 72: 228–35.

Stockwell, D. R. B., and D. P. Peters. 1999. "The GARP Modelling System: Problems and Solutions to Automated Spatial Prediction." *International Journal of Geographical Information Systems* 13: 143–58.

Stoms, D. M., F. W. Davis, and C. B. Cogan. 1992. "Sensitivity of Wildlife Habitat Models to Uncertainties in GIS Data." *Photogrammetric Engineering and Remote Sensing.* 58: 843–50.

Sturrock, R. N., S. J. Frankel, A. V. Brown, P. E. Hennon, J. T. Kliejunas, K. J. Lewis, J. J. Worrall, and A. J. Woods. 2011. "Climate Change and Forest Diseases." *Plant Pathology* 60: 133–49.

The Nature Conservancy. 1997. *Designing a Geography of Hope: Guidelines for Ecoregional-Based Conservation in The Nature Conservancy.* Arlington, VA: The Nature Conservancy.

Thomas, J. W., ed. 1979. *Wildlife Habitats in Managed Forests: The Blue Mountains of Oregon and Washington.* USDA Forest Service, Agriculture Handbook Number 553. Washington, DC: US Government Printing Office.

Thomas, L., S. T. Buckland, E. A. Rexstad, J. L. Laake, S. Strindberg, S. L. Hedley, J. R. B. Bishop, T. A. Marques, and K. P. Burnham. 2010. "Distance Software: Design and Analysis of Distance Sampling Surveys for Estimating Population Size." *Journal of Applied Ecology* 47: 5–14.

Thomson, J. R., E. Fleishman, R. Mac Nally, and D. S. Dobkin. 2006. "Comparison of Predictor Sets for Species Richness and the Number of Rare Species of Butterflies and Birds." *Journal of Biogeography* 34: 90–101.

Verner, J., M. L. Morrison, and C. J. Ralph, eds. 1986. *Wildlife 2000: Modeling Habitat Relationships of Terrestrial Vertebrates.* Madison: University of Wisconsin Press.

Vierling, K. T., L. A. Vierling, W. A. Gould, S. Martinuzzi, and R. M. Clawges. 2008. "LiDAR: Shedding New Light on Habitat Characterization and Modeling." *Frontiers in Ecology and the Environment* 6: 90–98.

White, G. C., and K. P. Burnham. 1999. "Program MARK: Survival Estimation from Populations of Marked Animals." *Bird Study* 46(supplement): 120–38.

Williams, B. K., J. D. Nichols, and M. R. Conroy. 2002. *Analysis and Management of Wildlife Populations.* San Diego, CA: Academic Press.

Wolman, A. G. 2006. "Measurement and Meaningfulness in Conservation Science." *Conservation Biology* 20: 1,626–34.

Zar, J. H. 1996. *Biostatistical Analysis.* 3rd ed. Upper Saddle River, NJ: Prentice-Hall.

8

Marine and freshwater conservation planning: From representation to persistence

Natalie C. Ban, PhD; Stephanie Januchowski-Hartley, PhD; Jorge Alvarez-Romero, PhD; Morena Mills, PhD; Robert L. Pressey, PhD; Simon Linke, PhD; and Debora de Freitas, PhD

Most of the science and practice of conservation planning has focused on terrestrial environments (Vance-Borland et al. 2008), just as this book has, with similarly threatened marine and freshwater systems receiving less attention. In contrast to land, anthropogenic changes to marine and freshwater systems are less easily observed and, until recently, the phrase "out of sight, out of mind" aptly applied to the conservation of these systems. As human impacts on marine and freshwater systems have become increasingly apparent (e.g., fisheries collapses), there has been an increasing realization that these ecosystems require protection, better resource management, and conservation (Nel et al. 2009; Sala and Knowlton 2006). Biodiversity in marine and freshwater is declining alarmingly, and humans are starting to feel the effects of reductions in ecosystem services (Worm et al. 2006). Increased conservation of marine and freshwater systems is therefore necessary, and a key part of these efforts will be conservation planning.

The goals of this chapter are to provide an overview of established good practices in marine and freshwater conservation planning, highlight distinctive characteristics of marine and freshwater systems important for their conservation, and place these systems in the broader framework of conservation planning (Margules and Pressey 2000; Pressey and Bottrill 2009). We highlight elements of marine and freshwater systems that differ both from each other and from terrestrial systems, and discuss how these differences influence approaches to planning. Also, despite some differences between aquatic and terrestrial systems, many of the principles, lessons, tools, and techniques for terrestrial conservation planning described in other chapters of this book can be adapted to marine and freshwater systems. We also pinpoint some emerging and cutting-edge conservation issues and practices, but the predominant focus of this chapter is on established good practices.

We focus this chapter on systematic conservation planning for biodiversity at a regional scale. Regions can be variously defined as natural regions (or ecoregions), administrative regions, or combinations thereof (Ferrier 2002). However they are defined, regions provide broad spatial context for conservation actions in particular areas. Although a main focus of this book is on planning for focal species at finer scales of analysis, planning systematically for marine and freshwater ecosystems primarily has been done at broader scales (at the level of an ecoregion, state, province, or country). This chapter discusses regional planning and relates it to finer scales where possible. Our chapter is concerned primarily with aspects of the planning process that promote the representation and persistence of biodiversity (using known and mapped components of biodiversity as surrogates), rather than being based upon focal species per se. As described by Chan et al. (chapter 2), the social and economic aspects of planning are also critical for effective design and application of conservation areas, and should be considered alongside the elements presented here. Our chapter should also be read together with chapters that describe emerging issues in conservation related to ecosystem services (Chan et al., chapter 2) and climate change (Bachelet, chapter 13).

In this chapter, we first provide background on systematic conservation planning. Second, we compare and contrast marine and freshwater systems with their terrestrial counterparts, justifying why a separate chapter on aquatic systems is warranted. Third, we give a short history of marine and freshwater conservation planning. Fourth, as the main focus of this chapter, we summarize best practices for marine and freshwater conservation planning. We use the systematic conservation planning stages outlined by Pressey and Bottrill (2009) to address important considerations for marine and freshwater systems, and expand beyond these stages to include best practices for multiple conservation actions, and integrating planning for terrestrial, marine, and freshwater realms. Finally, we finish with an overall synthesis and highlight some remaining challenges.

Systematic conservation planning

Systematic conservation planning is a framework for achieving explicit and quantitative conservation objectives with limited resources (Margules and Pressey 2000). Characteristics of systematic conservation planning include efficient use of limited resources to achieve explicit conservation objectives (e.g., related to biodiversity, ecosystem services, or livelihoods), defensibility and accountability in the face of competition for natural resources, and flexibility in accommodating opportunities and constraints (Margules and Pressey 2000).

Systematic conservation planning is a synthetic and constantly evolving discipline. Conservation planning must weave together ideas and techniques from the mathematical, physical, biological, social, economic, and political sciences. Approaches to conservation planning are being progressively refined. Description of the planning process has evolved from the initial list of stages by Margules and Pressey (2000) to include social aspects by Pressey and Bottrill (2009). See table 8.1 for the most recent systematic conservation planning stages that are applicable at all scales.

The use of decision support tools has become an important aspect of systematic conservation planning (see also Williams, chapter 12). These computer-based software systems can analyze known and mapped elements of biodiversity and provide options for areas where conservation objectives could be achieved.

Table 8.1. Description of stages in systematic conservation planning (SCP)

SCP stages	Description of stages, from Pressey and Bottrill (2009)
Scoping, context and stakeholders	
1. Scoping and costing the planning process	Decisions are necessary on the boundaries of the planning region, the composition and required skills of the planning team, the available budget, necessary funds in addition to those available, and how each step in the process will be addressed, if at all.
2. Identifying and involving stakeholders	Important stakeholders include those who will influence or be affected by conservation actions arising from the planning process, or be responsible for implementing those actions. Different groups of stakeholders will need to be involved in different ways in specific stages of planning.
3. Describing the context for conservation areas	The planning team describes the social, economic, and political setting for conservation planning, identifying the types of threats to natural features that can be mitigated by spatial planning and the broad constraints on, and opportunities for, conservation actions
Goals	
4. Identifying conservation goals	May begin with agreement on a broad vision statement for the region that is then progressively refined into qualitative goals about biodiversity (e.g., representation, persistence), ecosystem services, livelihoods, and other concerns.
Data	
5. Collecting data on socioeconomic variables and threats	Relevant spatially explicit data will include variables such as tenure, extractive uses, costs of conservation, and constraints and opportunities to which planners can respond. Will also involve predictions about the expansion of threatening processes.
6. Collecting data on biodiversity and other natural features	The planning team will collect spatially explicit data on biodiversity that include representation units (e.g., vegetation types), focal species, and ecological processes. This may extend to ecosystem services (e.g., maintenance of water flows, carbon sequestration).
Objectives	
7. Setting conservation objectives	Involves interpreting goals to define quantitative conservation objectives for each spatial feature (e.g., 2,000 ha of vegetation type 16,500 individuals of each species) and, where necessary, qualitative objectives related to configuration, past disturbance, and other criteria. Spatial data helps visualize and track progress toward reaching the stated goals in number 4 above.
Assessment and application	
8. Reviewing current achievement of objectives	Remote data, and perhaps also field surveys, are used in this stage to estimate the extent to which objectives have already been achieved in areas considered to be adequately managed for conservation.
9. Selecting additional conservation areas	With stakeholders, this stage requires decisions about the location and configuration of additional conservation areas that complement the existing ones in achieving objectives. Factors influencing decisions will include costs, constraints on, and opportunities for, effective conservation.
10. Applying conservation actions to selected areas	Application of conservation actions requires a variety of technical analyses and institutional arrangements to ensure that selected areas are given the most feasible and appropriate conservation management and that areas are prioritized for action when resources are limited
Monitoring and review	
11. Maintaining and monitoring conservation areas	Activities ensure that individual areas are managed to promote the long-term persistence of the values for which they were established. This involves explicit management objectives and monitoring to ensure that management actions are effective.

Note: Feedbacks exist among many of the stages; Pressey and Bottrill (2009) provide an illustration of these feedbacks, which are too many to depict here. Shaded stages are those that, in particular, require different considerations for marine and freshwater systems compared to terrestrial ecosystems. We address these stages in this chapter.

Key examples are Marxan (http://www.uq.edu.au/marxan/), C-Plan (http://www.edg.org.au/free-tools/cplan.html), and Zonation (http://www.helsinki.fi/bioscience/consplan/software/Zonation/index.html).

Although much emphasis in conservation planning has been on the design of reserve networks, a wide variety of actions for protection, management, and restoration are relevant. Increasingly, multiple objective planning frameworks have been applied whereby conservation management objectives have been examined alongside other objectives (e.g., renewable energy, transportation, or coastal hazard mitigation). With this type of planning, there has been a surge of online decision support tools that allow planners to visualize multiple scenarios, including such climate factors as sea-level rise, when considering priorities for conservation (Robbins et al. 2011). Therefore, when we discuss "conservation areas," we refer to areas where the range of interventions, from no-take areas to restoration of degraded systems, might be applied (Sarkar et al. 2006).

Comparison of marine, freshwater, and terrestrial systems

The science and application of conservation in terrestrial systems have greatly influenced thinking about conservation planning. The applicability of this largely terrestrial-based theory to marine and freshwater conservation areas depends, in part, on the commonalities between these systems. Differences in ecological and evolutionary processes do exist, stemming from physical characteristics and processes as well as types and patterns of human impacts (Carr et al. 2003).

Differences between terrestrial and aquatic systems tend to be relative rather than absolute (Avery 2003), and most generalizations about these differences come with exceptions. Nonetheless, compared to marine and freshwater realms, movements of energy, materials, individuals, and so on in terrestrial systems tend to be more "closed." Connectivity between terrestrial populations relies mainly on adjacency and corridors that are discrete and highly predictable (Rosenberg et al. 1997). Overall, terrestrial connectivity is less prevalent and extensive than in marine or freshwater systems. In terrestrial systems, dispersal and recruitment occur mainly at local scales. Unlike aquatic systems, long-distance dispersal of terrestrial propagules or young is relatively uncommon and is generally restricted to wind dispersed spores, bacteria, viruses, pollen, seeds, and some insects (Kinlan and Gaines 2003). Unlike aquatic systems, intraspecific variation between terrestrial populations is common, and wide spatial separation of terrestrial life-history stages is uncommon.

Important characteristics of marine systems

Environments in the sea are temporally more variable than on land and have more extensive flows of energy and materials. Connectedness between populations is determined by currents and patterns of dispersal. Parts of the water column constitute habitat for different marine species, but there is also a great deal of vertical movement of species and life-history stages. Ocean currents move vast volumes of water continuously. Some currents are consistent, having roughly the same patterns year-round. Others change with the seasons, and others are unpredictable. Some species are associated with these currents, or with fronts at their edges, and move as their habitat moves. Therefore, it becomes important to explicitly consider such dynamic environments in conservation planning (Carr et al. 2003; Game et al. 2009).

Where cold, deep currents reach continental shelves, their water masses are forced upward. These upwelling areas bring nutrients from the deep sea to the surface. The high levels of nutrients create very productive regions, which support sea life of all kinds, from phytoplankton and zooplankton to fishes, birds, and mammals. Similarly, where water masses collide (e.g., warm and cold currents), they form fronts, and frontal regions can have features similar to those of upwellings. Upwellings and fronts are some of the most productive parts of the ocean, and hence are very important for both biodiversity conservation and fisheries (Roff and Evans 2002). In general, upwellings are spatially stable, but fronts are not. Some species, such as tunas, follow the fronts. Remotely sensed data, such as sea-surface temperature from satellite imagery, can indicate the locations of frontal systems, and these in turn can be incorporated into conservation planning (Hobday and Hartmann 2006).

Ocean dynamics affect the connectivity of populations as well as their demographics and genetics. A large proportion of marine organisms have early planktonic stages, with larvae floating or swimming for varying lengths of time, and some dispersing long distances. Planktonic phases, coupled with water movements, cause habitats and ecosystems to be much more connected in marine than in terrestrial systems, and recruitment might be decoupled from parental populations. Furthermore, many marine species have multiple life-history stages, which often involve extensive movements among quite different habitats (ontogenetic shifts), adding further connectivity to marine systems (Avery 2003). For example, most fish species have a larval phase, followed by settlement of juveniles in a specific habitat. Upon adulthood, some fishes move to yet other habitats. Migrations and aggregations of marine species during these movements can occur in predictable areas in association with physical (e.g., temperature) or biological (e.g., food) factors; although movement corridors can be related to currents, the corridors can be spatially diffuse. Protecting one marine species implies protecting all of these habitats and ensuring that organisms can safely move between them. Some species require safe passage through a wide variety of habitats and travel up to 18,000 km in their lifetimes (Robbins et al. 2011). Salmon, for example, start their lives as fry in freshwater streams, move downstream to estuarine areas to grow, migrate to the open ocean as adults, and then return to their natal streams to spawn and die (Dittman and Quinn 1996).

Important characteristics of freshwater systems

Freshwater systems are variably open to upstream–downstream, land–water, and groundwater–surface transfers of energy and materials (Pringle 2001). Springs, isolated wetlands, and many lakes fall at the closed end of the spectrum. Physical and biological processes are defined and strongly influenced by four types of connectivity: lateral (continuity between floodplains and rivers, including wetlands and riparian zones); longitudinal (transport of nutrients and sediments and migration and dispersal of invertebrates, fish, and birds along rivers); temporal (regimes of events such as rainfall and flooding); and vertical (linkages between groundwater and surface water). Evolutionary processes are maintained, in part by these connections, as well as by animal behavior (Allan and Flecker 1993), and include genetic differentiation of stocks of migratory fish species and long isolation between species such as freshwater crayfish in high-altitude headwaters.

Recruitment in freshwater populations can be local, but many taxa are highly mobile within and between life-history stages. Species persistence depends on population processes across a wide range of spatial scales, and migratory species connect disparate habitats over great distances (Sedell et al. 1990). Areas occupied by different life-history stages can be quite dissimilar and widely separated for some species. Some species are dependent on both freshwater and terrestrial habitats, and on both freshwater and marine environments. Conservation planning for mammals such as otter and beaver in North America, amphibians such as frogs and toads, and some riparian birds and waterfowl, require systematic conservation planning across freshwater and terrestrial realms. Similarly, anadromous (e.g., salmon) and amphidromous (e.g., stream gobies) fish species require access to both fresh and salt waters, further necessitating considerations of connectivity between these realms when planning for conservation.

A brief history of marine and freshwater conservation planning

Marine and freshwater systems need conservation planning because they are threatened by human activities and are not well protected. Currently, only 1.17% of the world's ocean surface is formally protected (Toropova et al. 2010), compared to 12.9% of the world's land area (IUCN and UNEP-WCMC 2010). The percentage of the world's freshwater ecosystems protected is poorly known. The most commonly cited figure is the estimate from the World Database on Protected Areas (WDPA) that 1.5% of "lake systems" are protected (Chape et al. 2005). Pittock et al. (2008) explained that many spatial datasets lack the resolution to describe the global distribution and extent of freshwater systems. However, it is known that freshwater ecosystems have lost a greater proportion of their species and habitats than those of land or oceans (Abell et al. 2007; Pittock et al. 2008).

Oceans and estuaries are overfished (Pauly et al. 1998), suffer extensively from the effects of invasive species, and are increasingly polluted (Crain et al. 2009; Dudgeon et al. 2006). Marine top predators have declined by more than 90% since 1950 (Myers and Worm 2003), anoxic ("dead") zones are increasing (Diaz and Rosenberg 2008; Dybas 2005), massive coral bleaching is occurring (Carpenter et al. 2008), and climate change is affecting the ocean's chemistry (Orr et al. 2005) and altering major biophysical patterns and processes (Harley et al. 2006).

Freshwater systems are similarly affected. The Millennium Ecosystem Assessment (2005) summarized extensive losses of wetlands and described freshwater ecosystems as being over-used, underrepresented in protected areas, and having the highest proportion of species threatened with extinction. Key threats include dams and water withdrawals that alter natural flow regimes; modifications of riparian and in-stream habitats; and impacts on water quality from sediments, nutrients, and toxins, and invasive species. Direct habitat manipulation and destruction of freshwater ecosystems has resulted from agriculture practices and urban expansion. Subsistence, recreational, and commercial fisheries place further stress on declining species and alter ecosystem functioning (Dudgeon et al. 2006; Saunders et al. 2002). Already, climate change is affecting freshwater ecosystems through altered hydrologic regimes and water chemistry, and increased vulnerability to salinization. Some of the best known examples of related effects on fauna are changes to upstream fish migrations (Daufresne and Boe 2007), reduced fish stocks and productivity (Casselman 2002), reduced species diversity (Abell et al. 2007; Dudgeon et al. 2006), and altered community composition (Wrona et al. 2006).

Marine conservation planning

Marine protected areas (MPAs) are one of the most effective tools for ocean conservation (Halpern and Warner 2002) and have been implemented worldwide to manage fisheries and protect habitats (Boersma and Parrish 1999). Most marine conservation planning theory and practice has been directed toward MPA network design, and hence this topic is a focus of this chapter. However, we recognize that MPAs offer only a partial solution to marine conservation (Allison et al. 1998) and need to be complemented by both spatial and nonspatial measures that abate threats occurring outside MPA boundaries. These complementary measures include fisheries stock assessments, gear restrictions, temporary closures, land management to reduce land-based pollution, and management of invasive species. Ideally, MPAs will be implemented as integrated networks comprising areas with different protection levels that operate cooperatively to meet regional-scale objectives (WCPA/IUCN 2007). In addition, the link between terrestrial, marine, and freshwater systems is increasingly being recognized and incorporated into marine conservation planning (Halpern, Ebert, et al. 2009; Tallis et al. 2008).

Forms of MPAs, such as traditional sacred places or fisheries protection zones, have long been part of traditional societies' management of marine resources. Systematic planning for marine conservation is, however, a much more recent endeavor. No-take areas were first cited as a method of protecting fish populations by Beverton and Holt (1957) and, in 1992, the first comprehensive guidelines for establishing MPAs were prepared (Kelleher and Kenchington 1992). The oldest MPA that still exists and meets the International Union for Conservation of Nature definition is the Royal National Park in New South Wales, Australia, designated in 1879 (MPA News 2002). Beginning in the 1990s, efforts to apply a systematic approach to marine conservation planning began increasing. The first of these took place in the Florida Keys National Marine Sanctuary, USA (Bohnsack 1997; Leslie 2005). Marine conservation planning is gaining sophistication. Initial emphasis was on the protection of complementary and representative habitats and general design criteria (e.g., location, minimum size, and number of MPAs). Considerations have since expanded to explicitly include connectivity between areas (Almany et al. 2009), climate change (Game et al. 2008), land-based threats (Tallis et al. 2008) and the ecological processes that sustain marine biodiversity (Lombard et al. 2007), as well as socioeconomic and political constraints (Ban and Klein 2009). The most commonly cited success story of marine conservation planning is the rezoning of the Great Barrier Reef in Australia, which increased the total area of no-take areas (reserves) from 4.5% to 33%, and addressed many of the considerations outlined above (Fernandes et al. 2005).

Freshwater conservation planning

Some of the earliest published work on systematic conservation planning covered wetlands. The beginning of the field has been linked to two publications by Jamie Kirkpatrick in 1983 (Pressey 2002). One of these (Kirkpatrick and Harwood 1983) demonstrated that an iterative method employing complementarity could be applied to wetlands in Tasmania. A similar application was developed independently for wetlands in coastal New South Wales (Margules et al. 1988), and Pressey and Nicholls (1989) used the same data to show that iterative methods were more efficient at representing wetland plant species than the earlier scoring approaches. Noss et al. (2002) included classified stream reaches in their conservation plan for the Greater

Yellowstone Ecosystem. These early studies focused on representation of species and habitats, and none incorporated hydrological connectivity.

In 2007, some key studies for systematic conservation planning in freshwater systems were published, starting with a new framework for planning freshwater reserves (Abell et al. 2007), and an important study in a very data-poor region, the Madre de Dios Basin in the Amazon (Thieme et al. 2007). Since then, assessments of freshwater biodiversity values and protection of freshwaters through reserves and other management actions have gained momentum and importance within the conservation community. It is widely accepted that protected area networks have fallen short of their potential to conserve freshwater biodiversity (Abell et al. 2007; Pittock et al. 2008). Although terrestrial protected areas have provided some (often incidental) protection to freshwaters, it has been difficult to maintain the integrity of freshwater ecosystems within their boundaries.

Linke et al. (2007) built upon existing national assessments in Australia, such as the Assessment of River Condition (Norris et al. 2001) and the Wild Rivers project (Stein et al. 2002). They were the first researchers to simultaneously combine irreplaceability (conservation value of river sections and their catchments) with ecological condition and vulnerability (likelihood that condition will change negatively without intervention). Their work resulted in the first integrated framework for the conservation of rivers in Australia, providing a platform for prioritizing management and restoration actions. Along with Linke et al. (2007), Moilanen et al. (2008) considered connectivity in their conservation assessment of streams. They, however, not only found an intuitive way to include upstream and downstream connectivity, but also implemented this in the preexisting software package Zonation (Moilanen 2007). Following Moilanen et al. (2008), similar connectivity solutions have been found for the Marxan software (Esselman and Allan 2011; Hermoso et al. 2011). These examples demonstrate how freshwater conservation planning can be operationalized within two well-maintained software packages, both developed originally for terrestrial applications (Moilanen et al. 2005; Possingham et al. 2000). Further applications of systematic conservation planning methods have been used for palustrine wetlands by Ausseil et al. (2010) and for rivers in several parts of the world (Heiner et al. 2011; Moilanen et al. 2008; Rivers-Moore et al. 2011).

Advancing beyond systematic planning for surface waters, Michel et al. (2009) provided the first systematic conservation planning application using groundwater biota as conservation features. Using the protocols for the assessment and conservation of aquatic life in the subsurface (PASCALIS) dataset described by Deharveng et al. (2009), they compared hot-spot strategies to a complementarity-based approach. Including spatial constraints in the reserve design algorithm, their approach highlighted that appropriate methods can yield efficient results, even if the background data is sparse. Michel et al. (2009) also showed that aquifer complexes can be used as a spatial framework for conservation, and that human influences are a critical component of spatial prioritization approaches for groundwater. Collectively, the studies outlined in this section are a solid foundation for best-practice conservation planning for freshwaters (Linke et al. 2011).

Best practices for marine and freshwater conservation planning

In table 8.1 (based on Pressey and Bottrill [2009]), we highlight aspects of marine and freshwater conservation planning that are most different from, or are additional to, considerations for terrestrial systems. These are stages one (scoping and costing the planning process), five (collecting data on socioeconomic variables and threats), six (collecting data on biodiversity and other natural features), seven (setting conservation objectives), and nine (selecting additional conservation areas). We discuss these stages in detail below, and add sections on planning for multiple conservation actions and integrating terrestrial and aquatic planning. Building on current practice, we demonstrate key differences and similarities between terrestrial, marine, and freshwater planning, and the ways in which these differences have shaped approaches to prioritization. Best-practice statements are global generalizations, so there will be exceptions for particular regions. Approaches to planning will also be influenced by data availability, human activities, socioeconomic context, size of planning regions, and governance and jurisdictional issues.

Stage 1: Scoping and costing the planning process

Some specific considerations for marine and freshwater systems are necessary to define the boundaries of the planning region. Other steps in this part of the process, such as deciding on the planning team and budget, do not differ among terrestrial, marine, and freshwater systems, and hence will not be addressed here.

Marine boundaries of the planning region

Boundaries between parts of the ocean with different characteristics are generally diffuse, especially due to the open nature of ecosystems and communities. In principle, marine planning regions would be aligned with bioregions or ecoregions that are relatively homogeneous with respect to physical and biological patterns and processes (Spalding et al. 2007). In practice, marine planning regions are sometimes defined by bathymetry (e.g., division of communities based on depth), coastal features (e.g., bays or estuaries), or the influence of freshwater. Because of the connected nature of marine systems, definition of the planning region should consider the scales of connectivity related to, for example, larval dispersal and migrations. Planning at a bioregional scale is therefore particularly important in marine systems because planners can protect processes that extend over large areas and are spatially and temporally dynamic.

Marine governance and cultural systems are additional considerations in defining marine planning regions, and planners have to weigh these issues against ecological considerations when deciding on regional boundaries (Mills et al. 2011; see sidebar "Delineating the planning region: An example from Kimbe Bay, Papua New Guinea"). For example, planning across multiple jurisdictions can involve high transaction costs, whereas working within boundaries of single governance units such as provinces or traditionally owned inshore waters can facilitate management, enforcement, and legislation of restrictions on resource use. Many

Delineating the planning region: An example from Kimbe Bay, Papua New Guinea

Delineating the planning region—the geographic domain within which areas are evaluated and compared as candidates for conservation action—is among the first steps in any conservation planning exercise (Pressey and Bottrill 2009). The planning exercise by The Nature Conservancy (TNC) in Kimbe Bay (Papua New Guinea), within the Bismarck ecoregion and Kimbe-Witu seascape (Green et al. 2007), illustrates considerations that influence decisions about planning region boundaries. There is no optimal approach to defining the boundaries of a planning region, and planners will face trade-offs between the respective advantages of various criteria (Mills et al. 2010).

TNC started identifying planning regions in the Coral Triangle by defining ecoregions and seascapes. Ecoregions incorporated biological attributes only at a coarse scale, while the seascapes used finer-scale information thought to be suitable for management (Green and Mous 2008). Seascapes were delineated specifically for future conservation planning purposes (Green and Mous 2008).

The Kimbe Bay Marine Protected Area Network planning region is at TNC's smallest scale, used specifically for MPA network planning purposes. This planning region was designed based on both biophysical and socioeconomic considerations (Green et al. 2009). The inner and outer boundaries of the planning region were determined based on biological considerations, including coastal habitats and islands that are important nesting sites for birds and turtles. In contrast, the eastern and western boundaries were based on provincial and community boundaries. These different considerations were used to facilitate implementation and future management (Green et al. 2009).

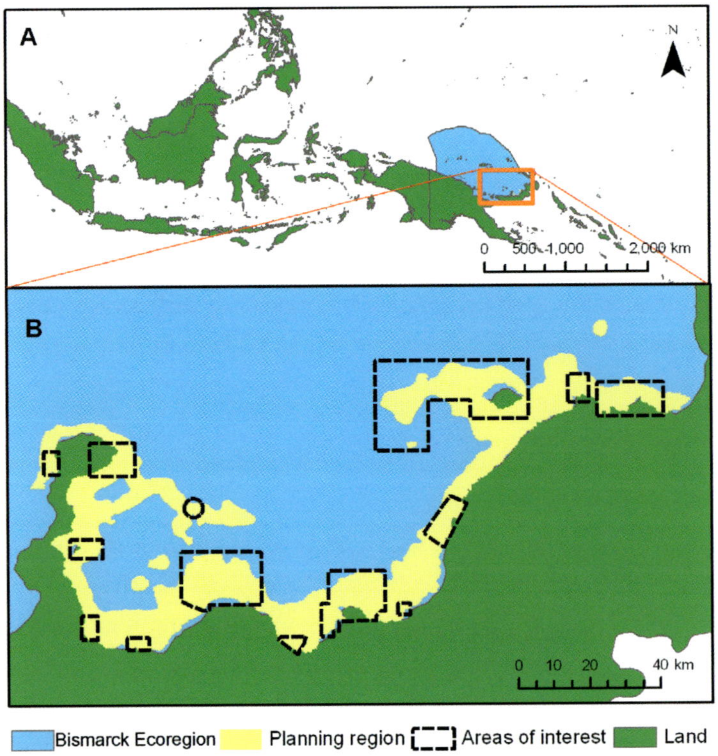

Figure 8.1. Ecoregion and planning region boundaries of Kimbe Bay. A. Bismarck ecoregion. B. Kimbe Bay planning region and areas of interest where attention for conservation was focused. Figure from The Nature Conservancy.

of the case studies of successfully implemented planning exercises cover single governance units (e.g., Great Barrier Reef Marine Park or California State Waters).

Freshwater boundaries of the planning region

For freshwater conservation planning, catchment boundaries define hydrologically sensible planning regions that incorporate linkages from headwaters to river mouths. Catchment boundaries can also be identified more precisely than the often fuzzy boundaries of bioregions or ecoregions commonly used in terrestrial planning. Ideally, the planning region should follow catchment boundaries and integrate terrestrial ecoregional boundaries to account for ecological processes linking both realms (e.g., Amis et al. 2009). When catchment and ecoregional boundaries do not align, planners should aim to account for the entire catchment extent, even if it extends beyond the ecoregional boundary. If planning exercises are restricted by jurisdictional boundaries, such as states or local government areas, planners should acknowledge the outside influence of truncated catchments. Even though catchments define the hydrological reality for freshwater conservation planning, the political reality, important for effective management, is defined by jurisdictions and ownership parcels. Better approaches to integrating these two ways of defining planning regions are needed.

Stage 5: Collecting data on socioeconomic variables and threats

Socioeconomic considerations are discussed in general by Chan et al. (chapter 2). Different types of socioeconomic considerations and data typically used and available for planning in marine and freshwater systems are outlined below. An example incorporating social elements into marine conservation planning from British Columbia is provided in chapter 2 (sidebar "Incorporating social elements into marine conservation planning: An example from British Columbia, Canada"). A key advantage of using systematic approaches is that complementarity-based planning is able to minimize impacts on stakeholders such as fishermen, farmers, indigenous peoples, or foresters. However, some basic difficulties remain in applying these tools for conservation. For example, the impacts of conservation on all stakeholders cannot be minimized simultaneously (Adams et al. 2010), therefore, trade-offs are inevitable, and the relative impacts of different management actions on diverse stakeholder groups are difficult to estimate.

Marine socioeconomic data

Marine areas are usually common property under government jurisdiction, although lack of management responsibility in areas outside national boundaries, in the high seas, is an issue for effective conservation (Ardron et al. 2008). In developed countries, private ownership of marine areas is rare, and ownership boundaries, especially offshore, are usually less distinct and are managed by harvest regulations or through use rights rather than tenure. In many developing countries, community ownership and management responsibility of inshore marine waters is common. Establishment of MPAs rarely implies acquisition of the seabed or water column (usually areas already under government jurisdiction), but does have opportunity costs for diverse stakeholders. Understanding tenure boundaries—for example, the distribution of fishing grounds—in marine areas is challenging because data on these boundaries is scant, and collecting such data can induce disputes. Tenure boundaries can be formal and mapped (e.g., customary tenure in Fiji), formal but unmapped (e.g., customary tenure in Melanesia; Johannes 2002), or informal (e.g., along the southeast coast of Brazil; Begossi 2006). Fishing grounds will also change in response to a range of temporal and environmental factors such as season, salinity, and lunar cycles.

Commonly, socioeconomic data is incorporated into conservation planning as a "cost" (see Chan et al., chapter 2). Decision support tools are then used to achieve conservation objectives while minimizing costs by avoiding areas of high costs where there are spatial options for representing species, habitats, and other features. Ban and Klein (2009) reviewed four approaches to including socioeconomic data in marine conservation planning. First, and most commonly, a uniform cost or planning unit area is used as a proxy for human use. The rationale is that, if objectives can be achieved in the smallest possible area, people would be less affected by conservation actions than if a larger area was designated. Second, opportunity costs are commonly used, usually represented by catches or catch per unit effort of one or more fisheries. Third, some planning exercises have combined seemingly disparate representations of human uses as a cost layer. For example, Green et al. (2009) combined community interest, fish spawning areas, cultural sites, adjacency to conservation areas, dive sites, and shipping channels, among others, to create a cost layer. Finally, measures of naturalness, ecological impact, or threat that represent opportunities or constraints for conservation management also have been used as costs.

Data that informs conservation planners about opportunities and constraints for conservation action increasingly is recognized as crucial. Application of conservation actions ultimately depends on people's support, and support and success can be increased if planners consider information on characteristics such as leadership, local preferences (e.g., areas a community wants to see protected), social capital, political stability, and strength of governance. Opportunities for conservation are reflected by such variables as willingness of people to participate in planning, effective governance of resources, and management infrastructure. Constraints on conservation actions include political instability and conflicts over resource ownership. Explicit use of such data for planning has been more common in developing countries where community support for the implementation of MPAs is crucial (e.g., Kimbe Bay, Papua New Guinea; Green et al. 2009). Ideally, consistent, fine-scale information on opportunities and constraints would be collected as part of extensive census surveys, although interpreting such data for conservation planning presents some difficulties. Alternatives to censuses are special-purpose interviews, which are inevitably limited in spatial coverage.

Data on human-use patterns, coupled with information on the impact of those activities on specific ecosystems, can be used to map threats. Mapping of human activities in the sea and on adjacent land, and the impacts of stressors associated with these activities on specific ecosystems, can be informed by modeling and empirical or experimental data, as well as expert opinion. The first global mapping of marine threats was done by Halpern et al. (2008), supported by data elicited from experts (Halpern et al. 2007). Similar exercises also are now being applied at regional scales, partly to inform conservation actions (Ban and Alder 2008; Halpern, Kappel, et al. 2009; Selkoe et al. 2009).

Freshwater socioeconomic data

Rivers are often in public tenure, so direct acquisition costs for conservation are seldom considered in freshwater planning. However, management costs can be significant, and establishment of conservation management on adjacent lands can present the same constraints as in terrestrial planning. Catchments encompass multiple stakeholder groups competing for finite natural resources. In particular, management of large catchments must deal with complexities of access and ownership, as well as multiple, overlapping jurisdictions and many levels of governance.

Social and economic datasets have only rarely been incorporated into freshwater planning (but see sidebar "Freshwater conservation: An example of planning with management costs and threats in north Queensland"). Exercises in freshwater planning that explicitly considered people and their activities have generally been limited to assessments of condition or future threat (Linke et al. 2007; Nel et al. 2007). These have considered threats in catchments typical of terrestrial planning such as agriculture, urbanization, deforestation, and road density, and some direct threats to fresh waters such as damming, pollution, weeds, and invasive fish. Rarely have measured or predicted flow regime modifications or sediment, nutrients, and pollutants been explicitly accounted for in systematic conservation plans for fresh waters. As in any planning exercises, information is needed in freshwater systems on temporal trajectories of threats so that planners can anticipate and preempt future problems. In addition, consideration should be given to habitats that are more or less vulnerable to threats (Hermoso et al. 2011; Pringle 2001). Land use—urban, industrial, or agricultural—and its associated demands and disturbances are key impacts on freshwater biodiversity (Pringle 2001). More generally, freshwater planning would benefit from additional information on the costs of establishing (for example, Januchowski-Hartley et al. 2011) and maintaining conservation management, and opportunities and constraints related to ownership and governance.

Freshwater conservation: An example of planning with management costs and threats in north Queensland

Invasive species are a major threat to freshwater ecosystems. Common methods used to identify management priorities for invasive species consider the value of biodiversity and ecosystems, but rarely account for costs associated with management actions. Existing methods range from "best guesses" by managers to identifying priorities more explicitly with, for example, use of geographic information systems to overlay environmental attributes (e.g., vegetation cover or soil types), and decision support tools that use scoring approaches to identify priorities based on a range of criteria. We refer to these methods as nonsystematic because they are not based on explicit objectives and do not evaluate the combined contributions of management actions toward those objectives.

A study in the Tully-Murray catchment in north Queensland, Australia (figure 8.2), demonstrated the benefits of using a systematic approach and accounting for management costs to prioritize actions for invasive species management in wetlands and stream reaches (Januchowski-Hartley et al. 2011). The target invasive species was olive hymenachne (*Hymenachne amplexicaulis*). Olive hymenachne is a robust rhizomatous perennial grass causing major impacts on wetlands, streams, native freshwater species, and agricultural production in northern Australia. The objectives of the study were to minimize (1) management costs, (2) spread of this invasive species, and (3) likelihood of reinvasion. Management costs were identified for three types of management actions (spray, amphibious vehicle + spray, and re-vegetation). The type of action used is dependent on the physical conditions and the density of olive hymenachne. Cost estimates were based on time requirements, travel, and number of workers. The study analyzed two scenarios that differed in their consideration of upstream connectivity of wetlands and stream reaches, and the potential for reinfestation (figure 8.2). The difference in direct management cost was negligible (about $20,000 AUD), with scenario 2 being slightly more expensive. However, the solution for scenario 2 had 50% more upstream connections and, therefore, a much greater probability of reinfestation of managed areas and much larger long-term cost. The solution for scenario 1 contained more upstream planning units in the southeast and northeast corners of the study area (indicated by the three black bolded boxes) than the solution for scenario 2. This difference reduced the potential for reinfestation in scenario 1 and lowered the long-term management cost.

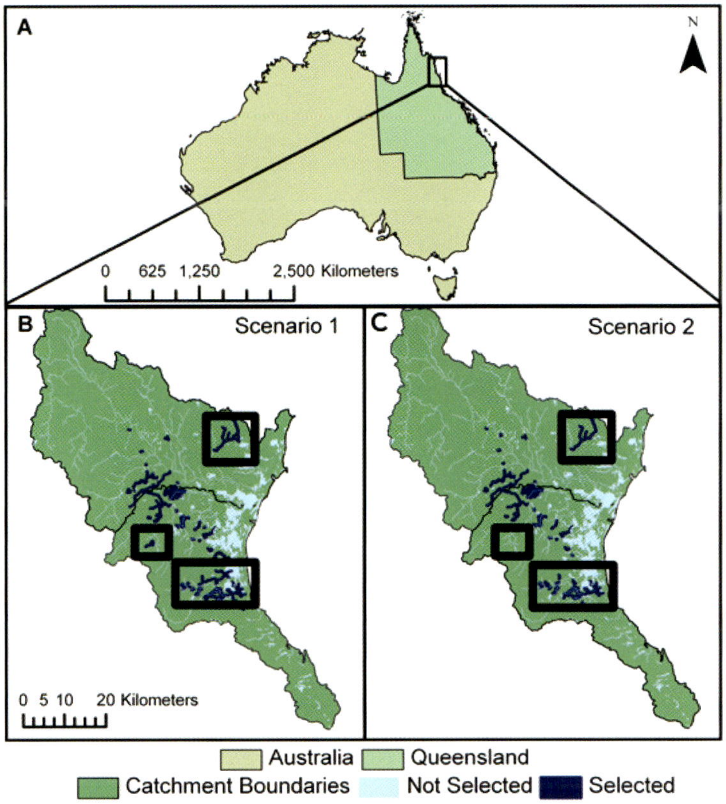

Figure 8.2. The Tully-Murray catchment study area. (A) Location of catchment study area in north Queensland, Australia. (B) Scenario 1 identified priority areas for management, considering connectivity. Specifically, this scenario selected areas for management while attempting to minimize the number of upstream connections to unmanaged areas, thereby reducing potential for reinfestation. (C) Scenario 2 identified priority areas, ignoring connectivity. Figure from Januchowski-Hartley et al. (2011).

A common impediment to systematic conservation planning is the difficult transition between design (stage 9 in table 8.1) and application of actions (stage 10; Pressey and Bottrill 2009). For this transition to be effective, much of the groundwork must have been done in earlier stages of the process, in particular, so that key stakeholders are informed and have a sense of ownership of the process. Freshwater planning is challenging because longitudinal, lateral, and vertical connectivity involves many users of natural resources. Yet these users often have distant connections to streams, lakes, wetlands, and groundwater of immediate concern for conservation, but not necessarily to the users. For example, a wide range of stakeholders will have to be involved if the proposed approach of Abell et al. (2007) is applied (which they call "three hierarchical zones"). Abell et al. (2007) divide freshwater protected areas into three zones—all with different management objectives and levers. The first zone is the freshwater focal area, the location of a specific freshwater feature requiring protection (e.g., richness or endemism hot spots, or the entire habitat of a range-restricted species or assemblage). The second zone is a critical management zone, where management is essential for maintaining functionality of a focal area (e.g., a wetland essential for regulating downstream water flows). The third zone is the catchment management zone, the entire upstream catchment of a critical management zone. The aim of this zone is to have whole-of-catchment best management practices that encourage actions such as maintenance of riparian zones, setting wastewater standards, and encouraging minimal impacts from other land uses such as agriculture and mining. The latter two zones would not necessarily be small or localized. Therefore, for approaches such as the three hierarchical zones to be effective, involvement and support is required from a range of stakeholders that affect, or are affected by, both surface and groundwater.

Stage 6: Collecting data on biodiversity and other natural features

Spatial coverage of biological data is typically patchy and biased for any realm, but more so for aquatic than terrestrial ecosystems. Mapping of terrestrial habitats, such as vegetation types, soil landscapes, geological units, and topographic categories as surrogates for biodiversity patterns, is far more extensive and detailed than for marine and freshwater realms. Narrow endemism is relatively common in terrestrial organisms, and the distributions of many terrestrial species are well studied and predictable (Avery 2003; Carr et al. 2003).

Below we describe the kinds of data typically available for marine and freshwater conservation planning. We assume that the goals of the planning process involve representation and persistence of biodiversity, and hence we subdivide the section below into planning for biodiversity patterns and planning for biodiversity processes. By patterns, we mean the snapshots of biodiversity, such as species occurrences and maps of ecosystems that are conventionally used in systematic conservation planning. By processes, we mean the many kinds of dynamics, across a wide range of temporal and spatial scales, on which the persistence of biodiversity depends. Examples are the dynamics of populations in patches of habitat, movements of organisms between patches of habitat, patch dynamics of disturbance and resources, migration, and ongoing diversification of lineages.

Data on biodiversity patterns

Marine biodiversity patterns

Patchiness and spatial biases of data on species distribution and abundance are accentuated in marine systems because of high ecological variability and difficulties involved in field surveys. Marine biodiversity data is most commonly concentrated around easily accessed areas such as coastal population centers and research stations. Very little is known about deeper marine habitats, especially those below 30 m (the recreational scuba diving limit). Data on marine species is mostly restricted to well-known taxonomic groups, including marine mammals, seabirds, sea turtles, fish, macroinvertebrates, seagrass, and macroalgae. Therefore, available data does not represent marine biodiversity per se, but rather serves as surrogates for biodiversity (Roberts et al. 2003). In these respects, data limitations are similar to those in terrestrial systems, but sparseness and both taxonomic and spatial biases are accentuated in the sea.

Knowledge of geographic patterns of ecosystem diversity, especially for offshore and deep environments, is relatively poor. Habitat classification maps rely more on geophysical features and are less extensive and detailed than on land (see sidebar "Great Barrier Reef Marine Park: An example of planning for representation"). A hierarchical ecological approach using abiotic factors was outlined by Zacharias and Roff (2000). Ardron (2002) outlined an approach for modeling benthic complexity using GIS. In another example, Airamé et al. (2003) defined marine habitats for conservation planning in relation to several physical and biological variables. These were sediment types (soft and hard) from sampling surveys, depth intervals defined as shoreline, photic zone (0–30 m), upper continental shelf (30–100 m), lower continental shelf (100–200 m), continental slope (> 200 m), and dominant assemblages of primary producers mapped from surveys and aerial photographs (giant kelp, surfgrass, and eelgrass). Using similar methodology, a report on marine regional assessment methods for The Nature Conservancy by Ferdaña (2006) includes a section entitled "Building a Benthic Habitat Model as Surrogates for Ecosystem-Scale Targets." Shoreline habitat types are also commonly incorporated as linear features to represent associated species and communities. These have been classified based on landform, substratum, exposure, slope, and vegetation across the intertidal zone. In intertidal marine areas, finer-resolution data has been found to better represent spatial variability in biodiversity (Banks et al. 2005).

Great Barrier Reef Marine Park: An example of planning for representation

The Great Barrier Reef Marine Park was the first regional-scale marine park to be systematically planned and implemented. The planning region covers 344,400 km^2. The planning process reconfigured multiple zonings that included a connected system of no-take areas (NTAs) that adequately and comprehensively represented the bioregions within the Great Barrier Reef region (figure 8.3A). Designed to ensure both representation and persistence of biodiversity, the planning exercise applied theoretical design principles discussed in the marine conservation literature, defining both quantitative and qualitative objectives (Fernandes et al. 2005). Over 70 bioregions (30 reef and 40 nonreef) were identified, and planners set objectives of at least 20% representation of the extent of each bioregion in NTAs (Lewis et al. 2003).

Qualitative objectives included rules for representation of bioregions across longitudinal and latitudinal gradients (Fernandes et al. 2005) to capture as much of their unmapped biophysical heterogeneity as possible. Qualitative design criteria determining size and spacing of NTAs were intended to ensure the persistence of ecological and evolutionary processes within the marine park. For example, where possible, NTAs were to be at least 20 km along their smallest dimension. Detailed information on the connectivity of areas for larvae of fish, corals, and other organisms was not, and is still not, available. However, recent analyses indicate that the size and spacing of NTAs should be adequate to ensure larval connectivity. NTAs connected by less than 50 km are likely to be connected demographically for most marine species. This should allow for the persistence of metapopulations because local extinctions of populations can be offset by recruitment from adjacent populations (Almany et al. 2009).

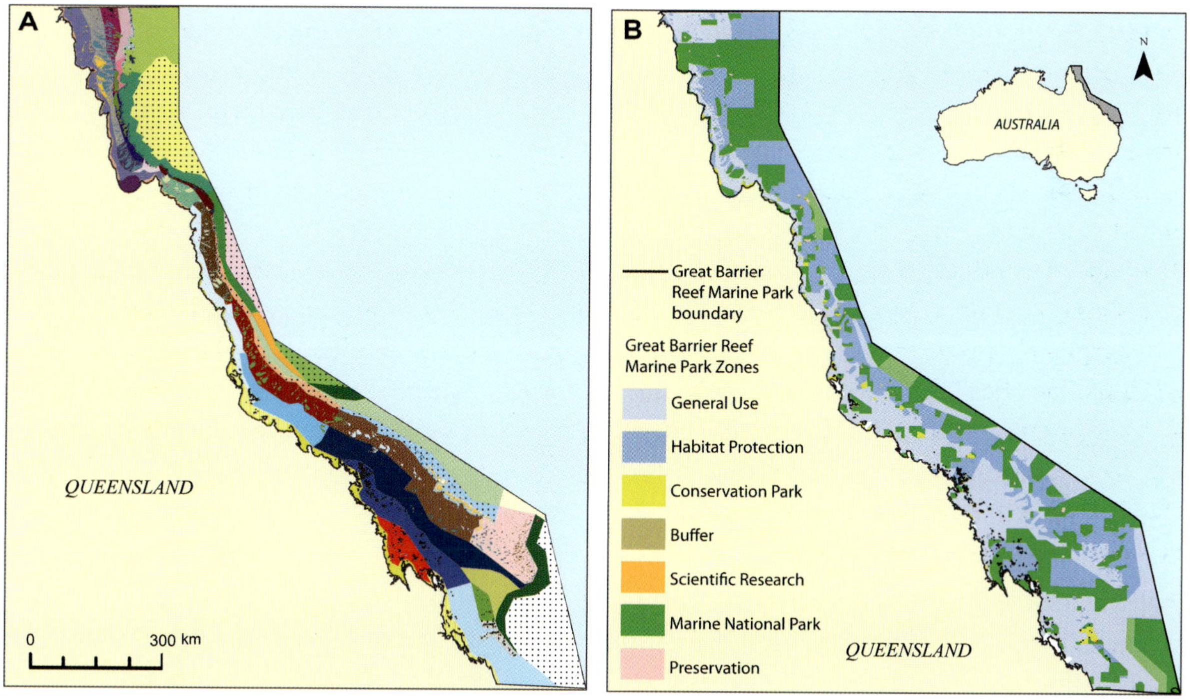

Figure 8.3. Bioregions and zoning of the Great Barrier Reef Marine Park. (A) Reef and nonreef bioregions identified within the Great Barrier Reef Marine Park. Please refer to the Great Barrier Reef Marine Park Authority for legends (http://www.gbrmpa.gov.au/__data/assets/pdf_file/0011/17300/nonreef-bioregions-in-the-gbrmp-and-gbrwh.pdf). (B) Great Barrier Reef Marine Park zones. For more information, refer to http://www.gbrmpa.gov.au. © Commonwealth of Australia (Great Barrier Reef Marine Park Authority).

To minimize the costs of management actions associated with the rezoning of the Great Barrier Reef Marine Park, there was extensive consultation with stakeholder groups and the community. Spatial data was collected as part of this endeavor, including the opportunity costs of NTAs to commercial fisheries and the values of areas to different interest groups. This data was incorporated into the selection of management areas not only to minimize costs of imposing stronger fishery regulations in many areas (figure 8.3B), but also to strengthen compliance and support for regulations. For example, places of special aesthetic, cultural, and historical value, as well as ecological values, were targeted for maintenance by NTAs. To minimize enforcement costs and maximize compliance, fewer and larger NTAs and other protection zones were preferred. Adjacent land uses were also considered when choosing areas for conservation to maximize the natural integrity of the conservation areas and minimize their exposure to land-based threats (Fernandes et al. 2005).

As well as representing different types of habitats and species, marine conservation planning should, where possible, incorporate areas and habitats important for the persistence of species at particularly vulnerable life-history stages. Examples are spawning aggregations, nursery areas, and roosting and nesting sites. For instance, MPA design for the Prince Edward Islands in the Southern Ocean included data on seabird and elephant seal foraging areas (Lombard et al. 2007). Where data on vulnerable life-history stages is not available, it can be modeled. Such habitat modeling is quite common for marine mammals, and examples include modeling cetacean habitats in the California Current (Reilly et al. 1997), a species-habitat model for large whale species developed by Gregr and Trites (2001) for coastal British Columbia, and a predictive model for fin whales in the Mediterranean (Laran and Gannier 2008). Demersal fish species have similarly been modeled (Leathwick et al. 2008). Where data is not sufficient for modeling, scientific experts or local communities can provide knowledge.

The tridimensional distribution of marine species and ecosystems also has important implications for marine conservation planning. Although some planktonic organisms and larvae are subject to lateral transport by ocean currents, others are active swimmers that move vertically as well as laterally. Many benthic and sessile marine species, such as corals, have an early pelagic phase. Yet the ecological linkages between pelagic and benthic environments are poorly understood, as are the influences on marine biodiversity patterns of both vertical and lateral ecological and oceanographic processes. Compilation of data on the tridimensional distribution of organisms is just beginning and has yet to be incorporated into marine planning (Game et al. 2009).

Freshwater biodiversity patterns

Limitations of data on freshwater biodiversity are related to spatial and taxonomic biases in species records, limited information on the distribution and abundance of most species, and poor knowledge of habitat requirements, including changes in habitat associations between different life-history stages. Data on freshwater biodiversity is concentrated at stream crossings, boundaries of protected areas, and human population centers. Data on taxa also is biased toward vertebrate and plant species occurring along and within streams and rivers. Wetlands and subterranean environments are some of the most poorly sampled freshwater ecosystems (Abell et al. 2007; Dudgeon et al. 2006), and algae, diatoms, crustaceans, and macroinvertebrates are poorly sampled taxonomic groups.

Similarly to terrestrial and marine systems, it is impossible to directly measure or map all facets of biodiversity in freshwater ecosystems. Therefore, available data on particularly well-known taxonomic groups and ecosystems is used as surrogates for conservation planning (Noss et al. 2002; Sowa et al. 2007). Detailed studies about the conservation needs of freshwater-dependent species have seen little application. One example includes a conservation assessment of the Madison Valley in Montana, where habitat for boreal toad (*Bufo boreas boreas*), Columbia spotted frog (*Rana luteiventris*), cutthroat trout (*Oncorhynchus clarki lewisi*), fluvial Arctic grayling (*Thymallus arcticus montanus*), and riparian birds were used as biodiversity surrogates (Brock et al. 2006).

Most freshwater conservation plans (Nel et al. 2009; Thieme et al. 2007) and gap analyses (e.g., Sowa et al. 2007) rely heavily upon coarse-filter surrogate methods (e.g., habitat classifications or geophysical characteristics; Thieme et al. 2007). Thieme et al. (2007) derived classes of stream reaches and drainage basins focusing on ecological functionality. They used several hydrogeomorphic characteristics to define classes: system type (e.g., headwater versus pass-through basin, small versus large streams, or floodplain type), elevation, slope, geology, vegetation, and hydrology (e.g., surface runoff or river discharge). The resulting classes were used as surrogates to represent common species and communities in Madre de Dios Basin, Brazil. Despite the heavy reliance on coarse-filter surrogates for freshwater conservation planning, their effectiveness has not been assessed. Therefore, although habitat mapping and classification are important to conservation planning, data on species occurrence, distribution, or abundance should be used wherever possible to complement habitat mapping (Nel et al. 2009).

With a few exceptions (Ausseil et al. 2010; Michel et al. 2009), freshwater systematic conservation planning has focused on patterns (and less so on processes) of river ecosystems. Usually wetlands and groundwater ecosystems are either omitted, or included only for their contribution to ecosystem processes (Nel et al. 2011). Nel et al. (2011) were unable to map wetlands confidently, and, therefore, wetlands were only recognized for their functional role in the persistence of river biodiversity; it is not known whether explicit inclusion of wetlands for their biodiversity would have changed the results. Where data exists, riverine, wetland, and groundwater biodiversity patterns should all be included because they are not adequate surrogates for one another. However, including degrees of connectivity, interdependence, and likely overlap of species composition between different freshwater ecosystems is a major challenge. Further research is needed to account for connectedness of freshwater systems because the persistence of their biodiversity depends on managing connectivity to maintain biological processes and mitigate threats.

Data on biodiversity processes

Marine biodiversity processes

Most marine planning has focused on representation of biodiversity patterns rather than the persistence of processes on which biodiversity ultimately depends (Pressey et al. 2007). As in terrestrial systems, marine ecological and evolutionary processes are often poorly represented by conventional spatial biodiversity surrogates. In conservation planning, some marine processes are occasionally considered, with information on minimum and maximum dispersal distances for well-known species guiding objectives for size and spacing of MPAs. Replication is generally applied as a generic design strategy to spread the risk of disturbances, for example, from storms and coral bleaching. Aggregation sites of species of special interest and other potentially vulnerable and important areas are targeted when data is available.

Connectivity, fluctuations in resources, and disturbances are important processes for planners to consider in marine systems, just as they are in freshwater and terrestrial systems. Marine connectivity is a rapidly expanding field, with advances being made on investigating larval dispersal through hydrodynamic modeling (Kool et al. 2010) and genetic parentage analysis (Jones et al. 2009). These advances can be used to improve design criteria for marine protected areas (Almany et al. 2009; see section on stage 9). We have discussed resource dynamics related to upwellings and ocean fronts in the section comparing marine and terrestrial systems. Examples of disturbance processes in marine systems include strong cyclonic waves that break fragile and sessile species such as sponges and corals in shallow water, and can remove many species from stretches of shoreline. Coral bleaching, often associated with prolonged high water temperatures, is another type of disturbance. Conservation planning should take disturbances into account so that individual MPAs or networks of MPAs have minimal likelihood of being completely affected by single disturbance events (Allison et al. 2003). An MPA system, for example, can be designed to minimize the probability of coral bleaching across the reefs it contains (Game et al. 2008).

Few datasets exist on marine processes (see sidebar "Planning for marine ecological processes: An example from the Southern Ocean"), although spatial and temporal datasets are available for some processes. Oceanographic studies have provided broad-scale data on ocean circulation. Insights into other important processes are available from satellite imagery. Sea-surface temperature is monitored with several types of imagery (e.g., moderate resolution imaging spectroradiometer [MODIS], sea-viewing wide field-of-view sensor [SeaWiFS], and advanced very high resolution radiometer [AVHRR]) to identify, for example, upwellings, high-temperature anomalies related to risk of coral bleaching, and fronts between cold and warm water. Similarly, some plankton detected with satellite imagery through ocean color (e.g., MODIS and SeaWiFS), provide a proxy for ocean productivity and, when combined with temperature and other data, allow the characterization of primary productivity dynamics (Behrenfeld and Falkowski 1997). Time-series data on these phenomena allows assessments of their temporal and spatial predictability in different areas, which in turn can inform the design of static and dynamic MPAs.

Planning for marine ecological processes: An example from the Southern Ocean

The importance of processes such as migration, metapopulation dynamics, and diversification of lineages to maintain biodiversity has long being been recognized (Cowling et al. 1999; Lombard et al. 2007; Pressey et al. 2007; Rouget et al. 2006), but their explicit incorporation into marine planning is still uncommon. One viable and practical option is to represent spatially processes of interest so that they can be considered explicitly in conservation planning. Following this approach, Lombard et al. (2007) mapped and used information on marine processes during the design of the Prince Edward Islands Marine Protected Area in the Southern Ocean. One of the goals of the planning process was to reduce bycatch of albatross and petrels associated with longline fishing. Another goal was to protect key ecological processes that link marine and terrestrial environments, in particular, the nutrient cycle driven by birds and seals, which sustains the unique terrestrial and coastal biodiversity of the islands.

The Prince Edward Islands are influenced by three major fronts, or areas where different water masses meet: the Sub-Antarctic Front (SAF) to the north of the islands, the Antarctic Polar Front (APF) to the south, and the southern SAF (SSAF) between the other two. Fronts are associated with high productivity, including high biomass of plankton and fish, and drive key ecological processes that sustain the marine and coastal biodiversity of the region, including the input of marine-derived nutrients to the islands by pelagic foragers.

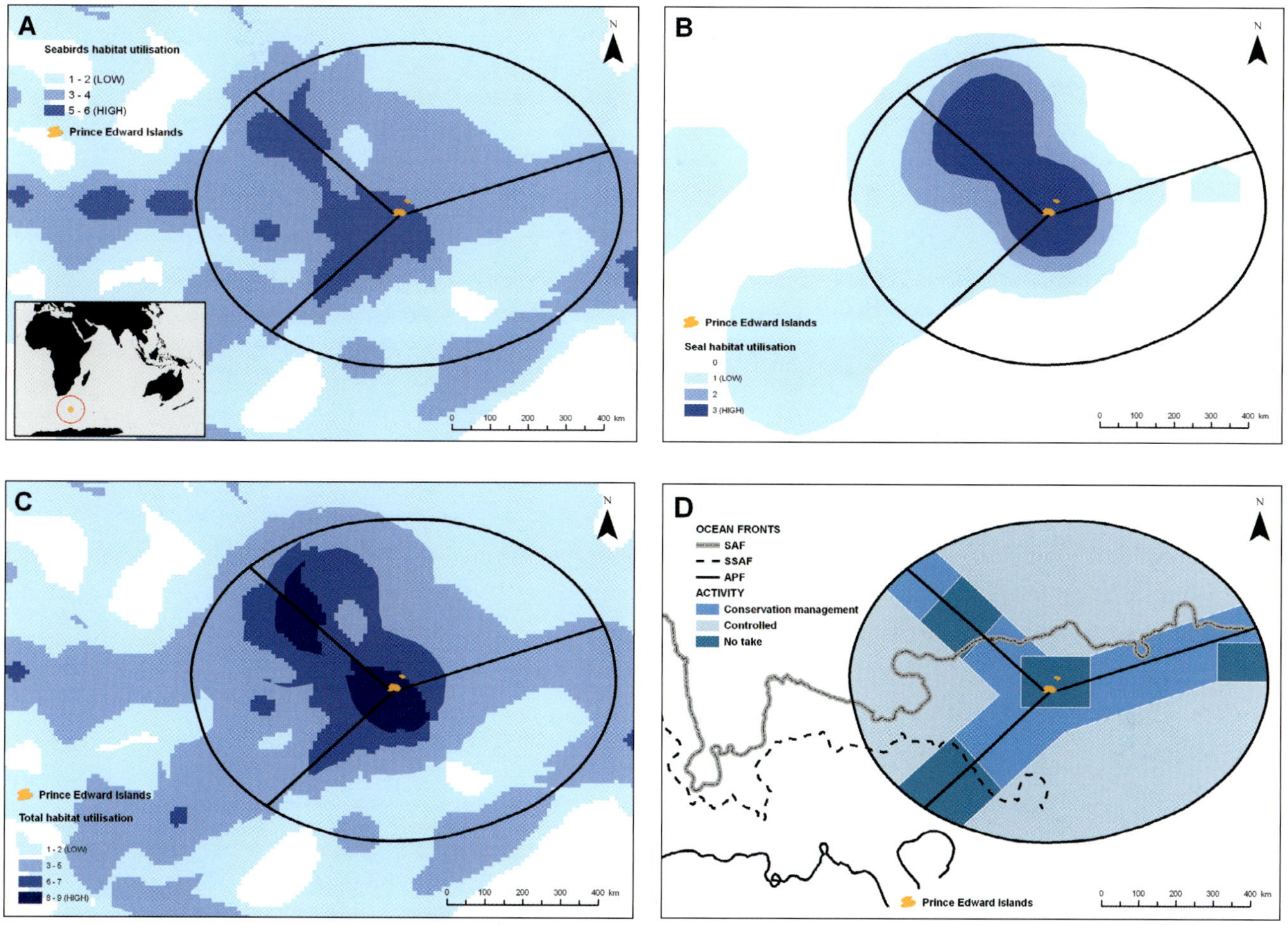

Figure 8.4. Prince Edward Islands. (A) Seabird habitat utilization, (B) seal habitat utilization, (C) integrated habitat utilization, and (D) boundaries of the proposed marine protected area. Figure adapted from Lombard et al. (2007).

Ocean areas used by pelagic foragers were identified from satellite tracking of seal and albatross. Habitat utilization was then calculated using kernel density estimators. The results were polygons containing specified percentages (i.e., 50%, 51–75%, and 76–95%) of densely clustered satellite points and representing different levels of utilization (figures 8.4A and 8.4B). An integrated habitat utilization map (figure 8.4C) was constructed by overlaying the maps of both species and identifying areas of low usage (minimum records of only one species) to high usage (maximum records of both species). This map was then used to delineate major axes of foraging movements. In addition, the average position of fronts was calculated based on subsurface (200-m depth) temperature data points derived from cruises (1989–2005), converted into raster grids and contoured. Four core reserves (dark blue) were delineated based on benthic habitats, major water bodies, and species richness. These were then connected with straight lines to incorporate the three major movement axes of pelagic foragers and simultaneously maximize coverage of the average position of oceanic fronts within conservation zones. Boundaries of the proposed conservation management and no-take zones were drawn as straight lines, and boundaries between zones were placed on exact degrees and minutes to facilitate navigation and compliance (figure 8.4D).

Work underway to incorporate oceanographic dynamics into conservation planning has used time-series data retrospectively. This has strengths and shortcomings. On the positive side, retrospective data allows planners to assess how well-established MPAs have achieved objectives related to dynamics over time (e.g., Ban et al. 2012) and to demonstrate how design of MPAs must change to accommodate dynamics (e.g., Grantham et al. 2011). In the latter case, individual time slices of variables of interest were analyzed as separate data layers, each with its own objective. The shortcoming of these retrospective assessments is that planning must ultimately be for the future. Forward-looking assessments of dynamics must therefore rely on simulations into the future, and these inevitably involve considerable uncertainty.

Freshwater biodiversity processes

Systematic conservation planning for freshwaters, like that for terrestrial and marine environments, has been strongly focused on representation of biodiversity pattern and done a poor job of accounting for processes (but see Roux et al. 2002). In freshwater ecosystems, as in marine and terrestrial systems, ecological and evolutionary processes are often poorly represented by conventional spatial surrogates for biodiversity. The following sidebar, "Large-scale freshwater conservation planning in a data-poor region: Madre de Dios Basin, Amazon," gives an example of planning for freshwater processes with limited data.

Large-scale freshwater conservation planning in a data-poor region: Madre de Dios Basin, Amazon

The Madre de Dios Basin region, covering approximately 160,000 km^2, is defined by the drainage basins of the Madre de Dios and Orthon Rivers of Peru and Bolivia in the southwest Amazon (figure 8.5). The Madre de Dios planning exercise addressed two key conservation planning issues: multiple zones and connectivity processes. This region is characterized by a near-complete lack of biological and physical data.

To address data gaps, Thieme et al. (2007) proposed an approach based on definition of coarse-filter freshwater habitats. They used newly available spatial and remote sensing datasets, including space-borne optical and radar observations, and new techniques of spatial data analysis to generate subbasin, stream, floodplain, and wetland habitat classifications (figure 8.6).

The coarse-filter habitat representation approach focused on protecting the most intact examples of each wetland and floodplain type while maximizing longitudinal and lateral connectivity through a network of conservation areas and multiple-use zones.

The proposed multiple-use management zones complement existing reserves as well as plans for additional terrestrial reserves. The objective for representation was at least 20% of each freshwater habitat type. Combined with design for connectivity, this resulted in the identification of 86 subbasins requiring new or additional protective management. Overall, about 11% (about 17,000 km^2) of the region was nominated for some level of protection. Of these areas, 44% were recommended as Level I (representation of areas currently lacking protection), 29% were recommended as Level II (connectivity

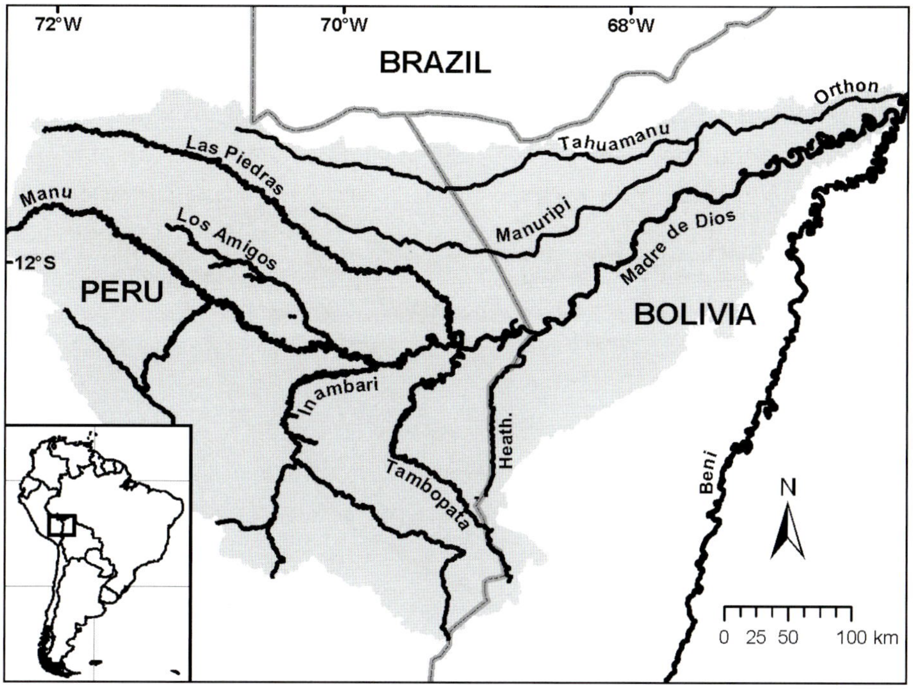

Figure 8.5. Major rivers within the drainage basin (gray) of the Madre de Dios and Orthon. From Thieme et al. (2007).

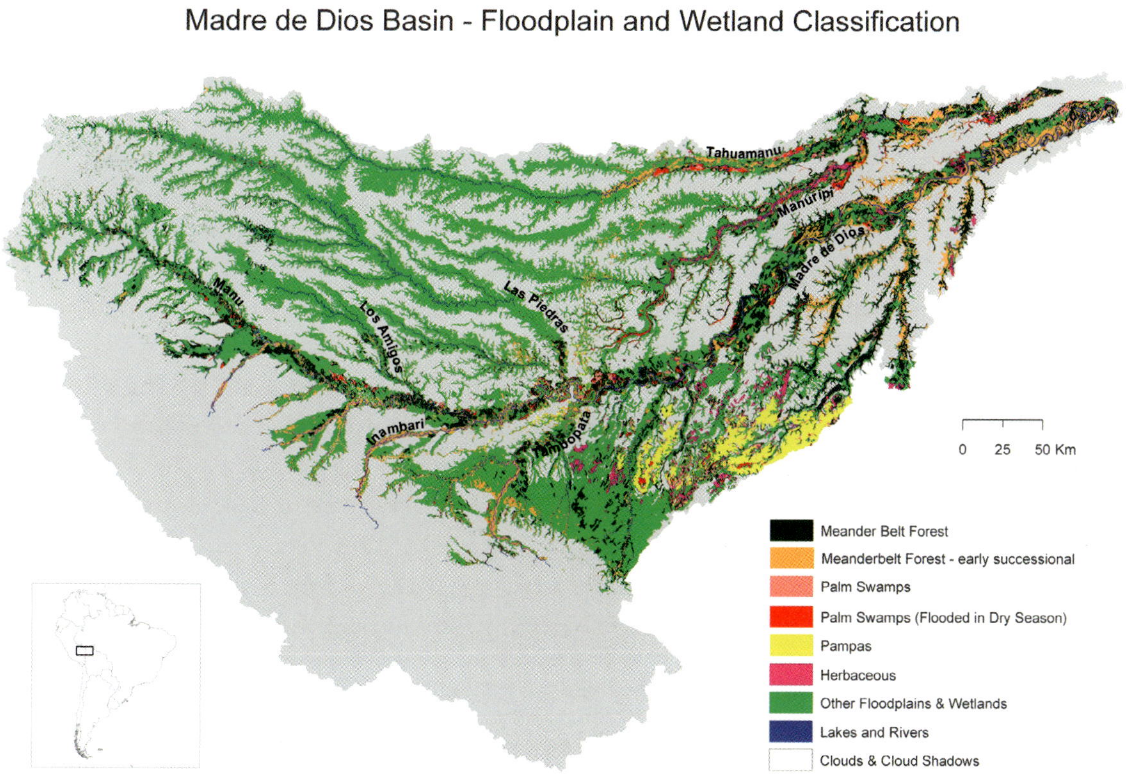

Figure 8.6. Floodplain and wetland classification. This map depicts current and relict river floodplains and other poorly drained lands below 400-m elevation that are likely to support wetland ecosystems. Classification based on Landsat ETM+ (optical multispectral), JERS-1 (radar) and digital elevation data from the Shuttle Radar Topography Mission (SRTM). From Thieme et al. (2007).

areas that overlap with indigenous lands) and 27% were recommended as Level III (degraded areas or potential threat mitigation zones that meet representation and connectivity goals; figure 8.7).

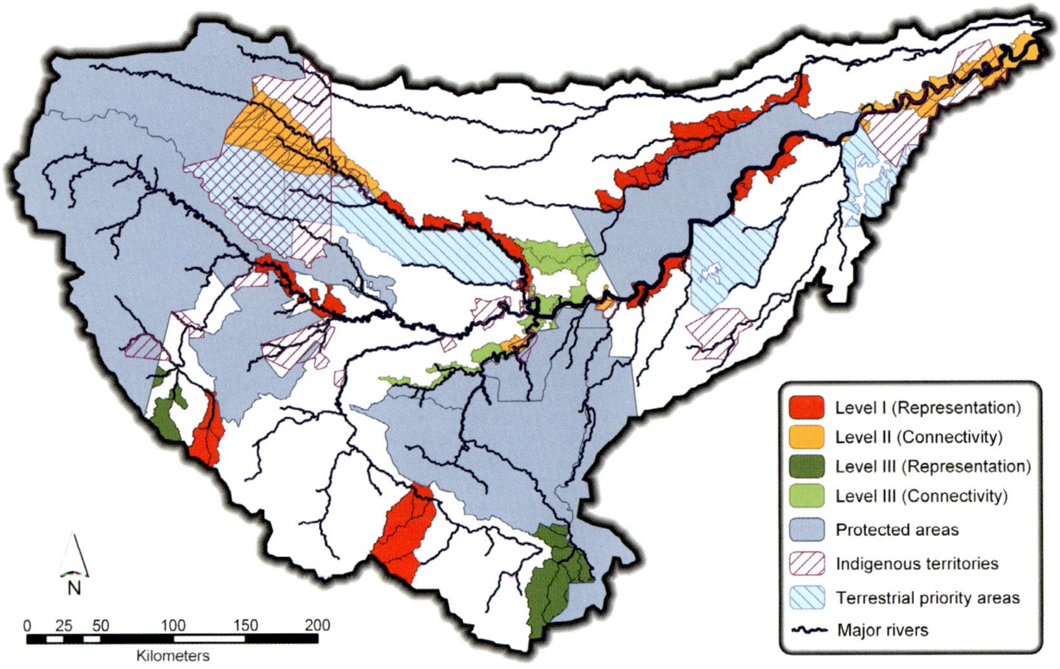

Figure 8.7. Preliminary integrated conservation plan for the freshwater and terrestrial ecosystems of the Madre de Dios Basin. Level I contains relatively intact areas currently lacking protection, Level II areas overlap with indigenous lands, and Level III includes degraded areas or potential threat mitigation zones. From Thieme et al. (2007).

The most obvious difference between planning in the terrestrial and freshwater realms is the connected nature of rivers and wetlands (Pringle 2001). Many important processes in freshwater systems are related to the four kinds of connectivity discussed above: lateral (e.g., filling of floodplain wetlands and associated breeding of waterbirds; movements of fish from river channels onto inundated floodplains), longitudinal (e.g., movements of diadromous and amphidromous fish), temporal (e.g., variations in flows and temperature within and between years), and vertical (e.g., flows between surface waters and groundwater; population dynamics of groundwater-dependent fauna). Other processes of interest to conservation planners include patch dynamics within streams related to movement of bed material and dead trees or changes in river course, and diversification of lineages in discrete, distinctive habitats such as high-altitude cool headwaters. Empirical data is becoming available on minimum and maximum dispersal distances of migratory species such as diadromous salmon in the Pacific Northwest of the United States or amphidromous Gobiidae species in the Pacific. However, this data has not to our knowledge been incorporated into planning for freshwaters or across freshwater and marine systems. Maintaining connectivity is particularly important for migratory species that need protection of widely separated spawning and adult habitats, and free passage between them.

Stage 7: Setting conservation objectives

Objectives for biodiversity patterns

Quantitative objectives for representing biodiversity patterns can be set in different ways. The most common approach is to identify objectives as thresholds that need to be reached before a system of conservation areas is considered complete (Pressey et al. 2003). A less common approach, but one deserving more attention, is to define benefit or utility functions (e.g., Arponen et al. 2005) that relate the amount of a feature protected (e.g., extent of a vegetation type or abundance of individuals) to the benefit associated with that feature. We limit our discussion here to objectives as thresholds. These can be defined in terms of extent, number of occurrences, probabilities of occurrence, or population sizes, and modified according to knowledge of habitat requirements, for example, to integrate habitat suitability and connectivity (e.g., Carroll et al. 2004). Ideally, objectives for biodiversity features are set in relation to estimates of previous extents or abundances (e.g., pre-European extents of vegetation types in Australia) and scaled to reflect relative need for protection, for example, in relation to threat (Pressey et al. 2003). Below we discuss some special considerations and guidelines for setting objectives for marine and freshwater systems.

Marine biodiversity pattern objectives

Ideally, all marine habitats and biogeographic zones should be adequately represented in a system of marine conservation areas (Ballantine 1997) to maximize the protection of species associated with them (Roberts et al. 2003). Locating marine reserves in both core areas and transitions between biogeographic zones might also be needed to protect genetic diversity, accommodate physical and biological shifts due to climate change, and confer resistance and resilience to shifts to undesirable states (Roberts et al. 2003).

How much of each marine habitat, species, or other feature should be protected depends on many considerations (Pressey et al. 2003; Rondinini and Chiozza 2010), but there is a general agreement about representing at least 20%, or ideally 30% to 60%, of each broad habitat type in high-protection zones within each marine bioregion (Airamé et al. 2003; Boersma and Parrish 1999; Bohnsack 2000; Roberts and Hawkins 2000; Roberts et al. 2003; Ward et al. 1999). These percentages are considered necessary to support fisheries production and safeguard biodiversity over the long term (Halpern and Warner 2003; Roberts et al. 2003). However, higher representation levels would be warranted to increase the likelihood of including unmapped features if planning is based only on habitat-level surrogates for biodiversity, without additional information on species, breeding sites, and other localized features (Airamé et al. 2003; Ward et al. 1999). Larger objectives are also justified for more threatened or degraded features because, by their nature, they are more in need of protective management (The Ecology Centre, University of Queensland 2009). Larger percentages (> 35%) of coastal waters in reserves might also be required to adequately protect species with longer dispersal distances (Botsford et al. 2003). Incomplete knowledge of marine biodiversity justifies a precautionary approach, ensuring representation of cross-shelf variation in depth and substratum and latitudinal gradients (The Ecology Centre 2009).

Freshwater biodiversity pattern objectives

Approaches to setting objectives for fresh waters are akin to those used in terrestrial and marine ecosystems, with consideration necessary for the multiple components of freshwater systems, including river reaches, riparian corridors, wetlands, subcatchments, entire catchments, and subterranean areas (groundwater). Types of quantitative objectives used to represent biodiversity pattern in fresh waters include the number of occurrences of a particular habitat type (e.g., Weitzell et al. 2003), species or species assemblages present; percentage total area or length of a specific wetland, subcatchment or river type; and a fixed number of habitat kilometers. For example, Hermoso et al. (2011) set objectives to select 10 subcatchments, roughly equating to 70 km of habitat for each species for which data was available. This approach could be particularly useful in helping conservation planners and stakeholders appreciate the length of river requiring conservation actions.

Most commonly, objectives for freshwater systems are set using a uniform percentage—that is, the same target for all freshwater species and habitat types (Moilanen et al. 2008; Nel et al. 2007; Thieme et al. 2007). However, such a blanket percentage rule makes little sense and is not recommended for any conservation plan (Pressey et al. 2003). For example, for a very common, secure feature, 20% representation could be unnecessary, but a rare and threatened feature could need 100% representation to persist.

Objectives for processes

Marine biodiversity process objectives

Processes have been incorporated into conservation planning in four ways: (1) incidentally by considering only biodiversity pattern, (2) with generic design criteria, (3) through process-specific design criteria, or (4) by setting objectives for spatial components associated with processes (Pressey et al. 2007). Generic criteria are guidelines about size and spacing of protected areas. They are intended to enhance the persistence of features in a system of conservation areas, for example, by ensuring connectivity among conservation areas (see sidebar "Great Barrier Reef Marine Park: An example of planning for representation"). Generic criteria might inform the size of individual protected areas to enable self-recruitment, and/or distances between reserves to maximize larval connectivity (Palumbi 2004). Reserve size, for example, can be based on the "ocean neighborhoods" of species (Palumbi 2004). Biophysical guidelines about reserve design were used during the design of the Great Barrier Reef Marine Park in Australia to ensure, among other objectives, the connectivity of no-take areas and minimization of edge effects (Fernandes et al. 2005). A retrospective assessment found that 98% of the protected reefs would receive high degrees of larval subsidies from protected reefs in adjacent no-take areas (Almany et al. 2009).

Quantitative objectives can be set for features that are dynamic in time and space. For example, an objective for net primary productivity identified from satellite imagery might be to have a minimum extent of high-productivity areas in a system of marine reserves at any time. This would mean that areas of high productivity could occur in different MPAs at different times. A related objective might be to protect a percentage of "refugia"—a region where conditions have remained relatively constant over time compared to surrounding regions—or areas such as upwellings with high productivity year-round. Analogous objectives could be set for patchy disturbances, such as a maximum extent within the MPA system or any single MPA at any time, high connectivity to areas likely to serve as sources of larvae after disturbance, or protection of refugia, in this case, areas with very infrequent and/or low-intensity disturbances (Ban et al., 2012).

Freshwater biodiversity process objectives

Connectivity is the most crucial attribute in freshwater systems (Pringle 2001). The need to retain connectivity in conservation planning for freshwaters has mainly been addressed by selecting longitudinally connected conservation areas. Hermoso et al. (2011) modified Marxan software to specifically include longitudinal distance through virtual connections between river reaches. This approach allowed them to avoid selecting isolated planning units and forced inclusion of closer upstream areas. Marxan thereby prioritized whole subbasins.

Stage 9: Selecting additional conservation areas

This is the stage where a spatial solution to the conservation design problem is developed. Such a solution, or multiple options for solutions, must synthesize information from all the previous stages, in particular, analyzing data on socioeconomic variables and threats on biodiversity, and addressing objectives for biodiversity features, while acknowledging the contributions of any established conservation areas. Decision support tools are commonly used in this stage (see Chan et al., chapter 2), and stakeholders should be actively engaged in decision making, perhaps by using software tools to identify conservation options and negotiate resolution of these options. Ideally, the types of conservation areas to be implemented (e.g., fully protected area or partially protected area) also are decided, although in reality this is often done after the general spatial prioritization is complete.

The main advantage of regional-scale design is that systems of conservation areas can be defined that are more than the sums of their parts. Design considerations important for marine and freshwater systems have been briefly outlined in the section on stage 7: setting conservation objectives. These design elements include complementarity and connectivity between individual conservation areas (e.g., size or spacing). The design stage aims to resolve spatial options for conservation, considering variables such as cost, previous and present extractive uses, and threats. However, no matter how well executed, this stage produces only notional systems of conservation areas that inevitably will have to change as application of actions proceeds (stage 10 of the systematic conservation planning process [see table 8.1], not described here).

Conservation design is becoming increasingly sophisticated in marine and freshwater systems. Representation was the initial focus of early conservation planning (e.g., Bedward et al. 1992). Recent emphasis has shifted to also including processes, for example, through design criteria (Fernandes et al. 2005) or process-specific objectives (Ban et al. 2012). It has also evolved from ignoring spatially variable costs (e.g., opportunity costs), to explicitly including them in decision support tools (Ban and Klein 2009). Because we have discussed the main design considerations elsewhere (see stage 7), below we highlight some frontiers in design for marine and freshwater systems.

Emerging frontiers in conservation area design

Frontiers in marine conservation area design

Here we briefly highlight three emerging frontiers in marine conservation area design: (1) dynamic protected areas to track dynamic processes, (2) connectivity, and (3) management costs. First, dynamic protected areas are those that can be moved in time and/or or space (Game et al. 2009). Their use is common in fisheries management, and is increasingly called for as a marine protected area strategy but to our knowledge has yet to be implemented. For example, if a particular frontal system which varies in location is highly productive and vulnerable to exploitation, a moveable protected area could track this feature as it shifts. With real-time data from satellites on the location of the front, one or more temporary protected areas could be progressively moved to track its location (Hobday and Hartmann 2006). The frontier lies in implementing this idea and testing additional environmental correlates that could facilitate definition of such moveable areas through satellite imagery. A second frontier concerns connectivity. Our understanding of the connectedness of marine systems is continually improving (e.g., Almany et al. 2007), and to date this research has resulted in guidelines for MPA design. The extent to which these guidelines apply to different marine ecosystems, however, is unknown, and continually improving connectivity data (e.g., from parentage analysis) might lead to refinements or change our approaches to marine conservation design. Finally, there is increasing research on, and interest in, better understanding the costs of ongoing management of marine conservation areas. However, management costs are rarely considered in the planning process. A recent study found that no-take areas are less expensive to manage than multiple-use areas, and that larger areas are cheaper to manage per unit area than small areas up to a point (Balmford et al. 2004). If one objective is to minimize management costs, then the explicit incorporation of ongoing management costs has the potential to change marine conservation design considerations.

Frontiers in freshwater conservation area design

Three emerging frontiers offer promise for the persistence of species and ecosystems in hydrologically connected landscapes: (1) integrating conservation plans to link rivers, wetlands, and subterranean aquatic ecosystems; (2) maintaining or restoring connectivity between realms; and (3) accounting for the costs of conservation actions. First, the integration of conservation plans between rivers and other freshwater ecosystems will require methods that account for connections between them. The persistence of biodiversity in these ecosystems depends on managing connectivity to maintain biological processes and mitigate threats that propagate between them (Linke et al. 2011). Second, and complementary to methods to maintain connections between freshwater ecosystems, research is emerging to integrate connectivity between terrestrial, freshwater, and marine realms (below). Finally, there is increasing research on, and interest in, better understanding the costs of conservation actions, including ongoing management (e.g., Januchowski-Hartley et al. 2011). Explicitly accounting for initial and ongoing management costs can encourage practitioners to invest public funds more transparently and accountably. Accurate estimates of conservation costs can refine approaches to setting conservation priorities, identify the actual costs required to achieve objectives, and even influence policy by demonstrating the full costs of new initiatives.

Planning for multiple conservation actions

Strict reservation of large areas of sea or freshwater ecosystems and their associated catchments is unlikely because of the large costs of acquisition and management. In addition, the preferences and diverse needs of different stakeholders commonly limit the extent of strict reservation. Consequently, some areas important for biodiversity will remain outside any system of strict reserves. This reality, and the connected nature of marine and freshwater systems, makes it critical for planners to consider appropriate management of nonreserved areas to achieve conservation objectives.

Most exercises in conservation design identify generic conservation areas, leaving the spatially explicit application of specific actions for "later" (sometimes never), if there is no effective transition from stage 9 (selection) to stage 10 (applying actions). Explicitly planning for multiple actions in the selection stage allows planners to consider the most appropriate and feasible spatial combinations of different actions. However, even with this approach, planners should expect to alter their designs as actions proceed. Tools such as Marxan with Zones (Watts et al. 2009) that explicitly incorporate multiple actions cannot preempt all the mistakes and surprises that emerge with close investigation of individual areas on the ground or in the water. Ongoing adaptation and adjustment of designs are therefore an inevitable part of the action stage (stage 10, see table 8.1).

Multiple marine conservation actions

Off-reserve conservation is relatively common in marine systems through the multiple-use model in which areas under strict protection (no-take areas) are buffered by other conservation zones that allow varying types and extents of extractive use. Despite this, with some notable exceptions, systematic methods have generally not been applied to define multiple actions, being largely restricted to identifying priorities for generic MPAs or strict no-take areas.

Good practice in marine conservation planning is to design multiple actions over the whole seascape, rather than planning only for generic MPAs. This is referred to as marine spatial planning (Ardron et al. 2008; Douvere 2008), or planning for multiple management objectives (see Beck et al. 2009), akin to managing the "matrix" in terrestrial systems (e.g., Donald and Evans 2006). Although this type of planning is well established in some regions of the world, including Europe, it is just now gaining momentum in other regions, such as in the United States. The management actions considered can include many fisheries management tools: seasonal closures for species, minimum size limits, gear restrictions, total allowable catches, and others (Walters and Martell 2004). In this way, representation of conservation features can be extended beyond that within high-protection zones, and an integrated zoning system can contribute to sustaining functional seascapes and ecological processes (e.g., maintaining connectivity). Multiple management actions allow planners to address the threats that specific activities pose in multiple-use zones and consider the management arrangements needed to mitigate those threats. Multiple actions also allow for the explicit incorporation of management actions suited to different socioecological contexts (Ostrom 2007).

Zoning can also be used to minimize conflict among users (Agardy 2003). For example, diving is incompatible with many types of fishing, and these uses can be separated by zoning. Identifying areas important for fishing, and designating these as permitted fishing areas while also identifying areas of high priority for conservation, can be one way to identify win-win areas for protection (Ban and Vincent 2009). The decision support tool Marxan with Zones, released in 2009 (Watts et al. 2009), provides a tool for selecting multiple zones (Klein et al. 2010). Ideally, knowledge on the relative effectiveness of different management actions to protecting biodiversity, as well as their relative costs, would be incorporated into such software. An approach to evaluating and planning for conservation that considers multiple management actions is especially important in countries where governance and social context make extensive permanent no-take areas infeasible (Johannes 2002; Mills et al. 2011).

Multiple freshwater conservation actions

However ideal it might be for the protection of freshwater ecosystems to conserve entire catchments, it is seldom feasible. Instead, spatial mixtures of actions across catchments, such as those proposed by Abell et al. (2007; see stage 5), are more realistic. Relatively small areas of particular interest could be declared "freshwater focal areas," with extractive uses likely to be restricted, similar to terrestrial protected areas. In addition, Abell et al. (2007) suggest that "critical management zones" upstream of the focal areas would be managed to maintain the key processes, including water quality and quantity. Examples of critical management zones are fish spawning habitats, stream reaches used for spawning migrations, and areas that would yield large volumes of sediment if disturbed. Thus, when considering multiple actions in catchments, freshwater and terrestrial conservation strategies overlap, and all tools to manage the matrix become relevant, as described in other chapters. For example, if the addition of effluent to a stream is the main threat to species of concern in a downstream wetland, remediation measures such as effluent settlement ponds would be effective conservation measures.

Planning to integrate terrestrial, marine, and freshwater objectives

Numerous ecosystem processes link marine, freshwater, and terrestrial realms and contribute to the maintenance of conservation features. For example, major river inflows sustain high-productivity areas in the sea and also provide migration pathways for anadromous fish. Conservation areas are rarely isolated from processes occurring beyond their boundaries, and impacts from external sources are perceived to be greater in marine and freshwater than terrestrial regions. In coastal marine and freshwater systems, threats posed by activities in adjacent catchments can be extensive. In particular, the major externally derived threats to marine reserves include chemical pollution, sedimentation, and exotic invasive species. These same externally derived threats can also have major consequences for freshwater reserves and fresh waters that incidentally fall within terrestrial protected areas (Pringle 2001). Integrated conservation planning across realms is therefore very important for long-term effective conservation.

In general, marine planning exercises have focused on mapping and incorporating threats occurring within the marine environment (e.g., fishing, shipping, and mining) and few have attempted to tackle integrated land-sea planning (but see The Nature Conservancy 2002 and sidebar "Integrating realms in conservation planning: Cross-system threats in the Pacific Northwest Coast ecoregion, USA"). Consideration of land-based threats is usually limited to those originating in the coastal zone (e.g., coastal populations). Existing land-sea planning exercises (e.g., Darimont et al. 2010; Hazlitt et al. 2010; Tallis et al. 2008; The Nature Conservancy 2002) demonstrate how the locations and configurations of conservation areas can change in both the land and the sea as a consequence of this integration, and highlight that, at finer scales of conservation planning, there is promise that a more detailed understanding of the needs of species that rely on multiple realms, such as anadromous fish, amphibians, and even birds, can serve to identify priority areas that encompass elements of those realms (Alvarez-Romero et al. 2011). They also show the trade-offs that planners will encounter in the process of integration (Klein et al. 2010). There is therefore a need for a framework to carry out land-sea planning, such as one recently proposed by Alvarez-Romero and colleagues (2011). They emphasize that making links between terrestrial, freshwater, and marine conservation planning will require substantial institutional and policy changes in the management of natural resources.

Integrating realms in conservation planning: Cross-system threats in the Pacific Northwest Coast ecoregion, USA

Modification and degradation of terrestrial ecosystems (e.g., agriculture, logging, and urbanization) can have significant downstream impacts on both freshwater and marine biodiversity (Abell et al. 2007; Beck and Odaya 2001). Similarly, sea-based activities such as fishing can have upstream effects by reducing the input of marine-derived nutrients into freshwater and terrestrial systems (Gresh et al. 2000). These are known as cross-system threats. A study in the Pacific Northwest Coast ecoregion illustrates how inclusion or exclusion of cross-system threats can result in very different configurations of conservation areas (Tallis et al. 2008). Potential conservation areas were identified using Marxan, and the best solution (i.e., the set of planning units that achieved conservation objectives in the least amount of area and with minimum threats) was selected under each of three threat scenarios: no threats considered, only within-realm threats considered (e.g., only threats to marine areas from marine sources), and cross-system threats considered.

To account for cross-system threats, Tallis et al. (2008) mapped the extent of Columbia River plume into the sea in two ways. First, they used data on long-term variability in sea-surface temperature (SST; 10-year values of mean annual temperature variance derived from AVHRR satellite images), with higher variances indicating more influence of the river plume (figure 8.8A). Second, the scientists used CORIE—a salinity-based plume model—to project the average annual plume extent by identifying areas with low salinity ($\leq$ 31.0 psu) as a proxy for freshwater influence of the river (figure 8.8B). The marine influence of eight smaller river plumes was also mapped based on temperature and salinity. Sites that were influenced by river plumes using these two methods were given a freshwater influence factor, which was translated to a higher "cost" (i.e., function of area and threat levels) in the selection of conservation areas (figure 8.8C). The purpose of this cost was to minimize exposure to land-based threats when selecting areas to achieve biodiversity objectives. The influence of rivers served as a proxy for these land-based threats.

Differences in the set of planning units selected for potential conservation areas under the three threat scenarios are in figure 8.9. The first scenario (8.9A), which does not account for threats, produced more compact and larger priority areas. The second scenario (8.9B) identified priority areas that were smaller and more dispersed, constrained by the occurrence of threats directly affecting terrestrial

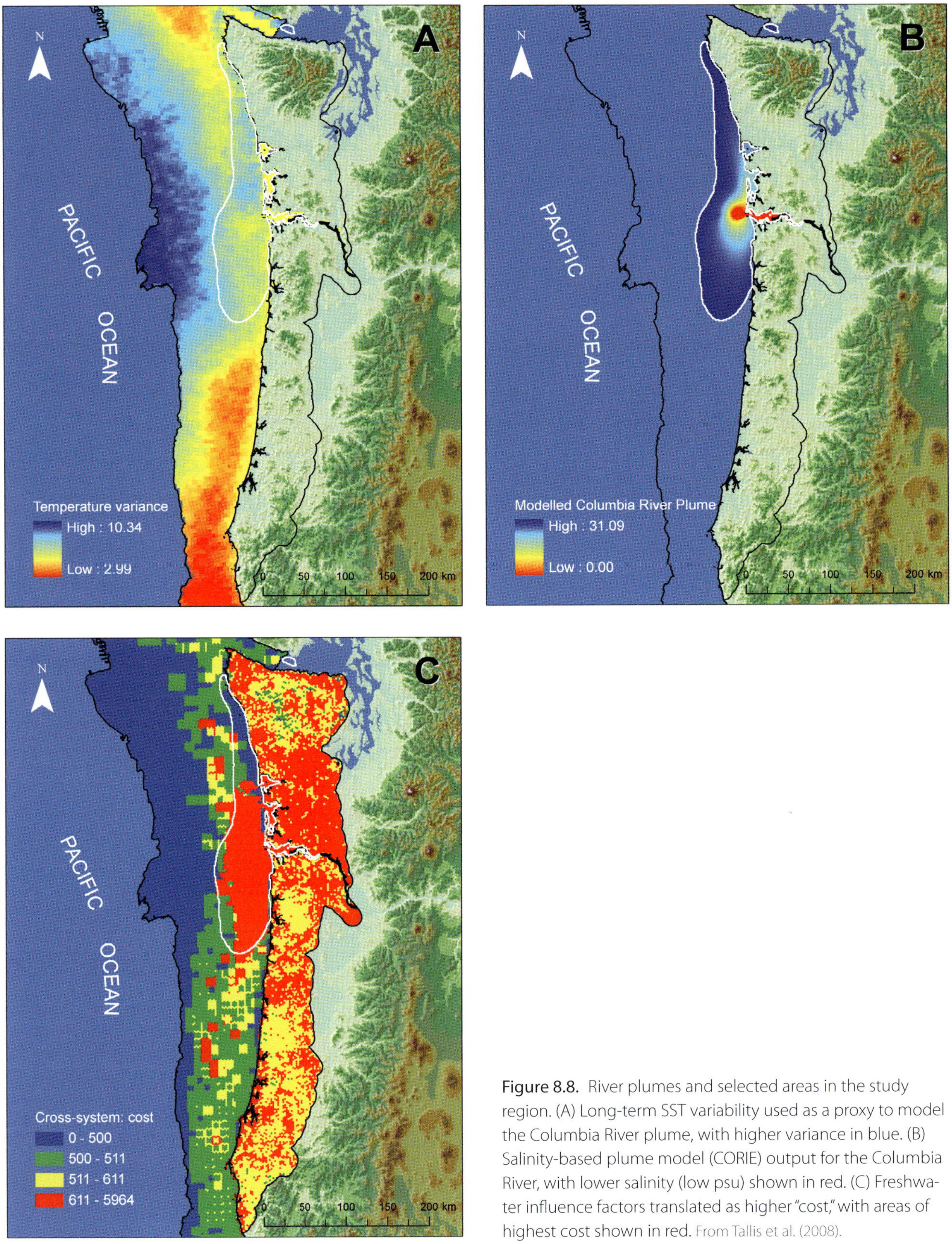

Figure 8.8. River plumes and selected areas in the study region. (A) Long-term SST variability used as a proxy to model the Columbia River plume, with higher variance in blue. (B) Salinity-based plume model (CORIE) output for the Columbia River, with lower salinity (low psu) shown in red. (C) Freshwater influence factors translated as higher "cost," with areas of highest cost shown in red. From Tallis et al. (2008).

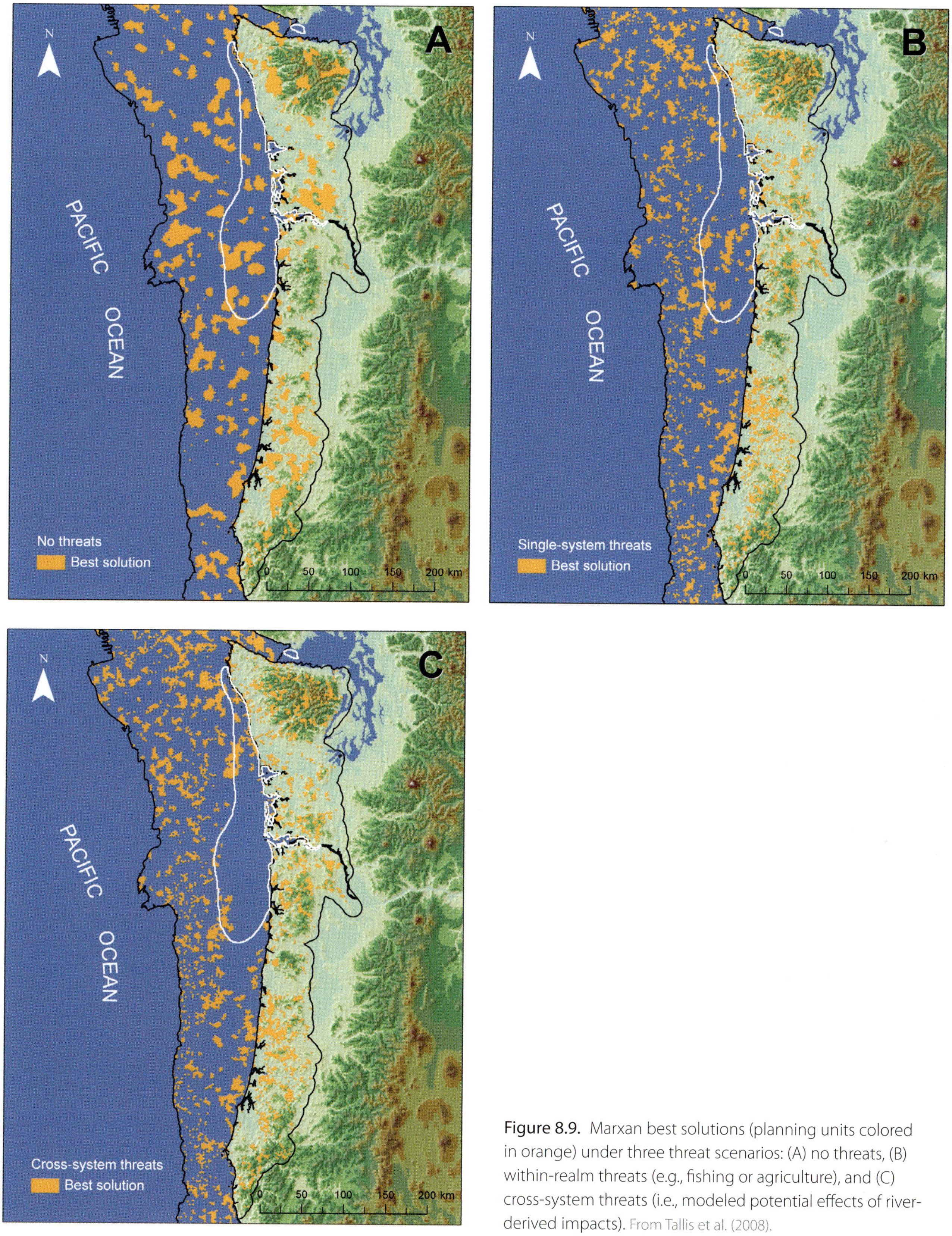

Figure 8.9. Marxan best solutions (planning units colored in orange) under three threat scenarios: (A) no threats, (B) within-realm threats (e.g., fishing or agriculture), and (C) cross-system threats (i.e., modeled potential effects of river-derived impacts). From Tallis et al. (2008).

and marine ecosystems. In the third scenario (8.9C), when cross-system threats were incorporated, coastal marine areas within the river plumes, and potentially affected by river-derived pollutants, were generally not selected. Accounting for cross-system threats, such as land-based pollution, can significantly alter the design of marine conservation areas.

Inappropriate land management in catchments continues to degrade even the best-designed MPAs (e.g., Great Barrier Reef Marine Park) through increased input of sedimentation and nutrients into rivers (Great Barrier Reef Marine Park Authority 2009). Configuring MPAs to maintain land-sea processes and mitigate threats from land will lead to MPAs that better achieve conservation objectives (Roberts et al. 2003). Land-sea planning allows planners to (1) mitigate or avoid cross-system threats and (2) promote the persistence of ecological process across realms (Tallis et al. 2008). For example, no-take areas can be selected where there is minimal impact from any present or likely emerging threats (e.g., major river discharges from modified catchments, proximity to industrial developments, or coastal infrastructure such as ports). This approach was taken during the rezoning of the Great Barrier Reef Marine Park (Fernandes et al. 2005).

At least two types of challenges lie ahead for integrating conservation planning across realms. First are the technical challenges of integrating data and approaches from the three realms. Second, and possibly more difficult, are the shifts in attitudes and priorities among politicians, policy makers, practitioners, and scientists required to understand that attempting to solve conservation problems in isolation is both unrealistic and inefficient (Linke et al. 2011). Thus, further developing the methodological foundations for integrated land-sea conservation, and effectively communicating these findings to decision makers, are important contributions to be made by conservation planners and scientists (Alvarez-Romero et al. 2011).

Summary and challenges

This chapter focuses on examining special considerations for conservation planning in marine and freshwater systems. Some stages of the planning process, such as identifying stakeholders and describing the planning context, apply similarly to all realms. However, some special considerations are necessary for planning in marine and freshwater ecosystems. In particular, connectivity is a dominant consideration for marine and freshwater planning, especially as conservation planners move from representing biodiversity patterns to also focusing on biodiversity processes. The extensive connectedness of marine and freshwater systems highlights the need to move beyond single-action and single-realm planning toward a more integrated consideration of processes, threats, and actions.

Although we have outlined some important considerations in marine and freshwater conservation planning, much remains to be refined and developed. Challenges currently being researched include improving the link between planning and implementation, better integrating considerations of social variables and ecosystem services into the planning process (see Chan et al., chapter 2), addressing climate change (see Bachelet, chapter 13), improving planning for multiple management objectives and actions incorporating ecological modeling into spatially explicit conservation planning (Christensen et al. 2009), and addressing uncertainty. The latter is particularly important because we will never have complete information about current and future biodiversity distributions and human uses, and as planners attempt to include climate science and data such as sea-level rise and storm surge into planning processes, the need to accurately portray uncertainty will only increase. Therefore, identifying robust strategies for conservation in the face of uncertainty is paramount.

Acknowledgments

We wish to thank Amanda Lombard and Zach Ferdaña for generously sharing information and data for sidebars "Planning for marine ecological processes: An example from the Southern Ocean" and "Integrating realms in conservation planning: Cross-system threats in the Pacific Northwest Coast ecoregion, USA," respectively.

References

Abell, R., D. Allan, and B. Lehner. 2007. "Unlocking the Potential of Protected Areas for Freshwaters." *Biological Conservation* 134: 48–63.

Adams, V. M., R. L. Pressey, and R. Naidoo. 2010. "Opportunity Costs: Who Really Pays for Conservation?" *Biological Conservation* 143: 439–48.

Agardy, T. 2003. "An Environmentalist's Perspective on Responsible Fisheries: The Need for Holistic Approaches." In *Responsible Fisheries in the Marine Ecosystem,* edited by M. Sinclair and G. Vladimarsson, 65–85. Rome: Food and Agriculture Organization of the United Nations.

Airamé, S., J. Dugan, K. D. Lafferty, H. Leslie, D. McArdle, and R. R. Warner. 2003. "Applying Ecological Criteria to Marine Reserve Design: A Case Study from the California Channel Islands." *Ecological Applications* 13: S170–S184.

Allan, J. D., and A. S. Flecker. 1993. "Biodiversity Conservation in Running Waters." *Bioscience* 43: 32–43.

Allison, G. W., S. D. Gaines, J. Lubchenco, and H. P. Possingham. 2003. "Ensuring Persistence of Marine Reserves: Catastrophes Require Adopting an Insurance Factor." *Ecological Applications* 13: S8–S24.

Allison, G. W., J. Lubchenco, and M. H. Carr. 1998. "Marine Reserves Are Necessary but not Sufficient for Marine Conservation." *Ecological Applications* 8: S79–S92.

Almany, G. R., M. L. Berumen, S. R. Thorrold, S. Planes, and G. P. Jones. 2007. "Local Replenishment of Coral Reef Fish Populations in a Marine Reserve." *Science* 316: 742–44.

Almany, G., S. Connelly, D. Heath, J. Hogan, G. Jones, L. McCook, M. Mills, R. Pressey, and D. Williamson. 2009. "Connectivity, Biodiversity Conservation, and the Design of Marine Reserve Networks for Coral Reefs." *Coral Reefs* 28: 339–51.

Alvarez-Romero, J., R. L. Pressey, N. C. Ban, K. Vance-Borland, C. Willer, C. J. Klein, and S. D. Gaines. 2011. "Integrated Land-Sea Conservation Planning: The Missing Links." *Annual Review of Ecology, Evolution and Systematics* 42: 381–409.

Amis, M. A., M. Rouget, M. Lotter, and J. Day. 2009. "Integrating Freshwater and Terrestrial Priorities in Conservation Planning." *Biological Conservation* 142: 2,217–26.

Ardron, J. 2002. "GIS Recipe for Determining Benthic Complexity: An Indicator of Species Richness." In *Marine Geography: GIS for the Oceans and Seas,* edited by J. Breman, 169–75. Redlands, CA: Esri Press.

Ardron, J., K. Gjerde, S. Pullen, and V. Tilot. 2008. "Marine Spatial Planning in the High Seas." *Marine Policy* 32: 832–39.

Arponen, A., R. K. Heikkinen, C. D. Thomas, and A. Moilanen. 2005. "The Value of Biodiversity in Reserve Selection: Representation, Species Weighting, and Benefit Functions." *Conservation Biology* 19: 2,009–14.

Ausseil, A. G. E., W. Lindsay Chadderton, P. Gerbeaux, R. T. Theo Stephens, and J. R. Leathwick. 2010. "Applying Systematic Conservation Planning Principles to Palustrine and Inland Saline Wetlands of New Zealand." *Freshwater Biology* 56: 142–61.

Avery, R. 2003. "Marine and Terrestrial Conservation Planning—How Different Are They?" In *Conserving Marine Environments: Out of Sight Out of Mind,* edited by P. Hutchings and D. Lunney, 18–40. Sydney: Royal Zoological Society of New South Wales.

Ballantine, B. 1997. *Design Principles of Systems of "No-Take" Marine Reserves. The Design and Monitoring of Marine Reserves.* Vancouver: Fisheries Centre, University of British Columbia.

Balmford, A., P. Gravestock, N. Hockley, C. J. McClean, and C. M. Roberts. 2004. "The Worldwide Costs of Marine Protected Areas." *Proceedings of the National Academy of Sciences* 101: 9,694–97.

Ban, N., and J. Alder. 2008. "How Wild Is the Ocean? Assessing the Intensity of Anthropogenic Marine Activities in British Columbia, Canada." *Aquatic Conservation: Marine and Freshwater Ecosystems* 18: 55–85.

Ban, N. C., and C. J. Klein. 2009. "Spatial Socio-Economic Data as a Cost in Systematic Marine Conservation Planning." *Conservation Letters* 2: 206–15.

Ban, N. C., R. L. Pressey, and S. Weeks. 2012. "Conservation Objectives and Sea Surface Temperature Anomalies in the Great Barrier Reef." *Conservation Biology* 26: 799-809.

Ban, N. C., and A. C. J. Vincent. 2009. "Beyond Marine Reserves: Exploring the Approach of Selecting Areas where Fishing Is Permitted, Rather than Prohibited." *PLoS ONE* 4: e6258.

Banks, S. A., G. A. Skilleter, and H. P. Possingham. 2005. "Intertidal Habitat Conservation: Identifying Conservation Targets in the Absence of Detailed Biological Information." *Aquatic Conservation: Marine and Freshwater Ecosystems* 15: 271–88.

Beck, M., J. Ferdania, K. Kachmar, P. Morrison, and P. Taylor. 2009. *Best Practices for Marine Spatial Planning.* Arlington, VA: The Nature Conservancy.

Beck, M. W., and M. Odaya. 2001. "Ecoregional Planning in Marine Environments: Identifying Priority Sites for Conservation in the Northern Gulf of Mexico." *Aquatic Conservation: Marine and Freshwater Ecosystems* 11: 235–42.

Bedward, M., R. L. Pressey, and D. A. Keith. 1992. "A New Approach for Selecting Fully Representative Reserve Networks: Addressing Efficiency, Reserve Design and Land Suitability with an Iterative Analysis." *Biological Conservation* 62: 115–25.

Begossi, A. 2006. "Temporal Stability in Fishing Spots: Conservation and Co-Management in Brazilian Artisanal Coastal Fisheries." *Ecology and Society* 11: 5.

Behrenfeld, M. J., and P. G. Falkowski. 1997. "Photosynthetic Rates Derived from Satellite-Based Chlorophyll Concentration." *Limnology and Oceanography* 42: 1–20.

Beverton, R. J. H., and S. J. Holt. 1957. *On the Dynamics of Exploited Fish Populations.* London: Chapman and Hall.

Boersma, P. D., and J. K. Parrish. 1999. "Limiting Abuse: Marine Protected Areas, a Limited Solution." *Ecological Economics* 31: 287–304.

Bohnsack, J. 1997. "Consensus Development and the Use of Marine Reserves in the Florida Keys National Marine Sanctuary." In *Protected Areas in Resource-Based Economies: Sustaining Biodiversity & Ecological Integrity*, Conference Proceedings of the 14th Annual General Meeting of the Canadian Council on Ecological Areas, 66–70.

———. 2000. "A Comparison of the Short-Term Impacts of No-Take Marine Reserves and Minimum Size Limits." *Bulletin of Marine Science* 66: 635–50.

Botsford, L. W., F. Micheli, and A. Hastings. 2003. "Principles for the Design of Marine Reserves." *Ecological Applications* 13: S25–S31.

Brock, B., E. Atkinson, C. Groves, A. Toivola, T. Olenicki, and L. Craighead. 2006. *A Wildlife Conservation Assessment of the Madison Valley, Montana.* Bozeman, MT: Wildlife Conservation Society, Greater Yellowstone Program.

Carpenter, K. E., M. Abrar, G. Aeby, R. B. Aronson, S. Banks, A. Bruckner, A. Chiriboga, J. Cortes, J. C. Delbeek, L. DeVantier, G. J. Edgar, A. J. Edwards, D. Fenner, H. M. Guzman, B. W. Hoeksema, G. Hodgson, O. Johan, W. Y. Licuanan, S. R. Livingstone, E. R. Lovell, J. A. Moore, D. O. Obura, D. Ochavillo, B. A. Polidoro, W. F. Precht, M. C. Quibilan, C. Reboton, Z. T. Richards, A. D. Rogers, J. Sanciangco, A. Sheppard, C. Sheppard, J. Smith, S. Stuart, E. Turak, J. E. N. Veron, C. Wallace, E. Weil, and E. Wood. 2008. "One-Third of Reef-Building Corals Face Elevated Extinction Risk from Climate Change and Local Impacts." *Science* 321: 560–63.

Carr, M. H., J. E. Neigel, J. A. Estes, S. Andelman, R. R. Warner, and J. L. Largier. 2003. "Comparing Marine and Terrestrial Ecosystems: Implications for the Design of Coastal Marine Reserves." *Ecological Applications* 13: S90–S107.

Carroll, C., R. E. Noss, P. C. Paquet, and N. H. Schumaker. 2004. "Extinction Debt of Protected Areas in Developing Landscapes." *Conservation Biology* 18: 1,110–20.

Casselman, J. 2002. "Effects of Temperature, Global Extremes, and Climate Change on Year-Class Production of Warmwater, Coolwater, and Coldwater Fishes in the Great Lakes Basin." In *Fisheries in a Changing Climate,* edited by Nature A. McGinn, 39–60. Bethesda, MD: American Fisheries Society.

Chape, S., J. Harrison, M. Spalding, and I. Lysenko. 2005. "Measuring the Extent and Effectiveness of Protected Areas as an Indicator for Meeting Global Biodiversity Targets." Philosophical Transactions of the Royal Society B: Biological Sciences 360: 443–55.

Christensen, V., Z. Ferdaña, and J. Steenbeek. 2009. "Spatial Optimization of Protected Area Placement Incorporating Ecological, Social and Economical Criteria." *Ecological Modelling* 220: 2,583–93.

Cowling, R. M., R. L. Pressey, A. T. Lombard, P. G. Desmet, and A. G. Ellis. 1999. "From Representation to Persistence: Requirements for a Sustainable System of Conservation Areas in the Species-Rich Mediterranean-Climate Desert of Southern Africa." *Diversity and Distributions* 5: 51–71.

Crain, C. M., B. S. Halpern, M. W. Beck, and C. V. Kappel. 2009. "Understanding and Managing Human Threats to the Coastal Marine Environment." *Annals of the New York Academy of Sciences* 1,162: 39–62.

Darimont, C. T., H. M. Bryan, S. M. Carlson, M. D. Hocking, M. MacDuffee, P. C. Paquet, M. H. H. Price, T. E. Reimchen, J. D. Reynolds, and C. C. Wilmers. 2010. "Salmon for Terrestrial Protected Areas." *Conservation Letters* 3: 379–89.

Daufresne, M., and P. Boe. 2007. "Climate Change Impacts on Structure and Diversity of Fish Communities in Rivers." *Global Change Biology* 13: 2,467–78.

Deharveng, L., F. Stoch, J. Gibert, A. Bedos, D. Galassi, M. Zagmajster, A. Brancelj, A. Camacho, F. Fiers, and P. Martin. 2009. "Groundwater Biodiversity in Europe." *Freshwater Biology* 54: 709–26.

Diaz, R. J., and R. Rosenberg. 2008. "Spreading Dead Zones and Consequences for Marine Ecosystems." *Science* 321: 926–29.

Dittman, A., and T. Quinn. 1996. "Homing in Pacific Salmon: Mechanisms and Ecological Basis." *Journal of Experimental Biology* 199: 83–91.

Donald, P. F., and A. D. Evans. 2006. "Habitat Connectivity and Matrix Restoration: The Wider Implications of Agri-Environment Schemes." *Journal of Applied Ecology* 43: 209–18.

Douvere, F. 2008. "The Importance of Marine Spatial Planning in Advancing Ecosystem-Based Sea Use Management." *Marine Policy* 32: 762–71.

Dudgeon, D., A. Arthington, M. Gessner, Z. Kawabata, D. Knowler, C. Lévêque, R. Naiman, A. Prieur-Richard, D. Soto, and M. Stiassny. 2006. "Freshwater Biodiversity: Importance, Threats, Status and Conservation Challenges." *Biological Reviews* 81: 163–82.

Dybas, C. 2005. "Dead Zones Spreading in World Oceans." *Bioscience* 55: 552–57.

Esselman, P. C., and J. D. Allan. 2011. "Application of Species Distribution Models and Conservation Planning Software to the Design of a Reserve Network for the Riverine Fishes of Northeastern Mesoamerica." *Freshwater Biology* 56: 71–88.

Ferdaña, Z. 2006. "Case Study: Pacific Northwest Coast (PNWC) Ecoregion Offshore Classification Methodology." Ecoregional Assessment Toolkit, http://science.natureconservancy.ca/centralinterior/docs/ERAtoolbox/7/Standard7_casestudy_PNWC.pdf.

Fernandes, L., J. Day, A. Lewis, S. Slegers, B. Kerrigan, D. Breen, D. Cameron, B. Jago, J. Hall, D. Lowe, J. Innes, J. Tanzer, V. Chadwick, L. Thompson, K. Gorman, M. Simmons, B. Barnett, K. Sampson, G. De'ath, B. Mapstone, H. Marsh, H. Possingham, I. Ball, T. Ward, K. Dobbs, J. Aumend, D. Slater, and K. Stapleton. 2005. "Establishing Representative No-Take Areas in the Great Barrier Reef: Large-Scale Implementation of Theory on Marine Protected Areas." *Conservation Biology* 19: 1,733–44.

Ferrier, S. 2002. "Mapping Spatial Pattern in Biodiversity for Regional Conservation Planning: Where to from Here?" *Systematic Biology* 51: 331–63.

Game, E. T., H. S. Grantham, A. J. Hobday, R. L. Pressey, A. T. Lombard, L. E. Beckley, K. Gjerde, R. H. Bustamente, H. P. Possingham, and A. J. Richardson. 2009. "Pelagic Protected Areas: The Missing Dimension in Ocean Conservation." *Trends in Ecology & Evolution* 24: 360–69.

Game, E. T., E. V. E. McDonald-Madden, M. L. Puotinen, and H. P. Possingham. 2008. "Should We Protect the Strong or the Weak? Risk, Resilience, and the Selection of Marine Protected Areas." *Conservation Biology* 22: 1,619–29.

Grantham, H. S., E. T. Game, A. T. Lombard, A. J. Hobday, A. J. Richardson, L. E. Beckley, R. L. Pressey, J. A. Huggett, J. C. Coetzee, C. D. van der Lingen, S. L. Petersen, D. Merkle, and H. P. Possingham. 2011. "Accommodating Dynamic Oceanographic Processes and Pelagic Biodiversity in Marine Conservation Planning." *PLoS ONE* 6: e16552.

Great Barrier Reef Marine Park Authority. 2009. "Great Barrier Reef Outlook Report 2009." Great Barrier Reef Marine Park Authority, Townsville. http://www.gbrmpa.gov.au/corp_site/about_us/great_barrier_reef_outlook_report.

Green, A., P. Lokani, S. R. G. Sheppard, J. Almany, S. Keu, H. Aitsi, J. Warku Karvon, R. Hamilton, and G. Lipsett-Moore. 2007. *Scientific Design of a Resilient Network of Marine Protected Areas, Kimbe Bay, West New Britain, Papua New Guinea.* The Nature Conservancy Pacific Island Countries Report No. 2/07. http://www.nature.org/ourinitiatives/regions/asiaandthepacific/papuanewguinea/placesweprotect/scientific_design_of_a_resilient_network_of_marine_protected_areas_kim.pdf.

Green, A., and P. Mous. 2008. "Delineating the Coral Triangle, Its Ecoregions and Functional Seascapes Version 5.0." The Nature Conservancy Coral Triangle Program Report 1/08. pp. 1–44. http://www.conservationgateway.org/file/delineating-coral-triangle-its-ecoregions-and-functional-seascapes.

Green, A., S. E. Smith, G. Lipsett-Moore, C. Groves, N. Peterson, S. Sheppard, P. Lokani, R. Hamilton, J. Almany, J. Aitsi, and L. Bualia. 2009. "Designing a Resilient Network of Marine Protected Areas for Kimbe Bay, Papua New Guinea." *Oryx* 43: 488–98.

Gregr, E. J., and A. W. Trites. 2001. "Predictions of Critical Habitat for Five Whale Species in the Waters of Coastal British Columbia." *Canadian Journal of Fisheries and Aquatic Sciences* 58: 1,265–85.

Gresh, T., J. Lichatowich, and P. Schoonmaker. 2000. An Estimation of Historic and Current Levels of Salmon Production in the Northeast Pacific Ecosystem: Evidence of a Nutrient Deficit in the Freshwater Systems of the Pacific Northwest. *Fisheries* 25: 15–21.

Halpern, B. S., C. M. Ebert, C. V. Kappel, E. M. P. Madin, F. Micheli, M. Perry, K. A. Selkoe, and S. Walbridge. 2009. "Global Priority Areas for Incorporating Land-Sea Connections in Marine Conservation." *Conservation Letters* 2: 189–96.

Halpern, B. S., C. V. Kappel, K. A. Selkoe, F. Micheli, C. Ebert, C. Kontgis, C. M. Crain, R. Martone, C. Shearer, and S. J. Teck. 2009. "Mapping Cumulative Human Impacts to California Current Marine Ecosystems." *Conservation Letters* 2: 138–48.

Halpern, B. S., K. A. Selkoe, F. Micheli, and C. V. Kappel. 2007. "Evaluating and Ranking the Vulnerability of Global Marine Ecosystems to Anthropogenic Threats." *Conservation Biology* 21: 1,301–15.

Halpern, B. S., S. Walbridge, K. A. Selkoe, C. V. Kappel, F. Micheli, C. D'Agrosa, J. F. Bruno, K. S. Casey, C. Ebert, H. E. Fox, R. Fujita, D. Heinemann, H. S. Lenihan, E. M. P. Madin, M. T. Perry, E. R. Selig, M. Spalding, R. Steneck, and R. Watson. 2008. "A Global Map of Human Impact on Marine Ecosystems." *Science* 319: 948–52, doi: 10.1126/science.1149345.

Halpern, B. S., and R. R. Warner. 2002. "Marine Reserves Have Rapid and Lasting Effects." *Ecology Letters* 5: 361–66.

Halpern, B., and R. Warner. 2003. "Matching Marine Reserve Design to Reserve Objectives." *Proceedings of the Royal Society of London, Series B: Biological Sciences* 270: 1,871–78.

Harley, C. D. G., A. R. Hughes, K. M. Hultgren, B. G. Miner, C. J. B. Sorte, C. S. Thornber, L. F. Rodriguez, L. Tomanek, and S. L. Williams. 2006. "The Impacts of Climate Change in Coastal Marine Systems." *Ecology Letters* 9: 228–41.

Hazlitt, S. L., T. G. Martin, L. Sampson, and P. Arcese. 2010. "The Effects of Including Marine Ecological Values in Terrestrial Reserve Planning for a Forest-Nesting Seabird." *Biological Conservation* 143: 1,299–303.

Heiner, M., J. Higgins, X. Li, and B. Baker. 2011. "Identifying Freshwater Conservation Priorities in the Upper Yangtze River Basin." *Freshwater Biology* 56: 89–105.

Hermoso, V., S. Linke, J. Prenda, and H. Possingham. 2011. "Addressing Longitudinal Connectivity in the Systematic Conservation Planning of Fresh Waters." *Freshwater Biology* 56: 57–70.

Hobday, A. J., and K. Hartmann. 2006. "Near Real-Time Spatial Management Based on Habitat Predictions for a Longline Bycatch Species." *Fisheries Management and Ecology* 13: 365–80.

IUCN and UNEP–WCMC (The International Union for Conservation of Nature and the United Nations Environment Programme–World Conservation Monitoring Centre). 2010. *The World Database on Protected Areas (WDPA): January 2010.* Cambridge: United Nations Environment Programme–World Conservation Monitoring Centre.

Januchowski-Hartley, S. R., P. Visconti, and R. L. Pressey. 2011. "A Systematic Approach for Prioritizing Multiple Management Actions for Invasive Species." *Biological Invasions* 13: 1,241–53.

Johannes, R. E. 2002. "The Renaissance of Community-Based Marine Resource Management in Oceania." *Annual Review of Ecology and Systematics* 33: 317–40.

Jones, G., G. Almany, G. Russ, P. Sale, R. Steneck, M. van Oppen, and B. Willis. 2009. "Larval Retention and Connectivity among Populations of Corals and Reef Fishes: History, Advances and Challenges." *Coral Reefs* 28: 307–25.

Kelleher, G., and R. A. Kenchington. 1992. *Guidelines for Establishing Marine Protected Areas: A Marine Conservation and Development Report.* Gland Switzerland: International Union for Conservation of Nature.

Kinlan, B. P., and S. D. Gaines. 2003. "Propagule Dispersal in Marine and Terrestrial Environments: A Community Perspective." *Ecology* 84: 2,007–20.

Kirkpatrick, J., and C. Harwood. 1983. "Conservation of Tasmanian Macrophytic Wetland Vegetation." *Proceedings of the Royal Society of Tasmania* 117: 5–20.

Klein, C. J., C. Steinback, M. Watts, A. J. Scholz, and H. P. Possingham. 2010. "Spatial Marine Zoning for Fisheries and Conservation." *Frontiers in Ecology and the Environmen* 8: 349–53.

Kool, J. T., C. B. Paris, S. Andréfouët, and R. K. Cowen. 2010. "Complex Migration and the Development of Genetic Structure in Subdivided Populations: An Example from Caribbean Coral Reef Ecosystems." *Ecography* 33: 597–606.

Laran, S., and A. Gannier. 2008. "Spatial and Temporal Prediction of Fin Whale Distribution in the Northwestern Mediterranean Sea." *ICES Journal of Marine Science: Journal du Conseil* 65: 1,260–69.

Leathwick, J., A. Moilanen, M. Francis, J. Elith, P. Taylor, K. Julian, T. Hastie, and C. Duffy. 2008. "Novel Methods for the Design and Evaluation of Marine Protected Areas in Offshore Waters." *Conservation Letters* 1: 91–102.

Leslie, H. M. 2005. "A Synthesis of Marine Conservation Planning Approaches." *Conservation Biology* 19: 1,701–13, doi:10.1111/j.1523-1739.2005.00268.x.

Lewis, A., S. Slegers, D. Lowe, L. Muller, L. Fernandes, and J. Day. 2003. "Use of Spatial Analysis and GIS Techniques to Re-zone the Great Barrier Reef Marine Park." Coastal GIS Workshop, July 7–8, 2003. University of Wollongong, Australia. http://www.gbrmpa.gov.au/__data/assets/pdf_file/0004/6178/lewis_et_al_final03.pdf.

Linke, S., R. Pressey, R. Bailey, and R. Norris. 2007. "Management Options for River Conservation Planning: Condition and Conservation Revisited." *Freshwater Biology* 52: 918–38.

Linke, S., E. Turak, and J. Nel. 2011. "Freshwater Conservation Planning: The Case for Systematic Approaches." *Freshwater Biology* 56: 6–20.

Lombard, A. T., B. Reyers, L. Y. Schonegevel, J. Coopers, L. B. Smith-Adao, D. C. Nel, P. W. Froneman, I. J. Ansorge, M. N. Beste, C. A. Tosh, T. Strauss, T. Akkers, O. Gon, R. W. Leslie, and S. L. Chown. 2007. "Conserving Pattern and Process in the Southern Ocean: Designing a Marine Protected Area for the Prince Edward Islands." *Antarctic Science* 19: 39–54.

Margules, C., A. Nicholls, and R. Pressey. 1988. "Selecting Networks of Reserves to Maximise Biological Diversity." *Biological Conservation* 43: 63–76.

Margules, C. R., and R. L. Pressey. 2000. "Systematic Conservation Planning." *Nature* 405: 243–53.

Michel, G., F. Malard, L. Deharveng, T. Di Lorenzo, B. Sket, and C. De Broyer. 2009. "Reserve Selection for Conserving Groundwater Biodiversity." *Freshwater Biology* 54: 861–76.

Millennium Ecosystem Assessment 2005. *Ecosystems and Human Well-Being: Synthesis.* Washington, DC: World Resources Institute.

Mills, M., S. D. Jupiter, R. L. Pressey, N. C. Ban, and J. Comley. 2011. "Incorporating Effectiveness of Community-Based Management in a National Marine Gap Analysis for Fiji." *Conservation Biology* 25: 1,155–64.

Mills, M., R. Weeks, R. L. Pressey, S. Foale, and N. C. Ban. 2010. "A Mismatch of Scales: Challenges in Planning for Implementation of Marine Protected Areas in the Coral Triangle." *Conservation Letters* 3: 291–303.

Moilanen, A. 2007. "Landscape Zonation, Benefit Functions and Target-Based Planning: Unifying Reserve Selection Strategies." *Biological Conservation* 134: 571–79.

Moilanen, A., A. M. Franco, R. I. Early, R. Fox, B. Wintle, and C. D. Thomas. 2005. "Prioritizing Multiple-Use Landscapes for Conservation: Methods for Large Multi-Species Planning Problems." *Proceedings of the Royal Society B: Biological Sciences* 272: 1,885–91.

Moilanen, A., J. Leathwick, and J. Elith. 2008. "A Method for Spatial Freshwater Conservation Prioritization." *Freshwater Biology* 53: 577–92.

MPA News. 2002. "Results from the Reader Challenge: Which MPA is the Oldest?" *MPA News* 3: 5–6.

Myers, R. A., and B. Worm. 2003. "Rapid Worldwide Depletion of Predatory Fish Communities." *Nature* 423: 280–83.

Nel, J., D. Roux, R. Abell, P. Ashton, R. Cowling, J. Higgins, M. Thieme, and J. Viers. 2009. "Progress and Challenges in Freshwater Conservation Planning." *Aquatic Conservation: Marine and Freshwater Ecosystems* 19: 474–85.

Nel, J., D. Roux, G. Maree, C. Kleynhans, J. Moolman, B. Reyers, M. Rouget, and R. Cowling. 2007. "Rivers in Peril inside and outside Protected Areas: A Systematic Approach to Conservation Assessment of River Ecosystems." Diversity and Distributions 13: 341–52.

Nel, J. L., B. Reyers, D. J. Roux, N. Dean Impson, and R. M. Cowling. 2011. "Designing a Conservation Area Network that Supports the Representation and Persistence of Freshwater Biodiversity." *Freshwater Biology* 56: 106–24.

Norris, R., I. Prosser, B. Young, P. Liston, N. Bauer, N. Davies, F. Dyer, S. Linke, and M. Thoms. 2001. *The Assessment of River Condition (ARC). An Audit of the Ecological Condition of Australian Rivers.* Final Report Submitted to the National Land and Water Resources Audit Office. http://www.anra.gov.au/topics/water/pubs/river_assessment/river_assessment.pdf.

Noss, R., C. Carroll, K. Vance-Borland, and G. Wuerthner. 2002. "A Multicriteria Assessment of the Irreplaceability and Vulnerability of Sites in the Greater Yellowstone Ecosystem." *Conservation Biology* 16: 895.

Orr, J. C., V. J. Fabry, O. Aumont, L. Bopp, S. C. Doney, R. A. Feely, A. Gnanadesikan, N. Gruber, A. Ishida, and F. Joos. 2005. "Anthropogenic Ocean Acidification over the Twenty-First Century and Its Impact on Calcifying Organisms." *Nature* 437: 681–86.

Ostrom, E. 2007. "A Diagnostic Approach for Going Beyond Panaceas." *Proceedings of the National Academy of Sciences* 104: 15,181–87.

Palumbi, S. R. 2004. "Marine Reserves and Ocean Neighborhoods: The Spatial Scale of Marine Populations and Their Management." *Annual Review of Environment and Resources* 29: 31–68.

Pauly, D., V. Christensen, J. Dalsgaard, R. Froese, and F. Torres, Jr. 1998. "Fishing Down Marine Food Webs." *Science* 279: 860–63.

Pittock, J., L. Hansen, and R. Abell. 2008. "Running Dry: Freshwater Biodiversity, Protected Areas and Climate Change." *Biodiversity* 9: 30–8.

Possingham, H. P., I. R. Ball, and S. Andelman. 2000. "Mathematical Methods for Identifying Representative Reserve Networks." In *Quantitative Methods for Conservation Biology,* edited by S. Ferson, and M. Burgman, 291–305. New York: Springer-Verlag.

Pressey, R. 2002. "The First Reserve Selection Algorithm: A Retrospective on Jamie Kirkpatrick's 1983 Paper." *Progress in Physical Geography* 26: 434–41.

Pressey, R. L., and M. C. Bottrill. 2009. "Approaches to Landscape- and Seascape-Scale Conservation Planning: Convergence, Contrasts and Challenges." *Oryx* 43: 464–75.

Pressey, R. L., M. Cabeza, M. E. J. Watts, R. M. Cowling, and K. A. Wilson. 2007. "Conservation Planning in a Changing World." *Trends in Ecology and Evolution* 22: 583–92.

Pressey, R. L., R. M. Cowling, and M. Rouget. 2003. "Formulating Conservation Targets for Biodiversity Pattern and Process in the Cape Floristic Region, South Africa." *Biological Conservation* 112: 99–127.

Pressey, R. L., and A. O. Nicholls. 1989. "Efficiency in Conservation Evaluation: Scoring versus Iterative Approaches." *Biological Conservation* 50: 199–218.

Pringle, C. 2001. "Hydrologic Connectivity and the Management of Biological Reserves: A Global Perspective." *Ecological Applications* 11: 981–98.

Reilly, S., P. Fiedler, K. Forney, and J. Barlow. 1997. "Cetacean Habitats in the California Current: Partitioning Oceanographic and Geographic Patterns." International Whaling Commission Working Paper. SC/49/022.

Rivers-Moore, N. A., P. S. Goodman, and J. L. Nel. 2011. "Scale-Based Freshwater Conservation Planning: Towards Protecting Freshwater Biodiversity in KwaZulu-Natal, South Africa." *Freshwater Biology* 56: 125–41.

Robbins, J., L. D. Rosa, J. M. Allen, D. K. Mattila, E. R. Secchi, A. S. Friedlaender, P. T. Stevick, D. P. Nowacek, and D. Steel. 2011. "Return Movement of a Humpback Whale between the Antarctic Peninsula and American Samoa: A Seasonal Migration Record." *Endangered Species Research* 13: 117–21.

Roberts, C., and J. Hawkins. 2000. "Fully-Protected Marine Reserves: A Guide." Washington, DC: World Wide Fund Endangered Seas Campaign.

Roberts, C. M., S. Andelman, G. Branch, R. H. Bustamente, J. C. Castilla, J. Dugan, B. S. Halpern, K. D. Lafferty, H. Leslie, J. Lubchenco, D. McArdle, H. P. Possingham, M. Ruckelshaus, and R. R. Warner. 2003. "Ecological Criteria for Evaluating Candidate Sites for Marine Reserves." *Ecological Applications* 13: S199–S214.

Roff, J. C., and S. M. J. Evans. 2002. "Frameworks for Marine Conservation—Non-hierarchical Approaches and Distinctive Habitats." *Aquatic Conservation: Marine and Freshwater Ecosystems* 12: 635–48.

Rondinini, C., and F. Chiozza. 2010. "Quantitative Methods for Defining Percentage Area Targets for Habitat Types in Conservation Planning." *Biological Conservation* 143: 1,646–53.

Rosenberg, D. K., B. R. Noon, and E. C. Meslow. 1997. "Biological Corridors: Form, Function, and Efficacy." *Bioscience* 47: 677–87.

Rouget, M., R. M. Cowling, A. T. Lombard, A. T. Knight, and I. H. K. Graham. 2006. "Designing Large-Scale Conservation Corridors for Pattern and Process." *Conservation Biology* 20: 549–61.

Roux, D., F. de Moor, J. Cambray, and H. Barber-James. 2002. "Use of Landscape-Level River Signatures in Conservation Planning: A South African Case Study." *Conservation Ecology* 6 (2): 6. http://www.consecol.org/vol6/iss2/art6.

Sala, E., and N. Knowlton. 2006. "Global Marine Biodiversity Trends." *Annual Review of Environment and Resources* 31: 93–122.

Sarkar, S., R. L. Pressey, D. P. Faith, C. R. Margules, T. Fuller, D. M. Stoms, A. Moffett, K. A. Wilson, K. J. Williams, P. H. Williams, and S. Andelman. 2006. "Biodiversity Conservation Planning Tools: Present Status and Challenges for the Future." *Annual Review of Environment and Resources* 31: 123–59.

Saunders, D. L., J. J. Meeuwig, and A. C. J. Vincent. 2002. "Freshwater Protected Areas: Strategies for Conservation." *Conservation Biology* 16: 30–41.

Sedell, J. R., G. H. Reeves, F. R. Hauer, J. A. Stanford, and C. P. Hawkins. 1990. "Role of Refugia in Recovery from Disturbances: Modern Fragmented and Disconnected River Systems." *Environmental Management* 14: 711–24.

Selkoe, K., B. Halpern, C. Ebert, E. Franklin, E. Selig, K. Casey, J. Bruno, and R. Toonen. 2009. "A Map of Human Impacts to a 'Pristine' Coral Reef Ecosystem, the Papah Naumoku Kea Marine National Monument." *Coral Reefs* 28: 635–50.

Sowa, S., G. Annis, M. Morey, and D. Diamond. 2007. "A Gap Analysis and Comprehensive Conservation Strategy for Riverine Ecosystems of Missouri." *Ecological Monographs* 77: 301–34.

Spalding, M. D., H. E. Fox, G. R. Allen, N. Davidson, Z. A. Ferdaña, M. A. X. Finlayson, B. S. Halpern, M. A. Jorge, A. L. Lombana, and S. A. Lourie. 2007. "Marine Ecoregions of the World: A Bioregionalization of Coastal and Shelf Areas." *BioScience* 57: 573–83.

Stein, J. L., J. A. Stein, and H. A. Nix. 2002. "Spatial Analysis of Anthropogenic River Disturbance at Regional and Continental Scales: Identifying the Wild Rivers of Australia." *Landscape and Urban Planning* 60: 1–25.

Tallis, H., Z. Ferdana, and E. Gray. 2008. "Linking Terrestrial and Marine Conservation Planning and Threats Analysis." *Conservation Biology* 22: 120–30.

The Ecology Centre, University of Queensland. 2009. Scientific Principles for Design of Marine Protected Areas in Australia: A Guidance Statement. http://www.ecology.uq.edu.au/docs/Scientific_Principles_MPAs_c6.pdf.

The Nature Conservancy. 2002. *Chesapeake Bay Lowlands Ecoregional Plan.* Anchorage, AK: The Nature Conservancy.

Thieme, M., B. Lehner, R. Abell, S. Hamilton, J. Kellndorfer, G. Powell, and J. Riveros. 2007. "Freshwater Conservation Planning in Data-Poor Areas: An Example from a Remote Amazonian Basin (Madre de Dios River, Peru and Bolivia)." *Biological Conservation* 135: 484–501.

Toropova, C., I. Meliane, D. Laffoley, E. Matthews, and M. Spalding, eds. 2010. *Global Ocean Protection: Present Status and Future Possibilities.* Brest, France: Agence des aires marines protegees, Gland, Switzerland, Washington, DC and New York: International Union for the Conservation of Nature World Commission on Protected Areas, Cambridge: United Nations Environment Programme–World Conservation Monitoring Centre, Arlington, VA: The Nature Conservancy, Tokyo: United Nations University, New York: Wilderness Conservation Society. http://data.iucn.org/dbtw-wpd/edocs/2010-053.pdf.

Vance-Borland, K., D. Roux, J. Nel, and B. Pressey. 2008. "From the Mountains to the Sea." *Conservation Biology* 22: 505–7.

Walters, C. J., and S. J. D. Martell. 2004. *Fisheries Ecology and Management.* Princeton, NJ: Princeton University Press.

Ward, T., M. Vanderklift, A. Nicholls, and R. Kenchington. 1999. "Selecting Marine Reserves Using Habitats and Species Assemblages as Surrogates for Biological Diversity." *Ecological Applications* 9: 691–98.

Watts, M. E., I. R. Ball, R. S. Stewart, C. J. Klein, K. Wilson, C. Steinback, R. Lourival, L. Kircher, and H. P. Possingham. 2009. "Marxan with Zones: Software for Optimal Conservation Based Land- and Sea-Use Zoning." *Environmental Modelling & Software* 24: 1,513–21.

WCPA–IUCN (World Commission on Protected Areas–International Union for Conservation of Nature). 2007. *Establishing Marine Protected Areas Networks: A Guide for Developing National and Regional Capacity for Building MPA Networks.* Nontechnical Summary Report. World Commission on Protected Areas–International Union for Conservation of Nature. http://www.medpan.org/_upload/893.pdf.

Weitzell, R. E., M. L. Khoury, P. Gagnon, B. Schreurs, D. Grossman, and J. Higgins. 2003. *Conservation Priorities for Freshwater Biodiversity in the Upper Mississippi River Basin.* NatureServe and The Nature Conservancy. http://www.natureserve.org/library/uppermsriverbasin.pdf.

Worm, B., E. B. Barbier, N. Beaumont, J. E. Duffy, C. Folke, B. S. Halpern, J. B. C. Jackson, H. K. Lotze, F. Micheli, S. R. Palumbi, E. Sala, K. A. Selkoe, J. J. Stachowicz, and R. Watson. 2006. "Impacts of Biodiversity Loss on Ocean Ecosystem Services." *Science* 314: 787–90.

Wrona, F., T. Prowse, J. Reist, J. Hobbie, L. Lévesque, and W. Vincent. 2006. "Climate Change Effects on Aquatic Biota, Ecosystem Structure and Function." *AMBIO: A Journal of the Human Environment* 35: 359–69.

Zacharias, M. A., and J. C. Roff. 2000. "A Hierarchical Ecological Approach to Conserving Marine Biodiversity." *Conservation Biology* 14: 1,327–34.

9

Identification and mapping of habitat cores

Richard Church, PhD

Conservation planning requires good spatial data and a basic understanding of what type of habitat is required to support a given species, ensemble, or community. Thus, a critical step in conservation planning is often the development of a map of suitable habitat across a target landscape (see Dickson et al., chapter 7). The next key step is using this map of suitable habitat to identify and map the core habitat required to meet conservation objectives or to be used as an input to subsequent analysis, like population modeling and corridor planning.

Defining core habitat

"Core habitat" is a commonly used but context-dependent concept, somewhat difficult to define and measure in concrete terms. The word "core" can be defined as the basic or most important part of something, and in a conservation context, core habitat is often estimated subjectively by expert opinion (Craighead and Cross 2007; see also Johnston et al., chapter 14).

The term "core" or "core habitat" has been in common use in conservation since the 1970s, when the United Nations Educational, Scientific, and Cultural Organization (UNESCO) Man and the Biosphere Programme included the term "core" in its statutory framework. The general criteria for designation of a site as a biosphere reserve includes "a legally constituted core area or areas devoted to long-term protection, according to the conservation objectives of the biosphere reserve, and of sufficient size to meet these objectives." The term has generally been used in this sense ever since: core areas are defined in the context of a particular conservation project, although the objectives and size are often not explicitly stated. For example, Tilson et al. (2010) defined a core habitat block for a tiger conservation landscape as "big enough for five adult tigers." Identifying an area as core also generally includes some sense of protection.

A more general definition has been used by forestry biologists, who defined the term to mean a forest interior area free of edge effects, where an edge effect is a function of the surrounding habitat (Zipperer 1993). In other cases, core habitat has been defined in terms of its purpose or values. For example, the Natural Heritage and Endangered Species Program (2004) defines core habitats as those areas that contain the most critical sites for biodiversity protection in that state: they "represent the state's most viable rare plant and animal populations and include exemplary communities and aquatic habitats." Within this context, core represents a desired protection status. Similarly, Noss (1987) described core areas as places "where the conservation of biodiversity, ecological integrity, wilderness, or similar values takes precedence over other values and uses." Core has also been described by its degree of habitat alteration: "A distinguishing characteristic of core areas is limited human access—that is, low road density or, ideally, roadlessness" (Noss et al. 1999). In the context of reserve design, a core is an area that is managed so that human use is restricted and nature can operate in an undisturbed fashion (Soule and Terborgh 1999). National parks or other protected areas are often described as core areas. "Core" can also be defined in terms of demographics, referring to the contiguous area of land necessary to support an entire population, or a "viable" population, or even a subunit of a metapopulation of a given species.

In most conservation planning, then, "core" has been used to reflect a sense of intact habitat that is unaffected by human disturbance or other neighboring habitat. Thus, this chapter will examine the approaches that can be used to identify and map core habitat based upon a broad definition of core as "an area or patch of relatively intact habitat that is sufficiently large to support more than one individual of a species." Identification and mapping of the core is a critical prelude to identification of important buffer areas and connectivity issues.

The term "relatively intact" is context- and scale-related. At a very broad scale, even a highway through hundreds of acres of habitat may leave it relatively intact. Thus, Yellowstone Park is generally considered core habitat on a regional scale for many species, even though it is intertwined with busy highways and lesser roads. At the scale of the park itself, though, individual cores for different species may be defined in areas remote from the highways. The definition above also lacks a sense of time. In many cases, the conservation goal may be to support or maintain a family unit of a given species throughout their lives or throughout seasons of the year (including migratory routes). At broader scales, a core might maintain an entire population of that species for many generations.

Unfortunately, most definitions do not address how to determine where core exists, or how much is needed to meet a specific conservation objective. Table 9.1 gives several examples of defining core habitat based upon rules/metrics that have been set by a conservation team or study. These range from the lesser prairie-chicken—where core is defined as an area of suitable habitat containing at least two home ranges that are continuously occupied—to the marten—where core habitat involves at least 30–50 square kilometers containing at least 75% suitable habitat with an overstory canopy closure of at least 75%. Also note that the presence of roads can significantly degrade habitat (grizzly bear) or represent a boundary for a species (Coachella lizard). Many of the rules or standards given in table 9.1 can easily be applied to geographical data to produce a map of those areas that meet the definition of "core." It is important to emphasize that "core" is commonly defined with minimum size restrictions. That is, it is an area of suitable, contiguous habitat that exceeds a given minimum number of acres.

Decisions about how many animals are desired to persist for how long into the future need to be made on a case-by-case basis. Determining a core that can support a viable population is discussed by Stanton and Akçakaya (chapter 11); for many species this requires a regional scale of analysis. In many cases however,

Table 9.1. Examples of standards for defining core habitat

Species	Definition of core habitat	Citation
Mountain quail	• A contiguous area of habitat of medium to high quality that has an area greater than two home ranges in size • In continuous use by the species successfully enough to produce offspring that disperses	Timossi et al. (1995a)
Marten	• 30 to 50 square km, 75% of which is in suitable stands (overstory of at least 40% cedar, spruce, pine that has a canopy closure > 75%)	Watt et al. (1996)
Coachella lizard	• Shall contain populations of sufficient size to be considered viable, independent of others • Core cannot be fragmented by roads or development • Core has intact processes, including sand source and delivery system for the lizard • Each contains a sand source	
Prairie chicken	Core habitat as patches of suitable habitat (mixed grass prairie, sandhill prairie, tallgrass prairie, sand sagebrush or shinnery) that are • either more than 2,000 ha in area or between 500 ha to 2,000 ha in area; and • no more than 10 km from another patch of at least 500 ha in size.	Hagen et al. (2004)
Grizzly bear	• Suitable habitat: based upon road density, building density, land cover, and amount of forest and grassland • Core areas: contiguous areas > 250 square km with a habitat effectiveness > 3.0	Walker and Craighead (1997)
	• An area of secure grizzly bear habitat within a bear management unit that is at least 0.31 miles from any road, motorized trail, or nonmotorized high-use trail that is open during the active bear season	Schwartz et al. (2005)
Veery	• Deciduous/mixed forests • Core habitat must be contiguous forest that is at least 80 m from an edge; core areas average greater than 100 ha	Herkert (1995)
Northern goshawk	• A home range of at least 2,000 ha, of which 1,000 ha are core • Core habitat is defined as habitat that is medium- to high-quality	Timossi et al. (1995b)
Black bear, river otter, moose (Vermont)	• Patches of unfragmented landscape • Large core: > 10,000 acres • Medium core: > 1,500 to 10,000 acres • Small core: 0–1,500 acres	Austin et al. (2006)
California spotted owl	• Core area delineated around mixed conifer old-growth stands that can serve as nest sites where the surrounding 600 acres has a canopy closure of at least 50%	Gerrard (2006)
Tiger populations in Sumatra	• Core comprises the highest quality habitat > 200 square km • Provided that the area has a high proportion of occupation based upon radio-collar GPS data	Linkie et al. (2006)
Salmon	Biologically suitable (capable of supporting life phase function) that is • temporally persistent (spatially fixed and has a high potential for habitat rejuvenation), • located within or adjacent to a migratory corridor, • accessible most of the time (not dependent on low return-interval events to facilitate access), and • cumulatively large (supports sizable portion of population).	Martin and Benda (2004)

Source: Richard Church.

defining a core for a smaller number of animals persisting for shorter time frames can be an effective conservation goal at a subpopulation level, or even a smaller group at a local level of analysis. Core habitat at these scales can contribute to supporting a viable population at a broader scale.

The starting point: A map of suitable habitat

The process of identifying core habitat begins with a map of suitable habitat for the species or community in question. To recap Dickson et al. (chapter 7), habitat mapping for terrestrial species begins with the collection of spatial data that can be used to identify the extent of a specific habitat type. Such data is collected and mapped in vector or raster format. Each cell or polygon represents a specific factor such as vegetation cover, soil type, land use, or distance to the nearest road. From the data available, it is necessary to estimate the extent of useable habitat for the species or community of interest across a target area or region. This, in turn, requires an estimate of just what is necessary to support a given species. For example, suppose that the goal is to identify black-footed ferret habitat. Black-footed ferrets prey on prairie dog colonies (Hoff and Bevers 1998). Developing a map of prairie dog colonies that are not subject to prairie dog abatement programs would be a good start to identifying potential suitable habitat for the black-footed ferret.

Another approach commonly used in mapping habitat involves the use of a wildlife habitat relationship (WHR) model. WHR models are developed by experts who have estimated what habitat elements are needed for a given species (examples of such models are presented by Johnston et al., chapter 14). For example, the California spotted owl is found in late-seral stage old-growth stands of mixed conifers (Gerrard 2006). Therefore, suitable habitat for the spotted owl can be pinpointed by identifying areas in which old-growth mixed conifer forests exist. Habitat for a species may also be determined by using species presence or species presence/absence data, when available. A number of modeling approaches—ecological niche factor analysis, generalized additive model (GAM), generalized linear model (GLM), and principal components analysis (PCA)—have been developed to predict suitable habitat (Guisan and Zimmermann 2000). Many of these approaches have been covered in detail in chapter 7. The current chapter assumes that a map of suitable habitat has been developed. The objective here is to identify and map core habitat.

General approaches to modeling habitat range from detailed scoring of habitat values to a simple habitat/nonhabitat dichotomy. Figure 9.1 presents a habitat suitability map for the San Joaquin kit fox for the eastern portions of Contra Costa and Alameda Counties in California (Gerrard et al. 2001). The habitat suitability value of each raster cell is based upon a number of environmental variables (e.g., land cover and the presence of active agriculture) and has been defined in terms of a ratio-valued suitability score ranging from -20 (not suitable or even dangerous for the fox) to 4 (highly suitable). It is common to define suitability for a value range of 1 to 10 (Craighead and Cross 2007); however, the use of a score and its value range should reflect habitat utility and may include negative values as well. More information on ratio-scaled suitability scores can be found in Church and Murray (2009), and a cautionary note on such rankings is given by Wolman (2006).

Another example of a suitability map is given in figure 9.2, which depicts habitat suitability for the California spotted owl. This second example is a map of the percentage of canopy closure for a region of mature forest stands. The higher the percentage of canopy closure, the more suitable a location is for the spotted owl. These two maps depict habitat suitability in terms of a range from poor suitability to high suitability.

A simpler form of a habitat suitability map involves classifying each spatial unit as either suitable or not. An example of this third form is depicted in figure 9.3A, where each cell is classified as either being suitable habitat or part of a matrix of other lands (McGarigal and Marks 1995). It is possible to derive this third form of habitat suitability map from the second type of suitability map, as is often done.

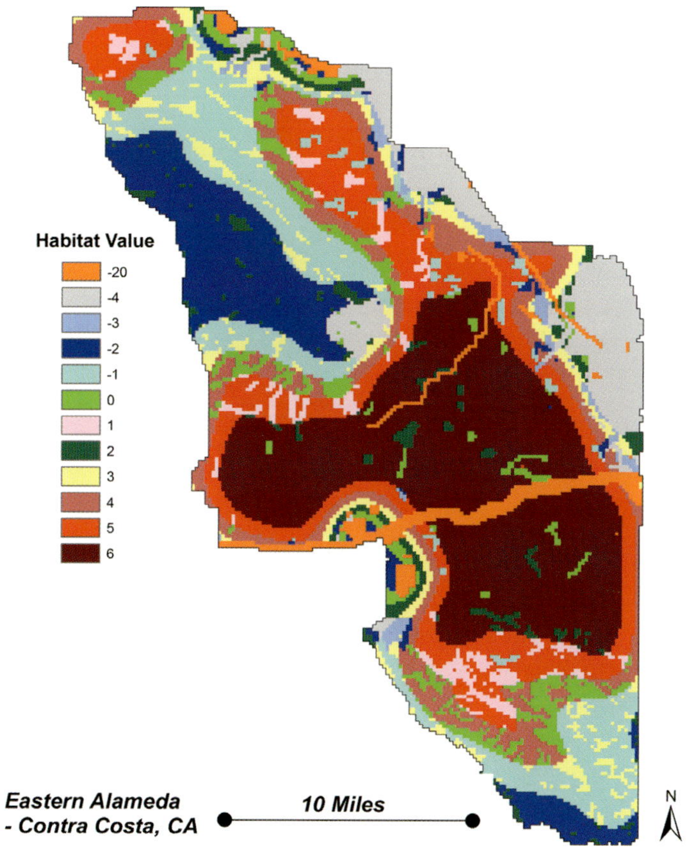

Figure 9.1. Habitat suitability map for the San Joaquin kit fox. Derived from Gerrard et al. (2001).

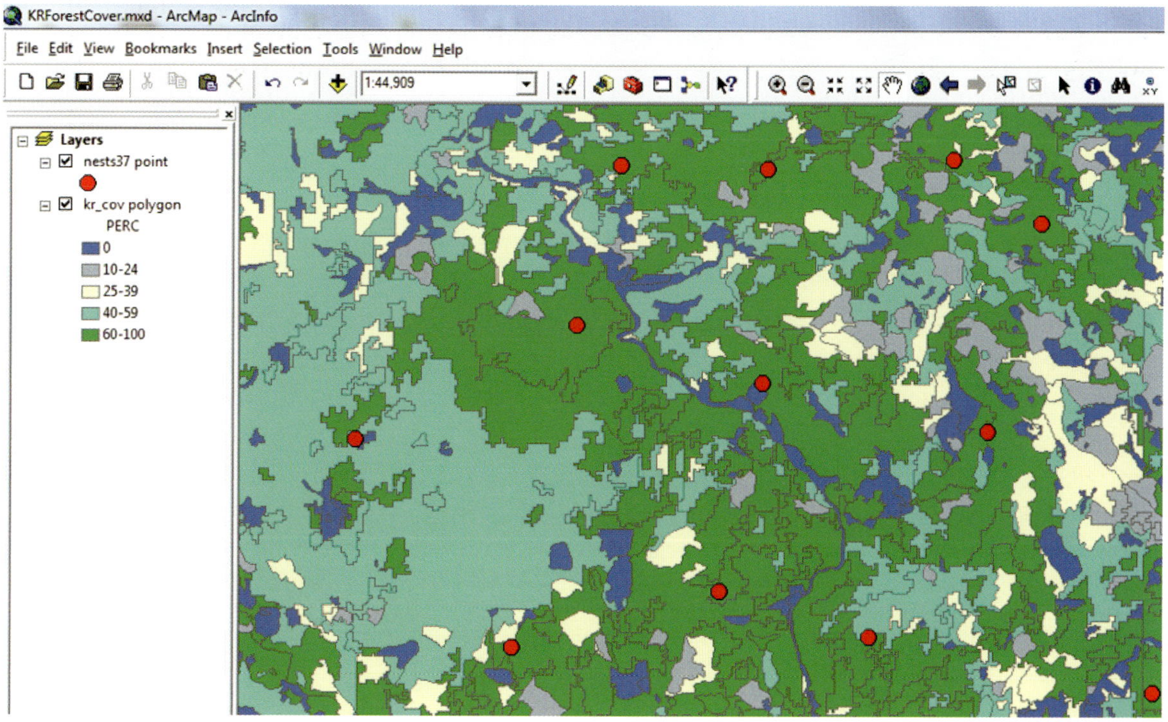

Figure 9.2. A habitat suitability map for the California spotted owl in the Kings River ranger district of the Sierra National Forest. The red circles depict known owl nest sites. Data from the US Forest Service.

Figure 9.3A. Habitat and landscape matrix. Figure from Richard Church.

Basic approaches to determining core habitat

For many conservation planning applications, the mapped range of habitat values can be used to delineate core habitats for purposes of protection, management, or analyses of connectivity. Johnson et al. (chapter 14) provides one example in which the most favorable habitats for each season are considered cores: a threshold of habitat value is chosen and all habitat above the threshold is considered favorable. They provide another example in which all suitable habitat for wolverines is considered core for purposes of examining the effects of predicted climate change.

One variation of using a habitat threshold approach is to consider all habitat above a certain value and then map contiguous blocks of that habitat to attain a desired area. The result is an irregularly shaped patch of the highest-value habitat. Depending upon the ecology of the species in question, and the characteristics of the intervening habitat (less preferred, unusable, avoided), a core area can be defined as all the habitat within the outer boundaries of the high-value habitat, or only the high-value habitat itself, or some other measure. In this fashion, grizzly bear core habitat in Yellowstone was delineated in blocks of 50 km^2 or larger, and 250 km^2 or larger for purposes of analyzing connectivity (Craighead et. al. 2005). In reality, most of the habitat was used by grizzlies to some extent, but it was assumed that the lower-value habitats were used more for travel to reach areas with greater food resources. Fifty km^2 represents an average home range (over all habitat values) for a single female, and 250 km^2 represents habitat for several family groups.

Dickson et al. (chapter 7) discuss an occupancy modeling approach for modeling habitat. In this example, any habitat that a given species occupies could be considered core; alternatively, a threshold of probability of occupancy could be selected to identify core habitat.

The next sections present five different approaches for defining core habitat: buffer-defined core, density-defined core, patch-defined core, population-defined core, and territory-separated core.

Buffer-defined core

One of the most common definitions of core is a contiguous area of good habitat that falls entirely within a habitat patch (McGarigal and Marks 1995). Two such habitat patches are surrounded by matrix cells in figure 9.3A. The matrix cells surrounding a habitat patch are either dangerous to the species of interest or lack sufficient quality to be suitable habitat. This means that patch cells on the edge of the patch are not fully surrounded by good habitat. We can define the core to be those cells that are surrounded by suitable habitat and are at least some minimum specified distance to any edge. We call this "buffer-defined" core habitat. Without loss of generality, suppose that the buffer distance is equivalent to one cell width. Using this definition of core, figure 9.3B depicts the core as those cells that are surrounded by habitat and are at least one cell distance away from the edge of the patch and potentially dangerous matrix areas.

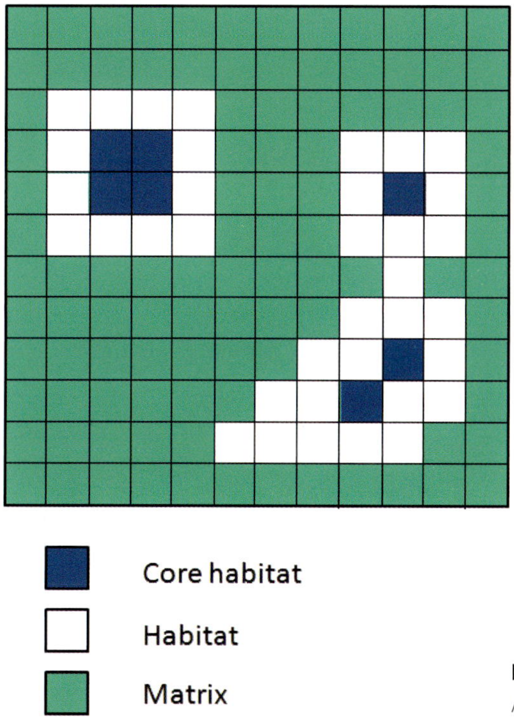

Figure 9.3B. Buffer-defined core areas generated by a program such as Patch Analyst. Figure from Richard Church.

Several approaches can be used to generate this type of map:

1. Develop a special-purpose, stand-alone computer program, Arc Macro Language (AML) script, Python/Visual Basic code, and so on.
2. Use functions provided in the grid tool box found in Esri ArcGIS software or tools found in Geographic Resources Analysis Support System (GRASS) or some other GIS software.
3. Use a specially designed application, for example, Patch Analyst.

Figure 9.3B is an example generated by using Patch Analyst (Elkie et al. 1999). Other programs have been developed to calculate core habitat as well. Most notable is FRAGSTATS (McGarigal and Marks 1995). The definition of a core cell generated in FRAGSTATS for a one-cell buffer is shown in figure 9.3C.

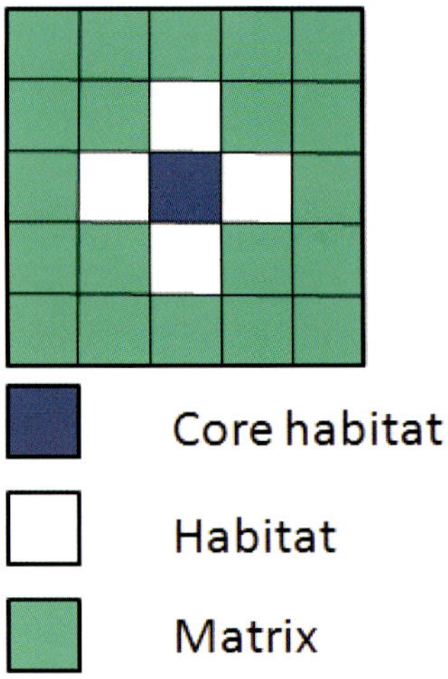

Figure 9.3C. Depicting the definition of "core" used in FRAGSTATS. Figure from Richard Church.

FRAGSTATS checks the neighboring cells in each of the cardinal directions (east, west, north, and south) to identify core. Using this definition of "core" (one-cell buffer in the cardinal directions), the FRAGSTATS program estimates core for our two-patch example given in figure 9.3D. The difference between what Patch Analyst and FRAGSTATS will classify as core results from the algorithm used in the program code and not the definition of the buffer.

Figure 9.3D. "Core" as defined by the FRAGSTATS program. Figure from Richard Church.

In another example, the two patches are only one cell apart, as in figure 9.4A. Girvetz and Greco (2007) suggest that these two patches may really be one, depending on scale and the quality of the cells separating them.

Figure 9.4A. Two patches that are close together. Figure from Richard Church.

The two patches can be morphed together using rules in a program called PatchMorph to generate one patch, as shown in Figure 9.4B. However, this process may also affect which cells are determined to be core.

Figure 9.4B. An example of using PatchMorph software. Figure from Richard Church.

For example, figure 9.4C depicts core for the morphed patch using the definition of "core" given in Patch Analyst. Note here that some cells now depicted as core were originally classified as matrix land, not habitat. Whether this approach is right or wrong is debatable. For instance, if some of the gap cells between the two patches are indeed habitat and were wrongly classified, then the morphing process may have identified the real patch and the actual core. However, the gap cells may be appropriately classified, and the two patches should not be merged. There is also an issue of scale. Even if the habitat quality of the "gap" cells is not high, it may not play much of a role in defining habitat when the cell size is smaller than the size needed by the species. Thus, morphing may be reasonable when such gaps are small in relation to the size of the total habitat and its use. Overall, the lesson to learn here is that codes such as Patch Analyst, FRAGSTATS, and PatchMorph, as well as other software packages or scripts, should be used with caution, with an understanding of the underlying rules and definitions employed in the codes and their appropriateness with respect to the species of concern. The computer-generated cores should be ground-truthed in the field whenever possible to determine their validity.

Figure 9.4C. Core defined on two morphed patches. Figure from Richard Church.

Density-defined core

A density-defined core is one that is defined by a measurement of some important habitat characteristic. For example, the US Forest Service has been developing protection plans for the California spotted owl in the Sierra Nevada (Gerrard 2006). The spotted owl is principally found in mixed conifer stands of old-growth forests. Nest sites are usually found in locations of dense stands and high canopy cover. Figure 9.2 presents a simple habitat suitability map for the spotted owl based upon the percentage of canopy cover found in each polygon. This map presents the percent canopy cover in five broad classes: 0–9%, 10–24%, 25–39%, 40–59%, and 60–100%. These broad classes were defined based upon the classes of vegetation/forest stands present in the ranger district. The known nest sites for the 2003 season are represented as large red dots. The spotted owls tend to nest in areas having a canopy closure that exceeds 60%, and their home range tends to have an average canopy closure of 50% or greater. In general, one can think of the areas that are colored blue or light yellow as the matrix lands, and the areas colored aqua or dark green as preferred habitat. Although the map clearly delineates areas that are preferred by the owls, it is a complex pattern. It is difficult to discern exactly where owls might nest. For example, some nests are on the edge between two canopy closure classifications. Even though many habitat suitability maps are produced in the form presented here, it can be difficult to get a sense of where core might exist.

Simple habitat suitability maps do not necessarily convey complex issues. For example, owls are found in a variety of canopy closure levels during foraging, and one can think of the nest site being located by the owl within a complex mixture of stands of various ages, where the preponderance of stands are dense, old growth. Owl biologists have suggested that the average of what is found within a 600-acre (242.8-hectare) area adjoining the nest site is the best indicator of habitat suitability for a nesting area. That is, the suitability of a nest site is not only a function of the stand in which the nest is located but a function of what surrounds the nest up to a circular region of 242.8 hectares (about 2.4 kilometers squared of land, with a radius of 879 meters). Thus, the map of each polygon or cell misses the situational elements of what surrounds that polygon or cell. That is, a habitat suitability map may need to not only represent the value of the habitat in a specific land unit (intrinsic value) but also represent the values of surrounding units (neighboring value) (Gerrard et al. 2001). We can take a map of habitat values such as that given in figure 9.2, and generate another map of area-based averages. For example, we can use the Focal Statistics tool in ArcGIS software, using the circle option to calculate the average of all cells within a radius of 879 meters of each cell.

Thus, for each cell, the new map layer indicates the average of cell values in a 242.8 hectare circle around and including that cell. Essentially, this would be a map of cell neighborhoods (intrinsic and situational values), and it represents a density-defined measurement. For this example, we have further classified the average canopy closure values into three ranges: 0–49%, 50–59%, and 60–100%. This map is given in figure 9.5, along with the known location of owl nest sites. In figure 9.2, we see that the owls tended to select nest sites at locations with a canopy closure of 60% or higher in stands of old-growth mixed conifers; from figure 9.5, we see that the nest sites were also in areas that have a 600-acre average canopy closure of 50% or higher.

There are two important points to be made here: first, a habitat map is often more complex than that depicted as examples in manuals (e.g., Patch Analyst or FRAGSTATS, where the landscape is classified as either habitat or matrix), and second, neighborhood values may be as important as the intrinsic value of a given land unit in depicting a habitat map.

Figure 9.6 presents the entire Kings River Ranger District. Forest biologists defined the core of the habitat to be those areas that equaled or exceeded a 600-acre average of 60% canopy closure, and the habitat to be those areas in which the canopy closure was 50% or greater. Note that the core areas depicted in figure 9.6 vary in size and shape, but at any cell within the core, the average canopy closure equals or exceeds 60% for a circular region around that site. This means that core elements may be dependent upon habitat outside of the core to fully function as core.

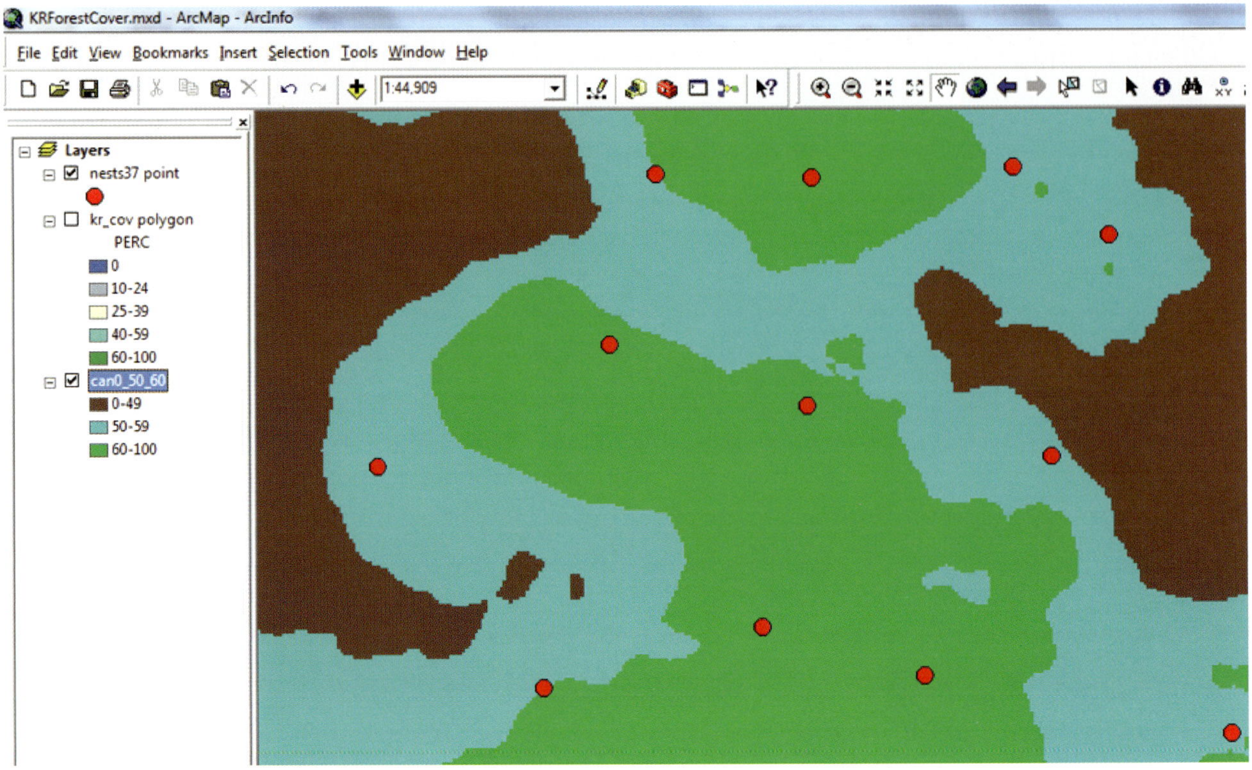

Figure 9.5. Spotted owl nest sites mapped with canopy-cover averages. Data from the US Forest Service.

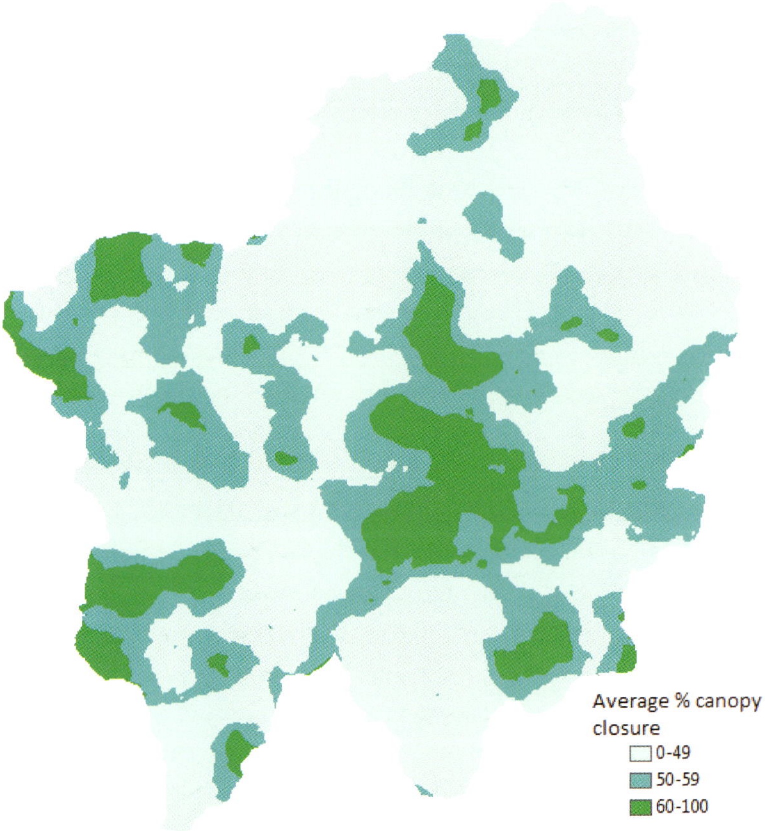

Figure 9.6. Core areas of the Kings River Ranger District of the Sierra National Forest are depicted in dark green. Data from the US Forest Service.

Patch-defined core: Growing a patch

Another method to define core areas of habitat is based in part on the fact that owls and other species generally forage as efficiently as possible. They will tend to avoid dangerous areas and visit areas with greater density of prey. What this means is that the territory surrounding a nest or den may not be an ideal circle, but is large enough to support the species' needs, and varies in shape due to local changes in resources and cover. In a completely homogeneous range, the use by the owl or other animal might be quite circular. Similarly, in a heterogeneous landscape, given the variability of cover on the landscape, the shape of the territory is likely to be as compact as possible to efficiently utilize locally concentrated resources.

An example is a study of the San Joaquin kit fox (SJKF) by Gerrard et al. (2001). The objective of the study was to identify feasible sites (patches) large enough and of suitable quality to support a mated pair of kit foxes. In other cases, the objective may be to determine suitable sites or planning units that would support some other demographic unit of a species (e.g., herd or pack) using this method. The basis for the work on the SJKF was the habitat suitability map given in figure 9.1, in which each cell was valued in terms of a ratio-scored habitat value (Gerrard et al. 2001). Suitability values ranged from +6 (high quality) to a -20 (not suitable/dangerous). The dark band across the map south of the center is associated with an interstate highway. Undisturbed grasslands scored a +6, mixed habitat (e.g., oak woodland) scored 1 through 5, and areas scoring negative values were regarded as places that a kit fox would avoid. Even highly suitable regions contain local pockets that would be avoided by a fox. The scoring approach was developed so that the value of a patch (a connected set of raster cells) was the sum of the suitability scores of each of the cells within the patch. To support a mated pair of kit foxes, the team determined that a minimum of 4 square miles (1,036 ha or 2,560 acres) of the best-quality habitat was required. The habitat suitability map was based upon a cell size of 4 hectares. This meant that 256 cells of the best-quality habitat were needed to support a mated pair. This equates to 256 cells at +6 value points per cell, which equals 1,536 habitat-value points. Any feasible patch that could support a mated pair needs at least 1,536 habitat-value points. For instance, if the habitat was rated at +4 throughout, then a feasible patch would require 1,536/4 = 384 hectares—a larger size due to a lower average suitability. The scoring function can be applied to any delineation of area to determine whether that area could support a mated pair of kit foxes; that is, it needs to have a sum of scores that equals or exceeds 1,536 value points. It is important to note that a cell of -20 in a patch would have to be compensated by more than 3 cells of best-quality habitat (+6), just to balance out at a net of zero. One of the features in the kit fox habitat that was scored a -20 was urban area, which would be avoided by a kit fox, making its territory larger and day-to-day hunting less efficient because the fox would have to move in more circuitous routes.

Given a map of habitat values, and knowing that a feasible patch of core habitat needs to contain at least 1,536 habitat quality points to support a mated pair of kit foxes, we can now turn our attention to generating feasible patches. The basic idea is that habitat, and especially fragmented habitat, is often mixed and not homogeneously good. Core habitat for the kit fox, in this context, must be those areas in which we can draw relatively compact patches where the sum of habitat-value points within those patches exceeds 1,536. This approach can be used for any species, given values of habitat classes and an area or total value size to achieve. Keep in mind that the total habitat-value goal is just an estimate. It is based upon many assumptions, such as any cell rated as a 4 (due to its classification through remotely sensed imagery) is identical to any other cell rated 4. Land-cover classification is discussed by Olenicki (chapter 4) and model assumptions by Johnston et al. (chapter 14). The goal of 1,536 habitat-value points, however, is a useful estimate that can be achieved and mapped in a GIS.

To address the problem of turning a map of habitat values into feasible core patches, Church et al. (2003) developed a patch generation technique called the patch growing process (PGP). This process can be started in virtually any place on the map with a search for a feasible patch that includes a starting or seed cell. Scale is important; it is assumed here that the cell size in relationship to needed habitat size is quite small, say, 1% or less. This means that 100s of cells will be selected in generating a patch. The objectives of the PGP are fourfold:

1. To generate good representative patches about a seed cell
2. To keep the length of the final perimeter of the patch low
3. To maintain a tendency to select the most suitable cells as it grows
4. To generate patches that meet the expert's criteria for feasibility

The basic process of the PGP can be described as follows. Because the patch is derived around a seed cell, it is necessary to set a minimum standard for a cell to be a seed: it would be nonsensical to start defining a patch in an area in which there is no suitable habitat. When starting at a given seed cell, we will automatically expand the starting footprint to either a 3 × 3 set of cells, 5 × 5 set of cells, or some minimum compact footprint. Selecting an odd numbered size allows us to center the starting footprint on the seed cell. Once this small, compact square footprint is defined, the rest of the process is determined by growth.

Growth of a patch occurs on the perimeter. We can start by identifying all cells outside the small patch that share at least one edge with a patch cell and place them on a list. The order of this list is then randomized. As we add cells to the list, we also generate a composite suitability score for that cell associated with selecting it to be added to the patch:

$CS_i = Suit_i + N \times e_i$

where $Suit_i$ is the suitability value for cell i, e_i is the number of edges that cell i shares with the existing patch, and N is the weight attached to sharing edges with the existing patch.

We now have a list of cells, ordered randomly, that are adjacent to the present boundary of the patch. Each of these cells has a score that is based upon the suitability of the cell and the number of edges that the cell shares with the current patch boundary. This list is then ordered from highest value to lowest value.

Next, we select the top X percent of the cells on this list to add to the patch. If the patch is now large enough in terms of value points, we stop, because we have defined a feasible patch. If the patch as defined so far is not large enough in terms of the value-points total, we repeat the process by developing a new list of cells sharing an edge with the new patch perimeter in randomized order, calculate the composite suitability value for adding each cell to the patch, and so on.

The process is very simple, but quite powerful. The major driving force in adding cells on the perimeter of the growing patch is to add the highest quality cells that exist on the perimeter. But, other cells may be needed to fill in a jagged edge, when a cell outside the patch shares three or four edges with the patch, or when the perimeter does not contain many high-valued cells. If the values of X and N are high, then the shape of the patch is more important, and it will be a compact, box-like shape centered about the seed cell. With lower values of X and higher values of N, the patch will tend to grow in regions of higher-valued habitat and have a smooth perimeter; with low values of N and X, the process will select those cells with as high a suitability as possible, which could result in a very jagged, lengthy perimeter. Thus, the process can be tuned as appropriate so that the habitat patches are realistic and can support an intended function.

Three important points should be made before giving an example of this process. First, the purpose of randomizing the order of the list of prospective patch cells is to eliminate any bias associated with how the list of cells is generated. Because a computer program would process a list according to an approach that would be executed the same way at each iteration, unless the order is randomized, cells would have some type of order based upon the process (e.g., starting at 12 o'clock and proceeding in a clockwise direction). If all cells had the same composite suitability value, then without randomizing the order, the top of the ordered list would have those cells added to the list first, thus giving a bias toward growing in the north and easterly directions. Randomizing the order of the list after it is generated eliminates any bias associated with the list generation phase. The second point is that if the same cell is used as a starting seed several times, each resulting patch may be somewhat different. This is also due to the fact that the cells have essentially been added to the list in a randomized order. Thus, different representative patches may be generated from the same starting cell. The third point is that if a generated patch is too large when it reaches the needed quality points, then it can be eliminated from being used in any further analysis. Such an event would happen if the amount of useful habitat in a local region around the seed cell is almost nonexistent, and the lack of good habitat requires the process to grow a patch to such a size that it is not viable for an animal to occupy. Finally, to reiterate the premise given earlier, core habitat can exist only in those areas in which a viable patch can be generated, or when a contiguous area contains at least a minimum number of nonoverlapping viable habitat patches.

Figure 9.7 shows three different patches generated by the PGP on a portion of the SJKF suitability map. Note that the patch is sometimes bordered by poor-value habitat (patches B and C) and sometimes incorporates poor-value areas (patch A).

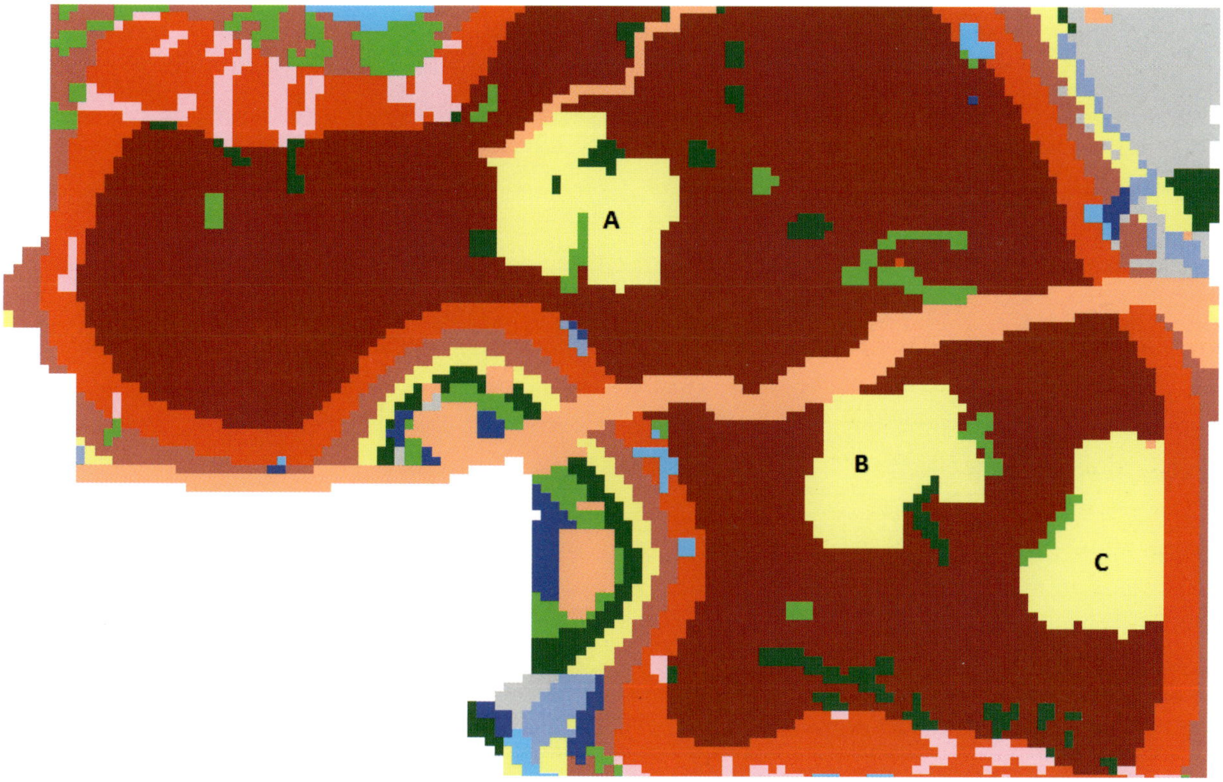

Figure 9.7. An example of three SJKF patches (A, B, and C) generated by PGP. Patch footprints are depicted in yellow. Patch A incorporates an area of poor-value habitat (green), while patches B and C border poor-value habitat. Data from the National Center for Ecological Analysis and Synthesis in a project funded by the National Science Foundation.

Figure 9.8 gives a composite view of the boundaries of 350 patches generated for the map given in figure 9.1. Each patch was generated by picking at random a seed cell that contained high-value habitat, so that the patches are equally scattered through the region. When an area has a high density of patch boundaries, it can be interpreted as potentially playing a role in many different patches. That is, it is core. When the density of patch boundaries is low, few if any definable, viable patches could be found, and the likelihood that such an area is core is low, even if it contains high-value habitat. As always, whenever possible, the core areas identified by modeling should be validated in the field. If the land-cover classification that the models are based upon is accurate and has been ground-truthed, then the cores that are identified should be reasonable estimates.

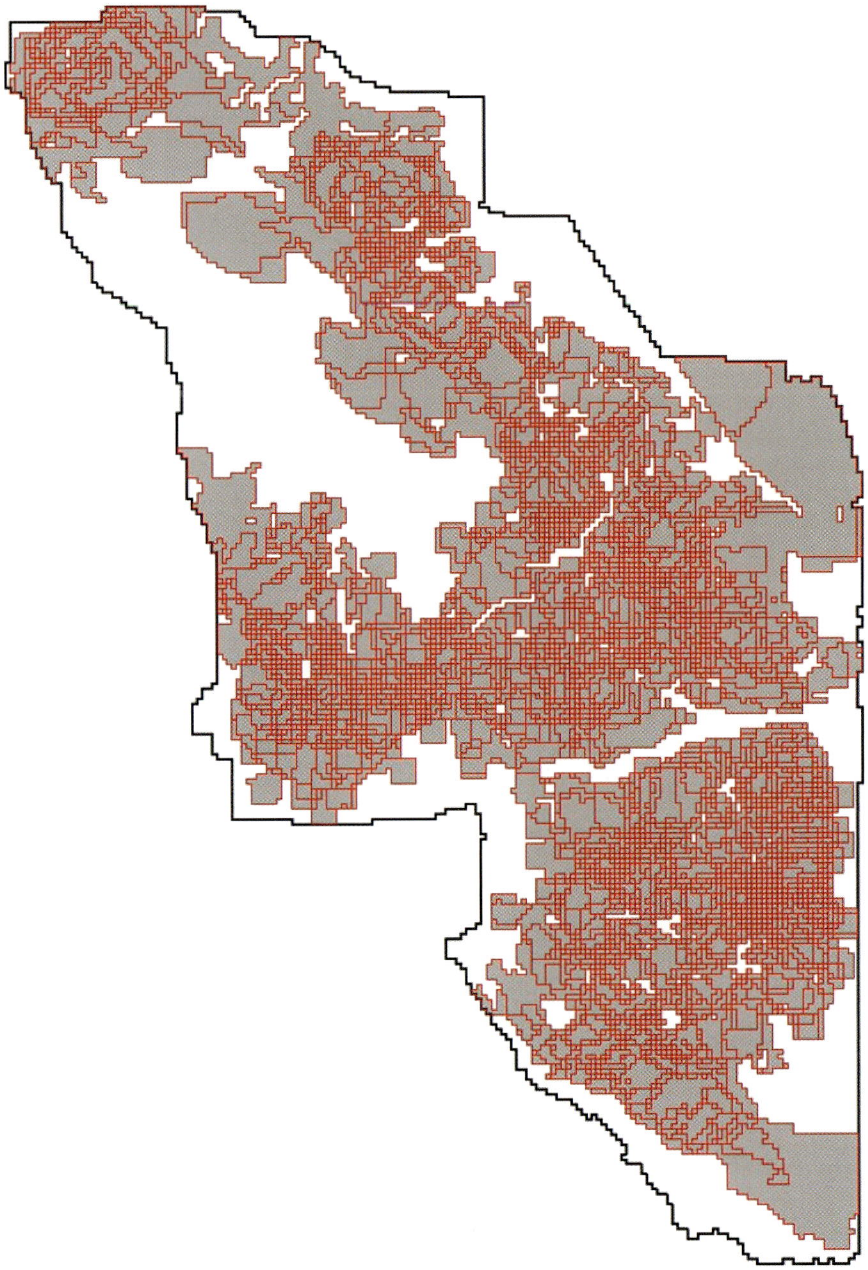

Figure 9.8. A composite of 350 feasible patches generated for the area presented in figure 9.1. The boundaries of patches are shown in red. Figure produced by Ross Gerrard and Richard Church in a project funded by the National Center for Ecological Analysis and Synthesis.

Estimating useful core for a population with overlapping territories

Many species, including some top predators often used as focal species, do not defend discrete territories but instead range widely, overlapping with other individuals. For these species, one commonly used metric for conservation planning is population density: for example, Yellowstone National Park had an average population density of one bear per 50 km^2 of habitat (Blanchard and Knight 1980). Such a bear density might be used to suggest that each bear requires 50 km^2 of good habitat in that region. In contrast, in the Arctic Alaska habitat in the Eastern Brooks Range, a bear required from 148 km^2 (Quimby and Snarski 1974) to 173 km^2 in high-quality habitat (Reynolds 1976), and in the Western Brooks Range, a bear required an average of 42–44 km^2 of habitat (Reynolds 1984). We can then consider core habitat to be that area that has a bear density that exceeds a certain minimum density. Species density can also be used to estimate the number of animals supported in habitat where there have been no census attempts. If density estimates are available for different habitat types, or for habitat classified as core versus noncore, population estimates can be made.

Demographic parameters are also used to characterize habitat as either "source" (where residents have a positive growth rate) or "sink" (where residents have a negative growth rate; see Doak 1995). Where such data is available, core habitat can be defined as only the source habitat or as an area that includes source and other habitats. For example, Schwartz et al. (2005) examined individual, temporal, and spatial covariates that potentially affected reproduction and survival of grizzly bears, and found that habitat containing whitebark pine influenced survival. Because of this, a source–sink dynamic was suggested for the Greater Yellowstone Ecosystem, with positive growth rates inside Yellowstone National Park and the Yellowstone Recovery Zone (an area within which a threatened or endangered population and its habitat is managed so as to recover the population), and negative growth rates outside the Yellowstone Recovery Zone.

Estimating useful core habitat as spatially distinct territories

In contrast to wide-ranging species, territorial species exclude other individuals from areas of habitat and thus space themselves more predictably across the landscape. Whether they defend a territory as a family unit, a pack, or a single individual, their cores can be calculated in terms of how much habitat each territorial unit requires.

It is a relatively easy task (using GIS functionality) to calculate the total area that is designated core habitat. If 2,560 acres (4 square miles) of suitable habitat are needed to support a mated pair of kit foxes, then dividing the total acreage of core habitat by 2,560 acres per mated pair will yield a very approximate estimate of the population that may be supported by the core habitat and surrounding buffer habitat. Figure 9.9 represents a hypothetical habitat range for the kit fox, a territorial species, containing 3,000 acres. If two hypothetical circular territories are superimposed on the raster map, circumscribing an area of 2,560 acres, some of the habitat within the circular region is not core habitat. The habitat might support two mated pair of kit foxes if the habitat quality outside of the two core areas is quite desirable as well.

Two issues are particularly relevant here: (1) no compact circular-like territory exists that is comprised of only core habitat, and (2) the fox territories appear to need or rely on habitat outside of a hypothetical circular core because the core habitat is distributed unevenly. Depending upon whether foxes can use noncore habitat effectively, the answer to how many mated pairs this patch will support is none or two, not one,

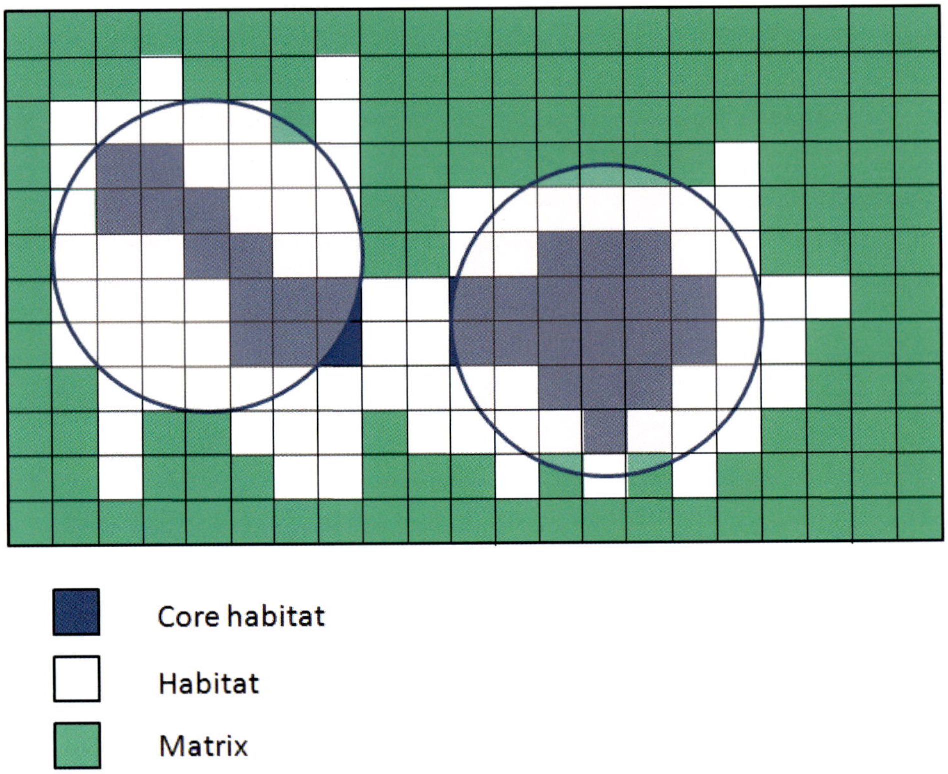

Figure 9.9. Estimating useful habitat. Figure from Richard Church.

which is the answer generated by the simple mathematical calculation. The confusion here arises due to the quantity and distribution of core and is perhaps one of the major research questions in habitat utilization. One method in modeling the effectiveness of a specific distribution of islands or patches of core is to consider the use of a population viability analysis (PVA) model, which is the subject of chapter 12. In what follows, we will discuss an alternative spatial analysis model for estimating the usefulness of core.

The above discussion seems to suggest that the definition of core may need to include other elements beyond those given in table 9.1. That is, even though a given raster cell may qualify as core because it contains desirable habitat and is far enough from a dangerous boundary, other circumstances surrounding the cell are important. Figure 9.6 presents a map of habitat suitability for the spotted owl as a function of average canopy-closure percentages.

This region has suitable habitat within a matrix of land cover that is less suitable, but is it all core? The presence of a nesting site is one of the key details in delineating what is core, but the spotted owl will forage in areas that are less suitable. With few exceptions, owls nested in a location with a canopy closure of 60–100% and where the canopy closure of a 600-acre circular region surrounding the nest site averaged 50% or higher (Gerrard 2006). Biologists defined core habitat as those areas that are nest sites and have a canopy closure of 60% or greater, which may partially rely on core buffer areas that average 50% or greater. Given this interpretation, core habitat would be the dark green areas in figure 9.6, and core buffer would be the aqua areas in the figure.

To define these core areas in terms of how many owls they can support (how "useful" they are) we can ask:

If a circular area of size R composed of core habitat is large enough to support a territorial species, then how many circular territories can be drawn within an area of core habitat without overlapping?

The size and shape of a core area may not be distributed so that a species can maximize the use of all the best habitat. There are two ways to view the issue of how many viable territories will fit into a core area, based upon the process under which a species will populate an area: (1) what is the "packed" arrangement of circular areas that effectively maximizes the number of territories that can be located within the core region (assuming dispersal is tight and just far enough), and (2) what is a more randomized, scattered arrangement of territories in which the gaps between any territories are too small to support an owl? A species might display some of both tendencies. A packed arrangement would represent the largest number of owls that could be supported, and a looser, somewhat random scattered pattern would represent the lowest estimate of what the core will support. The actual value is likely to be somewhere between these two estimates.

Random Maximum Scatter routine

We can develop a method for the determining the lower, scattered estimate as follows for circular territories of radius R:

Method: Random Maximum Scatter (assume a raster)

Step 1. From the core habitat, identify those raster cells that can serve as territorial centers. Define this as list Ω; set $k = 0$, where k represents the number of circular territories placed; set F = null, where F is the set of center locations selected; and go to step 2.

Step 2. Randomly select a cell j from list Ω. Place cell j on list F, increment k by 1, and proceed to step 3.

Step 3. Remove from list Ω all cells that are less than 2^*R distance away from cell j, and then proceed to step 4.

Step 4. If list Ω is empty, stop with set F containing k centers whose territories do not overlap. Otherwise, return to step 2.

This routine randomly selects cells to serve as territorial centers, as long as they do not overlap with any previously chosen centers (they must be at least 2^*R distance away). To prevent overlaps from occurring, once a cell has been selected, all remaining cells on the list Ω that are too close to the newly selected cell are eliminated from further consideration. A simple script or code can easily be written in Python, VBA, or other programming language. The key to applying this process is the development of the list Ω. The list can be generated by first using the Focal Statistics tool, which can be used to calculate the total amount of habitat within a circular area. For example, using a circle of radius R, we can total the number of cells having core habitat designation. Thus, we can produce a map depicting, for each cell, how much core habitat exists within a circular region centered at that cell. We can then select those cells, having 100% core habitat within the circular region, or some reasonable standard (e.g., 90% core). This yields a map layer that represents those cells that can serve as a territorial center—that is, the cells which comprise list Ω.

Forest biologists defined this standard for cells to serve as a territorial center for spotted owls (Gerrard 2006): within the area of suitable habitat of old growth, if the average canopy closure in a circular area about a given cell exceeds 50%, and if the cell itself has a canopy closure between 60–100%, then the cell can serve as a territorial center for spotted owls. Overall, there were 74,705 30-m cells that were considered feasible territorial centers in the Kings River Ranger District. Figure 9.10 presents a result of the Random Maximum Scatter routine when the circle radius was set to 789 meters. This routine placed 47 territories at random on the map, where no additional circular territories could be placed without overlapping with already selected locations. This routine took a second or two to execute and could be rerun several times to generate an expected maximum number of possible scattered territories.

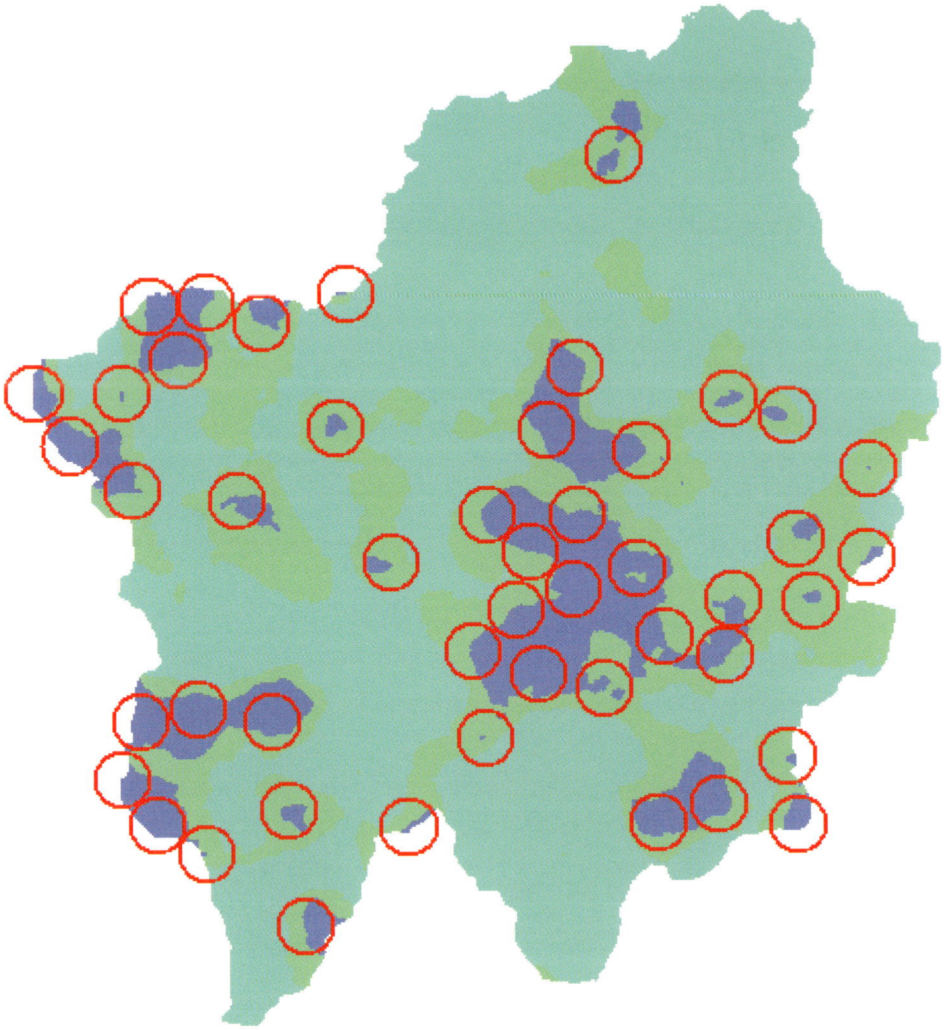

Figure 9.10. Forty-seven scattered territories generated by the Random Maximum Scatter method. Figure from Richard Church.

The Maximal Packing heuristic and the *r*-separation location model (RSLM)

Solving for a packed arrangement that contains the largest number of nonoverlapping circular territories requires a different approach. This particular problem can be formulated as a formal optimization model (see Williams, chapter 12) that is equivalent to the *r*-separation location model (RSLM). To do this, consider the following notation:

$\Omega = \{ j \mid \text{cell } j \text{ is a viable territory center} \}$, basically the same set generated at the beginning of the Maximum Scatter method

$$x_j = \begin{cases} 1, & \text{if cell } j \text{ is selected for a center of a circular territory} \\ 0, & \text{if not} \end{cases}$$

R = the radius of the circular territory
d_{ij} = the distance between cell i and cell j
Using this notation, we can define the following integer-linear programming problem:

$$\textit{Maximize} \quad z = \sum_{j \in \Omega} x_j$$

subject to:

1. $x_j + x_k \le 1$ for each $(j,k) \in \Omega$ and $d_{jk} < 2 * R$
2. $x_j \in \{0,1\}$ for each $j \in \Omega$

This model is quite simple conceptually. First, the objective is to select as many cells as possible from the set Ω. Constraints of type 1 ensure that any two selected cells must be at least 2*R distance apart. The constraint works as follows: if a pair of cells i and j are within a distance of 2*R there will be a constraint of the form $x_i + x_j \le 1$ that will prevent both of them being selected simultaneously. Thus, constraints of type 1 ensure that each selected territory center is at least 2*R distance from all other selected centers. The second constraint specifies that the decision variables must be discrete zero-one in value. This is a rather simple formulation and easily solved for small problems. The difficulty arises when there are thousands of cells that can serve as viable territorial centers.

It is important to note that this optimization problem is a location model that can be found in the location science literature (Church and Murray 2009). It is mathematically equivalent to the vertex packing problem on a graph, the RSLM (Erkut et al. 1996), or the anti-cover problem (Moon and Chaudhry 1984). It has been applied to forest management problems (Murray and Church 1997) as well as habitat analysis (for the spotted owl, see Gerrard 2006); for the sandhill crane, see Downes et al. 2008).

Although there are techniques that have been developed to increase the problem sizes for which the above model can be applied, solving it optimally requires sophisticated optimization software that is not readily available and usually requires knowledgeable personnel in optimization modeling. In an application of the RSLM model to analyze sandhill crane carrying capacity for an area in Ohio, Downes et al. (2008) relied on a software package called CPLEX. Because CPLEX is not readily available without an expensive license, there has been an attempt to develop an approach that is robust, fast, highly accurate, and inexpensive. One of the processes is described below:

Method: the Maximal Packing or RSLM heuristic

Step 1. From the core habitat, identify those raster cells that can serve as territorial centers. Define this as list Ω, set $k = 0$, set F = null, and go to step 2.

Step 2. Select from random a cell j from list Ω. Place cell j on list F, increment k by 1, and proceed to step 3.

Step 3. Remove from list Ω all cells that are less than 2^*R distance away from cell j. If Ω is empty, stop; otherwise proceed to step 4.

Step 4. If $k < 4$, find the cell in $j \in \Omega$ that minimizes the combined distances to the first k sites already in F. Make this the $k + 1^{st}$ site of set F, increment k to $k + 1$ and return to step 3. If $k \geq 4$, then go to step 5.

Step 5. If list Ω is empty, stop with set F containing k centers whose territories do not overlap; otherwise proceed to step 6.

Step 6. Find a cell in $j \in \Omega$ that minimizes the combined distances to the first four sites selected in set F. Make this the $k + 1^{st}$ site of set F, increment k to $k + 1$, and then proceed to step 7.

Step 7. Remove from list Ω all cells that are less than 2^*R distance away from cell j, and then return to step 5.

The RSLM heuristic detailed above first selects a "building block" cluster of four sites that are as close to as possible to each other. It begins with the random selection of the first site, and then identifies three additional sites that are as close as possible to the first selected site, where no overlap of the associated circular territories is allowed. Once it has selected the building block cluster, it then selects additional sites one at a time, ensuring that the added sites do not overlap with previously selected circular regions and that the combined distance to the building block cluster is minimized. This heuristic strategy mimics the dispersal strategy of a new juvenile individual that moves no further than necessary from the existing territories. This strategy tends to build very tight clusters of sites, with no territorial overlap, while maximizing the number of territories allocated to the landscape. In testing, this process has been found to be quite robust at finding a pattern of territories that yields an optimal or near optimal solution to RSLM. Because the starting position of the heuristic is generated randomly, it is possible that solutions to the RSLM heuristic will vary somewhat, depending upon which cell is selected at step 2. In practice, it has been found that the solutions do vary, but the overall number of located circular territories tends to fall within a very narrow band that is close to optimal. The RSLM heuristic can be applied to relatively large problems and takes less than a minute to execute on a computer with modest processing speed.

The RSLM heuristic applied to the Kings River landscape data produced the solution in figure 9.11 (Gerrard 2006). The packed pattern involves 62 territorial units. Thus, the core area for the Kings River Ranger District should support somewhere between 47 and 62 mated pairs of owls, depending upon their dispersal characteristics.

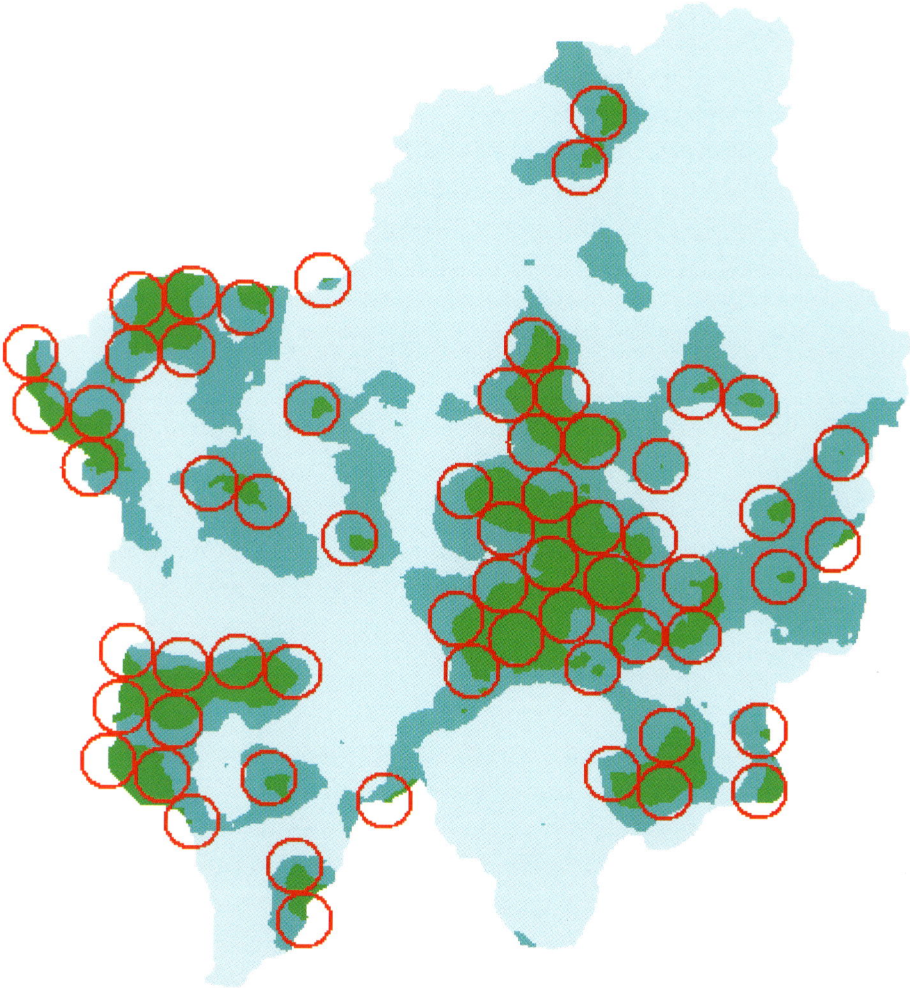

Figure 9.11. Sixty-two packed territories generated by the RSLM heuristic. Figure from Richard Church.

Future developments

Planning for the protection of natural landscapes in the face of human encroachment, increasing fragmentation, and climate change means that greater efforts will need to be made in modeling habitat and the impacts of change. Better-integrated methods are needed to support planning for the protection of focal species. For example, it is realistic to expect that modeling tools integrated with GIS will form the backbone of future work. As these tools evolve, tools such as Maximum Scatter and RSLM heuristic will become readily available as an add-on to GIS software. Just as FRAGSTATS can be used to measure many different metrics associated with habitat, core, and fragmentation, better tools and models for modeling and delineating core habitat will be integrated into decision-making aids (e.g., Marxan; Ball et al. 2009). It is also reasonable to expect that one will be able to move across scales using such tools as population viability analysis, habitat models, corridor location, reserve design models, and climate models.

It is also realistic that research will continue on the issue of delineating core habitat as a key element in conservation planning (Craighead and Cross 2007). The identification of core areas is a necessary first step before connectivity can be assessed across a landscape as discussed in the following chapter.

Acknowledgements

I would like to acknowledge Ross Gerrard as having prepared figures 1, 6, 7, and 8. I also want to acknowledge the support of the US Forest Service to the University of California–Santa Barbara for contacts under which the spotted owl habitat was studied, and the National Center for Ecological Analysis and Synthesis for support for modeling the San Joaquin kit fox habitat. Finally, I want to recognize the collaborations that I have had with Pete Stine, Mike Gilpin, and Ross Gerrard, who were instrumental in the development of several of the approaches discussed here.

References

Austin, J., K. Viani, and F. Hammond. 2006. "Vermont Wildlife Linkage Habitat Analysis: A GIS-Based, Landscape-Level Identification of Potentially Significant Wildlife Linkage Habitats Associated with State of Vermont Roadways." Vermont Agency of Transportation VTrans Research Advisory Council No. RSCH008-967.

Ball, I. R., H. P. Possingham, and M. Watts. 2009. "Marxan and Relatives: Software for Spatial Conservation Prioritisation. In *Spatial Conservation Prioritisation: Quantitative Methods and Computational Tools,* edited by A. Moilanen, K. A. Wilson, and H. P. Possingham, 185–95. Oxford: Oxford University Press.

Blanchard, B. M., and R. R. Knight. 1980. Status of Grizzly Bears in the Yellowstone System. *Transactions North American Wildlife and Natural Resource Conference* 45: 263–67.

Church, R. L., R. A. Gerrard, M. Gilpin, and P. Stine. 2003. "Constructing Cell-Based Species Territories Useful in Conservation Planning." *Annals of Association of American Geographers* 93 (4): 814–27.

Church, R. L., and A. T. Murray. 2009. *Business Site Location, Location Analysis, and GIS.* Hoboken, NJ: John Wiley & Sons.

Craighead, L., and B. Cross. 2007. *Identifying Core Habitat and Connectivity for Focal Species in the Interior Cedar-Hemlock Forest of North America to Complete a Conservation Area Design.* USDA Forest Service Proceedings RMRS-P-49. http://www.fs.fed.us/rm/pubs/rmrs_p049/rmrs_p049_281_296.pdf.

Craighead, L., B. Gilbert, and T. Olenicki. 2005. "Comments Submitted to the US Fish and Wildlife Service Regarding Delisting of the Yellowstone Grizzly Bear DPS." *Federal Register* 70 (221): 69,853–84. http://www.craigheadresearch.org/uploads/7/6/9/0/7690832/craighead_gilbert_olenicki_final_comments.pdf.

Doak, D. F. 1995. "Source-Sink Models and the Problem of Habitat Degradation: General Models and Applications to the Yellowstone Grizzly Bear." *Conservation Biology* 9: 1,370–79.

Downes, J. A., R. J. Gates, and A. T. Murray. 2008. "Estimating Carrying Capacity for Sandhill Cranes Using Habitat Suitability and Spatial Optimization Models." *Ecological Modeling* 214: 284–92.

Elkie, P., R. Rempel, and A. Carr. 1999. *Patch Analyst User's Manual.* TM-002. Thunder Bay: Ontario Ministry of Natural Resources, Northwest Science and Technology. http://www.rocchini.net/ecopae/mat_did/pa_manual.pdf.

Erkut, E., C. ReVelle, and Y. Ulkusal. 1996. "Integer-Friendly Formulations for the *r*-Separation Problem." *European Journal of Operational Research* 92: 342–51.

Gerrard, R. A. 2006. *Measuring Habitat Quality and Carrying Capacity for the California Spotted Owl Case Study Using Data from Sierra National Forest.* Interim Report Compiled for Sierra Nevada Research Center and Plumas-Lassen-Tahoe National Forests.

Gerrard, R. A., P. Stine, R. Church, and M. Gilpin. 2001. "Habitat Evaluation Using GIS: A Case Study Applied to the San Joaquin Kit Fox." *Landscape and Urban Planning* 52: 239–55.

Girvetz, E. H., and S. E. Greco. 2007. "How to Define a Patch: A Spatial Model for Hierarchically Delineating Organism-Specific Habitat Patches." *Landscape Ecology* 22: 1,131–42.

Guisan, A., and N. E. Zimmermann. 2000. "Predictive Habitat Distribution Models in Ecology." *Ecological Modeling* 135: 147–86.

Hagen, C. A., B. E. Jamison, K. M. Giesen, and T. Z. Riley. 2004. "Guidelines for Managing Lesser Prairie-Chicken Populations and Their Habitats." *Wildlife Society Bulletin* 32 (1): 69–82.

Herkert, J. R. 1995. "Status and Habitat Area Requirements of the Veery in Illinois." *The Auk* 112: 794–97.

Hoff, J., and M. Bevers. 1998. *Spatial Optimization for Managed Ecosystems.* New York: Columbia University Press.

Linkie, M., G. Chapron, D. J. Martyr, J. Holden, and N. Leader-Williams. 2006. "Assessing the Viability of Tiger Subpopulations in a Fragmented Landscape." *Journal of Applied Ecology* 43: 576–86.

Martin, D., and L. Benda. 2004. "Identifying Core Areas for Chinook Salmon." In *Development of Salmonoid Conservation Strategies,* edited by Douglas Martin and Lee Benda, chapter 4. Project no. T01426T, King County, Washington.

McGarigal, K., and B. J. Marks. 1995. "FRAGSTATS Spatial Analysis Program for Quantifying Landscape Structure." General Technical Report PNW-GTR-351. USDA Forest Service, Pacific Northwest Research Station.

Moon, I. D., and S. S. Chaudhry. 1984. "An Analysis of Network Location Problems with Distance Constraints." *Management Science* 30: 290–307.

Murray, A. T., and R. L. Church. 1997. "Facets for Node Packing." *European Journal of Operational Research* 101: 598–608.

Natural Heritage and Endangered Species Program. 2004. *Core Habitats of Sharon, Massachusetts.* Westborough: Division of Fisheries and Wildlife, Commonwealth of Massachusetts.

Noss, Reed F. 1987. "Protecting Natural Areas in Fragmented Landscapes." *Natural Areas Journal* 7 (1): 2–13.

Noss, Reed F., Eric Dinerstein, Barrie Gilbert, Michael Gilpin, Brian J. Miller, John Terborgh, and Steve Trombulak. 1999. "Core Areas: Where Nature Reigns." In *Continental Conservation,* edited by M. Soulé and J. Terborgh. Covelo, CA; Washington, DC: Island Press.

Quimby, R., and D. J. Snarski. 1974. "A Study of Fur-Bearing Mammals Associated with Gas Pipeline Routes in Alaska." In *Distribution of Moose, Sheep, Muskox and Furbearing Mammals in Northeastern Alaska,* edited by R. D. Jakimchuk, chapter 2. Arctic Gas Biological Report Series, Volume 6.

Reynolds, H. V. 1976. *Big Game Investigations. Movements and Population Discreteness of North Slope Grizzly Bears.* Federal Aid to Wildlife Restoration Project. Report W-17-6. Juneau, AK: Alaska Department of Fish and Game.

Reynolds, H. V. 1984. Grizzly Bear Population Biology in the Western Brooks Range, Alaska. In *Structure, Status, Reproductive Biology, Movement, Distribution, and Habitat Utilization of a Grizzly Bear Population,* edited by H. V. Reynolds and J. L. Hechtel, 2–19. Federal Aid to Wildlife Restoration Project. Final Report W-17-6. Juneau: Alaska Department of Fish and Game.

Schwartz, C. C., M. A. Haroldson, G. C. White, R. B. Harris, S. Cherry, K. A. Keating, D. Moody, and C. Servheen. 2005. "Temporal, Spatial, and Environmental Influences on the Demographics of Grizzly Bears in the Greater Yellowstone Ecosystem." *Wildlife Monographs* 161: 1–68. http://www.fws.gov/mountain-prairie/species/mammals/grizzly/Schwartz_et_al_2006a.pdf.

Soulé, M. E., and J. Terborgh. 1999. *Continental Conservation.* Washington, DC: Island Press.

Tilson, R. L., and P. J. Nyhus, eds. 2010. *Tigers of the World: The Science, Politics, and Conservation of* Panthera tigris, 2nd ed. Amsterdam and Boston: Elsevier/Academic Press.

Timossi, I. C., E. Woodard, and R. H. Barrett. 1995a. *Habitat Suitability Models for Use with ARC/INFO: Mountain Quail.* California Wildlife Habitat Relationships Technical Report No. 10. Sacramento: California Department of Fish and Game. http://nrm.dfg.ca.gov/FileHandler.ashx?DocumentVersionID=17456.

———. 1995b. *Habitat Suitability Models for Use with ARC/INFO: The Northern Goshawk.* California Wildlife Habitat Relationships Technical Report No. 14. Sacramento: California Department of Fish and Game. http://nrm.dfg.ca.gov/FileHandler.ashx?DocumentVersionID=14268.

Walker, R., and L. Craighead. 1997. *Analyzing Wildlife Movement Corridors in Montana Using GIS.* 1997. Environmental Sciences Research Institute. Proceedings of the 1997 International Esri Users Conference. http://proceedings.esri.com/library/userconf/proc97/proc97/to150/pap116/p116.htm.

Watt, W. R., J. A. Baker, D. M. Hogg, J. G. McNicol, and B. J. Naylor. 1996. *Forest Management Guidelines for the Provision of Marten Habitat, Version 1.0.* Sault Ste. Marie: Ontario Ministry of Natural Resources, Forest Management Branch.

Wolman, A. G. 2006. "Measurement and Meaningfulness in Conservation Science." *Conservation Biology* (20) 6: 1,626–34.

Zipperer, W. C. 1993. "Deforestation Patterns and Their Effects on Forest Patches." *Landscape Ecology* 8 (3): 177–84.

10

Assessing habitat connectivity

Peter H. Singleton; and Brad H. McRae, PhD

Animals move across landscapes to acquire food, reproduce, migrate, and shift in response to environmental change or other evolutionary pressures. In fact, most ecological processes happen within a spatial context that influences how those processes occur across a landscape. Examples include nutrient cycling, seed dispersal, range shifts of interacting species, and many others (Crooks and Sanjayan 2006). Landscape features, including patches of natural vegetation, agriculture, urban areas, and major highways, can influence how these processes proceed within an area (figure 10.1). Understanding how heterogeneous landscapes provide for or inhibit ecological flows at a variety of scales is important for responsible land management and conservation. The degree to which a landscape allows these flows to occur can be described in a broad sense as connectivity (see sidebar "Defining connectivity").

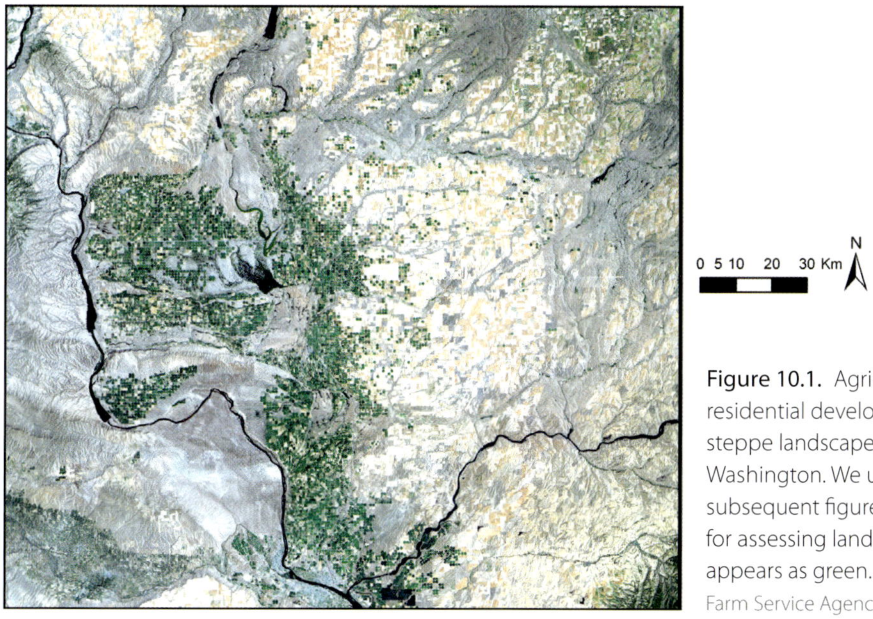

Figure 10.1. Agricultural land uses, major highways, and residential development create a fragmented shrub-steppe landscape in the central Columbia Plateau of Washington. We use this landscape as the basis of the subsequent figures showing different analysis approaches for assessing landscape permeability. Irrigated vegetation appears as green. Figure from US Department of Agriculture, Farm Service Agency, Aerial Photography Field Office.

Defining connectivity

Several definitions of "habitat connectivity" have been proposed, including: "the degree to which absolute isolation is prevented by landscape elements which allow organisms to move among patches" (Merriam 1984), "the degree to which the landscape impedes or facilitates movement among resource patches" (Taylor et al. 1993), and "the functional relationship among habitat patches owing to the spatial contagion of habitat and the movement responses of organisms to landscape structure" (With et al. 1997). All of these definitions imply that some feature or patch of interest is spatially related to other, similar features, and that intervening landscape characteristics influence the relationship between those features of interest. Precise measures of connectivity will depend on the specific process of interest. For example, connectivity for foraging movements of a Chelan mountainsnail (*Oreohelix* spp.), which may span a few meters of rocky hillside, is a very different concept from connectivity for maintaining metapopulation functions of wolves or grizzly bears (see Unnasch and Karl, chapter 3), which may span hundreds of kilometers.

Thus, connectivity assessment is a spatially explicit concept involving movement or flow. To help understand it, here are some important terms:

- *Barrier.* A landscape feature that obstructs the movement of an animal or ecological process
- *Corridor.* A linear, connected land area that joins habitat blocks or sources of ecological flows to provide opportunities for the movement of animals or ecological processes
- *Fracture zone.* An area of reduced landscape permeability between blocks of habitat or sources (see below) of ecological processes
- *Landscape permeability.* The quality of a heterogeneous land area to provide for passage of animals or other ecological functions (Singleton et al. 2002)
- *Least-cost corridor.* The set of map cells for which the least-cost path distance between two sources passing through the cell falls below a user-defined threshold
- *Least-cost distance, weighted distance, or cost-weighted distance.* The least accumulative cost distance (the sum of cell size times resistance of the cell) to the nearest source
- *Least-cost path.* The one-cell-wide path between two sources with the least accumulative cost distance (the sum of cell size times resistance of the cell for the cells along the path)
- *Resistance, friction, or cost surface.* A map of how much the habitat characteristics at each map cell facilitates or impedes the movement of an animal or ecological process
- *Sources or nodes.* Patches of habitat or origins of an ecological process that serve as the features between which linkages are modeled using least-cost distance or other approaches; not a source in the context of population source–sink dynamics; often referred to as "cores."

This chapter provides an introduction to spatial analysis concepts and methods for assessing habitat connectivity. We give an overview of commonly used methods for habitat connectivity assessment, summarize steps involved in conducting a connectivity assessment, and provide some suggestions for advancement of the field. A variety of analysis approaches are possible for assessing habitat connectivity, and as is often the case, important trade-offs exist between the level of information provided by the analysis and the complexity of that analysis. The most important message in this chapter may be that the best analysis approach is the simplest one that adequately answers the question at hand (for more on movement model assumptions, see Johnston et al., chapter 14).

This field is developing rapidly, with many new and innovative tools and analyses being published each month. In this chapter, we have chosen to focus on a few of the more commonly applied techniques. Our hope is to provide a starting point for conservation planners who can then refer to the most current scientific literature as well as the website accompanying this book, http://www.esri.com/conservationbook, to learn about the latest tools and techniques.

Why connectivity is important

The ability of animals to move at a variety of scales is important for the health and persistence of their populations. In particular, the study of metapopulation dynamics has demonstrated that the persistence of many species across broad landscapes is influenced by their ability to move and recolonize areas when local subpopulations become extinct (Levin 1974; McCullough 1996; Hanski and Gilpin 1997). Other examples of important ecological processes influenced by habitat connectivity include genetic exchange (Manel et al. 2003; Storfer et al. 2010), seed dispersal (Levey et al. 2005, 2008), and nutrient transport (Wipfli 2005). However, connectivity can also have a downside, with such processes as fire, insect outbreaks, and disease transmission also being associated with specific kinds of connectivity patterns (Haddad et al. 2011). The scientific literature about habitat connectivity is broad and deep. Bennett (2003), Crooks and Sanjayan (2006), and Hilty et al. (2006) all provide excellent surveys of the field.

One important area of concern today is maintaining or enhancing habitat connectivity so plant and animal communities can move in response to a changing climate (IPCC 2007). Substantial shifts in species ranges have occurred over the past century (Parmesan and Yohe 2003). Continuing and accelerating shifts are expected to be one of the consequences of climate change (see Bachelet, chapter 13). Providing the opportunity for organisms to shift in response to changing climatic and ecological conditions will be an important strategy for adapting to climate change (Heller and Zavaleta 2009; Krosby et al. 2010).

Awareness of habitat connectivity issues is growing among land-management agencies and the general public. This awareness is indicated by the Western Governors' Association Connectivity Initiative (http://www.westgov.org/initiatives/wildlife), state and regional connectivity planning efforts (see sidebar "Recent broad-scale connectivity assessments"), and conservation campaigns to raise awareness of habitat connectivity issues: for example, the Freedom to Roam (http://www.freedomtoroam.org) and Spine of the Continent (http://wildlandsnetwork.org/wildways/western-wildway) initiatives.

Recent broad-scale connectivity assessments

Some broad-scale landscape assessments that address connectivity include the following:

- Washington State Connected Landscapes Project (http://www.waconnected.org)
- California Essential Habitat Connectivity (http://www.dfg.ca.gov/habcon/connectivity/)
- Arizona Wildlife Linkage Assessment (http://www.azdot.gov/inside_adot/OES/AZ_Wildlife_Linkages/index.asp)
- Two Countries–One Forest (http://www.2c1forest.org/en/mainpageenglish.html)
- Yellowstone to Yukon (http://www.y2y.net)

Ecological and theoretical background

Structural versus functional connectivity

Habitat connectivity has two components: (1) the structural characteristics or arrangement of patches on the landscape (structural connectivity), and (2) the specific ecological processes that flow through the landscape (functional connectivity; Crooks and Sanjayan 2006). Measures of structural connectivity quantify the characteristics and spatial arrangement of elements in the landscape, for example, the amount and distribution of a species' habitat. Measures of functional connectivity address the way an ecological process, such as the movement of individual animals, is influenced by the arrangement of elements on the landscape. Structural connectivity is usually much easier to quantify than functional connectivity, but interpreting the ecological significance of structural connectivity patterns can be very difficult unless the analyst has a specific ecological function in mind. Many connectivity assessments have investigated the structural connectivity in a landscape and tried to relate those patterns to ecological functions (for example, Girvetz et al. 2008). Some analysts have used sophisticated simulation models that specifically represent the mechanisms involved in an ecological process to address functional connectivity (for example, Carroll and Miquelle 2006).

Some general modeling principles

A model is a systematic representation of something else. By definition, models are simplified approximations of reality, and that simplification is based on a series of assumptions about how a system works. In the context of ecology and natural resource conservation, models use formal (often mathematical) reasoning structures that allow us to investigate our assumptions about how the system works. Clearly understanding and communicating the assumptions that a model is built upon is critical for the fruitful application of that model.

The use of models for understanding wildlife habitat relationships has a rich and productive history (Scott et al. 2002; Franklin 2009; Millspaugh and Thompson 2009). Broadly, such models can be categorized as being data-driven empirical models (those that describe patterns from a specific dataset) or expert-based conceptual models (those based on a synthesis of the understanding of a process derived from a variety of sources, including the scientific literature, field observations, and professional judgment; see also Dickson et al., chapter 7). In most cases, models used for assessing habitat connectivity patterns have been conceptual. Although there are good examples of empirical models that have been used for evaluating connectivity patterns (for example, Shirk et al. 2010; Cushman et al. 2006; Braunisch et al. 2010; Short Bull et al. 2011), the data simply is not available to develop empirical models for many species of conservation concern. Given those limitations, there are five main objectives for using conceptual models for connectivity assessment (based on Morrison et al. 1998, 310):

1. To formalize or describe our current understanding about patterns of structural or functional connectivity within a landscape
2. To understand which environmental factors affect those connectivity patterns
3. To identify weaknesses in our understanding of those patterns
4. To predict connectivity patterns in new geographies or in the future
5. To generate testable hypotheses about the patterns of interest

Many analysts focus on the prediction objective, but these expert-based conceptual modeling approaches are often most beneficial for the things that we learn in the modeling process. The modeling framework forces us to formalize our understanding of a system, identify the weaknesses of that understanding, and identify testable hypotheses that can be used to advance our understanding of the system. Additionally, most of the assessment approaches we present are deterministic—they will always provide an answer (or identify a "best" linkage route through a landscape), even when it doesn't make ecological sense. It is up to the modelers and the assessment team to provide a rigorous ecological interpretation of the modeling procedure and results to ensure a useful product. As Morrison et al. (1998, 313) suggest, "models are like politicians: support them, use them, but don't unquestioningly trust them."

Questions to consider in choosing a connectivity assessment approach

The following questions are important to consider in choosing a connectivity assessment approach.

What is the specific question to be addressed?

Simply assessing habitat connectivity, without being specific about what is being connected, does not provide useful information for land-management and conservation decisions. The most important step in any scientific investigation is clearly articulating the question under investigation. Ecological questions related to landscape permeability can range from the very specific and fine scale, such as "Where along this 15-mile stretch of highway should crossing structures for painted turtles be located?" (Montana Department of Transportation et al. 2008) to the more general and broad scale, such as "Which portions of California are least impacted by human activities and how are they connected?" (Spencer et al. 2010). These are both worthwhile questions. The critical issue here is that the question needs to be clearly articulated so that the assumptions associated with the assessment can be identified.

What is the simplest approach that will provide the necessary information?

One of the biggest challenges in modeling is to find the best trade-off between complexity and uncertainty (Millspaugh et al. 2009). The best explanation of something is usually the explanation that makes the fewest assumptions (the principle of parsimony). Adding complexity to a model may add specificity (more complex models can provide more precise estimates), but it also adds uncertainty (the more complex a model is, the more likely some part of it will be wrong). This is particularly important for broad-scale spatial analysis projects in which the accuracy of available spatial datasets can be variable: errors multiply as those datasets are combined, and there can be substantial uncertainty regarding the effect of spatial pattern on the process of interest. In the context of landscape analysis, a good principle is to use the fewest GIS layers with the fewest classes that adequately address the question at hand. Remember that each additional class in a categorical GIS layer is an additional model parameter and an additional opportunity to get something wrong. Although he wasn't a spatial analyst, Albert Einstein provided excellent advice when he said, "Make everything as simple as possible, but not simpler."

What is the scale of the question?

The scale of the analysis needs to be carefully matched to the ecological process under consideration (see Unnasch and Karl, chapter 3). GIS datasets typically used for landscape analysis have important limitations that must be considered when using them to represent ecological process. For example, mapping landscape permeability patterns at a broad regional scale has little utility for species that are associated with specific fine-scale habitat features unless there are spatial data that depicts those habitat features with reasonable reliability and consistency. As with other topics in this book, broad-scale assessments can provide important information about ecological relationships and context, while finer-scale assessments may be more practical to implement. Sometimes multiple scales of analysis will be needed to proceed from an understanding of regional connectivity patterns to local implementation (e.g., see WHCWG 2010).

Is the model testable?

In connectivity assessment, models are used to project our understanding of how animals move onto a virtual landscape. The best models allow us to determine if that understanding (and the way it is represented in the model) is correct. One of the objectives of using a conceptual model should be the development of testable hypotheses. The process of hypothesis testing is the process by which science advances. It's always best to strive to formulate the modeling question and the model output in a manner that can be validated. For example, an analysis that suggests there is connectivity for cougars between mountain ranges (based on habitat) can be validated by comparing genetic distances among populations based upon DNA sampling (McRae et al. 2005).

Tools and techniques

Landscape permeability assessment methods range from relatively simple to relatively complex (figure 10.2). All of these methods summarize spatial patterns that influence flows or movement across the landscape. The intent is to distill the highly complex patterns present in heterogeneous landscapes into simplified representations that can be used to increase our understanding of ecological processes and communicate that information. More complex analysis approaches can provide more information, but require more assumptions, while simpler approaches require fewer assumptions, but provide less specific information. Careful consideration of the level of complexity required to address the question at hand is an important part of selecting a connectivity analysis approach. In the following sections, we describe some connectivity analysis approaches, starting with the simplest approaches that provide general information on landscape structure, and work through increasingly complex approaches that provide more specific information about ecological processes.

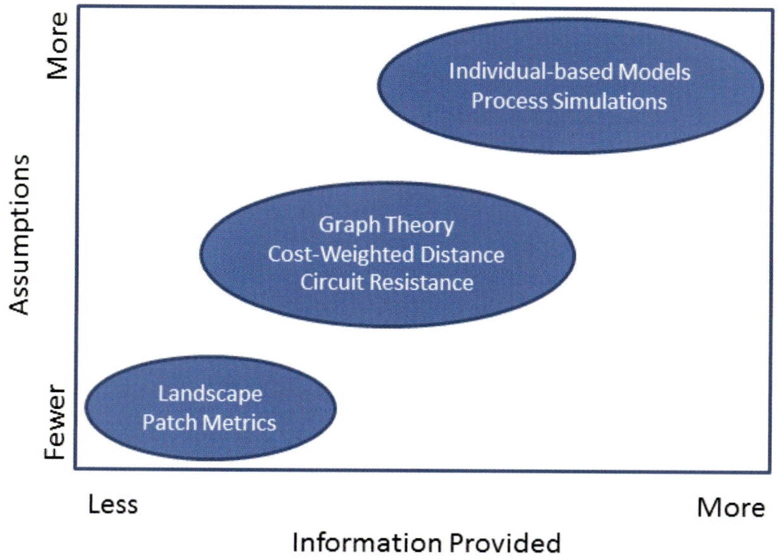

Figure 10.2. Connectivity analysis approaches range from the simple to the complex. In most cases, the choice of an analysis approach involves a trade-off between the number of assumptions required and the amount of information provided by an analysis. Figure from Peter H. Singleton.

Landscape patch metrics

Landscape patch metrics are summaries of the composition and configuration of elements in the landscape. These metrics have been the workhorses of landscape ecology for many years, and they have been widely applied to describe landscape pattern (McGarigal and Marks 1995). They are the simplest measures of structural connectivity within a landscape. Patch metrics provide very useful measures of landscape structure (including fragmentation) but are not direct measures of ecological functions. Common metrics that measure habitat fragmentation include nearest neighbor distance, patch size, patch core area, edge to area ratios, and their variants (McGarigal and Marks 1995); effective mesh size is another metric that has been developed specifically to assess connectivity patterns (Girvetz et al. 2008). Patch metrics generally are summarized across some landscape unit (e.g., mean patch size within a watershed), and are less spatially explicit than some other approaches. Landscape patch metrics are often compared to some benchmark like historic range of variability (Landres et al. 1999). Often, areas that are most divergent from historic conditions are those areas that have been most impacted by habitat fragmentation and associated barriers. Identifying these areas can be a useful first step in assessing habitat connectivity patterns (Rowland and Wisdom 2009). FRAGSTATS (McGarigal and Marks 1995) is a standard software package for calculating these metrics.

Graph theory

Graph theory is a collection of mathematical concepts that describes the relationships between things based on a representation of nodes (the things being connected) and edges (the links between the nodes). It has been applied broadly in a variety of disciplines, including particle physics, computer science, sociology, and ecology (Newman 2010). Graph theory provides a strong theoretical framework for understanding habitat patch networks. In fact, most landscape permeability assessment methods (for example, cost-distance, circuit resistance, and centrality methods) are applications of graph theory that use different mathematical methods for describing the relationships between landscape features or cells in a GIS raster layer.

Graph theory is particularly useful for characterizing complex sets of patches and linkages forming networks. Fortin and Dale (2005, 57–75) provide a good introduction to graph theory in the context of landscape ecology, and several reviews have recently been published (Laita et al. 2011; Rayfield et al. 2011; Galpern et al. 2011; Fall et al. 2007; Pinto and Keitt 2009; Minor and Urban 2008). Graph theory has been applied to ecological questions with increasing frequency in the past several years (e.g., Urban and Keitt 2001; Baggio et al. 2011; Bodin and Saura 2010; Cerdeira et al. 2010; Pascual-Hortal and Saura 2008). Spatial analysis tools that can be used to calculate metrics based on graph theory include Conefor Sensinode (Saura and Pascual-Hortal 2009; http://www.conefor.org), Linkage Mapper (http://www.circuitscape.org/linkagemapper), and the igraph package for R and Python (Hartvigsen 2011; http://igraph.sourceforge.net).

Cost-distance analysis

Cost-distance analysis is the most commonly applied spatial analysis technique for landscape connectivity assessment focused on conservation planning (Beier et al. 2011). In this case, "cost" refers to an energetic cost or mortality risk of moving through the landscape (figure 10.3). Cost-distance analysis starts with the creation of a resistance raster layer (figure 10.3a). Each cell in the resistance layer is assigned an index of resistance or "cost," usually determined by the habitat characteristics at that cell. For each cell in the output cost-distance surface, the cost-distance algorithm calculates the sum of the resistance index times the cell size for each cell in the least costly path to the closest source (for more detail, see the "Cost Distance algorithm" page under ArcGIS Spatial Analyst Desktop Help at http://webhelp.esri.com/arcgisdesktop/9.2/index.cfm?TopicName=welcome; we describe the analysis steps for conducting a connectivity assessment using cost-distance methods in the following section). The algorithm produces a continuous map of the cumulative

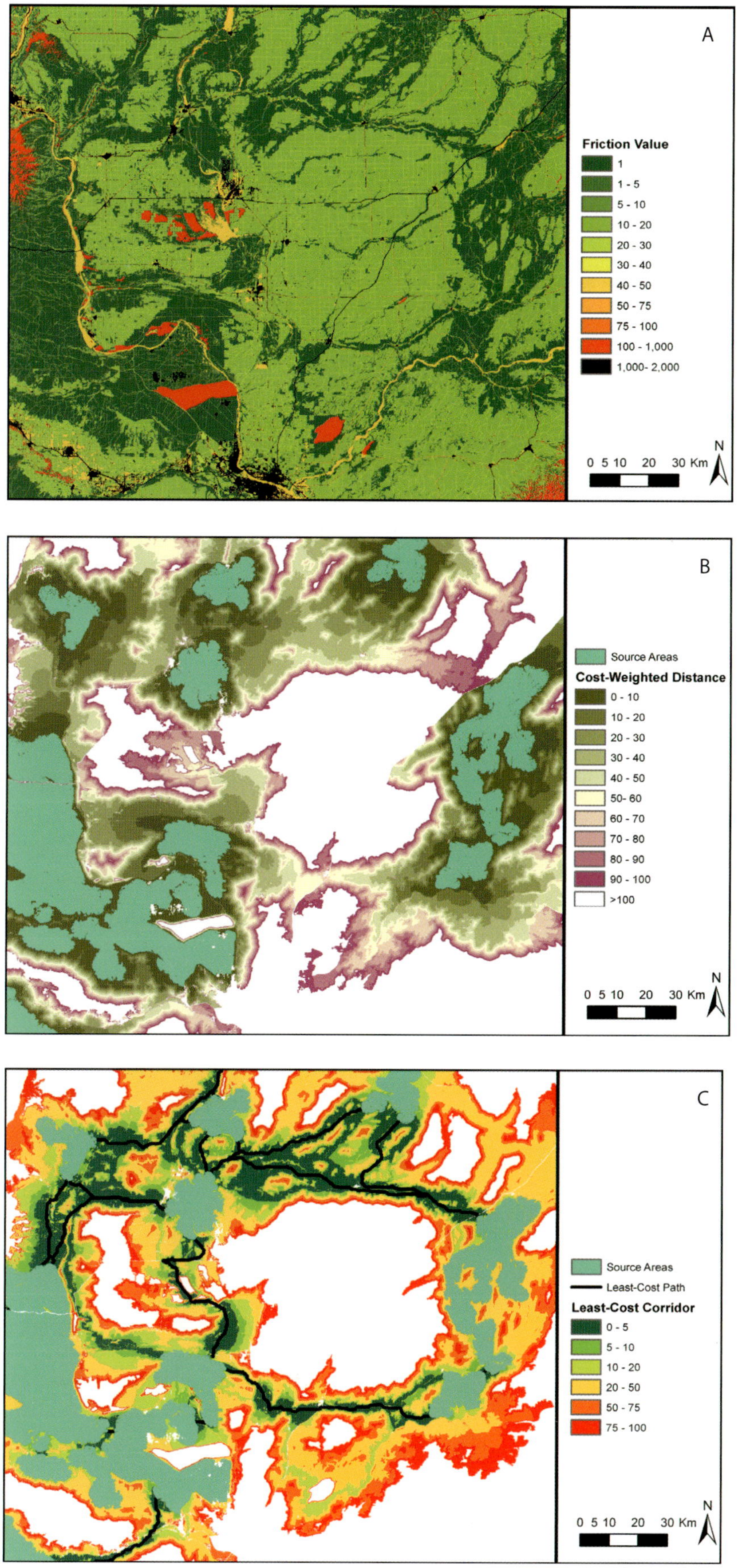

Figure 10.3. A cost-weighted distance analysis of landscape permeability for American badger in the Central Columbia Basin of Washington. This example is drawn from analysis conducted by the Washington Connected Landscapes Project (WHCWG 2010); (a) shows the resistance surface upon which the model was run; (b) shows the source (core) areas and the cost-weighted distance surface (in cost-weighted kilometers), which is a raster resulting from weighting the resistance between source areas; and (c) shows the least-cost path and normalized least-cost corridors (buffer of least-cost path in cost-weighted kilometers) between core areas. WHCWG (2010).

resistance between each cell on the map and the most accessible source area (figure 10.3b). This algorithm forms the basis for two useful map products: (1) Least-cost paths between two sources, or the set of adjacent cells with the minimum cumulative resistance between two sources. This identifies the least resistant one-cell-wide path between two sources. (2) Least-cost corridors, or the set of cells for which the cumulative cost-distance between two sources falls below one or more user-defined thresholds (figure 10.3c). Least-cost path and least-cost corridor analyses are described more extensively in the "Least-cost path and least-cost corridor" page under ArcGIS Spatial Analyst Desktop Help. Examples of least-cost path analyses are provided by Johnston et al. (chapter 14). ArcGIS tools for conducting cost-distance analysis are found in the Spatial Analyst Distance toolbox.

To interpret cost-distance analysis results, start by examining the resistance surface to get a feel for the distribution of habitat and barriers (figure 10.3a), then look at the cost-distance raster to evaluate the relative permeability across the landscape (figure 10.3b), and finally, examine the least-cost corridor results to find the best linkages between sources (figure 10.3c). Each one of these products provides unique information about the landscape, and considering them together provides a much more complete understanding of connectivity patterns than any one of these layers alone could. Cost-distance measures are best interpreted as relative indices of landscape permeability. The cost-weighted distance surface displays the relative resistance between each cell and the most accessible source, and is the most informative product in many ways. The least-cost paths alone are not particularly informative as geographic features because they are only one cell wide and do not provide any information on the permeability of the rest of the landscape. However, the least-cost distance can be a useful index of the degree of connectedness between two sources. These least-cost distance indices can be particularly informative when they are incorporated into graph representations of the landscape (e.g., Pascual-Hortal and Saura 2008).

Cost distance values can be very sensitive to small errors in the base GIS layers. The presence of a few connected cells with low resistance can dramatically reduce least-cost path values across a landscape. This can be particularly misleading if there are spatial errors in the base data (for example if there is a gap in the raster representation of a linear barrier like a highway), or if linear features are represented as being one cell wide (in which case, there will be a lower cost for crossing that feature where cells are arranged corner-to-corner than there will be for areas where the feature is straight and resistant cells are placed edge-to-edge; Rothley 2005). Using buffers of linear features can be helpful for avoiding this problem.

Cost-distance analysis can produce useful information on the relative resistance of the landscape to animal movement, and cost-distance landscape metrics can be conceptually linked to the movement cost and risk concepts of optimal foraging theory (Stephens and Krebs 1987). However, it does not predict specific animal movement paths through the landscape. Because cost-distance analysis represents cumulative landscape resistance between discrete start- and endpoints, the cost-distance results would only predict actual animal movement choices if a moving animal knew exactly where it wanted to go and only traveled through the very best habitat to get there. In cases where an animal has experience of the landscape, or there is only one good route through the landscape, this may be realistic. In many other cases, when an animal is exploring a new landscape (e.g., dispersing), or substantial portions of the landscape are equally good from the animal's perspective, centrality or circuit resistance may provide better representations of actual individual animal movements. However, centrality and circuit resistance pose unique challenges for mapping and interpretation. The important point here is that cost-distance analysis can provide a compelling representation of the relative permeability of different portions of a landscape, but it should not be interpreted as predicting specific movement routes through that landscape for individual animals.

Circuit theory

Another set of graph-based connectivity measures uses networks of electrical nodes connected by resistors as models for networks of populations, habitat patches, or grid cells connected by movement (McRae et al. 2008). Using circuits as models for ecological networks is based upon a simple idea: multiple pathways in electrical networks increase connectivity. Thus, distance metrics based on electrical connectivity are applicable to processes that increase in magnitude with increasing numbers of pathways (figure 10.4). Moreover, connections between circuit and random walk theories mean that resistance, current, and voltage can all be interpreted in terms of the behavior of random walkers (Doyle 1984; McRae et al. 2008). For example, circuit theory can be related to population genetic theory via random walk and coalescent theories, allowing predictions of gene flow in complex landscapes (McRae 2006; McRae and Beier 2007).

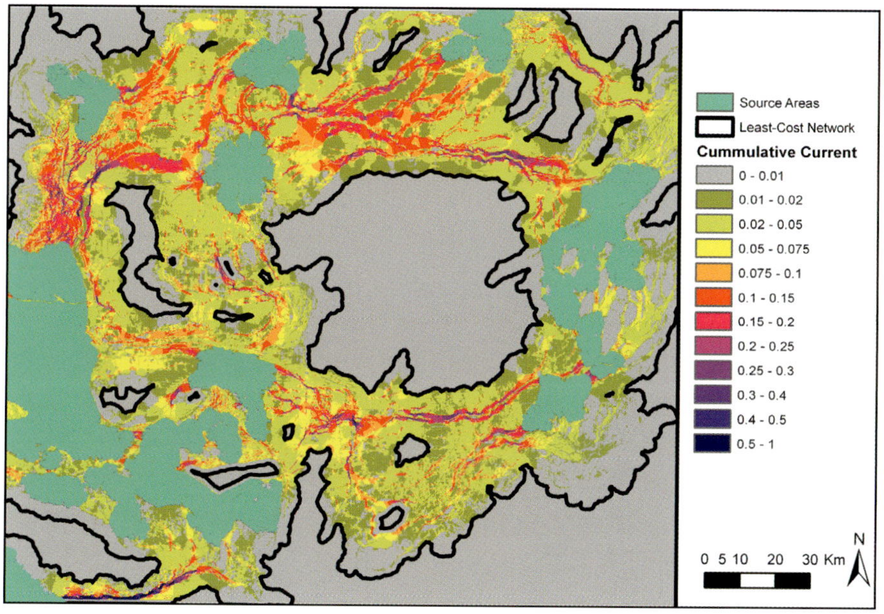

Figure 10.4. Circuit analyses complement least-cost path results by identifying important alternative pathways and "pinch points" where loss of a small area could disproportionately compromise connectivity. A promising application is restricting circuit analyses to networks identified by least-cost corridor analysis to take advantage of the strengths of both approaches (McRae and Shah 2009). This hybrid approach shows both the most efficient movement pathways and critical pinch points within them (in purple). WHCWG (2010).

Analyses based on cost-distance or circuit theory are unable to precisely predict real-world movement behavior or gene flow (Spear et al. 2010), but the two methods provide complementary perspectives on potential movement patterns. Corridors mapped using cost-distance analysis emphasize the "best" possible route through the landscape (the least-cost path or corridor), which would correspond to the route taken by animals with perfect or near-perfect knowledge of a landscape. In contrast, with circuit models, all possible paths contribute at least somewhat to connectivity, and they reflect movement routes taken by random walkers. Thus, circuit and least-cost models represent two extremes in assumptions about movement and connectivity. Least-cost analyses can show the safest and most direct movement routes or zones connecting source habitat areas; if a large portion of a landscape is likely to be developed, identifying such areas is important for efficient connectivity conservation. Circuit theory has the advantage of identifying and quantifying "pinch points"—constrictions in linkage areas that, if lost, could sever connectivity entirely. Such areas can be prioritized for early conservation action because options are limited. Circuit algorithms also integrate across all movement pathways to highlight alternative pathways for movement, which can lessen the effect of pinch points and provide redundant networks that are more robust to disturbance.

Circuit theory is unlikely to supplant least-cost corridor modeling for two reasons: it neither produces contiguous polygons running from one source area to another nor identifies the most efficient movement pathways connecting them—two qualities that are desirable in efforts to prioritize connectivity areas. But hybrid approaches, such as using cost-distance analyses to identify broader corridors or connectivity zones, and then using circuit theory to identify pinch points within those corridors, can take advantage of the strengths of both models. Circuitscape (McRae and Shah 2009) applies circuit theory to landscape connectivity analyses.

Centrality analysis

Centrality analysis is an emerging approach to connectivity modeling that focuses on how landscape elements contribute to connectivity among all or many locations across a landscape, not just individual pairs of locations (Freeman et al. 1991; Estrada and Bodin 2008). By analyzing how individual nodes, grid cells, habitat patches, or corridors contribute to providing connectivity across broader habitat networks, this class of methods seeks to identify those areas that act as gatekeepers of flow across landscapes (Carroll et al. 2011).

The difference between centrality and other methods is that centrality tallies results from analyses connecting many or all pairs of source and target nodes (sites or cells) on a landscape. For example, betweenness centrality (Freeman et al. 1991) identifies the shortest or least resistant paths connecting all pairs of nodes (or a subset of nodes) in a network, and sums the number of such shortest paths involving each intervening node. Measures of resistance between nodes can employ least-cost path, circuit theory, or other algorithms that are typically used to analyze connectivity between single pairs of habitat patches. This procedure identifies areas lying on a large proportion of the shortest paths in a network, the loss of which can disproportionately disrupt connectivity across the network as a whole. When applied to raster data, centrality analyses are often used to connect all grid cells (or many randomly chosen grid cells) in a landscape, eliminating the need to predefine discrete patches or source areas to connect (e.g., Carroll et al. 2011; Landguth et al. 2012). The Connectivity Analysis Toolkit developed by Carlos Carroll (Klamath Center for Conservation Research), Brad McRae (The Nature Conservancy), and Allen Brookes (University of Washington), Universal Corridor Network Simulator (UNICOR; Landguth et al. 2012), and the approach of Theobald et al. (2011) all use centrality analysis to identify important connectivity areas. The Connectivity Analysis Toolkit is available at http://www.klamathconservation.org/scienceblog/?page_id=138 and other sites.

Individual-based modeling

Individual-based (or agent-based) modeling is built on the understanding that landscape-scale ecological processes such as metapopulation dynamics, range shifts, or genetic exchange emerge from the behavior of a number of individuals, and the behavior of each of those individuals can be simulated in a manner that allows modelers to investigate the emergent processes of interest (Grimm and Railsback 2005; Johnston et al., chapter 14; figure 10.5). Individual-based models are mechanistic models, that is, they actually simulate the processes we are interested in (e.g., animal population dynamics or genetic exchange). They can also incorporate stochasticity, adding realism by acknowledging the role of random chance in ecological processes. Individual-based models are generally more focused on assessing functional connectivity than the methods described above. In this way, they can be substantially more informative than the landscape assessment approaches that we have presented so far. Using individual-based modeling, conservation planners can conduct population viability analysis, identify portions of the landscape that serve as population sources (figure 10.5a),

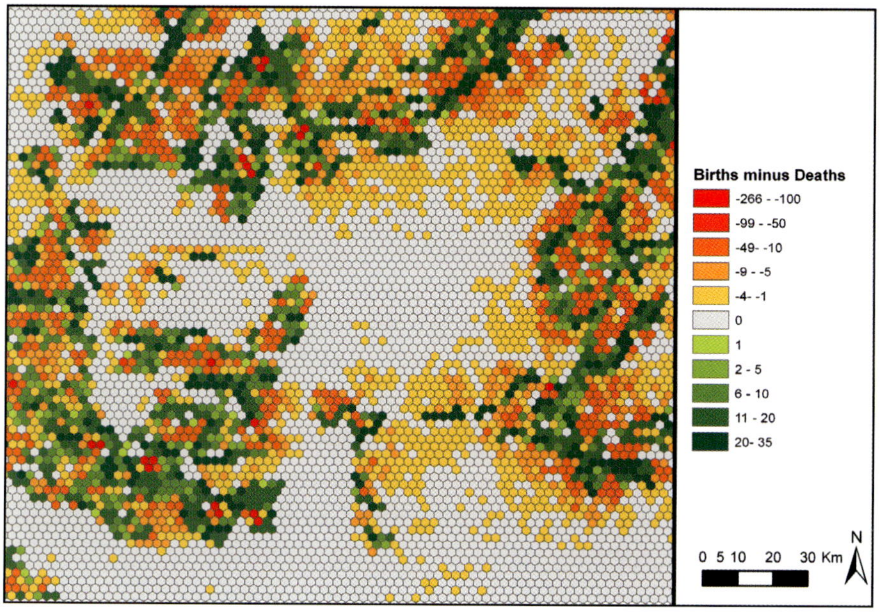

Figure 10.5a. The green areas indicate population sources (where births exceed deaths) and orange areas indicate population sinks (where deaths exceed births) on the landscape. WHCWG (2010).

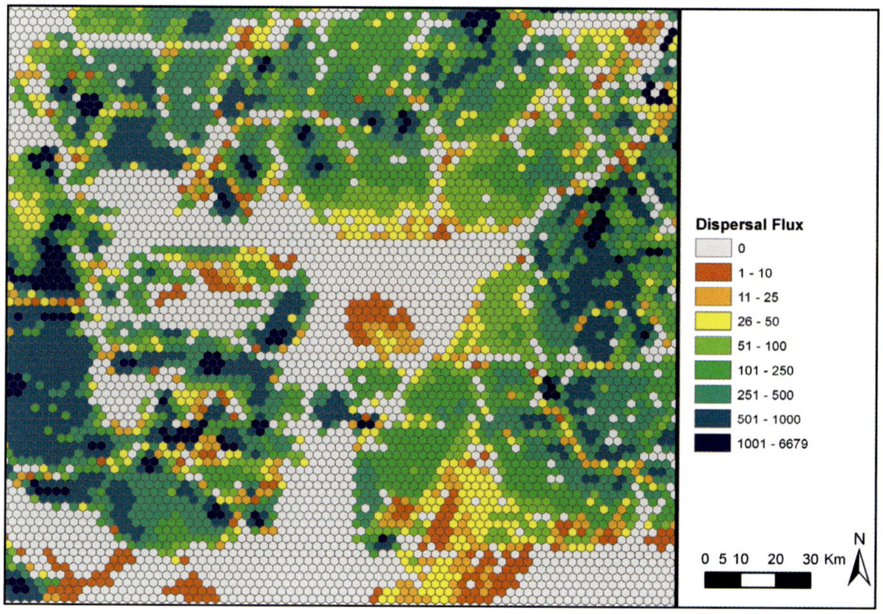

Figure 10.5b. HexSim can quantify simulated movement rates through a landscape by recording the number of times an animal passes through each hexagon during dispersal movements (the dispersal flux). WHCWG (2010).

identify the linkages between those sources (figure 10.5b), and estimate the relative contribution of different linkages to overall population viability—all in the same set of model runs (e.g., Heinrichs et al. 2010). However, individual-based models are also much more complex and require much more specific information to parameterize than the deterministic models described earlier.

Although individual-based models have been used for exploring ecological theory for some time, the application of these models for conservation planning has grown substantially in recent years (e.g., Johnston, 2013; Carroll and Miquelle 2006; Schumaker et al. 2004). Two tools that are currently available for individual-based modeling are HexSim (Schumaker 2011) and the Agent Analyst extension for ArcGIS (Johnston et al., chapter 14). HexSim, and its precursor PATCH, were specifically developed for population viability assessment and conservation planning applications (Schumaker 2011). Examples of the application of HexSim for conservation planning include population viability analysis for tigers (Carroll et al. 2006), and northern spotted owl recovery planning (USFWS 2011). Johnston and colleagues provide an example of using the Agent Analyst extension for modeling cougar movement in chapter 14.

Steps for applied connectivity assessment

The following discussion provides examples of steps that may be involved in a connectivity assessment using a cost-weighted distance approach. Other analysis methods (e.g., using circuit theory or centrality analysis) will proceed through similar steps, but we focus on this one example for brevity. Obviously, the exact steps needed to accomplish an assessment will be dependent on the specific goals and analysis approaches identified for the assessment. Note that the methods we focus on require defining core or source areas, but several emerging methods (Carroll et al. 2011; Landguth et al. 2012; Theobald et al. 2011) take a more gradient-based approach and do not connect discrete habitat patches.

Step 1. Convene an assessment team

Creating an inclusive analysis team leads to a broad sense of investment in the successful application of the results of the analysis. Often landscape modeling is perceived as a "black box" by regulators or other stakeholders in landscape planning. Inclusion in the modeling process is perhaps the best way to educate potential users about the process and develop a sense of commitment and investment in the application of the results (Beier et al. 2008). Assembling such a team will not always be possible, but it is an ideal worth striving for.

An ideal assessment team would include the following:

- Land managers
- Regulatory agency representatives
- Local species experts (biologists)
- Other stakeholders (NGOs and private landowners)
- Spatial analysts
- Communications specialists

Step 2. Identify the goals of the assessment team and objectives of the analysis

Goals are generally broad visions, and objectives are specific actions designed to accomplish those goals. Thus, there are two important steps at the start of an assessment. First, clearly define the goals of the team. Second, decide how to advance those goals through specific spatial analysis objectives. The goals of the team will determine the kinds of analyses that are most useful; however, the foundation of a successful landscape modeling effort depends on clearly defining the purpose and specific objectives of those analyses.

Step 3. Select the focal species or ecological characteristics to assess

The selection of focal species is addressed by Brock and Atkinson (chapter 6), but it is worth highlighting the special considerations for connectivity analysis. In particular, focal species selected for connectivity analysis need to be sensitive to the landscape barrier effects that are of management concern and that can be captured in modeling (Battisti and Luiselli 2011). For example, the Washington Wildlife Habitat Connectivity Working Group (WHCWG 2010) identified four types of threats that contributed to landscape barrier effects and used those as part of their focal species selection process. Those threats were (1) land clearing and vegetation removal, (2) development, (3) roads and traffic, and (4) presence of people or domestic animals.

An alternative to selecting a suite of focal species for analysis is to focus on the level of human disturbance or landscape integrity (Comer and Schulz 2007; Theobald et al. 2011). This approach has been applied for broad-scale linkage assessment in California (Spencer et al. 2010) and Washington (WHCWG 2010). Landscape integrity approaches focus on identifying and quantifying the portions of the landscape that are least altered by human activities. No assumptions are made about animals being associated with specific habitat conditions, only that substantial human alteration (for example, proximity to roads, human dwellings, agricultural landscapes, and so on) is likely to deter movement by animals. In general, landscape integrity approaches will be most informative for broad-scale, coarse-filter assessments that are not trying to address individual species or ecological communities. Landscape integrity approaches will often be most useful in conjunction with focal species analyses (Theobald et al. 2011; WHCWG 2010).

Step 4. Define the analysis area and scale

Defining a study area (extent) for connectivity analysis can be particularly challenging because connections to features outside of the area of emphasis are often important too. It's generally best to start by identifying the area you want to assess (e.g., a state or administrative unit), and then examine the surrounding areas to capture adjoining features that might be ecologically connected to the focal area.

Another important choice at the outset of analysis is determining the appropriate spatial resolution (grain) for the analysis. Smaller GIS raster cell sizes use up more computational resources (particularly processing time). Larger cells reduce resolution. A good guideline is to choose a resolution where important landscape features are represented by several cells. One approach used by the WHCWG for ecoregional analysis was to compile base data and the resistance surface at 30-m resolution, and then generalize that to 90-m resolution for the cost-distance analysis. This substantially reduced processing time (from days to hours) and also reduced the influence of single-cell misclassification errors in the cost-distance analysis.

At this step, it is also worthwhile to consider the temporal scale of the analysis. Is the analysis going to address existing conditions only, or are change scenarios going to be considered? There may be spatial data or model parameterization choices that are better suited for climate or land-use change scenarios.

Step 5. Compile the spatial data

The compilation of land-cover maps is discussed by Olenicki (chapter 4), but special considerations for broad-scale linkage assessments are worth highlighting. Linkage assessments will often require the compilation of datasets across management agencies, states, or even international borders. The structure and quality of those datasets can differ substantially across jurisdictions. In particular, such features as forest structure, tree species composition, and low-use road management status can be problematic. The data compilation phase is often the most time-consuming component of an assessment project. Once again, simpler is usually better! Compiling data from the fewest sources and categorizing it into the fewest classes to adequately address the question at hand will generally lead to more efficient analyses.

Step 6. Identify the areas to connect

Conducting a connectivity analysis usually requires identifying the features on the landscape to connect (commonly referred to as sources, cores, or nodes). Sources are most often identified in one of two ways: either they are modeled or they are based on previously defined landscape units. Modeling approaches for identifying source areas for connectivity analysis typically use habitat suitability modeling, followed by some spatial analysis that highlights areas with the appropriate amount and configuration of habitat (see Church, chapter 9). One approach for identifying source areas to connect consists of three steps: The first step is to use a habitat suitability model to evaluate the suitability of each individual pixel. As discussed by Dickson et al. (chapter 7) that habitat suitability model can be based on information in the scientific literature and other sources, or it may be empirically derived from species detection locations (see also Franklin 2009). For the second step, some sort of spatial moving-window algorithm (such as a focal sum or focal mean) may be used to identify the portions of the landscape where suitable habitat is abundant. For the third step, some rule set is applied (often incorporating a minimum threshold of habitat suitability and minimum polygon size) to select the portions of the landscape to serve as sources.

Known areas of species occurrence or land units with special management status (e.g., GAP conservation status designations) can also be used as sources for connectivity modeling. Known species occurrence areas can be ideal sources for connectivity modeling as long as the distribution of the species is relatively well-known across the analysis extent. However, for most species, especially at broad scales, survey efforts are often inconsistent, and source maps based only on detection data may be biased by survey effort. In such cases, known species-distribution polygons can be combined with sources identified from habitat suitability modeling to provide a consistent map for the entire analysis area. Analysts need to be particularly cautious about using administrative boundaries to identify source areas for connectivity analysis. For example, connections between wilderness areas in mountainous parts of the western United States may be important for some species, but those wilderness areas are often very heterogeneous. Natural landscape features or extreme ecotones can function as important barriers even within protected natural landscapes (e.g., Vignieri 2005).

When mapping source areas for linkage analysis, keep in mind that the number and distribution of source areas will have much more influence on the linkage model outcomes than the outer boundaries of the individual sources. Differences in the outer boundaries of source areas are unlikely to substantially change the resulting network, but the configuration of a connectivity network can be very sensitive to the addition or deletion of source areas.

Step 7. Generate a resistance surface

A resistance surface (also called a friction or cost surface) is a map of how much the characteristics in each cell facilitate or impede the movement of the focal species or process of concern. The generation of this surface is a common critical step for linkage assessment methods. It is essentially the inverse of a habitat suitability map—a habitat suitability map displays how "good" each area is for providing the habitat needs of an animal, but a resistance surface represents how "bad" an area is for an animal to move across. Note that an area can have poor habitat suitability (for example, it may not provide food or nesting opportunities) but still be relatively permeable to movement. Once a resistance surface is developed, patterns in that surface can be investigated using a variety of methods (including cost-distance analysis, circuit analysis, or individual-based modeling).

Most resistance estimates based on literature review and expert opinion involve compiling a suite of habitat layers (e.g.; land cover, roads, buildings, and topography), collapsing those layers into appropriate categories, assigning a resistance value to each category, then combining those resistance values into a single resistance estimate that represents the combination of habitat characteristics at that cell. There are important potential differences in dispersal or movement habitat and foraging or breeding habitat to consider when developing the modeling rules for classifying resistance values. Most studies of animal habitat selection have focused on the ways animals use habitat for foraging or breeding within their home ranges. Habitat selection behavior—or, more importantly for connectivity analysis, habitat avoidance behavior—may be very different for an animal that is making a long-distance movement. Understanding and quantifying animal movement patterns during long-distance movements is important for understanding habitat connectivity, but it presents substantial challenges for researchers. Several studies have used movement data (Driezen et al. 2007), landscape genetics data (Cushman et al. 2006; Epps et al. 2007; Perez-Espona et al. 2008), or resource selection functions (Chetkiewicz et al. 2006; Schadt et al. 2002; Kautz et al. 2006; Kindall and van Manen 2007) to develop and evaluate resistance measures. However, in most cases, analysts involved in connectivity assessments have had to base resistance estimates on a combination of literature review and expert opinion (e.g., WHCWG 2010). Spear et al. (2010) reviewed approaches for generating resistance surfaces for landscape genetics studies.

Step 8. Run the connectivity analysis

The next step is to extract the patterns of connectivity between the mapped sources based on the patterns of resistance represented in the resistance surface. Cost-distance analysis can be conducted in ArcGIS using the Cost Distance tool in the Distance toolbox. Specifying a maximum distance beyond which movement is unrealistic for the species can reduce processing time in large landscapes. The cost-weighted distance surface provides a tremendous amount of information on relative landscape permeability across the analysis area. It shows the cumulative resistance between each cell and the nearest source, so relative connectivity values across the entire analysis area can be visualized with this single layer. The cost-weighted distance surface alone may be an appropriate final product, but most connectivity assessment projects have gone on to identify linkage networks as outlined in the following steps.

Step 9. Investigate the source patch network and decide which patches to link

Identifying which source patches to connect is not a trivial problem. Identifying linkages in landscapes that are highly permeable will not be informative because the movement of animals is not going to be directed by barriers or areas with greater resistance, and the animals are likely to be essentially moving randomly (at least in relation to the connectivity model). Identification of linkages through areas that are highly impermeable is not informative either, because animals are unable to access those areas in the first place. The mid-range of permeability values are the most interesting; these are areas where the landscape is heterogeneous enough that barriers and differences in habitat characteristics are likely to constrain and direct animal movement. It is important to remember that linkage models will always identify a "best" linkage, even when it is not meaningful.

Several approaches are possible for identifying which source areas to model linkages between. Singleton et al. (2002) identified fracture zones (areas with reduced landscape permeability between sources) for forest carnivores by identifying the portions of the landscape that were 10–100 km weighted distance from the nearest source areas. Areas within the lower cost-weighted distance threshold (10 km cost-weighted distance) were termed "available habitat," and adjacent source polygons that fell within this distance were considered to be well connected to each other. Areas above the upper threshold (> 100 km cost-weighted distance) were considered to be too resistant to movement to be considered for connectivity modeling. Connectivity analyses were constrained to the portions of the landscape that fell within the lower and upper cost-weighted distance thresholds ("fracture zones"). This limited the number of modeled linkages while still capturing the important characteristics of the network being assessed.

Another approach was used by the Washington Connected Landscapes Project (WHCWG 2010), and implemented in Linkage Mapper. This approach used ArcGIS scripts written in Python to model linkages between pairs of source polygons that were adjacent in Euclidean or cost-weighted distance space and were closer than a specific upper limit in cost-weighted distance. The Singleton et al. (2002) approach worked very well for relatively simple regional networks with a small number of sources, but the automated scripts were necessary for the more complex networks with large numbers of sources assessed by the Washington Connected Landscapes Project.

Step 10. Run the corridor analysis

Once the analyst has identified which sources will be connected to which other sources, three steps may be involved in conducting the connectivity analysis for each source pair. These steps must be repeated for each source pair to be directly connected in the network. The first step is to create a fracture-zone polygon that defines the extent of the corridor analysis for that source pair. This polygon serves as an analysis mask for assessing the connectivity between the sources in the pair. It can be digitized by hand, or the Cost Allocation tool in the Distance toolbox can be used to systematically derive fracture-zone polygons from the cost-weighted distance surface created in step 8. Linkage Mapper uses bounding circles to constrain corridor analysis extents.

The second step is to run the least-cost corridor analysis in ArcGIS. This analysis involves using the Cost Distance tool (in the Distance toolbox) to create a cost distance surface for each source in the pair, and then adding the two cost distance surfaces into a least-cost corridor surface connecting the two sources. This produces a raster in which the value for each cell is the cost-weighted distance of the least costly path between the two sources that passes through that cell.

The third step is to post-process the least-cost corridor analysis results to highlight the areas within the fracture zone that provide the best connectivity between the two sources. Singleton et al. (2002) used the Spatial Analyst Slice tool in ArcGIS to identify the most permeable portions of fracture zones using 10-percentile area increments. This conveys the relativistic nature of the corridor delineation (e.g., which is the best 10%, the next best 10%, and so on), however, it can also be misleading if the best 10% is not much different from the next best 10%. It's also sensitive to the fracture-zone polygon delineation (i.e., larger polygons will have a larger best 10%).

An alternative to identifying fracture-zone polygons is the normalized least-cost corridor approach used by the Washington Connected Landscapes Project and implemented in Linkage Mapper (figure 10.3c). This approach normalizes least-cost corridors by identifying the least-cost path between each pair of sources, and then buffers that path by a species-specific cost-weighted distance. This highlights the most permeable portion of the landscape (the least-cost path), and also provides information on the landscape permeability adjacent to that path (i.e., narrower normalized least-cost corridors occur in landscapes where there are fewer alternative paths with similar resistance adjacent to the least-cost path). Delineating fracture zones works well for simple networks. However, it can become very cumbersome in complex networks with many pairs of source patches to connect.

Step 11. Compile and interpret the results

Summarizing the results of a complex landscape linkage assessment can be challenging. Assessment results include maps of the base data, resistance surfaces, connectivity assessment products (for example, cost-weighted distance surfaces and least-cost corridor maps), and documentation of the model. Tabular summaries of linkage analysis results should include a summary of the source areas that were connected, and a summary of the linkages identified between those sources. Tabular information on linkages can include Euclidean (point-to-point) distance, cost-weighted distance, circuit resistance, and the cost distance / Euclidean distance ratio. The cost distance / Euclidean distance ratio can be calculated in two ways: by taking the cost-weighted distance of the least-cost path divided by the straight-line distance between the two sources, or by taking the cost-weighted distance of the least-cost path divided by the unweighted length of the least-cost path. These ratios are particularly informative because they provide information on how resistant the linkage is to movement based on the habitat characteristics within the linkage (i.e., they standardize the effects of linkage length and help highlight which linkages have poor habitat or barriers). These corridor-quality statistics and others are automatically generated by Linkage Mapper.

Multivariate statistical methods such as hierarchical cluster analysis can be used to identify common connectivity patterns for multiple focal species or species groups. For example, hierarchical cluster analysis was used to identify three groups of focal species that had similar linkage patterns, or connectivity guilds, for the Washington Connected Landscapes Project (WHCWG 2010).

Step 12. Communicate the assessment findings

Successfully communicating the results of an assessment is perhaps the most important step in an effective linkage assessment. A primary goal for conducting connectivity analysis is to produce a product that will actually get used. Planning for communication should start at the very beginning of the process and should be conducted throughout: explicitly including social factors and stakeholder participation is essential for implementation (see Chan et al., chapter 2 and Pressey and Bottrill 2009). The final report needs to include the following supporting documentation:

- A clear statement of the goals of the assessment, and specific questions or objectives the GIS analysis was intended to address
- A detailed description of the methods and procedures used to conduct the analysis
- Detailed metadata for the base layers that were used in the analysis
- Guidance on the interpretation of the results of the analysis, including appropriate spatial scales of application and a discussion of the level of uncertainty associated with the assessment products

Distribution of digital products, including reports, base data, scripts, and GIS products, is increasingly important. This chapter emphasizes the technical aspects of connectivity assessment; a more detailed approach to communication and implementation is discussed in chapter 11.

Step 13. Have a plan for the next step

Plan ahead for what follows the completion of a connectivity assessment. Will finer-scale analysis be required to identify specific areas for planning or implementation to address connectivity? How will the assessment products be used to implement activities that address habitat connectivity? How can linkages be monitored to evaluate the effectiveness of the assessment approach or management activities?

Recommendations for future research

The science of landscape connectivity has advanced substantially during the last decade; however, important challenges remain. A few topics for future research are described below.

Continued development of better empirical methods and data for parameterizing permeability models.

A primary drawback of most linkage assessments is that they are based on conceptual models developed from expert opinion. Applications of landscape genetic techniques (Manel et al. 2003; Holderegger and Wagner 2008), resource selection functions (Chetkiewicz et al. 2006), and step selection functions (Frair et al. 2008) are important contributions to developing an empirical framework for assessing connectivity patterns. However, important challenges remain because information on habitat selection during infrequent dispersal movements is extremely difficult to collect, and the role of those movements in broad-scale demographic performance of populations can be difficult to determine (Lowe and Allendorf 2010).

Multispecies and multiprocess linkage optimization

Identifying networks that can provide connectivity values for the wide variety of ecological processes and species with different habitat requirements that are associated with them is an important challenge. Substantial work has been done on optimizing reserve placement (see Williams, chapter 12), but less work has been done on integrating entire networks for multiple species and ecological processes. Methods for integrating the results of multiple assessments into unified linkage networks that effectively serve connectivity requirements across species and ecological processes need to be developed.

Connectivity planning for climate change adaptation

Connectivity is one of the most commonly cited strategies for adapting to climate change (Heller and Zavaleta 2009). However, there are few examples of effective methods for developing linkage plans that address the changing ecological conditions associated with climate change (see Bachelet, chapter 13). Monitoring the effectiveness of conservation networks will be particularly important given expected ecological changes associated with climate change.

Model validation

Integration of adaptive management approaches for connectivity conservation and research will be essential for advancing the field. Connectivity assessments should strive to produce verifiable hypotheses, particularly if they involve enhancement or restoration activities. The integration of landscape genetics and movement studies for bears, marten, and sage grouse into the work being conducted by the Washington Wildlife Habitat Connectivity Working Group provides an excellent example of the integration of specific model validation into a multispecies, broad-scale connectivity assessment (WHCWG 2010).

Summary

Linkage assessment maps can provide a broad-scale vision to bring together a wide spectrum of stakeholders. These maps show the interconnectedness of large landscapes and can be very effective for fostering cooperation in landscape management and conservation planning. In this chapter, we provided an overview of basic concepts related to connectivity assessment, introduced some of the more common analytical tools, outlined the steps for a connectivity assessment according to best available practice, and highlighted some areas for future research. Important messages from this chapter are (1) it is important to clearly articulate the question to be addressed in each connectivity analysis, (2) analysis approaches range from simple to complex, and although more complex approaches may provide more information, they also require more assumptions, (3) keep things (in the words of Einstein) "as simple as possible, but no simpler" to address the question at hand, and (4) validation and communication of the results are important components of any connectivity assessment. As newer approaches are developed, they will be available on the website accompanying this book, http://www.esri.com/conservationbook.

References

Baggio, J. A., K. Salau, M. A. Janssen, M. L. Schoon, and O. Bodin. 2011. "Landscape Connectivity and Predator-Prey Population Dynamics." *Landscape Ecology* 26: 33–45.

Battisti, C., and L. Luiselli. 2011. "Selecting Focal Species in Ecological Network Planning Following an Expert-Based Approach: Italian Reptiles as a Case Study." *Journal for Nature Conservation* 19: 126–30.

Beier, P., D. R. Majka, and W. D. Spencer. 2008. "Forks in the Road: Choices in Procedures for Designing Wildland Linkages." *Conservation Biology* 22: 836–51.

Beier, P., W. Spencer, R. F. Baldwin, and B. H. McRae. 2011. "Toward Best Practices for Developing Regional Connectivity Maps." *Conservation Biology* 5: 879–92.

Bennett, A. 2003. *Linkages in the Landscape: The Role of Corridors and Connectivity in Wildlife Conservation.* Gland, Switzerland; Cambridge, UK: International Union for Conservation of Nature–The World Conservation Union. http://app.iucn.org/dbtw-wpd/edocs/FR-021.pdf.

Bodin, O., and S. Saura. 2010. "Ranking Individual Habitat Patches as Connectivity Providers: Integrating Network Analysis and Patch Removal Experiments." *Ecological Modelling* 221: 2,393–405.

Braunisch, V., G. Segelbacher, and A. H. Hirzel. 2010. "Modelling Functional Landscape Connectivity from Genetic Population Structure: A New Spatially Explicit Approach." *Molecular Ecology* 19: 3,664–78.

Carroll, C., B. H. McRae, and A. Brookes. 2011. "Use of Linkage Mapping and Centrality Analysis across Habitat Gradients to Conserve Connectivity of Gray Wolf Populations in Western North America." *Conservation Biology* 26: 78–87.

Carroll, C., and D. G. Miquelle. 2006. "Spatial Viability Analysis of Amur Tiger *Panthera tigris altaica* in the Russian Far East: The Role of Protected Areas and Landscape Matrix in Population Persistence." *Journal of Applied Ecology* 43: 1,056–68.

Cerdeira, J. O., L. S. Pinto, M. Cabeza, and K. J. Gaston. 2010. "Species Specific Connectivity in Reserve-Network Design Using Graphs." *Biological Conservation* 143: 408–15.

Chetkiewicz, C. L., C. C. St. Clair, and M. S. Boyce. 2006. "Corridors for Conservation: Integrating Pattern and Process." *Annual Review of Ecology, Evolution and Systematics* 37: 317–42.

Comer, P. J., and K. A. Schulz. 2007. "Standardized Ecological Classification for Mesoscale Mapping in the Southwestern United States." *Rangeland Ecology & Management* 60: 324–35.

Crooks, K. R., and M. Sanjayan. 2006. *Connectivity Conservation.* Cambridge: Cambridge University Press.

Cushman, S. A., K. S. McKelvey, J. Hayden, and M. K. Schwartz. 2006. "Gene Flow in Complex Landscapes: Testing Multiple Hypotheses with Causal Modeling." *The American Naturalist* 168: 486–99.

Doyle, P. G. 1984. "Random Walk on the Speiser Graph of a Riemann Surface." *Bulletin of the American Mathematical Society* 11: 371–77.

Driezen, K., F. Adriaensen, C. Rondinini, C. P. Doncaster, and E. Matthysen. 2007. "Evaluating Least-Cost Model Predictions with Empirical Dispersal Data: A Case-Study Using Radiotracking Data of Hedgehogs (*Erinaceus europaeus*)." *Ecological Modelling* 209: 314–22.

Epps, C. W., J. D. Wehausen, V. C. Bleich, S. G. Torres, and J. S. Brashares. 2007. "Optimizing Dispersal and Corridor Models Using Landscape Genetics." *Journal of Applied Ecology* 44: 714–24.

Estrada, E., and O. Bodin. 2008. "Using Network Centrality Measures to Manage Landscape Connectivity." *Ecological Applications* 18: 1,810–25.

Fall, A., M. J. Fortin, M. Manseau, and D. O'Brein. 2007. "Spatial Graphs: Principles and Applications for Habitat Connectivity." *Ecosystems* 10: 448–61.

Fortin, M. J., and M. R. Dale. 2005. *Spatial Analysis: A Guide for Ecologists.* Cambridge: Cambridge University Press.

Frair, J. L., E. H. Merrill, H. L. Beyer, and J. M. Morales. 2008. "Thresholds in Landscape Connectivity and Mortality Risks in Response to Growing Road Networks." *Journal of Applied Ecology* 45: 1,504–13.

Franklin, J. 2009. *Mapping Species Distributions: Spatial Inference and Prediction.* Cambridge: Cambridge University Press.

Freeman, L. C., S. P. Borgatti, and D. R. White. 1991. "Centrality in Valued Graphs—A Measure of Betweenness Based on Network Flow." *Social Networks* 13: 141–54.

Galpern, P., M. Manseau, and A. Fall. 2011. "Patch-Based Graphs of Landscape Connectivity: A Guide to Construction, Analysis and Application for Conservation." *Biological Conservation* 144: 44–55.

Girvetz, E. H., J. H. Thorne, A. M. Berry, and J. A. G. Jaeger. 2008. "Integration of Landscape Fragmentation Analysis into Regional Planning: A Statewide Multi-Scale Case Study from California, USA." *Landscape and Urban Planning* 86: 205–18.

Grimm, V., and S. F. Railsback. 2005. *Individual-Based Modeling and Ecology*. Princeton, NJ: Princeton University Press.

Haddad, N. M., B. Hudgens, D. Damschen, J. Levey, L. Orrock, J. J. Tewksbury, and A. Weldon. 2011. "Assessing Positive and Negative Ecological Effects of Corridors." In *Sources, Sinks, and Sustainability across Landscapes,* edited by J. Lui, V. Hull, A. Morzillo, and J. Weins, 475–506. Cambridge: Cambridge University Press.

Hanski, I. A., and M. E. Gilpin. 1997. "Metapopulation Biology: Ecology, Genetics, and Evolution." San Diego, CA: Academic Press.

Hartvigsen, G. 2011. "Using R to Build and Assess Network Models in Biology." *Mathematical Modelling of Natural Phenomena* 6: 61–75.

Heinrichs, J. A., D. J. Bender, D. L. Gummer, and N. H. Schumaker. 2010. "Assessing Critical Habitat: Evaluating the Relative Contribution of Habitats to Population Persistence." *Biological Conservation* 143: 2,229–37.

Heller, N. E., and E. S. Zavaleta. 2009. "Biodiversity Management in the Face of Climate Change: A Review of 22 Years of Recommendations." *Biological Conservation* 142: 14–32.

Hilty, J. A., W. Z. Lidicker, A. M. Merenlender. 2006. *Corridor Ecology: The Science and Practice of Linking Landscapes for Biodiversity Conservation.* Washington, DC: Island Press.

Holderegger, R., and H. H. Wagner. 2008. "Landscape Genetics." *Bioscience* 58: 199–207.

IPCC (Intergovernmental Panel on Climate Change). 2007. *Climate Change 2007: Synthesis Report. Contribution of Working Groups I, II, and III to the Fourth Assessment Report of the Intergovernmental Panel on Climate Change.* Geneva, Switzerland: IPCC.

Johnston, K. M., ed. 2013. *Agent Analyst: Agent-Based Modeling in ArcGIS.* Redlands, CA: Esri Press.

Kautz, R., R. Kawula, T. Hoctor, J. Comiskey, D. Jansen, D. Jennings, J. Kasbohm, F. Mazzotti, R. McBride, L. Richardson, and K. Root. 2006. "How Much Is Enough? Landscape-Scale Conservation for the Florida Panther." *Biological Conservation* 130: 118–33.

Kindall, J. L., and F. T. van Manen. 2007. "Identifying Habitat Linkages for American Black Bears in North Carolina, USA." *Journal of Wildlife Management* 71: 487–95.

Krosby, M., J. Tewksbury, N. M. Haddad, and J. Hoekstra. 2010. "Ecological Connectivity for a Changing Climate." *Conservation Biology* 24: 1,686–89.

Laita, A, J. S. Kotiaho, and M. Monkkonen. 2011. "Graph-Theoretic Connectivity Measures: What Do They Tell Us about Connectivity?" *Landscape Ecology* 26: 951–67.

Landguth, E. L., B. K. Hand, J. Glassy, S. A. Cushman, and M. A. Sawaya. 2012. "UNICOR: A Species Connectivity and Corridor Network Simulator." *Ecography* 35: 9–14.

Landres, P. B., P. Morgan, and F. J. Swanson. 1999. "Overview of the Use of Natural Variability Concepts in Managing Ecological Systems." *Ecological Applications* 9: 1,179–88.

Levey, D. J., B. M. Bolker, J. J. Tewksbury, S. Sargent, and N. M. Haddad. 2005. "Effects of Landscape Corridors on Seed Dispersal by Birds." *Science* 309: 146–48.

Levey, D. J., J. J. Tewksbury, and B. M. Bolker. 2008. "Modelling Long-Distance Seed Dispersal in Heterogeneous Landscapes." *Journal of Ecology* 96: 599–608.

Levin, S. A. 1974. "Dispersion and Population Interactions." *American Naturalist* 108: 207–28.

Lowe, W. H., and F. W. Allendorf. 2010. "What Can Genetics Tell Us about Population Connectivity?" *Molecular Ecology* 19: 3,038–51.

Manel, S., M. K. Schwartz, G. Luikart, P. Taberlet. 2003. "Landscape Genetics: Combining Landscape Ecology and Population Genetics." *Trends in Ecology and Evolution* 18: 189–97.

McCullough, D. R. 1996. *Metapopulations and Wildlife Conservation.* Washington, DC: Island Press.

McGarigal, K., and B. J.Marks. 1995. *FRAGSTATS: Spatial Pattern Analysis Program for Quantifying Landscape Structure.* Gen. Tech. Rep. PNW-GTR-351. Portland, OR: US Department of Agriculture, Forest Service, Pacific Northwest Research Station.

McRae, B. H. 2006. "Isolation by Resistance." *Evolution* 60: 1,551–61.

McRae, B. H., and P. Beier. 2007. "Circuit Theory Predicts Gene Flow in Plant and Animal Populations." *Proceedings of the National Academy of Sciences of the United States of America* 104: 19,885–90.

McRae, B. H., P. Beier, E. Dewald, Y. Huynh, and P. Keim. 2005. "Habitat Barriers Limit Gene Flow and Illuminate Historical Events in a Wide-Ranging Carnivore, the American Puma." *Molecular Ecology* 14: 1,965–77, doi: 10.1111/j.1365-294X.2005.02571.x.

McRae, B. H., B. G. Dickson, T. H. Keitt, V. B. Shah. 2008. "Using Circuit Theory to Model Connectivity in Ecology, Evolution, and Conservation." *Ecology* 89: 2,712–24.

McRae, B. H., and V. Shah. 2009. *Circuitscape User's Guide.* Santa Barbara: University of California. http://www.circuitscape.org.

Merriam, G. 1984. "Connectivity: A Fundamental Ecological Characteristic of Landscape Pattern." In *Proceedings of the First International Seminar on Methodology in Landscape Ecological Research and Planning,* edited by M. Ruzicka, T. Hrnciarova, and L. Miklos, 5–15. Roskilde Denmark: International Association for Landscape Ecology.

Millspaugh, J. J., R. A. Gitzen, D. R. Larsen, and F. R. Thompson. 2009. "General Principles for Developing Landscape Models for Wildlife Conservation." In *Models for Planning Wildlife Conservation in Large Landscapes,* edited by J. J. Millspaugh and F. R. Thompson, 1–31. Boston, MA: Academic Press.

Millspaugh, J. J., and F. R. Thompson. 2009. *Models for Planning Wildlife Conservation in Large Landscapes.* Boston, MA: Academic Press.

Minor, E. S., and D. L. Urban. 2008. "A Graph-Theory Framework for Evaluating Landscape Connectivity and Conservation Planning." *Conservation Biolog* 22: 297–307.

Montana Department of Transportation, Confederated Salish and Kootenai Tribes, and US Department of Transportation Federal Highway Administration. 2008. "US Highway 93 Ninepipe/Ronan Improvement Project: Final Supplemental Environmental Impact Statement and Section 4(f) Evaluation." FHWA-MT-EIS-95-01-FS. http://www.skillings.com/US93/PDF/US93_FSEIS/US93_FSEIS.pdf.

Morrison, M. L., B. G. Marcot, and R. W. Mannan. 1998. *Wildlife-Habitat Relationships: Concepts and Applications.* Madison, WI: University of Wisconsin Press.

Newman, M. E. J. 2010. *Networks: An Introduction.* Oxford: Oxford University Press.

Parmesan, C., and G. Yohe. 2003. "A Globally Coherent Fingerprint of Climate Change Impacts across Natural Systems." *Nature* 421: 37–42.

Pascual-Hortal, L., and S. Saura. 2008. "Integrating Landscape Connectivity in Broad-Scale Forest Planning through a New Graph-Based Habitat Availability Methodology: Application to Capercaillie (*Tetrao urogallus*) in Catalonia (NE Spain)." *European Journal of Forest Research* 127: 23–31.

Perez-Espona, S., F. J. Perez-Barberia, J. E. Mcleod, C. D. Jiggins, I. J. Gordon, and J. M. Pemberton. 2008. "Landscape Features Affect Gene Flow of Scottish Highland Red Deer (*Cervus elaphus*)." *Molecular Ecology* 17: 981–96.

Pinto, N., and T. H. Keitt. 2009. "Beyond the Least-Cost Path: Evaluating Corridor Redundancy Using a Graph-Theoretic Approach." *Landscape Ecology* 24: 253–66.

Pressey, R. L., and M. C. Bottrill. 2009. "Approaches to Landscape- and Seascape-Scale Conservation Planning: Convergence, Contrasts and Challenges." *Oryx* 43: 464–75.

Rayfield, B., M. J. Fortin, and A. Fall. 2011. "Connectivity for Conservation: A Framework to Classify Network Measures." *Ecology* 92: 847–58.

Rothley, K. 2005. "Finding and Filling the 'Cracks' in Resistance Surfaces for Least-Cost Modeling." *Ecology and Society* 10 (1): 4. http://www.ecologyandsociety.org/vol10/iss1/art4/.

Rowland, M. M., and M. J. Wisdom. 2009. "Habitat Networks for Terrestrial Wildlife: Concepts and Case Studies." In *Models for Planning Wildlife Conservation in Large Landscapes,* edited by J. J. Millspaugh and F. R. Thompson, 501–31. Boston, MA: Academic Press.

Saura, S., and L. Pascual-Hortal. 2009. "Conefor Sensinode 2.2: A Software Package for Quantifying the Importance of Habitat Patches for Landscape Connectivity." *Environmental Modelling & Software* 24: 135–39. http://www.conefor.org.

Schadt, S., F. Knauer, P. Kaczensky, E. Revilla, T. Wiegand, and L. Trepl. 2002. "Rule-Based Assessment of Suitable Habitat and Patch Connectivity for the Eurasian Lynx." *Ecological Applications* 12: 1,469–83.

Schumaker, N. H. 2011. HexSim User's Guide. Corvallis, OR: US Environmental Protection Agency, Environmental Research Laboratory. http://www.hexsim.net/guide/guide.pdf.

Schumaker, N. H., T. Ernst, D. White, J. Baker, and P. Haggerty. 2004. "Projecting Wildlife Responses to Alternative Future Landscapes in Oregon's Willamette Basin." *Ecological Applications* 14: 381–400.

Scott, J. M., P. J. Heglund, and M. L. Morrison. 2002. *Predicting Species Occurrences: Issues of Accuracy and Scale.* Washington, DC: Island Press.

Shirk, A. J., D. O. Wallin, S. A. Cushman, C. G. Rice, and K. I. Warheit. 2010. "Inferring Landscape Effects on Gene Flow: A New Model Selection Framework." *Molecular Ecology* 19: 3,603–19.

Short Bull, R. A., S. A. Cushman, R. Mace, T. Chilton, K. C. Kendall, E. C. Landguth, M. K. Schwartz, K. McKelvey, F. W. Allendorf, and G. Luikart. 2011. "Why Replication Is Important in Landscape Genetics: American Black Bear in the Rocky Mountains." *Molecular Ecology* 20: 1,092–107.

Singleton, P. H., W. L. Gaines, and J. F. Lehmkuhl. 2002. *Landscape Permeability for Large Carnivores in Washington: A Geographic Information System Weighted-Distance and Least-Cost Corridor Assessment.* Research Paper PNW-RP-549. Portland, OR: USDA Forest Service Pacific Northwest Research Station.

Spear, S. F., N. Balkenhol, M. J. Fortin, B. H. McRae, and K. Scribner. 2010. "Use of Resistance Surfaces for Landscape Genetic Studies: Considerations for Parameterization and Analysis." *Molecular Ecology* 19: 3,576–91.

Spencer, W. D., P. Beier, K. Penrod, K. Winters, C. Paulman, H. Rustigian-Romos, J. Strittholt, M. Parisi, and A. Pettler. 2010. *California Essential Habitat Connectivity Project: A Strategy for Conserving a Connected California.* Prepared for California Department of Transportation, California Department of Fish and Game, and Federal Highways Administration. http://nrm.dfg.ca.gov/FileHandler.ashx?DocumentID=18366.

Stephens, D. W., and J. R. Krebs 1987. *Foraging Theory.* Princeton, NJ: Princeton University Press.

Storfer, A., M. A. Murphy, S. F. Spear, R. Holderegger, and L. P. Waits. 2010. "Landscape Genetics: Where Are We Now?" *Molecular Ecology* 19: 3,496–514.

Taylor, P. D., L. Fahrig, K. Henein, and G. Merriam. 1993. "Connectivity Is a Vital Element of Landscape Structure." *Oikos* 571–73.

Theobald, D. M., S. E. Reed, K. Fields, and M. Soulé. 2011. "Connecting Natural Landscapes Using a Landscape Permeability Model to Prioritize Conservation Activities in the United States." *Conservation Letters* 5: 123–33.

Urban, D., and T. Keitt. 2001. "Landscape Connectivity: A Graph-Theoretic Perspective." *Ecology* 82: 1,205–18.

USFWS (US Fish and Wildlife Service). 2011. *Revised Recovery Plan for the Northern Spotted Owl,* Strix occidentalis caurina. Portland, OR: US Fish and Wildlife Service. http://www.fws.gov/oregonfwo/Species/Data/NorthernSpottedOwl/Recovery/.

Vignieri, S. N. 2005. "Streams over Mountains: Influence of Riparian Connectivity on Gene Flow in the Pacific Jumping Mouse (*Zapus trinotatus*)." *Molecular Ecology* 14: 1,925–37.

WHCWG (Washington Wildlife Habitat Connectivity Working Group). 2010. *2010 Washington Connected Landscapes Project: Statewide Analysis.* Olympia: Washington Departments of Fish and Wildlife, and Transportation. http://www.waconnected.org.

With, K. A., R. H. Gardner, and M. G. Turner. 1997. "Landscape Connectivity and Population Distributions in Heterogeneous Environments." *Oikos* 78: 151–69.

Wipfli, M. S. 2005. "Trophic Linkages between Headwater Forests and Downstream Fish Habitats: Implications for Forest and Fish Management." *Landscape and Urban Planning* 72: 205–13.

11

Conservation planning to ensure viability of populations and metapopulations

Jessica Stanton; and H. Reşit Akçakaya, PhD

An important (if not the ultimate) goal of conservation planning is to preserve natural populations—that is, to prevent the extinction of species. Assessing viability is the most direct way of measuring how close to this goal your conservation plan is, or will be in the future, when the planned conservation actions are in place. Viability is the likelihood that a species (or a population) will remain extant. Population viability analysis (PVA) is a method for calculating this likelihood, either under current conditions, or under assumed future changes.

At this stage in your conservation planning, you should have a good understanding of the landscape, what focal species you might use to represent ecosystems of conservation concern in your area, and the habitat available to those species. You even may have mapped out areas of suitable habitat and analyzed the connectivity between potential conservation areas. However, merely protecting areas of suitable habitat does not necessarily ensure that the amount and configuration of habitat patches will be sufficient for the long-term survival of the focal species. Likewise, even if sufficient amounts of suitable habitat are available, species may still decline due to threats unrelated to habitat loss, such as harvesting, introduced predators, competitors, or diseases. To analyze the impact of those threats, information about the species' demography also should be considered. PVA provides a method to link demographic data with habitat maps to address questions about the prospects for recovery, longevity, and resilience specific to your focal species in your landscape. PVA also can be particularly useful when evaluating different management options or proposed reserve design configurations, and for prioritizing research and data collection needs through the use of sensitivity analysis.

The ability to analyze habitat and landscape characteristics simultaneously with demography in a single model is an important aspect of PVA. Populations are often faced with multiple threats that affect their long-term viability, some of which may primarily impact habitat availability (such as logging or urban sprawl), and others that may have greater impacts on demography (such as reduced survival due to pollution or hunting). Combined, these different threats may have synergistic impacts on the population that may not be detected if considered singly. Similarly, some threats have effects that inherently impact both demography and habitat and cannot be adequately represented by a model that considers habitat availability but disregards demography or vice versa. For example, road building may fragment the landscape, decrease dispersal ability, and increase mortality for some species. Methods such as PVA allow you to model multiple impacts that may have complex effects on your focal species and provide an important complement to other analyses of the habitat, landscape connectivity, and reserve design.

A main focus of this book is on planning for focal species at finer scales of analysis, as discussed previously. Questions of viability, however, occur primarily at the population level, which for many focal species requires regional-scale analysis. Issues of scale are discussed in detail below, and in Unnasch and Karl (chapter 3). If you are working at a finer scale than the population extent, you can perhaps consider your conservation plan as a building block, or piece of a larger puzzle, which is population persistence. Consider the extent of your own analysis in the hierarchies of scale encompassed by the entire population or metapopulation. To assess viability, you may need to extrapolate the habitat maps of your focal species to a much broader region.

Current best practices

PVA is a process of predicting measures of population persistence, such as risk of population extinction or decline, chance of population recovery, expected future population size, and expected time to extinction. Most outputs from PVAs are probabilistic and are based on three variables: the amount of expected population change (e.g., 100% decline or total extinction, or partial decline), the probability that such a decline will occur, and the time frame over which the decline is expected to take place (Akçakaya 1992, 2000a). Thus, specific PVA results may include the following:

- Probability of extinction within the next 100 years
- Probability of a 50% decline by the year 2060
- Probability that there will be 50 or fewer mature individuals left in the population within the next 40 years
- Probability that the population will increase to 1,000 or more individuals by the year 2050

To predict these types of viability measures, PVA uses a variety of models (discussed below), which are parameterized with habitat and distribution data (e.g., occurrence locations and map layers describing habitat requirements) and demographic data (such as censuses, mark-recapture studies, surveys, and observations of reproduction and dispersal events). In addition, PVA relies on general information about the ecology of the species (e.g., life stages, age or size of first breeding, and so on), some of which may be based on expert opinion.

The best type of model to use depends on the conservation question being asked, the amount and type of available data, and the ecology of the species under consideration. The issue of formulating a specific question is addressed in detail below (see "Guidelines for application"). PVA is used often in the conservation and management of rare and threatened species, with two broad objectives. In the short term, the objective is to prevent extinction. In the longer-term, the objective is to promote conditions in which species not only remain extant but also retain their potential for evolutionary change without intensive management. With these objectives in mind, the specific questions addressed with a PVA may involve the following:

1. Assessing vulnerability: estimating the risk of extinction (an absolute measure of vulnerability), or the relative vulnerability of species and populations to extinction. These predictions can be used in ranking species (such as in the International Union for Conservation of Nature Red List of Threatened Species). When combined with such considerations as cost-effectiveness, social and cultural priorities, and taxonomic uniqueness, these results can also be used to set policies and priorities for allocating scarce conservation resources.
2. Impact assessment: quantifying the impact of human activities (such as urban sprawl and other types of land-use change; harvest, poaching, and other types of direct exploitation; pollution; and introduction of exotic species) by comparing results of models with and without the population-level consequences of the human activity. These assessments may involve past or current impacts (to assign responsibility) or projected future impacts (e.g., to evaluate proposed development plans).
3. Evaluating management options: predicting the likely responses of species to conservation and management actions, such as reintroduction, captive breeding, prescribed burning, weed control, habitat rehabilitation, or different designs for nature reserves or corridor networks.
4. Planning research: determining priorities for further data collection based on the sensitivity of model results to data uncertainties in model parameters.
5. Improving understanding: organizing the information and assumptions about a species or a population. PVA allows a structured process that makes the assumptions explicit and highlights the data deficiencies and uncertainties.

Although there is no single recipe to follow when doing a PVA, main components may include identification of the question, data collection, data analysis and parameter estimation, modeling and risk assessment, sensitivity analysis and refinement of the model, population monitoring and model evaluation, as well as the evaluation of any implemented conservation actions (see figure 11.1 and Akçakaya et al. 1999).

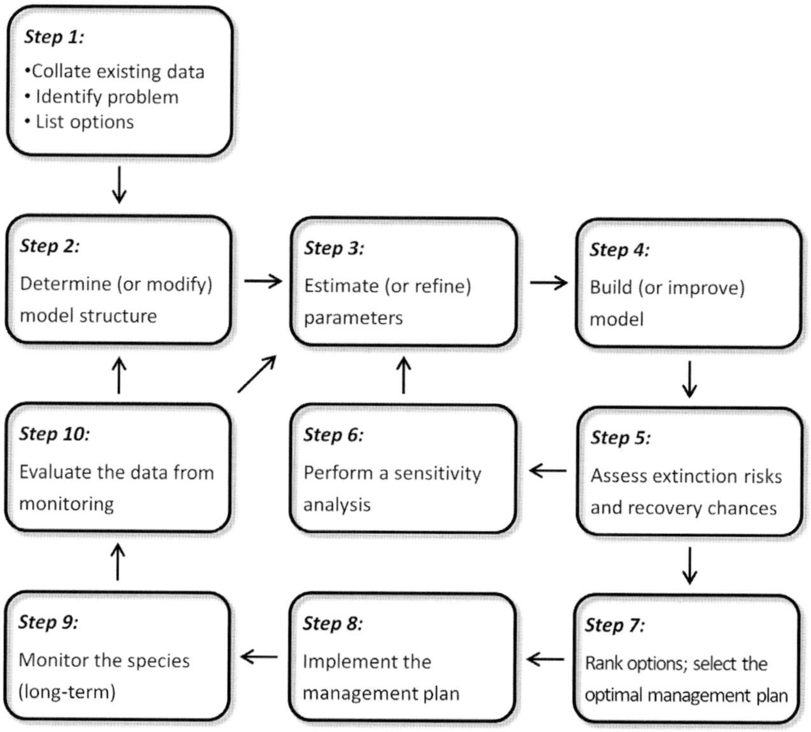

Figure 11.1. Main components of a population viability analysis. Figure from Akçakaya et al. (1999). Courtesy of Applied Biomathematics.

After formulating the specific questions, an important step is determining the type and structure of the model consistent with the question, the available data, and the ecology of the species. The types of models include occupancy models (see also Dickson et al., chapter 7), lattice (grid-based) models, age- or stage-structured metapopulation models, and individual- or agent-based models (see also Johnston et al., chapter 14). The details of these models are discussed by Akçakaya and Brook (2008), Akçakaya and Sjögren-Gulve (2000), DeAngelis and Mooij (2005), Grimm and Railsback (2005), Hanski (1994), and Sjögren-Gulve and Ebenhard (2000). A summary of the characteristics and examples of these model types are given in table 11.1. There are several software packages that implement these models. The most commonly used software programs include VORTEX (Lacy 1993) and the RAMAS library of programs (Akçakaya 1994, 2005; Akçakaya and Root 2003); others include ALEX (Possingham and Davies 1995), PATCH (Schumaker 1998), META-X (Grimm et al. 2004), and Agent Analyst (Johnston 2013). Of course, it is also possible to construct your own model from the ground up in any mathematical software package.

Table 11.1. Types of PVA models

Model type	Best use	Examples
Occupancy models	Equilibrium metapopulations High rates of local extinction and recolonization Limited demographic data Small, short-lived organisms (e.g., invertebrates) Large number of patches	Alpine Rhine valley amphibians (Gilioli et al. 2008) Saproxylic invertebrates (Ranius and Kindvall 2006) Heath fritillary butterfly (*Melitaea athalia*; Hodgson et al. 2009) Yellow-bellied marmot (*Marmota flaviventris*; Ozgul et al. 2006)
Lattice (grid-based) models	Relatively uniform or undisturbed landscapes Landscapes with continuous environmental gradients Locally abundant organisms	Serengeti grazers (Fryxell et al. 2005) *Helianthemum squamatum* (Quintana-Ascencio et al. 2009) *Aster kantoensis* (Shimada and Ishihama 2000) Asian water buffalo (Brook et al. 2006)
Demographically (age- or stage-) structured metapopulation models	Declining populations Locally abundant organisms Vertebrates and plants Large or dynamic landscapes Survival or fecundity depends on age or size Sufficient demographic data	Invasive tree *Syzygium jambos* (Brown et al. 2008) Woodland brown butterfly (Kindvall and Bergman 2004) Chinook salmon (Spromberg and Johnson 2008) Tree frog (Pellet et al. 2006) Eastern indigo snake (Breininger et al. 2004) Ovenbird (Larson et al. 2004) Tasmanian wedge-tailed eagle (Bekessy et al. 2009) Cougar (Cooley et al. 2009) Southern brown bandicoot (Southwell et al. 2008)
Individual-based models	Very small populations Abundant demographic and behavioral data Large-bodied, territorial species Modeling impact of genetic threats Determining emergent behaviors	*Micoures demerarae* (Brito and Fernandez 2000) *Panthera gombaszoegensis* (O'Regan et al. 2002) Community assembly (Hraber and Milne 1997) Giant panda (*Ailuropoda melanoleuca*; Li et al. 2003) Elk (*Cervus elaphus*; Murrow et al. 2009)

Source: Modified from Akçakaya and Brook (2008).

Guidelines for application

In this section, we outline some of the key issues to consider before approaching a PVA. Note that PVAs can become very complex because of the amount of data required, the number of variables involved, and the interaction between variables. Therefore, this chapter should be thought of as a guide to the types of information and data needed and as an overview of best practices. It is not a replacement for more thorough references, such as Akçakaya et al. (1999), Caswell (2001), Morris and Doak (2002), and Beissinger and McCullough (2002).

Formulation of goals and questions

Before you begin a PVA, it is important to clearly identify the question you wish to answer with your analysis. Outlining a realistic and useful question or set of questions is not trivial, and new questions may arise along the way as you begin to gather the data needed to construct your models. Likewise, you may find that the available data is not sufficient to address your original questions, and you may need to gather more data. Often, the process of constructing a PVA can make your data needs more clear and help identify where the critical areas of uncertainty or data deficiency lie. Examples of the types of questions that typically are addressed by a PVA include predictions about future abundances or assessing vulnerabilities, such as "What is the risk of local extinction?" or "What is the chance for recovery?"; questions related to assessing the impacts or current or future threats, such as "What is the population-level impact of harvesting?"; questions to evaluate competing management options, such as "Which management option will reduce the extinction risk the most for the least cost?"; and questions to guide future research or data collection, such as "Is the model more sensitive to uncertainty in adult survival or fecundity estimates?'"

The goals and the specific questions of a PVA, and the particular scenarios to be explored, should be developed in the course of a participatory process in which both experts and stakeholders outside of the professional scientific community provide input. Success of a PVA in contributing to positive conservation outcomes depends not only on its technical excellence but also on the acceptance and ownership of its results by the stakeholders and by the conservation agencies responsible for implementing the recommended actions. A broad stakeholder participation in the formative stages of a PVA project allows such acceptance.

Once your goals and questions are clearly articulated, you can begin to construct the model that will serve as the backbone of your PVA. The famous quote from statistician George E. P. Box (1979), "All models are wrong but some are useful," reminds us to think carefully about constructing a model that will be useful in addressing the questions you need it to address. Just as you would reach for a street map rather than a topographic map to plan a driving route, a model constructed for one type of conservation question or goal may or may not be much use to address a different type of question. Because a model is only a simplified representation of a real system, you must decide what type of model and which parameters are necessary to accurately represent the aspect of the system that is important to your goals. In addition to the nature of your conservation goals and management questions, additional considerations for deciding on what type of model to use include the amount and quality of data available as well as important ecological aspects of the focal species.

Habitat

Although other chapters in this volume cover predicting habitat (see Dickson et al., chapter 7), we briefly discuss it here, because it is an important aspect of viability analysis. In addition, habitat maps or models constructed for other conservation purposes may or may not be at the most appropriate spatial extent or resolution, or capture the characteristics of the landscape most useful in a PVA. A habitat map is the spatial backbone of a PVA and directly relates to many demographic parameters, such as the number and carrying capacity of subpopulations and dispersal rates. Therefore, most PVAs with spatial structure start from a map of suitable habitat for the modeled species.

Methods for estimating suitable habitat that are based on either statistical correlations between environmental variables and species occurrence locations (Guisan and Thuiller 2005; Franklin 2010) or known ecophysiological tolerances (Kearney and Porter 2009) are preferable to mapping habitat from expert opinion or the boundaries of coarse landform categories (such as wetland or forest). The benefits of these approaches include providing a means to incorporate a number of types of data, such as climatic variables, remote sensing, and digital elevation models, and accommodating interactions between predictor variables. Another advantage of modeling habitat suitability in either a correlation-based or ecophysiological-based framework is that it allows for projecting the model to new areas or future time steps. If possible, it is advisable to employ a few methods to construct the habitat model, and compare where they are similar and different (Elith et al. 2006).

Some care ought to be taken when constructing a habitat model to be used for a PVA. During the course of gathering the GIS layers and occurrence locations for constructing a habitat model, it is important to do the following:

1. Use a spatial scale that is relevant to the dispersal and patch size of the species (see below for more details on issues of scale).
2. Verify occurrence locations when constructing a correlation-based habitat model. Occurrence locations from publicly available biodiversity databases can be problematic at times if the coordinates given have an unknown radius of uncertainty. For instance, coordinates given may be the location of an actual sighting, or the center of a county or township where the species is reported to occur. Species misidentification is another issue that can affect the accuracy of habitat models (Lozier et al. 2009).
3. Select variables for the model that are biologically relevant to the species and that directly determine its occurrence through known ecophysiological mechanisms. Thus, indirect variables (e.g., altitude or topographic heterogeneity) should be avoided (Guisan and Zimmermann 2000).
4. Consider including climatic variables (derived from long-term weather data), land-use or land-cover variables (often derived from remote sensing products such as satellite images). When the model includes climatic variables that are predicted to change (see below), it is essential to exclude maps for variables that are proxies for climate (especially elevation). In addition, it is important to include the static maps (such as land-cover maps) in the model along with climate variables (especially if they interact with climate variables) or as binary masks (if there is no interaction; Stanton et al. 2012).
5. When using categorical variables (such as land cover), consider creating distance maps (e.g., distance to water) or percentage cover maps (e.g., percent of forest in a 120-m cell, based on 30-m cells of the original map), if such derived variables are considered to more directly affect the suitability of an area for the focal species.

6. Validate your model by using some form of data partitioning; this will help to avoid the risk of overfitting your model (see "Model parameters and assumptions" below).

Scale

An important consideration when constructing a model for a PVA is the spatial scale of your data, both in terms of the resolution and the extent (see also Unnasch and Karl, chapter 3). The scale should be relevant to the biology of the focal species, meaning it should relate to the patch size and dispersal distance. For example, too fine a resolution is needlessly data intensive; too coarse, and you may not get an adequate representation of the patch structure. A good guideline for vertebrate species is that the cell size of the raster map underlying the model should be about the same size or a bit smaller (but not much smaller) than the home range or territory size of the species.

Similarly, the spatial extent should be relevant to the dispersal distance of the focal species and the questions addressed by your PVA. Often, you may wish to construct a model for a subset of a species' total range, such as to assess the viability within a national park or system of reserve areas. A wide-ranging focal species can cause some misleading results. Consider a situation in which suitable habitat is located beyond the spatial extent of your model based on the political boundaries of the park or reserve. Subpopulations that lie on the border may be modeled as being truncated in size and, therefore, in carrying capacity. Individuals may also disperse into and outside of the park or reserve system. These issues can affect the outcome of the model in terms of estimated viability, time to decline or recovery, and the rate at which individual patches are colonized or go extinct within the metapopulation structure.

It is equally important to consider an appropriate timescale for your model, both in terms of resolution (i.e., the time step of the model) and extent (duration or time horizon). For most species in temperate regions, an annual time step is suitable for a PVA. However, this may be too coarse for other purposes, such as simulating dispersal behavior of individuals. The time horizon or duration of a model often follows from the formulation of goals. However, some technical issues must also be considered: population models projected for only a few years or generations may not show the longer-term trends inherent in the model. Models with very long duration will carry a high amount of uncertainty as the uncertainty is compounded through time. Also, model assumptions such as no noteworthy changes to the landscape or climatic conditions, or no adaptive changes or genetic drift, may be valid in the short term. However, landscape or genetic changes can be very difficult to predict over long timescales.

Model parameters and assumptions

The issues of scale mentioned above illustrate just a few of the common mistakes that can be made when constructing a population model. This is not meant as a discouragement; the process of constructing a model is an opportunity to think carefully about important dynamics of your focal species and all the factors that may be influencing the population in critical ways. It is important to be mindful of the explicit assumptions you are making in your model as well as the assumptions that are implied through model parameters.

For instance, mapping available habitat on the present landscape as a static variable implicitly assumes an unchanging landscape. Sometimes, because of lack of data (or lack of a crystal ball), we are forced to make assumptions that we know are not entirely accurate. By being aware of the assumptions you are making with your model, you can evaluate whether they seem reasonable or supportable within the context of the questions you are trying to address.

Finding the right balance between a detailed and realistic model, and a simple model that is well-supported by data, is very important. Simple models—for example, models with no age- or stage-based structure, with a single panmictic, closed population—are straightforward and understandable. However, how useful or accurate your model will be depends entirely on how well those assumptions describe the actual behavior of the population you are trying to describe (Lindenmayer et al. 2003) or the questions you are addressing. In some cases, a simple model may be adequate. However, models are overly simplified when they are missing important interactions between the population and the particular landscape, such as natural catastrophes, fragmentation, disease spread, or edge effects, when such factors substantially alter the trajectory of a population. But just as models can be too simple, they can also be overly complex by fitting many parameters to only a few data points (Ginzburg and Jensen 2004).

Model fitting

A good way to test the fit of a model is to use separate (ideally, independent) data for model building and for model validation. Separate training data and testing data can come from partitioning a single dataset spatially or temporally (while controlling for spatial or temporal autocorrelation). So, for example, a model of habitat suitability might be constructed using one-half of a landscape and then tested for accuracy by seeing how well it can predict the occurrences in the other half. This can be done with time-series data of population trends as well, when some portion of the time series is retained for evaluation. Using this approach, Brook et al. (2000) validated probabilistic predictions of PVAs and found that the predicted risk of population decline closely matched observed outcomes, there was no significant bias, population size projections did not differ notably from reality, and the predictions of five software packages they tested were highly concordant.

Demographic structure

"Demographic structure" refers to the categorizations of individuals within a population according to their age, size, sex, or physiological stage (such as immature, breeding, and so on). Models should have an age or stage structure if the demographic characteristics (fecundity, survival, growth, or dispersal) depend on the age, size, or other characteristics of the individual. These models (also referred to as "matrix models" because of the way the demographic rates can be organized in a matrix equation) are commonly used in conservation biology, and many techniques have been developed to parameterize them (Caswell 2001). The data that can be used for estimating the parameters of these models (e.g., age-specific survival rates and fecundities) include mark-recapture data (White and Burnham 1999) and population censuses in which individuals of different ages or stages are counted separately.

Models that lack age- or stage-based structure are called "scalar models" (they are also known as "count-based models" and "diffusion approximation"). Scalar models, which assume that all individuals in the population are functionally identical, are used in cases in which the only available data is a time series of total population size estimates. If the population being modeled has age structure, a scalar model of this population may overestimate the variability in the population size, and hence overestimate the risks faced by the population (Holmes 2004; Dunham et al. 2006). Simulations indicate that this bias increases as a function of the generation time of the species. Correcting the bias seems difficult, if not impossible, because it is not a simple function of generation time, and because any deviation of the initial age structure from the stable age structure adds uncertainty (Dunham et al. 2006).

Most demographic models of vertebrate populations include only females. In many cases, this is appropriate (as long as sex ratio is properly incorporated into fecundity estimates; see Akçakaya et al. 1999), because total reproduction is often limited by the number of breeding females. However, there are some cases in which it is necessary to model both males and females by developing matrix models with different stages or age classes for them. If, for example, the purpose of building a model is to evaluate the consequences of different hunting regimes, and only males (and perhaps only males over a certain age) are hunted, then the model obviously needs to have both male and female age classes (e.g., see Sezen et al. 2004). Regardless of the model objective, if males have higher mortality than females (causing a skewed sex ratio), and the mating system is (or is close to) monogamous, then the number of breeding females (and thus, overall population productivity) may be limited by the availability of males; in such cases, the model should include both females and males, in separate stages.

Density dependence

"Density dependence" refers to how the population growth depends on population size. In many species, both survival and fecundity decrease as a function of population size (i.e., more crowded populations have lower population growth). The mechanism for this is usually intraspecific competition for limited resources, such as food, light, and space (i.e., nesting or hiding sites). However, at very low populations, the reverse of this can also happen: because of factors collectively known as Allee effects, survival or reproduction may be lower when population size is very low (because of, for instance, difficulty in finding mates, or breakdown of social structures).

Density dependence can have important effects on extinction risks (e.g., Ginzburg et al. 1990), therefore, it is necessary to use caution in selecting the type of density dependence, and specifying its parameters. Most types of density dependence are modeled with two main parameters: the maximum population growth rate and carrying capacity. The maximum population growth rate (R_{max}) is the proportional increase in population size at low population densities, when the effects of intraspecific competition are negligible. This parameter can be estimated from the exponential growth-phase populations that become established in a new area. Another option is to regress growth rate (R) on population size, based on time-series data (e.g., see figure 11.2, from a PVA of the European bison, discussed below). Although this method can overestimate R_{max} when there is substantial measurement error, the results are nevertheless informative and can be used to estimate this parameter if supported by independent information (e.g., as in Kuemmerle et al. 2011). Carrying capacity (K) is the population size that can be supported by the available habitat and resources; it provides an important link between the suitability of habitat (see above) and population dynamics. For example, projected habitat loss is often translated as a decline in the population's carrying capacity in the population model. Thus, models that do not include such a habitat-related parameter (e.g., most density-independent models) may not be appropriate for such modeling threats as habitat loss. Estimating parameters for density dependence is a complicated process; for a discussion of these complexities, see Langton et al. (2002), Lande et al. (2002), and Saether et al. (2002), as well as the general modeling books mentioned in the introduction.

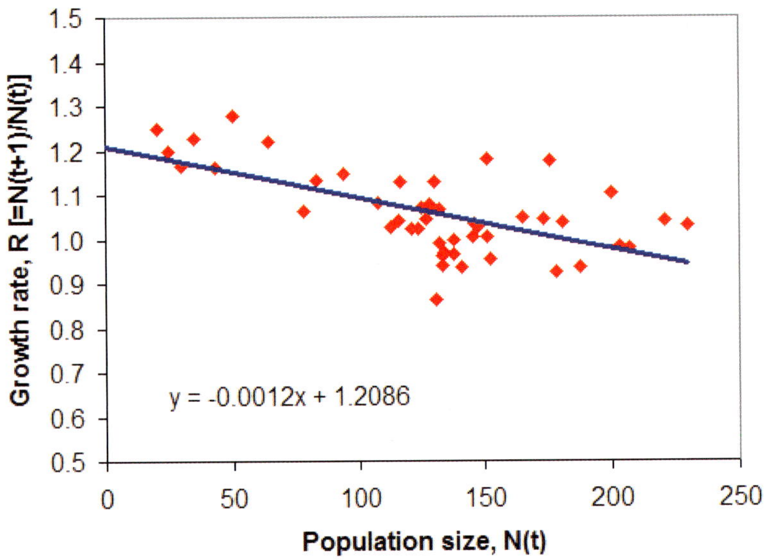

Figure 11.2. Estimation of R_{max} to model the density dependence, based on regression of growth rate on population size, derived from time-series data on population size of European bison (from Kuemmerle et al. 2011). The y-intercept is an estimate of R_{max}.
Figure from Kuemmerle et al. (2011).

Spatial structure

Most PVA models represent the landscape as discrete patches of suitable habitat, surrounded by a "matrix" that does not support populations but may allow dispersal. Each discrete habitable patch is assumed to support one biological population (sometimes called a subpopulation of a metapopulation). A biological population can be defined as a group of regularly interbreeding (i.e., panmictic) individuals. How these populations are configured (i.e., the number, size, shape, and location of these populations in the landscape) determines the model's spatial structure.

One approach to delineating populations is to identify clusters of cells in the habitat suitability map that have high values (high suitability). Clusters are identified based on a distance parameter called "neighborhood distance"; suitable cells within this distance of each other are grouped into one patch (see figure 11.3 for an example). Individuals (or territories) that are farther apart than the neighborhood distance are considered to be in different populations. This approach, combined with modeling and prediction of suitable habitat as described above, is used in habitat-based metapopulation models to delineate populations (Akçakaya 2000b; 2005).

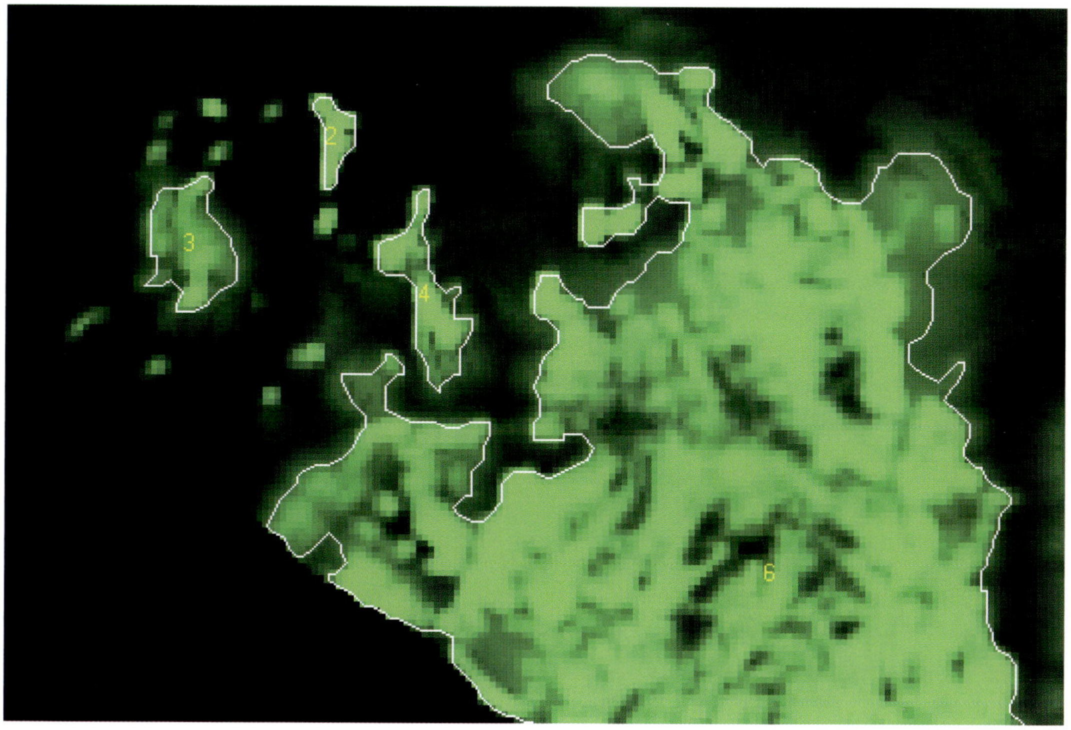

Figure 11.3. Detail of the spatial structure of the PVA model for the California gnatcatcher. Shades of green represent habitat suitability (the lighter the color, the higher the suitability); the white outlines are the approximate outer borders of patches, identified based on a neighborhood distance parameter of 300 m (Akçakaya and Atwood 1997). Cell (pixel) size of the underlying habitat map is 100 m. Each patch represents one population of the metapopulation model. In this small section of the study area, there are three small patches and part of a large patch (identified as "6"). The smallest patch ("2") consists of 70 cells. Smaller clusters of suitable cells are not identified as populations because they are too small to support a population. The location, size, and number of habitat patches based on habitat maps such as this one are then used to form the spatial structure of the metapopulation. Figure from Akçakaya and Atwood (1997).

Dispersal and connectivity

Dispersal is the movement of individuals in the landscape. Dispersal between populations allows extinct populations (unoccupied patches) to become recolonized, and decreases the probability of local extinction of extant but small populations ("rescue effect"). Thus, in many cases, increased dispersal increases viability. However, this is not always true (see Stacey et al. 1997; Beier and Noss 1998; Lecomte et al. 2004); in some cases, increasing dispersal by, for example, building or maintaining habitat corridors, may not be the best option. Whether such conservation actions are useful depends on many factors, including the behavior of the species in corridors, risk of spreading diseases or attracting predators, and cost of, and alternatives to, corridors (Akçakaya et al. 2006). Thus, defining connectivity as a conservation goal by itself may be counterproductive; instead, the role of dispersal should be studied in the context of the whole metapopulation, and in terms of its effect on the overall viability of the species in the landscape.

How connectivity is incorporated into a PVA depends on model type (see above). Parameters commonly used to model connectivity include dispersal rate (proportion of individuals moving from one habitat patch to another per unit time), and dispersal probability (probability of an individual moving from one patch or cell to another within a given time interval). The time step for these parameters should coincide with the time step of the model. Estimation of these parameters can be based on mark-recapture data. Relative rates of dispersal can be based on landscape-based approaches such as least-cost path (or friction) maps (Haines et al. 2006), gravity models (Beaudry et al. 2008) or circuit-theoretic methods (McRae et al. 2008). However, to be useful in a PVA, these relative measures must be combined with estimates of the actual number or proportion of individuals dispersing.

Variability and uncertainty

Variability is a pervasive characteristic of natural populations, and an important determinant of risk of extinction or decline in population models. Some variability in population sizes arises from the internal dynamics of the population; these include deterministic changes such as fluctuations in age composition due to departure from the stable age distribution, and random changes caused by demographic stochasticity. Demographic stochasticity is the sampling variation in the number of survivors and the number of offspring that occurs (even if survival rates and fecundities were constant) because a population is made up of a finite, integer number of individuals. Demographic stochasticity is often modeled by sampling the number of survivors from a binomial distribution and the number of offspring from a Poisson distribution (Akçakaya 1991). Relative to other factors, demographic stochasticity becomes more important at small population sizes.

Variability also is caused by fluctuations in environmental factors, causing stochastic changes in demographic rates such as survival, growth, fecundity, and dispersal. Environmental stochasticity often is modeled by selecting these demographic rates from random probability distributions, with specified mean, variance, and distribution type (such as normal, lognormal, beta, and so on). For examples of incorporating environmental variability into PVA models, see Akçakaya et al. (2004).

Estimating temporal variability of demographic rates for a PVA is often complicated because of measurement error (also called sampling variability). Temporal variability in any parameter (e.g., survival rate or fecundity) includes both natural variability and measurement error. For unbiased estimates, the components due to sampling and demographic stochasticity must be subtracted from total observed variance. Otherwise, variability may be overestimated, which may cause overestimated risks, as well as truncation (and, consequently, bias) in vital rates. There are several ways to obtain unbiased estimates:

- If survival rates are based on a mark-recapture analysis, see Gould and Nichols (1998), White et al. (2002), or the help file of Program MARK on how to remove demographic/sampling variance.
- If survival or fecundity estimates are based on age-structured census data, subtract the expected demographic variance from the observed variance (Akçakaya 2002).
- If the data is from a census of the total (or part of the) population, use methods discussed by Holmes (2001, 2004).
- If you have repeated or parallel estimates of the same population, you can use the covariance of the parallel measurements or the within-year variance estimates to subtract observation error from total observed variance (see Dunning et al. 2002; and Morris and Doak 2002, chapter 5).

An important consideration is the correlation among the demographic rates over time. This can be either different demographic rates in the same population (e.g., the correlation between survival and fecundity), or in different populations (e.g., the spatial correlation of population growth rate). Correlation increases the variability of the population size, and hence increases the risks of decline or increase (LaHaye et al. 1994). Thus, assuming no (or zero) correlation often gives results that are not precautionary.

Landscape change

Landscapes change according to seasons, climatic fluctuations (e.g., droughts and El Niño events), disturbances (e.g., fire and windthrow) and succession, as well as human impacts (e.g., urban sprawl, global climate change, and agricultural expansion; see Theobald, chapter 5, and Bachelet, chapter 13). For a species in such a dynamic landscape, these changes cause changes in the spatial structure of the population (e.g., populations may become more fragmented or more isolated), as well as in the characteristics of populations (e.g., habitat loss may mean a declining carrying capacity). These changes have important effects on the viability of the species, which depends on the interaction between landscape change (the pattern, scale, rate, and direction of habitat changes in size, structure, and quality) and the species' ecology (its ability to disperse between and function within the habitat patches).

For many species, habitat loss and global climate change are two threats that cause reduction, shifts, and fragmentation of their habitats. To adequately analyze these threats, a PVA should allow a dynamic spatial structure; in other words, it should simulate temporal changes in the location and number of populations that arises from habitat patches splitting, merging, appearing, and disappearing as the species' habitat changes (Akçakaya 2001; Akçakaya and Root 2003). PVA approaches that link a dynamic habitat to metapopulation models with dynamic spatial structure have been applied to study climate change (Keith et al. 2008; Anderson et al. 2009; Brook et al. 2009; Fordham et al. 2012), timber harvest, succession and natural disturbances (Akçakaya et al. 2004, 2005; Wintle et al. 2005).

Sensitivity analysis

Model parameters used in a PVA always contain some level of uncertainty. This uncertainty may be a consequence of measurement error, not knowing how different model parameters interact with one another (e.g., whether survival and fecundity covary), natural variability in the environment, or not knowing how parameters may change in the future. The relative effects of uncertainty in model parameters on model results can be evaluated by conducting a sensitivity analysis. A sensitivity analysis can help to identify important parameters in which small changes in estimated values or assumptions have relatively large effects on conclusions drawn from the model. Sensitivity analysis can also help to prioritize resource allocation for collecting field data because it can identify parameters whose more precise estimation has the largest effects on improving model confidence (Akçakaya and Sjögren-Gulve 2000). PVAs, particularly when constructed as spatially explicit demographic or individual-based models, can include complex interactions between multiple variables. Ideally, sensitivity analysis should include simultaneous tests of multiple parameters rather than one-at-a-time parameter tests (Chu-Agor et al. 2012), especially if spatial and nonspatial parameters may be interacting (Naujokaitis-Lewis et al. 2009).

Guidelines for implementing conservation using this approach

Evaluating conservation options

Population viability models naturally lend themselves to address questions related to available conservation management options. It may be that you wish to evaluate a proposed plan in terms of how effective it may be over time, or how the focal species might be expected to respond. You may have multiple management options on the table and wish rank to them in terms of expected effectiveness. For example, reserve design often involves selecting the network of reserve areas that will represent the greatest number of species. This approach does not examine whether the arrangement and connectivity of the selected reserve areas will be sufficient to support the long-term survival of any those species. By looking at the total species pool and selecting a subset of focal species that are representative of the range of dispersal abilities, life-history traits, and habitat preferences, and conducting a PVA for each focal species, you would be able to evaluate the proposed reserve designs with an eye toward long-term viability of the species represented. Within a PVA framework, each reserve can be evaluated as one or more patch or subpopulation for each focal species. Other examples of how PVA can inform conservation planning include simulating habitat corridor design by modeling increased dispersal between patches and simulating translocation or reintroduction by modeling a set number of individuals being moved or reintroduced to a patch.

When making decisions between competing management options, very often the final decision depends on a combination of both biological and nonbiological considerations. Biological considerations include those related to the long-term viability of the focal species (but see discussion in "Formulation of goals and questions" above). Nonbiological considerations include those that are related to the societal, cultural, or financial constraints or costs associated with each proposed action (see also Chan et al., chapter 2). Any proposed management option with unacceptably high costs or serious constraints should have been ruled out early in the decision-making process. PVA can be useful in ranking the remaining management options by providing a biologically based risk assessment criterion. Decision makers can then decide if it is more important to maximize viability for the species (subject to cost constraints) or minimize the costs (as long as the species viability is above some threshold).

An example of this kind of cost-effectiveness assessment is a study that evaluated various strategies for conservation of the European bison in the Carpathians (Kuemmerle et al. 2011). Once extinct in the wild, the European bison now exists in about 30, mostly small and isolated, populations in the fragmented landscapes of eastern Europe. Kuemmerle et al. (2011) developed a metapopulation model of this species in the Carpathian region. The spatial structure of the metapopulation was based on the distribution of potential habitat, as predicted by a habitat model (see "Habitat" above). Within-population dynamics were modeled using a stochastic, age-structured matrix model with density-dependent recruitment. Dispersal was modeled using a cost-surface (friction) approach that models the permeability of the background matrix for European bison dispersal. Friction values were based on the inverse habitat suitability value and four types of dispersal barriers: settlements, lakes, rivers, and roads. The linked habitat-metapopulation model was used to simulate the effects of three different conservation measures (reintroductions, wildlife overpasses, and antipoaching measures) on the viability of the metapopulation. These estimates were then combined with estimates of the cost of each conservation measure to rank these measures in terms of their cost-effectiveness.

Reintroductions appeared to be the most cost-effective approach for establishing a viable bison metapopulation, especially when coupled with wildlife overpasses to improve connectivity among herds. The study also identified the most promising regions for reintroductions and wildlife overpasses in the Carpathians. The study demonstrated that incorporating cost-effectiveness analyses into metapopulation viability analyses can elucidate the relative value (gain per unit cost) of conservation management options, allowing decision makers to choose cost-effective options to preserve large mammals.

Critical habitat

A conservation measure often proposed or required for a threatened species is the designation and protection of habitat critical to its survival. For example, the US Endangered Species Act mandates designating critical habitat based on the best available scientific data for each listed species. There are two main steps in critical habitat determination (Reed et al. 2006). The first is to characterize the habitat requirements of the species based on its ecology and life history, as discussed above. The second is to evaluate how different amounts and configurations of habitat affect the survival or recovery of the species. Areas of habitat are assessed for their contribution to the survival or recovery of the species. In this step, the appropriate criterion to use is species viability. In making critical habitat determination, different scenarios for the amount and configuration of habitat to place under protection are compared to each other in terms of the resulting viability of the species under each scenario. Reed et al. (2006) proposed a multilevel framework for designating critical habitat that is based on population viability analysis.

Climate change

Climate change has the potential to pose a threat to a great number of species, particularly those already being affected by habitat loss and fragmentation, hunting, pollution, disease, and competition from invasive species (Bachelet, chapter 13). PVA provides a method to examine the impact of climate change that goes beyond predicting net changes in amount and location of suitable habitat available under future climate scenarios. PVA models can be used to couple species demography and the changes in the landscape, allowing modelers to examine how a species' dispersal and colonization abilities will be able to keep pace with the rate of change. Recent work by Keith et al. (2008), Anderson et al. (2009), Brook et al. (2009), Aiello-Lammens et al. (2011), and Fordham et al. (2012) demonstrates how a species' risk of extinction under climate change depends on complex interactions between life-history characteristics and the landscape.

The method of linking PVA to future climate scenario models starts with generating or interpolating the climate predictions to create a time series of maps of available habitat. These maps provide the spatial basis of a metapopulation model. In addition to changing habitat suitability in time (and the resulting change in the spatial structure of the species' metapopulation), climate change can also affect other aspects of a species' ecology. These can be incorporated into a PVA through changes in average survival or fecundity of affected populations (if climatic conditions in those populations are predicted to become less suitable for the species), and in the variability of survival and fecundity (if the frequency or strength of extreme weather events such as hurricanes, droughts, or fires are predicted to increase in the future).

When constructing the dynamic habitat model under predicted future climate conditions, it is important to carefully consider how correlations between present environmental conditions may change in the future. For example, if elevation is shown to be an important predictor of habitat suitability for a species, it may be that the species is not responding to elevation per se, but rather the temperature and precipitation regimes associated with elevation—associations that may change with future climate conditions. Changing habitat in response to climate change often can result in a more fragmented spatial structure of the metapopulation. In such cases, the factors that affect survival of small populations (Allee effects and demographic stochasticity) and those that affect connectivity (dispersal rates) may be expected to play more important roles in the future.

A striking example of the interplay of spatial (habitat) and demographic factors involves the impacts of climate change on plant population viability in South African fynbos, one of the world's biodiversity hot spots. Based on habitat models of 234 plant species, Keith et al. (2008) selected three species representing contrasting patterns of distribution change: widespread range undergoing extensive contraction, restricted range contracting at the margins but with habitat suitability increasing in the core, and restricted range undergoing shift and fragmentation. Time series of habitat maps showing the projected change in the habitats of these species were linked to demographic models for two contrasting plant life histories dependent on fire: an obligate seeding shrub (all standing plants killed by fire, mature seeds retained in woody fruits and dispersed by wind); and a resprouting shrub (some plants survive fire, seeds released when mature and cached by ants). In both cases, seedling recruitment is cued to fire. The models also included recurring fires, seed dispersal, environmental and demographic stochasticity, and density dependence. The models were parameterized using demographic studies of Proteaceae and related species, and implemented in RAMAS GIS (Akçakaya 2005). The results indicated that the effect of climate change was dependent on complex interactions among the distribution pattern, life history, fire regime, and the evenness of density-dependence effects across life stages. The most robust generalization was that viability was more dependent on fire regime than other factors. Thus, changes in the spatial and temporal pattern of disturbances and extreme events may have more important consequences for species than gradual changes in average climatic values. Such effects and the complex interactions found in this study can only be explored with viability models that link habitat and demographic factors.

Cumulative impacts

An important advantage of PVA compared to alternative methods of assessment is that it can incorporate the effects of multiple impacts and conservation measures. Thus, a model can simulate effects of habitat loss (through decrease in carrying capacity of affected populations), harvest or poaching (through reduced survival rates in populations targeted), climate change (by changes in several model parameters, as discussed above), as well as conservation actions that are designed to mitigate against these threats, such as reintroductions, translocations, protected areas, habitat corridors, and habitat restoration and improvement. In some cases, different threats may exacerbate each other's impact. For example, habitat loss may result in smaller populations, which may make them more vulnerable to effects of hunting (e.g., due to Allee effects) or climate change. Thus, it is important to build PVAs that incorporate all known threats to the populations of the species, rather than focusing on one threat or one conservation measure.

Acknowledgments

This material is based in part on work supported by the National Aeronautics and Space Administration (NASA) under Grant No. NNX09AK19G issued through the NASA Biodiversity Program.

References

Aiello-Lammens, M., M. L. Chu-Agor, M. Convertino, R. A. Fischer, I. Linkov, and H. R. Akçakaya. 2011. "The Impact of Sea-Level Rise on Snowy Plovers in Florida: Integrating Geomorphological, Habitat, and Metapopulation Models." *Global Change Biology* 17: 3,644–54.

Akçakaya, H. R. 1991. "A Method for Simulating Demographic Stochasticity." *Ecological Modelling* 54: 133–36.

———. 1992. "Population Cycles of Mammals: Evidence for a Ratio-Dependent Prediction Hypothesis." *Ecological Monographs* 62: 119–42.

———. 1994. *RAMAS Metapop: Viability Analysis for Stage Structured Metapopulations.* Setauket, NY: Applied Biomathematics.

———. 2000a. "Population Viability Analyses with Demographically and Spatially Structured Models." *Ecological Bulletins* 48: 23–38.

———. 2000b. "Viability Analyses with Habitat-Based Metapopulation Models." *Population Ecology* 42: 45–53.

———. 2001. "Linking Population-Level Risk Assessment with Landscape and Habitat Models." *Science of the Total Environment* 274: 283–91.

———. 2002. "Estimating the Variance of Survival Rates and Fecundities." *Animal Conservation* 5: 333–36.

———. 2005. *RAMAS GIS: Linking Spatial Data with Population Viability Analysis.* Setauket, NY: Applied Biomathematics.

Akçakaya, H. R., and J. L. Atwood. 1997. "A Habitat-Based Metapopulation Model of the California Gnatcatcher." *Conservation Biology* 11: 422–34.

Akçakaya, H. R., and B. W. Brook. 2008. "Methods for Determining Viability of Wildlife Populations in Large Landscapes." In *Models for Planning Wildlife Conservation in Large Landscapes,* edited by J. J. Millspaugh and F. R. Thompson III, 449–71. Burlington, MA: Elsevier Academic Press.

Akçakaya, H. R., M. A. Burgman, and L. R. Ginzburg. 1999. *Applied Population Ecology.* Sunderland, MA: Sinauer Associates.

Akçakaya, H. R., M. A. Burgman, O. Kindvall, C. Wood, and P. Sjögren-Gulve, eds. 2004. *Species Conservation and Management: Case Studies.* New York: Oxford University Press.

Akçakaya, H. R., J. Franklin, A. D. Syphard, and J. R. Stephenson. 2005. "Viability of Bell's Sage Sparrow (*Amphispiza belli ssp. belli*): Altered Fire Regimes." *Ecological Applications* 15: 521–31.

Akçakaya, H. R., G. Mills, and C. P. Doncaster. 2006. "The Role of Metapopulations in Conservation." In *Key Topics in Conservation Biology,* edited by D. W. Macdonald and K. Service, 64–84. Malden, MA: Wiley-Blackwell.

Akçakaya, H. R., and W. T. Root. 2003. *RAMAS Landscape: Integrating Metapopulation Viability with LANDIS Forest Dynamics Mode.* Setauket, NY: Applied Biomathematics.

Akçakaya, H. R., and P. Sjögren-Gulve. 2000. "Population Viability Analyses in Conservation Planning: An Overview." *Ecological Bulletins* 9–21.

Anderson, B., H. R. Akçakaya, M. B. Araújo, D. A. Fordham, E. Martinez-Meyer, W. Thuiller, and B. W. Brook. 2009. "Dynamics of Range Margins for Metapopulations under Climate Change." *Proceedings of the Royal Society B: Biological Sciences* 276: 1,415–20.

Beaudry, F., P. G. deMaynadier, and M. L. Hunter. 2008. "Identifying Road Mortality Threat at Multiple Spatial Scales for Semi-Aquatic Turtles." *Biological Conservation* 141: 2,550–63.

Beier, P., and R. F. Noss. 1998. "Do Habitat Corridors Provide Connectivity?" *Conservation Biology* 12: 1,241–52.

Beissinger, S. R., and D. R. McCullough, eds. 2002. *Population Viability Analysis.* Chicago, IL: University of Chicago Press.

Bekessy, S. A., B. A. Wintle, A. Gordon, J. C. Fox, R. Chisholm, B. Brown, T. Regan, N. Mooney, S. M. Read, and M. A. Burgman, 2009. "Modelling Human Impacts on the Tasmanian Wedge-Tailed Eagle (*Aquila audax fleayi*)." *Biological Conservation* 142: 2,438–48.

Box, G. E. P. 1979. "Robustness in the Strategy of Scientific Model Building." In *Robustness in Statistics,* edited by R. L. Laurner and G. N. Wilkinson, 201–235. Elsevier Science & Technology Books.

Breininger, D. R., M. L. Legare, and R. B. Smith. 2004. "Eastern Indigo Snakes (*Drymarchon couperi*) in Florida: Influence of Edge Effects on Population Viability." In *Species Conservation and Management: Case Studies,* edited by H. R. Akçakaya, M. Burgman, O. Kindvall, C. C. Wood, P. Sjögren-Gulve, J. S. Hatfield, and M. A. McCarthy, 299–311. New York: Oxford University Press.

Brito, D., and F. A. S. Fernandez. 2000. "Metapopulation Viability of the Marsupial *Micoureus demerarae* in Small Atlantic Forest Fragments in South-Eastern Brazil." *Animal Conservation* 3: 201–9.

Brook, B. W., H. R. Akçakaya, D. A. Keith, G. M. Mace, R. G. Pearson, and M. B. Araujo. 2009. "Integrating Bioclimate with Population Models to Improve Forecasts of Species Extinctions under Climate Change." *Biology Letters* 5: 723–25.

Brook, B. W., D. M. J. S. Bowman, C. J. A. Bradshaw, B. M. Campbell, and P. J. Whitehead. 2006. "Managing an Endangered Asian Bovid in an Australian National Park: The Role and Limitations of Ecological-Economic Models in Decision-Making." *Environmental Management* 38: 463–69.

Brook, B. W., J. J. O'Grady, A. P. Chapman, M. A. Burgman, H. R. Akçakaya, and R. Frankham. 2000. "Predictive Accuracy of Population Viability Analysis in Conservation Biology." *Nature* 404: 385–87.

Brown, K. A., S. Spector, and W. Wu. 2008. "Multi-Scale Analysis of Species Introductions: Combining Landscape and Demographic Models to Improve Management Decisions about Non-Native Species." *Journal of Applied Ecology* 45: 1,639–48.

Caswell, H. 2001. *Matrix Population Models.* MA, USA: Sinauer Associates Sunderland.

Chu-Agor, M. L., R. Muñoz-Carpena, G. A. Kiker, M. E. Aiello-Lammens, H. R. Akçakaya, M. Convertino, and I. Linkov. 2012. "Simulating the Fate of Florida Snowy Plovers with Sea-Level Rise: Exploring Research and Management Priorities with a Global Uncertainty and Sensitivity Analysis Perspective." *Ecological Modelling* 224: 33–47.

Cooley, H. S., R. B. Wielgus, G. M. Koehler, H. S. Robinson, and B. T. Maletzke. 2009. "Does Hunting Regulate Cougar Populations? A Test of the Compensatory Mortality Hypothesis." *Ecology* 90: 2,913–21.

DeAngelis, D. L., and W. M. Mooij. 2005. "Individual-Based Modeling of Ecological and Evolutionary Processes." *Annual Review of Ecology, Evolution, and Systematics* 36: 147–68.

Dunham, A. E., H. R. Akçakaya, and T. S. Bridges. 2006. "Using Scalar Models for Precautionary Assessments of Threatened Species." *Conservation Biology* 20: 1,499–506.

Dunning, D. J., Q. E. Ross, S. B. Munch, and L. R. Ginzburg. 2002. "Measurement Error Affects Risk Estimates for Recruitment to the Hudson River Stock of Striped Bass." *The Scientific World Journal* 2: 238–53.

Elith, J., C. H. Graham, R. P. Anderson, M. Dudik, S. Ferrier, A. Guisan, R. J. Hijmans, F. Huettmann, J. R. Leathwick, A. Lehmann, J. Li, L. G. Lohmann, B. A. Loiselle, G. Manion, C. Moritz, M. Nakamura, Y. Nakazawa, J. McC. Overton, A. T. Peterson, S. J. Phillips, K. Richardson, R. Scachetti-Pereira, R. E. Schapire, J. Soberón, S. Williams, M. S. Wisz, and N. E. Zimmermann. 2006. "Novel Methods Improve Prediction of Species' Distributions from Occurrence Data." *Ecography* 29: 129–51.

Fordham, D. A., H. R. Akçakaya, M. Araújo, J. Elith, D. Keith, R. Pearson, T. D. Auld, C. Mellin, J. W. Morgan, T. J. Regan, M. Tozer, M. J. Watts, M. White, B. Wintle, C. Yates, and B. W. Brook. 2012. "Plant Extinction Risk under Climate Change: Are Forecast Range Shifts Alone a Good Indicator of Species Vulnerability to Global Warming?" *Global Change Biology* 18: 1,357–71.

Franklin, J. 2010. *Mapping Species Distributions: Spatial Inference and Prediction.* Cambridge: Cambridge University Press.

Fryxell, J. M., J. F. Wilmshurst, A. R. E. Sinclair, D. T. Haydon, R. D. Holt, and P. A. Abrams. 2005. "Landscape Scale, Heterogeneity, and the Viability of Serengeti Grazers." *Ecology Letters* 8: 328–35.

Gilioli, G., A. Bodini, J. Baumgartner, P. Weidmann, and J. Hartmann. 2008. "A Novel Approach Based on Information Theory to Rank Conservation Strategies: An Application to Amphibian Metapopulations." *Animal Conservation* 11: 453–62.

Ginzburg, L. R., S. Ferson, and H. R. Akçakaya. 1990. "Reconstructibility of Density Dependence and the Conservative Assessment of Extinction Risks." *Conservation Biology* 4: 63–70.

Ginzburg, L. R., and C. X. J. Jensen. 2004. "Rules of Thumb for Judging Ecological Theories." *Trends in Ecology and Evolution* 19: 121–26.

Gould, W. R., and J. D. Nichols. 1998. "Estimation of Temporal Variability of Survival in Animal Populations." *Ecology* 79: 2,531–38.

Grimm, V., H. Lorek, J. Finke, F. Koester, M. Malachinski, M. Sonnenschein, A. Moilanen, I. Storch, A. Singer, C. Wissel, and K. Frank. 2004. "META-X: Generic Software for Metapopulation Viability Analysis." *Biodiversity and Conservation* 13: 165–88.

Grimm, V., and S. F. Railsback. 2005. *Individual-Based Modeling and Ecology.* Princeton, NJ: Princeton University Press.

Guisan, A., and W. Thuiller. 2005. "Predicting Species Distribution: Offering More than Simple Habitat Models." *Ecology Letters* 8: 993–1,009.

Guisan, A., and N. E. Zimmermann. 2000. "Predictive Habitat Distribution Models in Ecology." *Ecological Modelling* 135: 147–86.

Haines, A. M., M. E. Tewes, L. L. Laack, J. S. Horne, and J. H. Young. 2006. "A Habitat-Based Population Viability Analysis for Ocelots (*Leopardus pardalis*) in the United States." *Biological Conservation* 132: 424–36.

Hanski, I. 1994. "A Practical Model of Metapopulation Dynamics." *Journal of Animal Ecology* 63: 151–62.

Hodgson, J. A., A. Moilanen, N. A. D. Bourn, C. R. Bulman, and C. D. Thomas. 2009. "Managing Successional Species: Modelling the Dependence of Heath Fritillary Populations on the Spatial Distribution of Woodland Management." *Biological Conservation* 142: 2,743–51.

Holmes, E. E. 2001. "Estimating Risks in Declining Populations with Poor Data." *Proceedings of the National Academy of Sciences of the United States of America* 98: 5,072–77.

———. 2004. "Beyond Theory to Application and Evaluation: Diffusion Approximations for Population Viability Analysis." *Ecological Applications* 14: 1,272–93.

Hraber, P. T., and B. T. Milne. 1997. "Community Assembly in a Model Ecosystem." *Ecological Modelling* 103: 267–85.

Johnston, K. M., ed. 2013. *Agent Analyst: Agent-Based Modeling in ArcGIS.* Redlands, CA: Esri Press.

Kearney, M., and W. Porter. 2009. "Mechanistic Niche Modelling: Combining Physiological and Spatial Data to Predict Species' Ranges." *Ecology Letters* 12: 334–50.

Keith, D. A., H. R. Akçakaya, W. Thuiller, G. F. Midgley, R. G. Pearson, S. J. Phillips, H. M. Regan, M. B. Araújo, T. G. Rebello. 2008. "Predicting Extinction Risks under Climate Change: Coupling Stochastic Population Models with Dynamic Bioclimatic Habitat Models." *Biology Letters* 4: 560–63.

Kindvall, O., and K.-O. Bergman. 2004. "Woodland Brown Butterfly (*Lopinga achine*) in Sweden: Viability in a Dynamic Landscape Maintained by Grazing." In *Species Conservation and Management: Case Studies*, edited by H. R. Akçakaya, M. A. Burgman, O. Kindvall, C. Wood, and P. Sjögren-Gulve, 171–78. New York: Oxford University Press.

Kuemmerle, T., K. Perzanowski, H. R. Akçakaya, F. Beaudry, T. R. van Deelen, I. Parnikoza, P. Khoyetskyy, D. M. Waller, and V. C. Radeloff. 2011. "Cost-Effectiveness of Different Conservation Strategies to Establish a European Bison Metapopulation in the Carpathians." *Journal of Applied Ecology* 48: 317–29.

Lacy, R. C. 1993. "VORTEX—A Computer-Simulation Model for Population Viability Analysis." *Wildlife Research* 20: 45–65.

LaHaye, W. S., R. J. Guitierrez, and H. R. Akçakaya. 1994. "Spotted Owl Metapopulation Dynamics in Southern California." *Journal of Animal Ecology* 63: 775–85.

Lande, R., S. Engen, and B.-E. Saether. 2002. "Estimating Density Dependence in Time-Series of Age-Structured Populations." *Philosophical Transactions of the Royal Society B* 357: 1,179–84.

Langton, S. D., N. J. Aebischer, and P. A. Robertson. 2002. "The Estimation of Density Dependence Using Census Data from Several Sites." *Oecologia* 133: 466–73.

Larson, M. A., F. R. Thompson, J. J. Millspaugh, W. D. Dijak, and S. R. Shifley. 2004. "Linking Population Viability, Habitat Suitability, and Landscape Simulation Models for Conservation Planning." *Ecological Modelling* 180: 103–18.

Lecomte, J., K. Boudjemadi, F. Sarrazin, K. Cally, and J. Clobert. 2004. "Connectivity and Homogenisation of Population Sizes: An Experimental Approach in *Lacerta vivipara*." *Journal of Animal Ecology* 73: 179–89.

Li, Y., G. Zhongwei, Y. Qisen, W. Yushan, and J. Niemelä. 2003. "The Implications of Poaching for Giant Panda Conservation." *Biological Conservation* 111: 125–36.

Lindenmayer, D. B., H. P. Possingham, R. C. Lacy, M. A. McCarthy, and M. L. Pope. 2003. "How Accurate Are Population Models? Lessons from Landscape-Scale Tests in a Fragmented System." *Ecology Letters* 6: 41–47.

Lozier, J. D., P. Aniello, and M. J. Hickerson. 2009. "Predicting the Distribution of Sasquatch in Western North America: Anything Goes with Ecological Niche Modelling." *Journal of Biogeography* 36: 1,623–27.

McRae, B. H., N. H. Schumaker, R. B. McKane, R. T. Busing, A. M. Solomon, and C. A. Burdick. 2008. "A Multi-Model Framework for Simulating Wildlife Population Response to Land-Use and Climate Change." *Ecological Modelling* 219: 77–91.

Morris, W. F., and D. F. Doak. 2002. *Quantitative Conservation Biology: Theory and Practice of Population Viability Analysis.* 1st ed. Sunderland, MA: Sinauer Associates.

Murrow, J. L., J. D. Clark, and E. K. Delozier. 2009. "Demographics of an Experimentally Released Population of Elk in Great Smoky Mountains National Park." *Journal of Wildlife Management* 73: 1,261–68.

Naujokaitis-Lewis, I. R., J. M. R. Curtis, P. Arcese, and J. Rosenfeld. 2009. "Sensitivity Analyses of Spatial Population Viability Analysis Models for Species at Risk and Habitat Conservation Planning." *Conservation Biology* 23: 225–29.

O'Regan, H. J., A. Turner, and D. M. Wilkinson. 2002. "European Quaternary Refugia: A Factor in Large Carnivore Extinction?" *Journal of Quaternary Science* 17: 789–95.

Ozgul, A., K. B. Armitage, D. T. Blumstein, D. H. Vanvuren, and M. K. Oli. 2006. "Effects of Patch Quality and Network Structure on Patch Occupancy Dynamics of a Yellow-Bellied Marmot Metapopulation." *Journal of Animal Ecology* 75: 191–202.

Pellet, J., G. Maze, and N. Perrin. 2006. "The Contribution of Patch Topology and Demographic Parameters to Population Viability Analysis Predictions: The Case of the European Tree Frog." *Population Ecology* 48: 353–61.

Possingham, H. P., and I. Davies. 1995. "ALEX—A Model for the Viability Analysis of Spatially Structured Populations." *Biological Conservation* 73: 143–50.

Quintana-Ascencio, P. F., I. Caballero, J. M. Olano, A. Escudero, and M. J. Albert. 2009. "Does Habitat Structure Matter? Spatially Explicit Population Modelling of an Iberian Gypsum Endemic." *Population Ecology* 51: 317–28.

Ranius, T., and O. Kindvall. 2006. "Extinction Risk of Wood-Living Model Species in Forest Landscapes as Related to Forest History and Conservation Strategy." *Landscape Ecology* 21: 687–98.

Reed, J. M., H. R. Akçakaya, M. Burgman, D. Bender, S. R. Beissinger, and J. M. Scott. 2006. Critical habitat. In *The Endangered Species Act at Thirty: Conserving Biodiversity in Human-Dominated Landscapes*, vol. 2, edited by J. M. Scott, D. D. Goble, and F. W. Davis, 164–77. Washington, DC: Island Press.

Saether, B.-E., S. Engen, F. Filli, R. Aanes, W. Schröder, and R. Andersen. 2002. "Stochastic Population Dynamics of an Introduced Swiss Population of the Ibex." *Ecology* 83: 3,457–65.

Schumaker, N. H. 1998. *A Users Guide to the PATCH Model.* Corvallis, OR: Environmental Protection Agency, Environmental Research Laboratory.

Sezen, Z., H. R. Akçakaya, and C. C. Bilgin. 2004. "Turkish Mouflon (*Ovis gmelinii anatolica*) in Central Anatolia: Population Viability under Scenarios of Harvesting for Trophy." In *Species Conservation and Management: Case Studies,* edited by H. R. Akçakaya, M. A. Burgman, O. Kindvall, C. Wood, and P. Sjögren-Gulve, 459–68. New York: Oxford University Press.

Shimada, M., and F. Ishihama. 2000. "Asynchronization of Local Population Dynamics and Persistence of a Metapopulation: A Lesson from an Endangered Composite Plant, *Aster kantoensis*." *Population Ecology* 42: 63.

Sjögren-Gulve, P., and T. Ebenhard. 2000. *The Use of Population Viability Analyses in Conservation Planning.* Wiley-Blackwell.

Southwell, D. M., A. M. Lechner, T. Coates, and B. A. Wintle. 2008. "The Sensitivity of Population Viability Analysis to Uncertainty about Habitat Requirements: Implications for the Management of the Endangered Southern Brown Bandicoot." *Conservation Biology* 22: 1,045–54.

Spromberg, J. A., and L. L. Johnson. 2008. "Potential Effects of Freshwater and Estuarine Contaminant Exposure on Lower Columbia River Chinook Salmon (*Oncorhynchus tshawytscha*) Populations." In *Demographic Toxicity: Methods in Ecological Risk Assessment,* edited by H. R. Akçakaya, J. D. Stark, and T. S. Bridges, 123–42. New York: Oxford University Press.

Stacey, P. B., V. A. Johnson, and M. L. Taper. 1997. "Migration within Metapopulations: The Impact upon Local Population Dynamics." In *Metapopulation Biology: Ecology, Genetics, and Evolution,* edited by I. Hanski and M. E. Gilpin, 267–91. San Diego, CA: Academic Press.

Stanton, J. C., R. Pearson, N. Horning, P. Ersts, and H. R. Akçakaya. 2012. "Combining Static and Dynamic Variables in Species Distribution Models under Climate Change." *Methods in Ecology and Evolution* 3: 349–57, doi:10.1111/j.2041-210X.2011.00157.x.

White, G. C., and K. P. Burnham. 1999. "Program MARK: Survival Estimation from Populations of Marked Animals." *Bird Study* 46: 120–39.

White, G. C., A. B. Franklin, and T. M. Shenk. 2002. "Estimating Parameters of PVA Models from Data on Marked Animals." In *Population Viability Analysis,* edited by S. R. Beissinger and D. R. McCullough, 169–90. Chicago, IL: University of Chicago Press.

Wintle, B. A., S. A. Bekessy, L. A. Venier, J. L. Pearce, and R. A. Chisholm. 2005. "Utility of Dynamic-Landscape Metapopulation Models for Sustainable Forest Management." *Conservation Biology* 19: 1,930–43.

12

Optimization models for reserve site selection and design

Justin C. Williams, PhD

Optimization models, once the domain of military logistics and business planning, are now also widely used for biological conservation. Margules and Pressey (2000) and Pressey and Cowling (2001) include reserve selection and design models (or "algorithms") as part of a multistage process for systematic conservation planning. Optimization models and algorithms may serve as stand-alone tools for reserve site selection and design, or be integrated within decision support systems, based on geographic information systems (GIS), that perform site selection along with data analysis and mapping. This chapter provides an overview of what optimization models are, and attempts to clarify their role in relation to (1) the way reserve site selection problems are described and formulated, (2) methods and algorithms that are used to select reserve sites, and (3) decision support systems that contain optimization models or algorithms as one of several components. In addition, to illustrate how simple optimization models can be used for reserve site selection and design in the real world, this chapter incudes a case study of North America's Inland Temperate Rainforest. This chapter is intended for conservation practitioners who have a modest (or no) background in mathematical optimization, operations research, and computer science.

A brief history of optimization methods for reserve site selection and design

Conservation scientists have grappled with the problem of how to (best) design nature reserves for at least the last 40 years. In the 1970s, Diamond (1975) and others proposed a set of geometric reserve design guidelines that suggested better versus worse reserve configurations, such as "a single large reserve is better than several small reserves" and "reserves that are close together are better than reserves that are far apart." These guidelines were based on the dynamics of migration and colonization and on species–area equilibrium relationships from island biogeography, and had as an underlying goal the long-term persistence of species. Although presented as flexible guidelines rather than rigid prescriptions, several of the guidelines—especially the "single-large-or-several-small" (SLOSS) guideline—spurred long-running debates. In time, something of a consensus emerged that the best way to design a reserve depends on what one is trying to protect within the context of available options for protection (see, for example, Soule and Simberloff 1986).

Within a few years, as declining biodiversity—particularly through species losses resulting from the degradation and loss of habitat—became an increasingly urgent problem, conservation scientists also asked the question, "Which areas of land or water should be set aside to have a reserve that is representative of as many species as possible?" Because conservation resources are typically very limited, this immediately was seen as a question of conservation efficiency. What is the "minimum reserve set"—the minimum number of reserve sites (or minimum cost) necessary to represent all species of concern? Or, alternatively, what is the maximum number of species that can be represented for a fixed budget? In response to these questions, conservation practitioners (e.g., Kirkpatrick 1983) developed systematic approaches for maximizing species richness or some other surrogate for biodiversity such as landscape heterogeneity, and for identifying the minimum reserve set or an approximation of it. These approaches used heuristic rules for site selection to find the smallest number of sites that altogether represented the species of interest (see, e.g., Csuti et al. 1997). For example, one could start by picking the site that contained the largest number of species. The second selected site would be the one that contained the largest number of species not already represented by the first site. Sites would continue being selected until all species were represented, with each additional site chosen to cover as many of the remaining unrepresented species as possible. This process, known as "greedy adding," was used in many of the early heuristics for reserve site selection (e.g., Margules et al. 1988).

Cocks and Baird (1989) were among the first to propose a formal mathematical optimization model for the selection of a set of sites for a reserve. Shortly after, it was recognized that the minimum reserve set problem was a version of the "set covering problem," a classic optimization problem in the field of operations research (Possingham et al. 1993; Underhill 1994). Once this connection was made, mathematical modelers and conservation scientists identified a wide variety of reserve site selection and design problems that could be formulated as optimization (minimization or maximization) models. Also, by the early 1990s, GIS, which facilitate the analysis and visualization of data, were becoming widely used in conservation planning (e.g., Smith et al. 1997 and references therein). The tandem and complementary use of GIS and optimization models for reserve selection and design would soon follow.

Three reviews discuss the evolution of systematic reserve design more extensively. Pressey (2002) recounts the historical development of several of the pioneering iterative heuristics developed for reserve site selection in the 1980s and early 1990s. Kingsland (2002a, 2002b) examines the historical forces that shaped the emergence of nature reserve design science, and investigates how problems of reserve design, which originated in ecology and conservation biology, have been reformulated to draw on the methods of operations research and decision science. Reviews of the many optimization models that have been formulated for reserve site selection and design can be found in ReVelle et al. (2002), Williams et al. (2004), Williams et al. (2005), Sarkar et al. (2006), and Haight and Snyder (2009).

Optimization models, solution methods, and decision support systems

Optimization models, solution algorithms, and decision support systems (DSSs) are some of the tools within the growing universe of computer-based methods for biological conservation planning and decision making. The Ecosystem-Based Management Tools Network database (http://www.smartgrowthtools.org/ebmtools/index.php) lists dozens of decision support tools for tackling various problems and challenges within this field. Below, the relationship of optimization models, algorithms, and DSSs to each other and to reserve site selection or design is discussed, and distinguishing features of each of these tools are explained.

Optimization models

A wide variety of mathematical models are used in conservation science. They tend to fall within two main categories, "descriptive/predictive" models and "prescriptive" models. The first category of models seeks to emulate dynamic biological and ecological systems, and may also try to predict their future states. Many models of this type are discussed elsewhere in this volume, particularly in chapters 7, 9, and 14. Most simulation models, including spatially explicit population models, are descriptive/predictive models. In contrast, prescriptive models, including optimization models, seek to identify a good (or best) course of action from among many possibilities. Here, "best" is defined relative to a stated objective used in the model, such as to minimize cost or maximize the number of species represented. Because models are (necessarily) approximations of reality, the best or optimal solution may not actually be the solution that is most appropriate for the real world. But, the intent is that the optimal solution will at least point toward what is best, or can help assess what is needed to protect a species and other conservation features.

Two relevant categories of optimization models are linear programming (LP) models and integer programming (IP) models. Although many conservation problems can be formulated as LPs, models for reserve site selection and design are typically of the IP variety. The primary difference between these two categories lies in their "decision variables," which represent the actions and decisions to be made. LPs contain only continuous decision variables, which represent continuous quantities such as the land area to allocate to a reserve, but IPs contain at least some variables that must take on integer values. Integer variables are used to represent whole number quantities, such as the number of times a particular species is represented in a reserve, or to represent a yes (1) or no (0) decision such as whether or not to select a particular site for the reserve. Unlike continuous variables, integer variables are not allowed to take on fractional values, such as 0.5, which are meaningless for what the integer variable represents. Although LPs—even very large ones—can be solved quickly using available software, integer restrictions tend to make IPs much more difficult to solve.

An optimization model, whether LP or IP, always follows directly from a statement—both precise and concise—of the conservation problem at hand. The model can frequently be written as a few mathematical statements. For example, the minimum reserve set problem can be stated as "minimize the number of sites selected such that each species is represented at least once in the reserve system." The optimization model that follows directly from this statement is a binary IP model that can be formulated as three mathematical expressions:

Minimum Reserve Set IP Model

$$\begin{array}{lll l}
\text{Minimize:} & \sum_{j \in J} Xj & & (1) \\
\text{Subject to:} & \sum_{j \in Mi} Xj \geq 1 & \text{for all species } i \in I & (2) \\
 & Xj = 0 \text{ or } 1 & \text{for all sites } j \in J & (3)
\end{array}$$

where i and I are the index and set of species; j and J are the index and set of candidate reserve sites; Mi is the set of candidate sites j that contain species i (the Mi sets can be derived from species presence/absence data); and Xj is a binary decision variable for site selection, where Xj takes on a value of 1 if site j is selected for the reserve system and takes on a value of 0 if site j is not selected. The first expression (1) is the *objective function,* which tells us what to optimize. Here we are minimizing the number of sites selected, which is calculated by summing up the values (either 0 or 1) of all the decision variables Xj. Expression (2) is a *constraint* enforcing the condition that we must represent each and every species within the reserve. This is achieved by selecting at least one site j in which each species i is known to be present. That is, the constraint says that the sum of the Xj values, summed over all sites that contain species i, must be at least 1. Expression (2) is in fact a collection of many separate constraints, one constraint for each species i. Expression (3) is the binary requirement for the decision variables, which stipulates that each site must be either selected or not selected.

As a numerical example of the minimum reserve set problem, consider the following hypothetical species presence/absence data with four species (A–D) and four candidate sites, where a one (1) indicates known species presence in the site and a zero (0) indicates absence (table 12.1).

Table 12.1. Example species presence data

Species	Site 1	Site 2	Site 3	Site 4
A	1	1	1	0
B	1	0	0	0
C	0	0	1	0
D	0	1	0	1

Source: Justin C. Williams.

The minimum reserve set IP model for this dataset is

$$\begin{array}{llllllll}
\text{Minimize:} & X1 + & X2 + & X3 + & X4 & & & (1) \\
\text{Subject to:} & X1 + & X2 + & X>3 & & \geq 1 & \text{for species A} & (2) \\
 & X1 & & & & \geq 1 & \text{for species B} & \\
 & & & X3 & & \geq 1 & \text{for species C} & \\
 & & X2 + & & X4 & \geq 1 & \text{for species D} & \\
 & \multicolumn{6}{l}{X1, X2, X3, \text{ and } X4 \text{ each must equal 0 or 1}} & (3)
\end{array}$$

What is the optimal solution to this problem? How many sites do we need to select? Observe that no site by itself is sufficient to represent all four species. It is also apparent that no combination of two sites covers all species. However, we can find a set of three sites that does, so three is our minimum number of sites. Because there are two different combinations of three sites that cover all species, (1, 2, 3) and (1, 3, 4), this problem has not one but two distinct optimal solutions (called "alternate optima"). While this small example was solved quickly by inspection, most large versions of this problem—with many species and many sites—require a systematic solution method (or algorithm) and a computer.

The minimum reserve set model as formulated in expressions (1) through (3) is generic in the sense that it can be applied to any geographical location as long as the user has a list of candidate sites, a list of species to protect, and species presence/absence data for the sites. This type of generality is true of virtually all model formulations. If the problem statement on which the model is based matches the user's problem, then the user need only supply the data. The conservation literature contains many generic and freely available model formulations that have been developed for a variety of reserve site selection and design problems (many are reviewed in ReVelle et al. 2002 and Williams et al. 2005).

Some researchers distinguish between reserve site selection models and reserve design models. Models in the site selection category, which includes the minimum reserve set model, do not account for spatial attributes, such as connectivity and compactness, in the selection of sites. Optimal solution to site selection problems are typically collections of disconnected sites with little spatial coherence. In contrast, models in the reserve design category employ spatial metrics such as inter-site distances and site adjacencies, and the resulting solutions are likely to be more spatially coherent. Reserve design models, though, tend to be more complex and computationally intensive than site selection models.

Solution methods

Although small reserve selection problems can sometimes be solved by inspection (as was done above), most realistically sized problems, involving perhaps hundreds or thousands of integer variables and constraints, require a computer. A computer is needed to check the large number of possible solutions, which grows exponentially with the number of decision variables. This can be illustrated by comparing two minimum reserve set problems of different sizes. In our example above with just four candidate sites (i.e., four binary or yes/no decision variables), there are $2^4 = 16$ possible ways to select sites for the reserve system, that is, 16 combinations of yes and no. (Some of these combinations are not feasible because they do not cover all species.) The entire solution space of 16 combinations or alternatives can be evaluated quickly for both feasibility and optimality. However, if we have a larger problem involving, say, 50 sites, the number of combinations grows (exponentially) to $2^{50} \approx 10^{15}$. Although site selection problems with 50 sites are not large problems in the real world, the solution space of roughly 10^{15} solutions is large—so large that exhaustive enumeration is likely to be impractical. Hence, a computer-based method is needed for efficiently finding the best solution without examining all combinations. Two broad categories of computer-based methods, exact methods and heuristic methods, have been developed for solving this type of optimization problem.

Exact solution methods

An exact method guarantees a mathematically optimal solution if it is carried out to completion. Depending on the type of problem (LP or IP), the problem's size (number of variables and number of constraints), and the speed of the computer, the run time may be short (seconds) or long (days). Linear programs, as a rule, solve quickly, and only extremely large LPs will encounter long run times. For IPs, however, solution times can vary widely, depending not only on problem size but also on how the model is formulated (e.g., see ReVelle 1993). For some IP models, even moderately-sized problems (e.g., 300 variables and 300 constraints) pose difficulties for exact methods, but other IP models can be solved as quickly or nearly as quickly as LPs of the same size.

Exact methods for solving problems modeled as LPs include the simplex algorithm and the interior barrier search algorithm. The most common exact method for solving IPs is the branch and bound method, which typically appears as an add-on to the simplex method. Other exact methods have been developed for specific types of IP problems, but such special-purpose methods have narrow applicability and are beyond the scope of this chapter. Most commercially available IP solvers (e.g., the IBM Corporation's CPLEX package) contain versions of both the simplex and the branch and bound algorithms. The depth of knowledge of linear programming one needs to successfully use LP and IP software packages varies from one package to another. Some have user-friendly interfaces that automate much of the preprocessing, such as the structuring of data and variables in the appropriate format, while other packages require more involvement by the user. Commercial optimization software can be expensive to purchase, although discounted academic licenses and free or low-cost demonstration versions of software often are available.

Heuristic solution methods

Although exact methods can guarantee optimal solutions, heuristic methods do not. This is the fundamental difference between the two types of methods. Heuristics are designed to find good solutions, and may happen upon the optimal solution, but can and often do terminate with suboptimal or "approximate" solutions. Furthermore, it may be impossible to discern how far away from true optimality a particular heuristic solution is. The advantage of heuristics is the speed with which they can find good and sometimes optimal solutions. For IP problems, especially large ones, heuristics are often much faster than exact methods. For LPs, on the other hand, heuristics offer no advantage because even large LPs can be solved quickly by exact methods. Heuristics are also used to solve nonlinear optimization problems that in many cases cannot be readily solved by exact methods (nonlinear models are models in which decision variables are multiplied together).

A wide variety of heuristic methods has been developed to solve optimization problems. Two simple iterative heuristics that have been used for reserve site selection are greedy adding (mentioned earlier) and "one-for-one swapping." Greedy adding, in which sites are selected one at a time based on their marginal contribution to species representation, was frequently used to solve the minimum reserve set problem, although Underhill (1994) demonstrated that it did not always find an optimal solution. The one-for-one swapping heuristic is typically applied to a slightly different site selection problem, the problem of maximizing the number of species that can be represented given a fixed number (N) of sites (Rosing et al. 2002; see also Church et al. 1996). One-for-one swapping starts with a candidate solution—any set of N selected sites—and checks to see if the solution would improve by swapping one of the selected sites for a site that was not selected. If the solution would improve (i.e., if more species would be represented), then the swap is made; otherwise, the swap is not made. The heuristic iteratively checks all possible swaps, updating the solution each time a swap is made, and continues until no further improvement can be realized.

The last 30 years have seen the rise of sophisticated "metaheuristic" methods for solving optimization problems. Metaheuristics employ simple heuristics, such as greedy adding and one-for-one swapping, within an overarching framework of rules that determine how and when the simple heuristics will be used. Metaheuristics have given researchers and practitioners a way to find solutions to problems that are not tractable by exact methods, and to find solutions that are superior to those found by simple heuristics. However, metaheuristics typically must be custom-coded for the problem at hand, requiring sufficient programming skills to implement. Three metaheuristics that have been used widely in operations research are simulated annealing, tabu search, and genetic or evolutionary algorithms (Osman and Kelly 1996). Which of these metaheuristics is the best? There is no simple answer. Although one metaheuristic may perform better than another for a particular type of problem, evidently, no metaheuristic is better than all others in all settings. Studies in which metaheuristics have been applied to reserve design include Csuti et al. (1997), simulated annealing; McDonnell et al. (2002), simulated annealing; Nalle et al (2002), simulated annealing, tabu search, and a genetic algorithm; Nalle and Arthur (2005), simulated annealing; and Mizumori et al. (2005), heuristic concentration (see also Rosing et al. 1999). Some DSSs for conservation planning incorporate heuristics and metaheuristics that a nonspecialist can access via a user-friendly interface. This is discussed in more detail below.

There has been a debate in the literature on whether one should use exact methods or (meta-) heuristics, given the pros and cons of each (Pressey et al. 1996; Rodrigues and Gaston 2002; Önal 2003; Sarkar et al. 2006). In fact, both types of methods have been, and continue to be, used in reserve design (see Moilanen and Ball 2009). Ultimately, the choice of exact methods versus heuristic methods depends on the user's preference for a guarantee of mathematical optimality versus the ability to find an approximate solution quickly. Good solutions found in "real time" may be necessary in, for example, an interactive planning session involving stakeholders and other interest groups. However, if time is less of an issue, the benefits (in cost savings or number of species protected) of the optimal solution relative to a second- or third-best solution may justify the longer computing time of an exact method.

Decision support systems

The need to generate, evaluate, and display alternative reserve plans rapidly has led to the development of computer-based DSSs. A well-designed DSS is an interactive software system that will help the user compile raw data, documents, personal knowledge, and models to identify and solve problems and make decisions. Although this is a fairly broad definition, DSSs are normally designed to provide decision support for specific problems, such as reserve planning and design.

A DSS for reserve planning and design typically works in the following way. The DSS first reads input data, then processes all of the data, and finally creates a reserve plan or design. At each step, the user may have options for interacting with and directing the DSS, and most DSSs can be run under different parameter settings to generate different reserve plans or designs. In the first step—formatting, reading, and analyzing input data—a DSS may interface with a GIS. GIS data utilized by a DSS may include species presence data; other site-based or polygon-based geospatial data, such as area, centroid coordinates, boundary length, and distance to the nearest road; and socioeconomic data, such as property values and human population densities. The GIS may also generate new data based on map overlays of other datasets. The second step—the processing step—is where optimization models and solution methods (or algorithms) would normally come into play, although not all DSSs for reserve design employ mathematical optimization. In the third step, the DSS may, through a GIS interface, create full-color maps showing which sites have been selected for the reserve system.

Because each of the DSS steps involves one or more complex and data-intensive tasks, the benefit of the DSS is that it provides a user-friendly and seamless interface so that the user does not need to be an expert in each task, nor take the time to set up the linkages among the components that carry out the tasks. Unfortunately, this convenience can render a DSS something of a "black box" if the user is not familiar with its components. For example, the user may wish to know how a DSS selects reserve sites—are heuristics or exact methods used? If a heuristic is employed, how does it work? The resulting solution may depend on the answer: two different heuristics may find different solutions to the same problem. Furthermore, some heuristics, if run twice, may find two different solutions for the same problem. To verify and validate the suitability of DSS output, the user may wish to compare the results of two or more DSSs for the same reserve site selection/design problem.

Although either heuristics or exact solvers can be embedded within (or linked to) a DSS, most DSSs that perform optimization incorporate heuristic or metaheuristic solvers instead of general-purpose exact solvers, such as the simplex and branch and bound methods for IPs. This may have to do with the relative difficulty of coding a simplex/branch and bound algorithm from scratch (versus the coding of metaheuristics), as well as with the flexibility computer programmers have for customizing metaheuristics for the particular type of problem addressed by the DSS. In addition, exact solvers in the form of commercial optimization software and other proprietary applications would likely require a licensing agreement for use in a DSS that was made available for distribution. Licensing would not be an issue for a (meta-) heuristic that was custom-coded by the DSS developer. Some DSSs in which the underlying optimization model has an explicit LP or IP form do incorporate exact solvers. Church et al. (2000), for example, report on a DSS for forest ecosystem management (RELMdss, developed for the US Forest Service) that uses LP models and incorporates the C-WHIZ commercial LP solver (Ketron Optimization). The CPLEX solver, too, can be linked to other software applications that are coded in C language, through its callable library.

The many DSSs available for conservation analysis and planning (as listed in the Ecosystem-Based Management Tools Network database) perform myriad tasks and are designed to answer a variety of conservation questions. The following two DSSs are designed for reserve site selection and planning:

- Marxan (Ecology Centre, University of Queensland, Australia, http://www.uq.edu.au/marxan)
- Zonation (Biodiversity Informatics and Conservation Group, University of Helsinki, Finland, http://www.helsinki.fi/bioscience/consplan/software/Zonation/index.html)

These DSSs are briefly discussed below, with emphasis given to their underlying optimization models and solution methods. As of this writing, the software and supporting documentation for Marxan and Zonation are freely available from their respective websites.

Marxan

Marxan evolved from an earlier DSS called SPEXAN, developed in the late 1990s (Ball 2000; see also Ball et al. 2009). (Another version of SPEXAN, called SITES, was also developed in the late 1990s for The Nature Conservancy [Andelman et al. 1999] but is no longer supported.) Although originally developed for marine ecosystems, Marxan can also be applied to terrestrial settings. The primary optimization model underlying Marxan can be found in McDonnell et al. (2002). This model is a nonlinear IP for the problem of selecting sites so as to minimize the total area (or cost) of the reserve system such that the realized attainment level of each "conservation value" (such as species or vegetation cover types) meets or exceeds a target specified by the user. This covering requirement is similar to the one used in the minimum reserve set problem. The Marxan model also contains a second objective function for minimizing the boundary length of the reserve, which promotes compactness. The user can choose from among several heuristic solution methods, including greedy adding of sites, greedy subtracting of sites, site swapping, and simulated annealing. Depending on which solution method is chosen, both the run time and the solution itself can be different, even given the same input values (i.e., conservation value targets and a boundary-length modifier weight). Simulated annealing takes longer than greedy adding but usually results in a better solution. Applications of Marxan for terrestrial conservation planning in western North America are reported in Chan et al. (2006) and Royle (2010).

Zonation

This DSS begins with the full landscape and iteratively removes sites (or grid cells) one at a time until only the desired reserve system remains. Sites are removed based on an evaluation of all remaining sites made by one of four site removal rules that the user selects (core area zonation, additive benefit function, target-based planning, or generalized benefit function). The site removal rule re-ranks all remaining sites in each iteration, and the worst remaining site—the site with the lowest ranking—is removed. A fifth rule, random removal, can also be used to create a baseline scenario for comparison. The user can decide when the right number of sites has been removed, based on updated calculations of the total conservation value of the remaining sites, at which time the site removal process stops. The solution method of Zonation is essentially a greedy subtracting heuristic. However, because the conservation problem that Zonation addresses has apparently not been formalized as a mathematical programming model (e.g., LP or IP) with an objective function and constraints (unlike Marxan), a determination of whether solutions are mathematically optimal cannot be made readily. An application of Zonation to forest conservation in Finland can be found in Lehtomaki et al. (2009).

Application: Site selection models for the North American Inland Rainforest

This section presents optimization models that were developed for a case study of habitat selection in the Inland Temperate Rainforest of western North America. Three relatively simple IP models were formulated to select sites to achieve area-based habitat targets as efficiently as possible (i.e., using minimum total area) and to compare the IP results to those of an insightful rule for selecting reserve sites. Although an integrated DSS was not used, IP software was used in tandem with ArcGIS data, analysis, and mapping capabilities. This study was carried out in conjunction with a conservation area design (CAD) developed for the Inland Temperate Rainforest at the Craighead Institute. Discussions of the context and the CAD can be found at the Institute's website (http://www.craigheadresearch.org), in Craighead and Cross (2007), at the University of Northern British Columbia website (http://web.unbc.ca/~wetbelt/2008-conference-Craighead-Conservation-Area-Design.html), and in Dellasala et al. (2011).

Background

In western North America, the Inland Temperate Rainforest or Interior Cedar-Hemlock Forest is a rare type of old-growth rainforest that occurs in wet mountain valleys in southeastern British Columbia and, to a lesser extent, northern Idaho, Washington, and Montana (Dellasala et al. 2011). The Inland Temperate

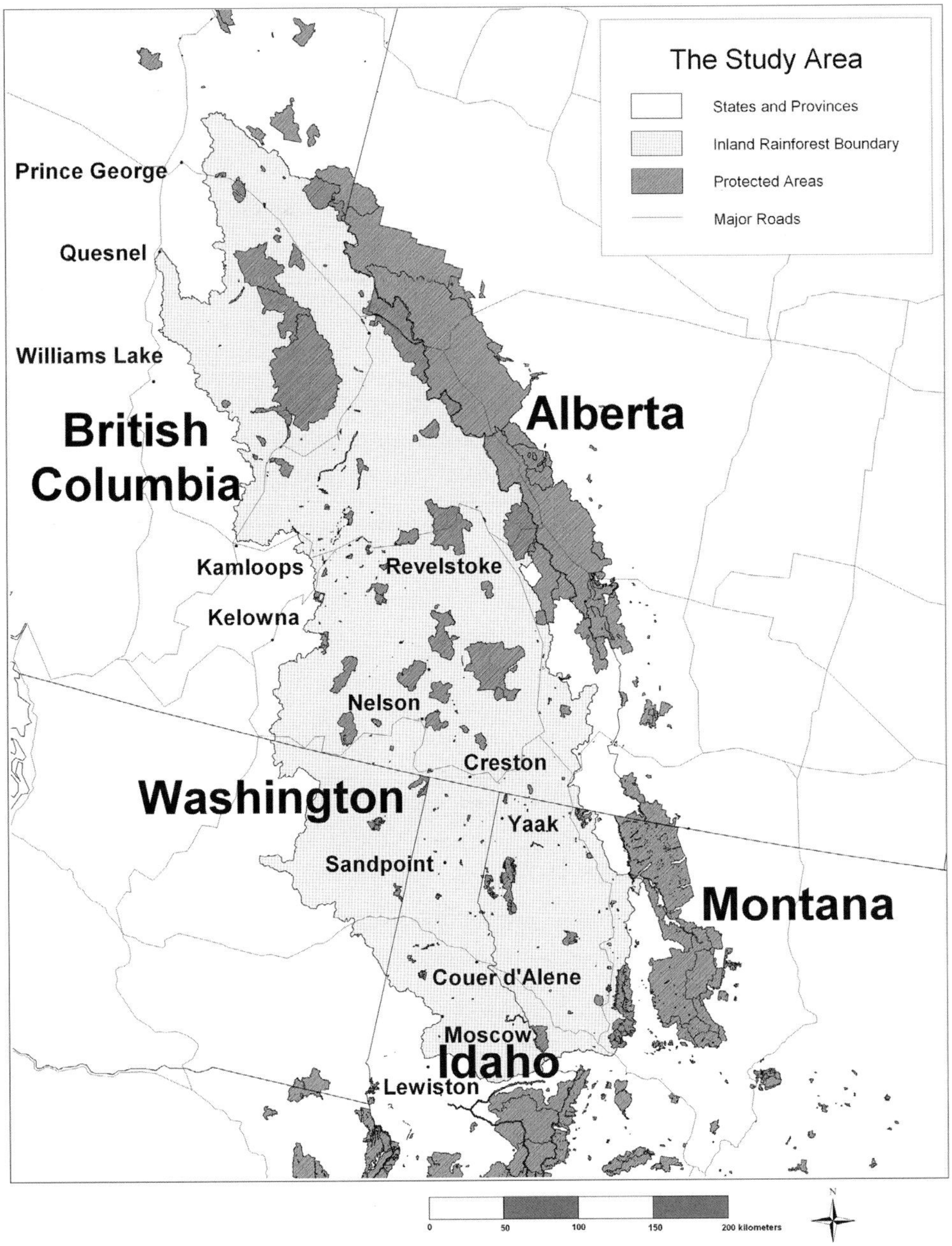

Figure 12.1. The Inland Temperate Rainforest. Protected areas from UNEP/IUCN (2008); DEM/hillshade from US Geological Survey; United States/Canada boundaries from ESRI Data & Maps, 2005, courtesy of ArcWorld Supplement; roads from ESRI Data & Maps, 2005, courtesy of TANA/GDT; streams/lakes/cities from ESRI Data & Maps, 2005, courtesy of ArcWorld; ITR boundary from Craighead Institute.

Rainforest is found at lower elevations and is interspersed with other habitat types within an area of about 194,824 km^2 (figure 12.1). The Inland Temperate Rainforest is distinguished by a predominance of western red cedar, western hemlock, and many typically coastal species of lichens. It provides critical habitat for many vertebrate species, including several at-risk species, such as grizzly bears, mountain caribou, and bull trout.

Using the general approach outlined in this book, researchers at the Craighead Institute developed a computer-based habitat suitability model for identifying high-quality and critical habitat (called "core" habitat) for each of six focal species, comprising five large carnivores—cougar (*Felis concolor*), grizzly bear (*Ursus arctos*), lynx (*Lynx canadensis*), wolf (*Canis lupus*), and wolverine (*Gulo gulo*)—as well as mountain caribou (*Rangifer tarandus caribou*). These species were chosen because they met most of the desired criteria for terrestrial focal species (e.g., large area requirements), they had been included in previous large-scale conservation plans for the region, and the data and scientific understanding were adequate for developing habitat suitability models (Craighead and Cross 2007). (The species were also chosen before more explicit approaches, as discussed by Brock and Atkinson [chapter 6] had been developed.) The basic assumption was that by maintaining suitable quality and quantity of habitat to support healthy populations of these six wide-ranging mammals, the Inland Temperate Rainforest ecosystem as a whole and the biodiversity of the region would be protected, which was the goal of the CAD.

The identification of core habitat provided a basis for selecting tracts of land in both Canada and the United States that would be recommended for protection in the Inland Temperate Rainforest CAD. It was not deemed practicable to protect all of the core habitats within the Inland Temperate Rainforest, so a conservation target, defined in terms of core habitat area, was determined for each focal species. The goal was to satisfy all targets simultaneously in the most efficient way possible, that is, with a minimum total number of sites. For this purpose we formulated three IP models to generate alternative solutions, including solutions in which the selected sites would (ideally) form large, connected areas of clustered sites. The resulting optimization problem was large (nearly 200,000 binary variables), and we wished to obtain exact optimal solutions in a short amount of time using a commercially available solver. For this reason, the IPs needed to be fairly simple.

Data

The Inland Temperate Rainforest region was partitioned into 194,824 1-km grid cells, which were the candidate selection sites for this study. Core cells were required to meet habitat quality thresholds, so only high-quality cells for a species were designated as core cells for that species. Nevertheless, 191,441 cells (98.3% of the Inland Temperate Rainforest) were designated as core habitat for at least one of the six focal species. These were the cells of primary interest, as the remaining cells had not been classified as core for any of the six species. Basically, our dataset was analogous to species presence/absence data in that a yes or no designation was made for every focal species and cell pair on the question of core habitat. Core habitat for caribou in Canada (British Columbia) and the United States were treated separately because different conservation targets were used to reflect differences in their at-risk status in the two countries. Table 12.2 lists the number of core cells for each species. Maps of core habitat for the individual focal species can be viewed at http://www.craigheadresearch.org/the-interior-cedar-hemlock-forest-of-north-america.html.

Table 12.2. Number of core habitat cells for each focal species

Focal species	Number of core habitat cells
Caribou total	77,295 (40.4% of all core cells)
BC Caribou	58,057 (30.3% of all core cells)
US Caribou	19,238 (10.1% of all core cells)
Cougar	170,426 (89.0% of all core cells)
Grizzly	116,333 (60.8% of all core cells)
Lynx	88,749 (46.4% of all core cells)
Wolf	136,798 (71.5% of all core cells)
Wolverine	80,220 (41.9% of all core cells)
Source: Justin C. Williams.	

Methods

Two methods were used to select core habitat: the "composite core" method developed by Lance Craighead, and IP models formulated by Justin Williams in consultation with Tom Olenicki and Lance Craighead. The Inland Temperate Rainforest study area contains existing protected areas such as Wells Gray Provincial Park and is adjacent to others, such as Banff National Park. The protected status of the areas was not factored into our study because we were primarily interested in identifying "clean slate" baselines for purposes of comparison. (Of course, decisions to protect additional habitat should take into account existing protected areas.)

Composite core method

The composite core method consists of a simple but effective rule for generating reserve designs. This approach resulted in six alternative designs for the Inland Temperate Rainforest. The first alternative ("1+," table 12.3, column 2), which is the largest in terms of total area, was generated by selecting all cells that were core for any of the six focal species. The second alternative ("2+," table 12.3, column 3) was generated by selecting only those cells that were core for at least two (any two) of the species. In the third alternative ("3+," table 12.3, column 4), cells that were core for any three or more of the species were selected, and so forth. In the sixth alternative ("6," table 12.3, column 7), which is the smallest in total area, only cells that were core for all six species were selected. Table 12.3 also lists the amount of core habitat selected for each focal species in each alternative; this is shown both as a percentage of the amount of core available for the species and as the number of cells selected. Observe that the six alternatives are "nested," in that solution 5+ contains solution 6, solution 4+ contains both solutions 5+ and 6, and so forth (figure 12.2).

Integer programming models

Three related optimization problems were defined and formulated as IP models. The first IP model (model 1) is a variation of the minimum reserve set model discussed earlier. The binary decision variables (X_j) are defined as before and indicate a 1 or 0 (yes or no) for the selection of each core cell j, where the set J is the set of all

Table 12.3. Results of the composite core method: six alternatives

Focal species	1 + Selected	2 + Selected	3 + Selected	4 + Selected	5 + Selected	6 Selected
Caribou	100% 77,295	~100% 77,273	96.6% 74,681	92.2% 71,296	72.7% 56,191	59.1% 45,701
Cougar	100% 170,426	83.1% 141,547	63.0% 107,293	55.2% 94,047	43.8% 74,604	26.8% 45,701
Grizzly	100% 116,333	98.9% 115,078	81.5% 94,760	72.4% 84,182	64.3% 74,776	39.3% 45,701
Lynx	100% 88,749	~100% 88,748	99.9% 88,697	97.4% 86,442	80.8% 71,704	51.5% 45,701
Wolf	100% 136,798	92.8% 126,935	78.3% 107,054	69.9% 95,566	54.7% 74,776	33.4% 45,701
Wolverine	100% 80,220	100% 80,220	93.3% 74,868	92.4% 74,084	84.2% 67,530	57.0% 45,701
Total cells selected	98.3% 191,441	77.7% 151,421	57.6% 110,197	50.3% 96,285	38.4% 74,776	23.5% 45,701
Dj(km) sum (mean)	10.526×10^6 (54.98)	5.940×10^6 (39.23)	2.907×10^6 (26.38)	2.403×10^6 (24.96)	1.791×10^6 (23.95)	0.940×10^6 (20.57)

Source: Justin C. Williams.

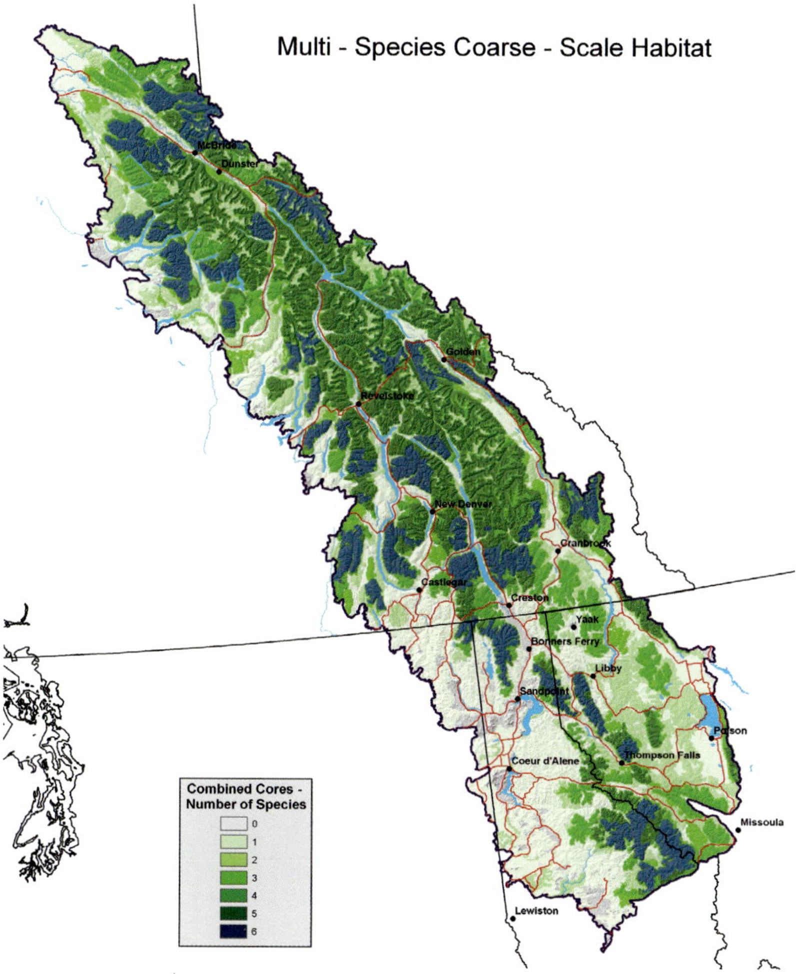

Figure 12.2. Composite cores. Cores are nested in terms of the number of focal species whose habitat is represented. Shining Mountains data from Ministry of Environment, Province of British Columbia; DEM/hillshade from US Geological Survey; United States/Canada boundaries from ESRI Data & Maps, 2005, courtesy of ArcWorld Supplement; roads from ESRI Data & Maps, 2005, courtesy of TANA/GDT; streams/lakes/cities from ESRI Data & Maps, 2005, courtesy of ArcWorld; ITR boundary from Craighead Institute; combined cores (six species) from Craighead Institute.

candidate core cells. The objective (expression 4) is also the same as before: to minimize the total number of core cells selected. The constraints (expression 5) are slightly different; instead of selecting just one core cell for each focal species *i*, we must select a specified target number (Ti) of core cells from the set of available core cells of each species (Ci). These constraints guarantee a minimum core habitat area for each of the focal species.

Minimize: $\sum_{j \in J} Xj$ (4)

Subject to: $\sum_{j \in Ci} Xj \geq Ti$ for all focal species $i \in I$ (5)

$Xj = 0$ or 1 for all sites $j \in J$ (6)

Model 1, as applied to the Inland Temperate Rainforest, has the following six core habitat constraints of type (5).

Subject to: $\sum_{j \in C\text{-}cari} Xj \geq T\text{-}caribou$ (5')

$\sum_{j \in C\text{-}coug} Xj \geq T\text{-}cougar$

$\sum_{j \in C\text{-}griz} Xj \geq T\text{-}grizzly$

$\sum_{j \in C\text{-}lynx} Xj \geq T\text{-}lynx$

$\sum_{j \in C\text{-}wolf} Xj \geq T\text{-}wolf$

$\sum_{j \in C\text{-}wver} Xj \geq T\text{-}wolverine$

The requirement in model 1 that core habitat meet or exceed a minimum target makes this model similar to two other reserve selection models: the "percent area" model (e.g., Pressey et al. 1996) and the Biodiversity Management Area Selection model (Davis et al. 1996). In these models, a specified percentage—such as 50%—of the available habitat area of each focal species must be selected.

The purpose of model 1 is to identify a solution with the smallest total area, which can then be used as a baseline of comparison for the composite core solutions and for the optimal solutions of the other two optimization problems. Because model 1 does nothing to promote potentially desirable spatial attributes, such as compactness and connectivity, and because its solutions are likely to be fragmented, a second optimization model was formulated. The second model promotes spatial proximity of the selected cells so as to improve the overall compactness of core habitat.

To characterize proximity, we examined the underlying clustering patterns of core habitat. The cluster analysis tool in ArcGIS was used to identify, for each species, discrete clusters of core habitat in the dataset. The centroid of each cluster was also identified. Thus, a suite of cluster centroids was identified for each focal species. The number of clusters (and centroids) was different for each species (table 12.4). In the case of cougar core habitat, which was fairly ubiquitous (89% of the total core), only two discrete clusters—and hence two centroids—were identified. In contrast, 44 centroids were identified for caribou, whose core habitat fell in many discrete pockets throughout the Inland Temperate Rainforest. For each core cell, we identified the centroid that was closest to it, using only those species for which the cell was core habitat. For example, if a cell was core for only cougars, grizzly bears, and wolves, its nearest centroid would be identified from among the centroids of clusters for these three species.

Table 12.4. Number of cluster centroids used for each focal species.

Focal species	Number of centroids
Caribou total	44
BC Caribou	32
US Caribou	12
Cougar	2
Grizzly	18
Lynx	35
Wolf	14
Wolverine	19
Source: Justin C. Williams.	

We then defined a coefficient Dj, which was the Euclidean distance from core cell j to its nearest centroid (see also Snyder et al. [2007], who use a similar distance coefficient for habitat selection). The objective of model 1 (expression 4) was modified to

$$\text{Minimize: } \sum_{j \in J} Dj\, Xj \qquad (7)$$

This objective function seeks to minimize the total number of distance-weighted cells selected. It was thought that this objective would promote compact clustering of selected habitat around naturally occurring centroids. This modified formulation, comprising objective (7) together with the habitat target constraints (5') and binary integer requirement (6), makes up model 2.

Solutions to model 2 are likely to have a total area larger than the minimum area needed to satisfy all targets, as identified by model 1. The reason is that the selection of two sites that are close to a centroid may add less to the objective function value than the selection of one distant site. So, model 2 would choose the former, whereas model 1 would choose the latter. Because the respective total areas of solutions to models 1 and 2 are likely to be different, they cannot be directly compared for assessing trade-offs. So, a third model—model 3—was formulated. Model 3 is the same as model 2 except that it contains one additional constraint:

$$\sum_{j \in J} Xj = |\mathrm{OM1}| \qquad (8)$$

Constraint (8) requires the total number of selected cells to equal $|OM1|$, which is the number of cells selected in the optimal solution to model 1. Model 3, then, not only encourages the clustering of selected cells and meets the core habitat targets for each species, but also uses no more than the minimum total area identified by model 1. When alternate optimal solutions exist for model 1 (which is very likely), model 3 identifies the one in which core habitat is most tightly clustered. If model 1 happens to have a unique optimum (very unlikely), this solution will be also be found by model 3.

Results and discussion

A core habitat target for each of the six focal species was specified. The targets ranged from 50% to 90% of the available core habitat area (table 12.5, column 2). The overall caribou target (80%) represented two separate targets, one for British Columbia (90%), and one for the United States (50%). These targets, used in all three IP models, were chosen so that alternate CADs reflecting high levels of habitat conservation could be explored. While the choice of targets is a judgment call and may be a point of contention for stakeholders, the choice should be defensible. Area-based targets would ideally reflect species' habitat needs based on population viability analysis. At the time of our study, this type of analysis was not available, so our targets reflected the opinion of researchers at the Craighead Institute, based upon available data. In the case of mountain caribou (an endangered caribou subspecies), a target of 90% in British Columbia was chosen to ensure almost complete habitat protection but also to allow more flexibility (compared to a target of 100%) for accommodating the needs of the other focal species. In the United States, only a remnant population of mountain caribou remains in northern Idaho, although large blocks of core habitat, not currently populated by mountain caribou, are located farther south. The target of 50% was chosen as a compromise to allow for the possibility of population reintroduction and recovery in the future.

Table 12.5. Habitat targets and optimal solutions for the three problems

Focal species	Core habitat targets	Model 1 optimum solution (figure 12.3)	Model 2 optimum solution (figure 12.4)	Model 3 optimum solution (figure 12.5)
Caribou total	80% 61,870	81.8% + 1,808	82.9% + 2,231	82.9% + 2,205
BC Caribou	90% 52,251	90.0% Same	90.0% Same	90.0% Same
US Caribou	50% 9,619	59.4% + 1,808	61.6% + 2,231	61.5% + 2,205
Cougar	50% 85,213	50.9% + 1,463	53.8% + 6,423	53.5% + 5,943
Grizzly	80% 93,066	80.0% Same	80.0% Same	80.0% Same
Lynx	70% 62,124	71.8% + 1,563	82.0% + 10,640	81.3% + 10,058
Wolf	50 % 68,399	54.7% + 6,469	62.5% + 17,043	62.2% + 16,713
Wolverine	85% 68,187	85.0% Same	85.6% + 495	85.8% + 677
Total cells selected	N/A	50.4% 96,520	50.7% 97,104 (+584)	50.4% 96,520
Dj distances (km) sum (mean)	N/A	2.989×10^6 (30.97)	2.157×10^6 (22.21)	2.161×10^6 (22.39)
Solution time (CPU seconds)	N/A	1.39	13.11	16.62

Source: Justin C. Williams.

IP model solutions

Each of the three optimization models was solved using the CPLEX (version 9.1) IP solver on a Dell laptop computer. The optimal solutions are listed in table 12.5, columns 3–5. The optimal solutions show that, although all targets were satisfied as expected, only some of the habitat targets were met exactly, and others were exceeded. In model 1, the targets for caribou in British Columbia, grizzly bears, and wolverines were met exactly, and in models 2 and 3, only the BC caribou and grizzly bear targets were met exactly. This might be expected because these percentage targets (80–90%) were higher than those of the other species (50–70%). Hence, by satisfying the two or three highest percentage targets exactly, the lower targets for the other species were exceeded. The model 1 optimal solution (figure 12.3) satisfied all targets through selection of just over 50% of all core cells (and just less than 50% of the entire region). The sum of *Dj* distances for this solution was 2.989×10^6 km, calculated from cell x- and y-coordinates from the Craighead Institute database. This resulted in a mean distance from selected cells to their nearest centroid of 30.97 km.

The model 2 optimal solution (figure 12.4), in which the sum of *Dj* distances was minimized, had a significantly lower total distance value of 2.157×10^6 km. The mean distance from selected cells to their nearest centroid in this solution was 22.21 km, which was 28% lower than in the model 1 solution, indicating a more compact and less fragmented reserve (compare figures 12.3 and 12.4). This was achieved with only a slight increase in the total area of the reserve (584 additional cells, less than 1% increase). While both the BC caribou and grizzly targets were met exactly in the model 2 solution, the other targets were exceeded by larger margins than they were in the model 1 solution. (We note that although model 2 minimizes the sum of selected cell-to-centroid distances, it does not take into account animal "movement costs," which might be high, due to barriers such as highways or poor intervening habitat, even when cell-to-cell distances are relatively short.)

The sum of *Dj* distances was also minimized in model 3 (figure 12.5), but the total number of selected cells was fixed at 96,520, the number identified in the model 1 solution. In the model 3 solution, the total distance value was 2.161×10^6 km, resulting in a mean distance from selected cells to their nearest centroid of 22.39 km, slightly higher than in model 2 but still much lower than in model 1, indicating a relatively compact reserve similar to the model 2 solution. The core habitat attainment levels in the model 2 and model 3 solutions were the same or similar for each species. Although the model 3 solution is technically an alternate optimum to the model 1 solution, it is arguably a better solution because not only is it more compact, but all of the species except BC caribou and grizzly bears have more selected core habitat than in the model 1 solution. This outcome demonstrates that covering problems such as this one frequently have (many) alternate optima (see Jacobi et al. 2007), and that some may be more desirable than others when additional criteria, such as compactness and coverage in excess of targets, are considered.

The computational burden of solving these problems was fairly small. The run times in central processing unit (CPU) seconds were 1.39, 13.11, and 16.62 for models 1, 2, and 3, respectively. The three problems were solved again with other habitat targets, and those run times were similar; the longest run time was 23 seconds, and most were less than 4 seconds. These results suggest that these three IP models are practical decision support tools for large conservation problems. Global optimal solutions were found quickly using an exact solver (CPLEX), and we did not need to resort to heuristic solvers or settle for approximate solutions. Although the problems contained nearly 200,000 binary variables (large by most standards), having only six or seven constraints—one constraint for each focal species plus an area constraint in model 3—helped make the problems computationally tractable.

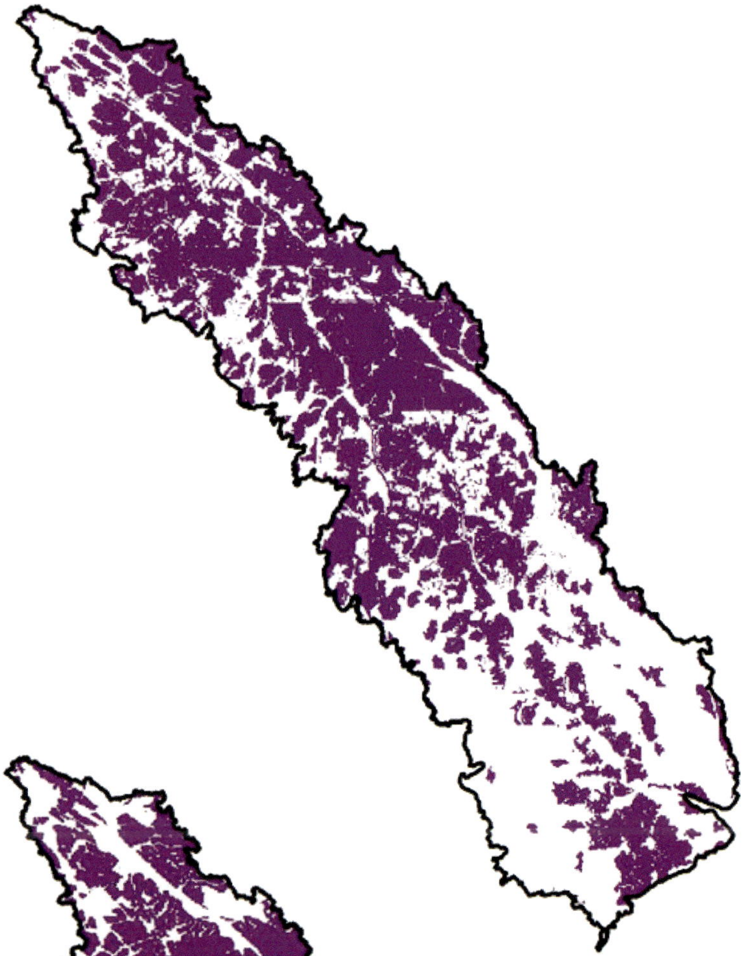

Figure 12.3. Optimal solution for IP model 1. Purple areas represent habitat cores for the targets specified in table 12.5. Shining Mountains data from Ministry of Environment, Province of British Colombia; DEM/hillshade from US Geological Survey; United States/Canada boundaries from ESRI Data & Maps, 2005, courtesy of ArcWorld Supplement; roads from ESRI Data & Maps, 2005, courtesy of TANA/GDT; streams/lakes/cities from ESRI Data & Maps, 2005, courtesy of ArcWorld; ITR boundary from Craighead Institute; optimized solution 1 from Craighead Institute.

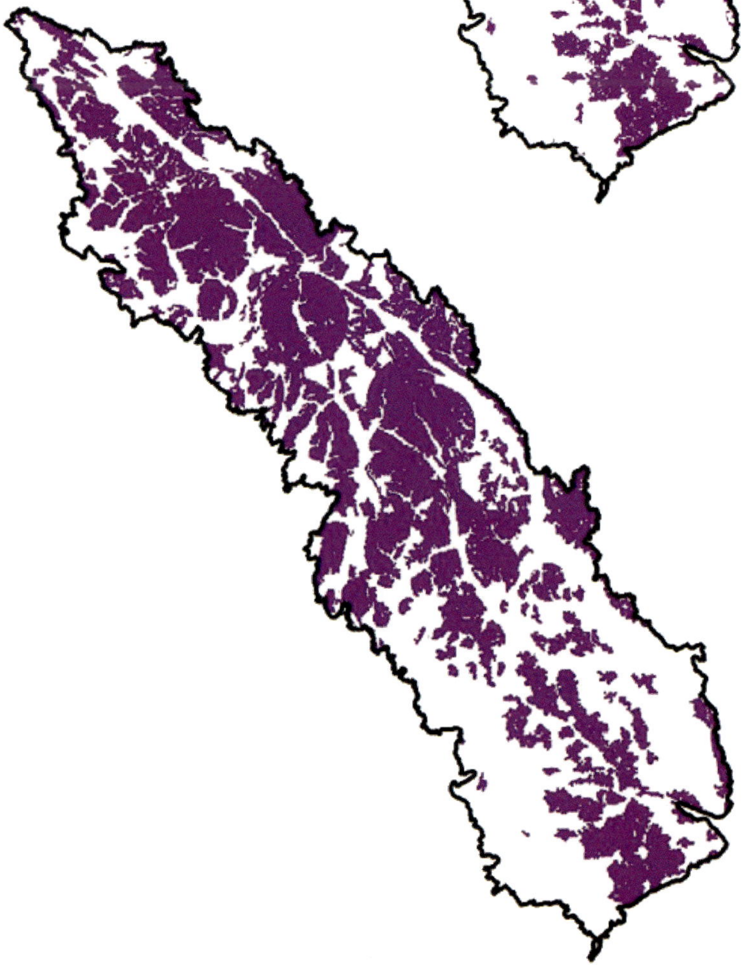

Figure 12.4. Optimal solution for IP model 2. Purple areas represent habitat cores for the targets specified in table 12.5. Shining Mountains data from Ministry of Environment, Province of British Colombia; DEM/hillshade from US Geological Survey; United States/Canada boundaries from ESRI Data & Maps, 2005, courtesy of ArcWorld Supplement; roads from ESRI Data & Maps, 2005, courtesy of TANA/GDT; streams/lakes/cities from ESRI Data & Maps, 2005, courtesy of ArcWorld; ITR boundary from Craighead Institute; optimized solution 2 from Craighead Institute.

Figure 12.5. Optimal solution for IP model 3. A compact solution with more core habitat for most species. Green areas represent habitat cores for the targets specified in table 12.5. Shining Mountains data from Ministry of Environment, Province of British Colombia; DEM/hillshade: US Geological Survey; United States/Canada boundaries from ESRI Data & Maps, 2005, courtesy of ArcWorld Supplement; roads from ESRI Data & Maps, 2005, courtesy of TANA/GDT; streams/lakes/cities from ESRI Data & Maps, 2005, courtesy of ArcWorld; ITR boundary from Craighead Institute; optimized solution 3 from Craighead Institute. Reprinted with permission from Dellasala et al. (2011).

Composite core solutions

The composite core method (described above) is a useful and intuitively appealing approach for generating reserve alternatives for the Inland Temperate Rainforest. The selected areas exhibit high levels of connectivity and compactness (figure 12.2), a result in part of natural overlap in core habitat of the six focal species. Furthermore, these alternatives were also found to be optimal with respect to IP models 1 and 3. Each composite core alternative resulted in some amount of core habitat for each species, as listed in table 12.3. When these amounts were used as the targets in models 1 and 3, the resulting optimal solutions turned out to be the same as the composite core solutions. Actually, this was necessarily the result. Consider, for example, the composite core 4+ solution (table 12.3). Here, all of the cells that provide core habitat for at least four species have been selected, 96,285 cells in total. Now, suppose that the corresponding percentage core habitat attainments (as listed in the 4+ column) were then used as targets in IP models 1 or 3. Would it be possible to satisfy these targets by selecting fewer than 96,285 cells? The answer turns out to be no. To see why, suppose that one of the cells in the 4+ solution was not selected. Then, at least four of the species would fall short of their targets by one cell. We would then need to select another cell to make up this deficit, but this new cell would (necessarily) be a cell that provided core for at most three species, so we would still fall short of the target for at least one species. This would force us to select a second new cell, but this would put us at 96,286 cells, one more than composite core solution. Hence, the composite core solution is optimal for models 1 and 3 if the composite core habitat attainments are used as targets in the models. Furthermore, with these targets, the composite core solution is the unique optimum for the models, because there is only one way to satisfy the targets.

The composite core solutions compared favorably with the IP solutions in terms of habitat coverage. For example, the composite core 4+ alternative and the solution for model 3 both selected just over 96,000 cells. But, the 4+ alternative provided significantly more habitat than the model 3 solution for five of the six focal species, all but grizzly bears. For grizzly bears, the 4+ alternative selected 72% of the available core habitat, falling short of the 80% target specified in model 3. This indicated that going from 72% to 80% for grizzly bears forced some level of sacrifice in the habitat selected for each of the other species, if total area was to remain (almost) constant. Conservation planners are likely to be interested in this type of trade-off. The sums of *Dj* distances were also calculated for the composite core solutions (bottom row of table 12.3). The mean cell-to-centroid distance in the 4+ solution was 24.96 km, indicating that 4+ was slightly less compact than the model 3 solution (22.39 km), although more compact than the model 1 solution (30.97 km). The 4+ and model 3 solutions present another interesting trade-off: when habitat targets change, one may gain or lose compactness even when the total area remains (almost) constant.

Conclusions

Model 3 provides a computationally tractable method for identifying the most efficient way to meet user-specified habitat targets while also promoting compactness. The advantage of the composite core method is that it identifies, without the need to run an IP model, the most efficient way to meet the targets dictated by its selection rule. The disadvantage of the composite core method is that the user has no flexibility in choosing habitat targets; the amount of habitat selected may not match what is needed. Models 1 and 3 can be used to identify the trade-offs resulting from the specification of different area targets for core habitat. These trade-offs involve total area, coverage in excess of targets, and reserve compactness. Much of the value of optimization is found in generating these trade-offs. The solution for model 3 (figure 12.5), for example, identified about 50% of the study area. Of this, about 17% was nonhabitat (rocks, ice, and open water). If subwatersheds identified as core habitat in part were included in their entirety, the total area of habitat identified as high conservation value totaled about 42% (see http://web.unbc.ca/~wetbelt/2008-conference-Craighead-Conservation-Area-Design.html, and Dellasala et al. 2011). This is similar in area to a number of other studies summarized by Noss et al. (2012).

The role of optimization models in conservation planning

Optimization models are primarily tools for decision support as opposed to decision making. Their purpose is not to dictate a course of action but to provide guidance, information, and insight for planning and decision making. To this end, optimization models provide several benefits for reserve site selection and design, in addition to finding mathematically optimal solutions. First, a model formulation provides a concise, exact statement of the problem. The model may not match perfectly what is needed in the real world, but it is transparent and is therefore open to criticism and possible improvement. Although the model's underlying mathematics may be inaccessible to nonspecialists, the purpose and function of each mathematical component (objective and constraints) can usually be explained in plain language. Second, optimization models are highly general in that they can be transferred from one setting to another, from one dataset to another, and potentially from one spatial scale to another. In addition, results are replicable; different users employing the same model and the same data will obtain the same result (although alternate optima may exist). Third, although the mathematically optimal solution may not be the best solution, it can establish absolute benchmarks or baselines for the criteria being modeled that can help put other (perhaps more practicable) alternatives in perspective.

However, results that take the form of a single optimal solution are unlikely to be satisfactory in conservation planning. Flexibility is called for—the ability to compare at least several reasonable alternatives. This is the fourth benefit of optimization models: the ability to generate a variety of alternative solutions. These alternatives may include not only an optimal solution but also other alternate optima (if they exist) or good suboptimal solutions that perform well with respect to other criteria not explicitly modeled.

Finally, optimization models can be formulated to address a wide array issues that may influence reserve design. For example, land-management decisions are often more complex and dynamic than simple binary choices for whether or not to include sites in a reserve. Planners and decision makers may wish to consider multiple options for how best to manage and protect each site (e.g., Williams et al. 2007; Orsi et al. 2011); and may wish to address uncertainty about species presence/absence (e.g., Camm et al. 2002) and about what the future holds with respect to encroaching development, climate change, and potential species loss (e.g., Costello and Polasky 2004; O'Hanley et al. 2007). A discussion of how the optimization models for the Inland Temperate Rainforest example (above) were used to help guide conservation efforts can be found in Dellasala et al. (2011).

References

Andelman, S. J., I. Ball, F. W. Davis, and D. M. Stoms. 1999. "Sites: An Analytical Toolbox for Ecoregional Conservation Planning." Manual prepared for The Nature Conservancy, 1–43. http://www.biogeog.ucsb.edu/projects/tnc/overview.html.

Ball, I. R. 2000. "Mathematical Applications for Conservation Ecology: The Dynamics of Tree Hollows and the Design of Nature Reserves." PhD diss., University of Adelaide, Adelaide, Australia.

Ball, I. R., H. P. Possingham, and M. E. Watts. 2009. "Marxan and Relatives: Software for Spatial Conservation Prioritization." In *Spatial Conservation Prioritization: Quantitative Methods and Computational Tools,* edited by A. Moilanen, K. A. Wilson, and H. P. Possingham, 185–95. Oxford: Oxford University Press.

Camm, J. D., S. K. Norman, S. Polasky, and A. R. Solow. 2002. "Nature Reserve Site Selection to Maximize Expected Species Covered." *Operations Research* 50 (6): 946–55.

Chan, K. M. A., M. R. Shaw, D. R. Cameron, E. C. Underwood, and G. C. Daily. 2006. "Conservation Planning for Ecosystem Services." *PLoS Biology* 4 (11): e379, doi: 10.1371/journal.pbio.0040379.

Church, R. L., A. T. Murray, M. A. Figueroa, and K. H. Barber. 2000. "Support System Development for Forest Ecosystem Management." *European Journal of Operational Research* 121 (2): 247–58.

Church, R. L., D. M. Stoms, and F. W. Davis. 1996. "Reserve Selection as a Maximal Covering Location Problem." *Biological Conservation* 76: 105–12.

Cocks, K. D., and I. A. Baird. 1989. "Using Mathematical Programming to Address the Multiple Reserve Selection Problem: An Example from the Eyre Peninsula, South Australia." *Biological Conservation* 49: 113–30.

Costello, C., and S. Polasky. 2004. "Dynamic Reserve Site Selection." *Resource and Energy Economics* 26: 157–74.

Craighead, L., and B. Cross. 2007. *Identifying Core Habitat and Connectivity for Focal Species in the Interior Cedar-Hemlock Forest of North America to Complete a Conservation Area Design.* USDA Forest Service Proceedings RMRS-P-49. Fort Collins, CO: US Department of Agriculture, Forest Service, Rocky Mountain Research Station.

Csuti, B., S. Polasky, P. H. Williams, R. L. Pressey, J. D. Camm, M. Kershaw, A. R. Kiester, B. Downs, R. Hamilton, M. Huso, and K. Sahr. 1997. "A Comparison of Reserve Selection Algorithms Using Data on Terrestrial Vertebrates in Oregon." *Biological Conservation* 80: 83–97.

Davis, F. W., D. M. Stoms, R. L. Church, W. J. Okin, and K. N. Johnson. 1996. "Selecting Biodiversity Management Areas." In *Sierra Ecosystem Project: Final Report to Congress, Vol. II, Assessments and Scientific Basis for Management Options,* edited by D. C. Erman, 1503–30. Davis, CA: University of California.

Dellasala, D. A., P. Alaback, L. Craighead, T. Goward, P. Paquet, and T. Sprybille. 2011. "Temperate and Boreal Rainforests of Inland North America." In *Temperate and Boreal Rainforests of the World: Ecology and Conservation,* edited by D. A. Dellasala, 82–110. Covelo, CA: Island Press.

Diamond, J. M. 1975. "The Island Dilemma: Lessons of Modern Biogeographic Studies for the Design of Nature Reserves." *Biological Conservation* 7: 129–46.

Haight, R. G., and S. A. Snyder. 2009. "Integer Programming Methods for Reserve Selection and Design." In *Spatial Conservation Prioritization: Quantitative Methods and Computational Tools,* edited by A. Moilanen, K. A. Wilson, and H. P. Possingham, 43–57. Oxford: Oxford University Press.

Jacobi, S., C. ReVelle, R. Pressey, and J. C. Williams. 2007. "Novel Operations Research Methods for Efficiently Determining Irreplaceable Sites for Conservation." *Environmental Modeling and Assessment* 12 (2): 91–103.

Kingsland, S. 2002a. "Creating a Science of Nature Reserve Design: Perspectives from History." *Environmental Modeling and Assessment* 7: 61–69.

———. 2002b. "Designing Nature Reserves: Adapting Ecology to Real-World Problems." *Endeavour* 26 (1): 9–14.

Kirkpatrick, J. B. 1983. "An Iterative Method for Establishing Priorities for the Selection of Nature Reserves: An Example from Tasmania." *Biological Conservation* 25: 127–34.

Lehtomaki, J., E. Tomppo, P. Kuokkanen, I. Hanski, and A. Moilanen. 2009. "Applying Spatial Conservation Prioritization Software and High-Resolution GIS Data to a National-Scale Study in Forest Conservation." *Forest Ecology and Management* 258: 2,439–49.

Margules, C. R., A. O. Nicholls, and R. L. Pressey. 1988. "Selecting Networks of Reserves to Maximize Biological Diversity." *Biological Conservation* 43: 63–76.

Margules, C. R., and R. L. Pressey. 2000. "Systematic Conservation Planning." *Nature* 405: 243–53.

McDonnell, M., H. Possingham, I. Ball, and E. Cousins. 2002. "Mathematical Methods for Spatially Cohesive Reserve Design." *Environmental Modeling and Assessment* 7 (2): 107–14.

Mizumori, M. M., C. S. ReVelle, and J. C. Williams. 2005. "The Maximal Multiple-Representation Species Problem Solved Using Heuristic Concentration." In *The Next Wave in Computing, Optimization, and Decision Technologies*, edited by B. L. Golden, S. Raghavan, and E. A. Wasil, 183–98. New York: Springer.

Moilanen, A., and I. Ball. 2009. "Heuristic and Approximate Optimization Methods for Spatial Conservation Prioritization." In *Spatial Conservation Prioritization: Quantitative Methods and Computational Tools*, edited by A. Moilanen, K. A. Wilson, and H. P. Possingham, 58–69. Oxford: Oxford University Press.

Nalle, D. J., and J. L. Arthur. 2005. "A Comparison of Fixed-Site and Non-Fixed-Site Approaches for Species Protection." *Environmental Modeling and Assessment* 10: 229–38.

Nalle, D. J., J. L. Arthur, and J. Sessions. 2002. "Designing Compact and Contiguous Reserve Networks with a Hybrid Heuristic Algorithm." *Forest Science* 48 (1): 59–68.

Noss, R. F., A. P. Dobson, R. Baldwin, P. Beier, C. R. Davis, D. A. Dellasala, J. Francis, H. Locke, K. Nowak, R. Lopez, C. Reining, S. C. Trombulak, and G. Tabor. 2012. "Bolder Thinking for Conservation." *Conservation Biology* 26 (1): 1–4, doi: 10.1111/j.1523-1739.2011.01738.x.

O'Hanley, J. R., R. L. Church, and K. Gilless. 2007. "Locating and Protecting Critical Reserve Sites to Minimize Expected and Worst Case Losses." *Biological Conservation* 134: 130–41.

Önal, H. 2003. "First-Best, Second-Best, and Heuristic Solutions in Conservation Reserve Site Selection." *Biological Conservation* 115: 55–62.

Orsi, F., R. L. Church, and D. Geneletti. 2011. "Restoring Forest Landscapes for Diversity Conservation and Rural Livelihoods: A Spatial Optimization Model." *Environmental Modeling & Software* 26: 1,622–38.

Osman, I. H., and J. P. Kelly. 1996. *Meta-Heuristics: Theory and Applications.* Boston: Kluwer Academic Publishers.

Possingham, H., J. Day, M. Goldfinch, and F. Salzborn. 1993. "The Mathematics of Designing a Network of Protected Areas for Conservation." In *Decision Sciences, Tools for Today*, edited by D. Sutton, E. Cousins, and C. Pierce, 536–45. Adelaide: Proceedings of the 12th Australian Operations Research Conference, ASOR.

Pressey, R. L. 2002. "The First Reserve Selection Algorithm—A Retrospective on Jamie Kirkpatrick's 1983 Paper." *Progress in Physical Geography* 26 (3): 434–41.

Pressey, R. L., and R. M. Cowling. 2001. "Reserve Selection Algorithms and the Real World." *Conservation Biology* 15 (1): 275–77.

Pressey, R. L., H. P. Possingham, and C. R. Margules. 1996. "Optimality in Reserve Selection Algorithms: When Does it Matter and How Much?" *Biological Conservation* 76: 259–67.

ReVelle, C. S. 1993. "Facility Siting and Integer-Friendly Programming." *European Journal of Operational Research* 65: 147–58.

ReVelle, C. S., J. C. Williams, and J. J. Boland. 2002. "Counterpart Models in Facility Location Science and Reserve Selection Science." *Environmental Modeling and Assessment* 7 (2): 71–80.

Rodrigues, A. S. L., and K. J. Gaston. 2002. "Optimization in Reserve Selection Procedures—Why Not?" *Biological Conservation* 107: 123–29.

Rosing, K. E., C. S. ReVelle, and D. A. Schilling. 1999. "A Gamma Heuristic for the p-Median Problem." *European Journal of Operational Research* 117: 522–32.

Rosing, K. E., C. S. ReVelle, and J. C. Williams. 2002. "Maximizing Species Representation under Limited Resources: A New and Efficient Heuristic." *Environmental Modeling and Assessment* 7 (2): 91–98.

Royle, K. M. 2010. "Exploring MARXAN as a Terrestrial Conservation Planning Tool." In *Ecosystem Based Management: Beyond Boundaries. Proceedings of the Sixth International Conference of Science and the Management of Protected Areas,* edited by S. Bondrup-Nielsen, K. Beazley, G. Bissix, D. Colville, S. Flemming, T. Herman, M. McPherson, S. Mockford, and S. O'Grady, 91–109. Science and Management of Protected Areas Association, Wolfville, NS.

Sarkar, S., R. L. Pressey, D. P. Faith, C. R. Margules, T. Fuller, D. M. Stoms, A. Moffett, K. A. Wilson, K. J. Williams, P. H. Williams, and S. Andelman. 2006. "Biodiversity Conservation Planning Tools: Present Status and Challenges for the Future." *Annual Review of Environment and Resources* 31: 123–59, doi: 10.1146/annurev.energy.31.042606.085844.

Smith, A. P., N. Horning, and D. Moore. 1997. "Regional Biodiversity Planning and Lemur Conservation with GIS in Western Madagascar." *Conservation Biology* 11 (2): 498–512.

Snyder, S. A., J. R. Miller, A. M. Skibbe, and R. G. Haight. 2007. "Habitat Acquisition Strategies for Grassland Birds in an Urbanizing Landscape." *Environmental Management* 40: 981–92.

Soule, M. E., and D. Simberloff. 1986. "What Do Genetics and Ecology Tell Us about the Design of Nature Reserves?" *Biological Conservation* 35: 19–40.

Underhill, L. 1994. "Optimal and Suboptimal Reserve Selection Algorithms." *Biological Conservation* 35: 85–87.

UNEP/IUCN (United Nations Environment Programme/International Union for Conservation of Nature). 2008. "World Database on Protected Areas." Produced by the United Nations Environment Programme–World Conservation Monitoring Centre (UNEP–WCMC) and the International Union for Conservation of Nature, World Commission on Protected Areas (IUCN WCPA). http://www.WDPA.org.

Williams, J. C., C. S. ReVelle, and S. A. Levin. 2004. "Using Mathematical Optimization Models to Design Nature Reserves." *Frontiers in Ecology and the Environment* 2 (2): 98–105.

Williams, J. C., C. S. ReVelle, and S. A. Levin. 2005. "Spatial Attributes and Reserve Design Models: A Review." *Environmental Modeling and Assessment* 10 (3): 163–81.

Williams, J. C., C. S. ReVelle, and W. Song. 2007. "Optimal Reserve Site Selection with Multiple Levels of Protection." *Environment and Planning B: Planning and Design* 34 (4): 725–39.

13

Integrating conservation planning with projected trends in climate change

Dominique Bachelet, PhD

Conservation organizations, federal agencies, and state legislators are all working on developing the ultimate list of climate change adaptation strategies—the magic cookbook that will help managers find the recipes to meet the climate change challenge from ocean bottom to mountaintop. Although the holy grail remains the reduction of global greenhouse gas emissions and other causes of anthropogenic stress, it is now understood that we have already committed our planet to a degree of warming that will affect our environment and our societies, even if we could somehow stop all anthropogenic emissions tomorrow. Therefore, land managers are preparing for a future world that will include higher sea levels, warmer ocean temperatures, more acidic seawater, more intense storms, warmer winter and nighttime temperatures, lower snowpacks, longer and drier summers, shortages in water resources, more frequent fires, more frequent pest outbreaks, new species assemblages, and expansion of exotic species. They will also face an increase in landscape fragmentation due to various human activities, as well as agricultural, industrial, and urban land-use pressure to produce more food, more energy, and more shelter for a growing and migrating world population. And of course, natural climate variability may exacerbate or mitigate the changes in climate driven by alterations in atmospheric chemistry and the reflectivity of our planet due to human activities.

As a representative of the National Park Service stated at the Ecological Society meeting in 2009, what is currently needed is an increase in everyone's tolerance to inevitable change. A recognition that change is constantly happening enhances the ability to seize emerging opportunities as the climate shifts. Such ability is the first essential component of strategies that conservation practitioners need to develop so as to meet upcoming challenges. For example, although projections of future sea levels show a dramatic loss of today's coastal areas, they also allow our society the opportunity to better engineer future coastlines and to consider such revolutionary concepts as rolling easements. Twenty percent of the world's population lives on a coast.

Mangroves and coastal wetlands are threatened around the world by housing and industrial development. Scientists have documented the danger of replacing natural vegetation by human constructions, and recent examples of catastrophic coastal erosion caused by tsunami or hurricanes have illustrated the point vividly. Today's objective should be designing new coastlines where natural vegetation is protected from development as sea level rises and that can buffer inland human structures from storm surges.

Pressey et al. (2005) defined conservation planning as the process of locating and designing conservation areas to promote the persistence of biodiversity in situ. Williams and Jackson (2007) warned that future warmer climates might promote no-analog communities and ecological "surprises" ill-suited to fit into spatially constrained conservation areas. Hannah et al. (2007) discussed the immediate need for additional protected areas to keep up with the shift of species ranges out of existing protected areas. Furthermore, conservation areas chosen through planning must be able to mitigate at least some of the proximate threats to biodiversity (Griffith et al. 2009). Ashcroft (2010) reminded readers of the uses and misuses of the term "refugia," separating macrorefugia based on large climate grids from microrefugia defined mostly by microclimatic conditions. They refined the distinction between habitat-based (ex situ) refugia—that are likely biased towards cooler locations (assuming uniform warming) and underestimate extinction risks in areas with barriers for dispersal or for species with limited dispersal ability—and climate stability-based (in situ) refugia that can exaggerate extinction risks for species with good dispersal ability but underestimate climate stress for species that are adapted to more climatically stable areas, and are thus more sensitive to small changes.

One of the most important steps in conservation planning is to identify what the ultimate goal is: is it the protection of one or more rare species, of a particular ecosystem or mosaic of ecosystems, or of an ecosystem function or service? Effective conservation planning includes the gathering and understanding of crucial information on the threats—in this case, climate change—to biodiversity (or other natural values) and on the vulnerability (a combination of sensitivity, adaptive capacity, and exposure) of the conservation areas (or species or ecosystem function) to those threats (Wilson et al. 2005; Pressey et al. 2005). Information gathering requires focus to be effective. Whether the emphasis is on an endangered species or an ecosystem function affects what data is collected and what tools can be used to learn more about the system. Once the information has been gathered and gaps in knowledge addressed, an action plan can be drawn up and applied. The success of this action plan should be evaluated regularly so that course corrections can be introduced and the effectiveness of the plan constantly improved.

In this chapter I introduce a framework to address climate change issues in conservation planning. I discuss projected trends in climate and the resulting simulated climate change impacts, including model result uncertainties and an acknowledgment of our incomplete understanding of biotic interactions and ecosystem processes, due in part to the lack of data. Finally, I propose a list a conservation planning strategies to prepare for future direct and indirect effects of climate change.

Frameworks for climate change awareness and a large-scale example

A straightforward framework to address climate change issues in conservation planning was developed by Cross and colleagues (Cross et al. 2012; figure 13.1).

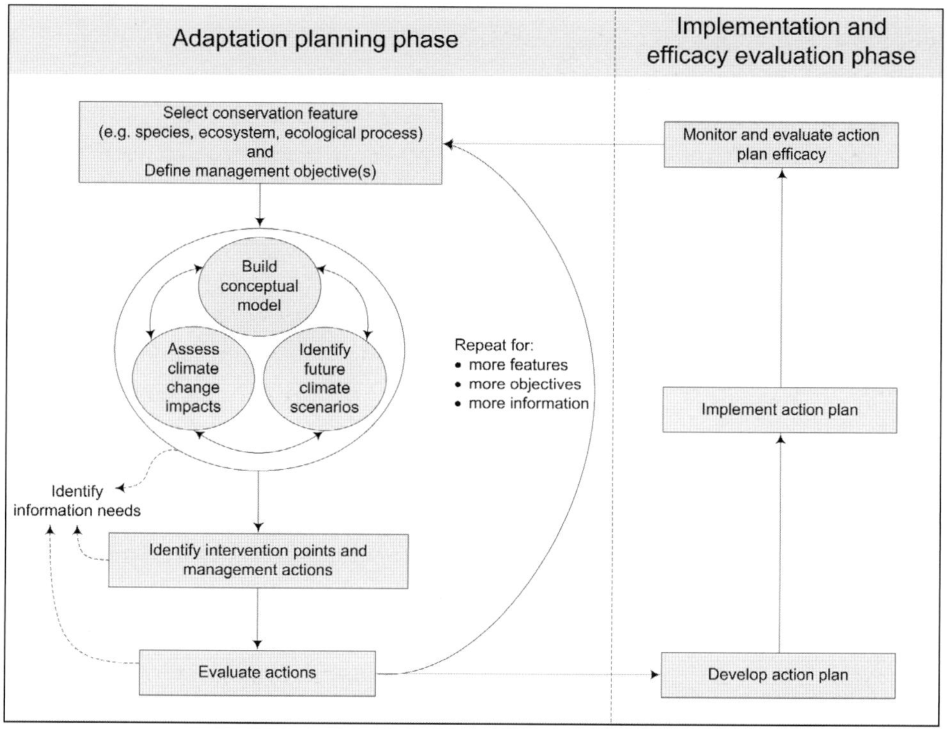

Figure 13.1. Framework for natural resource management and conservation to address climate change. Cross et al. (2012).

As a first critical step, a feature of interest or concern (species, ecosystem or its function, or area) is chosen in regard to climate change challenges. For example, Klausmeyer and Shaw (2009) chose Mediterranean ecosystems (see figure 13.2) as a feature of interest or concern because of their striking diversity of plants and the correlation of these ecosystems with a unique climate. They note that Mediterranean ecosystems represent "2% of our planet's land area containing 20% of the known vascular plant diversity with the second lowest level of land protection of all terrestrial biomes."

Once the feature has been identified, a clear management objective is necessary so that it is possible to evaluate when the objective has been met. The objective of Klausmeyer and Shaw (2009) was to identify which Mediterranean regions of the world needed the most immediate conservation action—thus a prioritized list of where new protected areas and connectivity pathways should be established. The Sierra Club has a catchy phrase for recommended action for conservation: "protect the best, remove the stress, restore the rest." In the case of the Mediterranean systems, the first order of business was to find out where "the best" was located.

According to Cross et al. (2012), the third step after identifying a feature and a practical management objective is to put together a conceptual model for the system of interest. Klausmeyer and Shaw (2009)

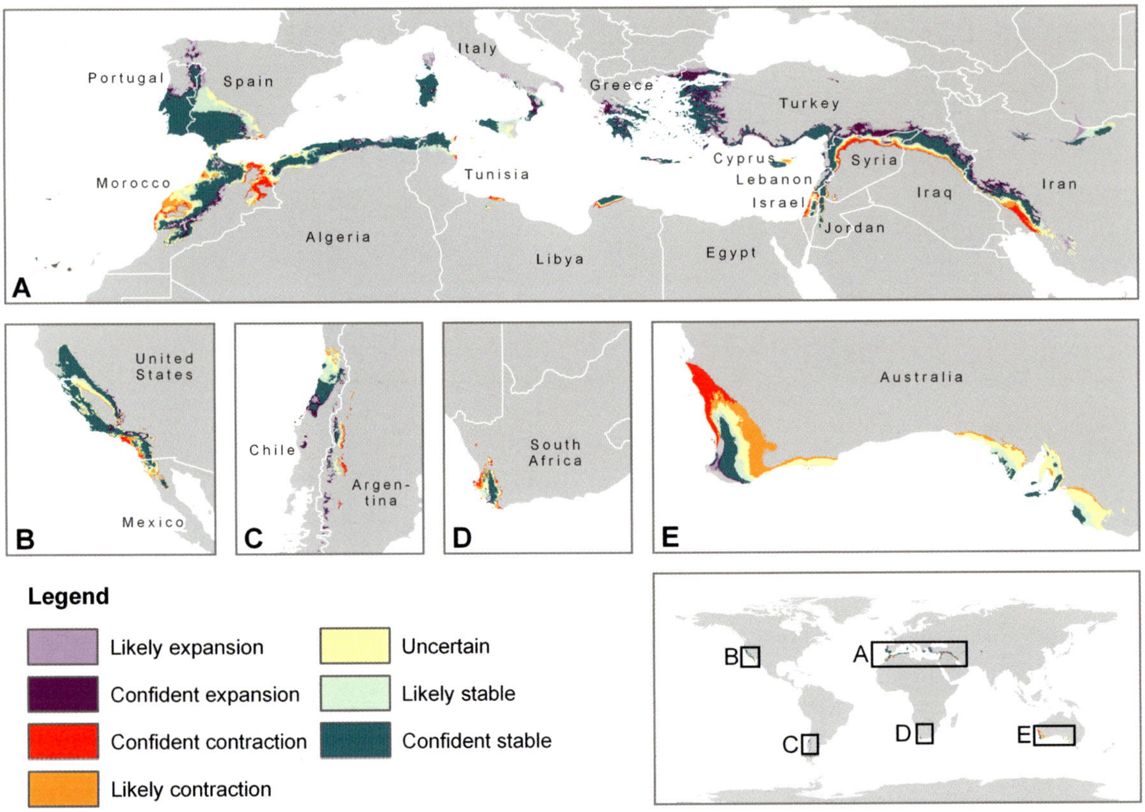

Figure 13.2. Projected status of the Mediterranean climate extent at the end of the 21st century under the A2 emissions scenario (see downloadable map at http://bit.ly/Lg9wXV).

defined the Mediterranean climate extent using climate indices defined earlier by Aschmann (1973). They then analyzed 136 projections of future climate from the World Climate Research Programme's Coupled Model Intercomparison Project phase 3 (CMIP3) multimodel dataset (Meehl et al. 2007). They came up with a map of the projected future area of Mediterranean climate extent, with various levels of confidence regarding their regional expansion or contraction. One of the areas with the most projected change was identified as the west coast of Australia, where the climate is projected to remain characteristic of Mediterranean lands in only 10% of the existing protected areas; in 17% of the current Mediterranean lands, climate definitely will shift to a new state.

The fourth step in the framework is to identify intervention points and management actions. In this case, apart from slowing the rate of climate change by reducing the amount of atmospheric greenhouse gases to give species more time to adapt, Klausmeyer and Shaw identified specific local strategies. Conservationists in this area should be encouraged to establish new protected areas, consider farmland areas for restoration, and modify current land-management practices to minimize the future impacts of warming.

The final steps include evaluating the feasibility of the suggested strategies, prioritizing and implementing them, and monitoring their effectiveness. These local actions should bring together landowners, provincial and national government representatives, conservation organizations, and farmers to determine a feasible plan of action, evaluate costs and available funding, and also possibly change policy to make the strategies more effective. Once a plan of action gets implemented, it is essential to monitor the site closely and evaluate the plan's efficacy as new data is collected.

Another recent framework for addressing strategies for climate change adaptation has been developed by conservation biologists working with the Yale School of Forestry and Environmental Sciences (see sidebar, "The Yale Framework").

The Yale Framework

The large number and diversity of models and data that can be applied to climate impact analyses and adaptation strategies can be confusing. Recognizing a need for clarity within this field, the Yale School of Forestry and Environmental Sciences convened a working group of leading conservation biologists, modelers, and policy makers to develop guidance for integrating climate change adaptation strategies into natural resource planning and policymaking.

In its most basic sense, the Yale Framework (http://www.databasin.org/yale) recommends that planners consider an optimal combination of six adaptation approaches at three distinct levels of ecological organization. The framework offers insights on how to do the following:

1. Establish a baseline of current environmental conditions.
2. Identify which ecological features will likely be most vulnerable to climate change.
3. Visualize the future needs of the most vulnerable species, ecosystems, and landscapes.

Adaptation approaches include the following:

1. What are the current patterns of biodiversity, and how can they be conserved? This constitutes the baseline of current assets and existing strategies aimed at ensuring the highest possible resilience to change.
2. What will the future patterns of biodiversity look like across a varied landscape? Models can be used to predict the manner in which climate change fundamentally influences the distribution of species, interactions among species, and the underlying ecological processes that support them.
3. What is the best way to maintain ecological processes? Changes in ecosystem-level processes can influence an ecosystem's ability to withstand and/or adapt to environmental change. Understanding these potential impacts is necessary for finding the right strategies to preserve ecological function.
4. How can ecological connectivity be best maintained or restored? Species may respond to changes in their habitats by shifting their geographic ranges, either longitudinally or by elevation. Maintaining or restoring landscape connectivity can help species adapt to a changing climate.
5. What areas are predicted to have relatively stable climate over the long term? When climate-related stressors change a species' habitat so quickly that migration is impossible, those species may benefit from the availability of uniquely situated areas within a specific habitat that offer relatively stable environmental conditions. It is important to identify these at least temporary havens, known as "refugia."

6. What on-the-ground environmental conditions are best suited to provide habitats for the native plant and animal species of a region, allowing them to persist in a climate-change era? Recent research suggests that conserving the underlying "ecological stage" or "enduring features" of a region (its geological, soil, and topographic features) is a solid approach for maintaining biodiversity under different climate impact scenarios.

The Yale Framework was constructed so that these six adaptation approaches can be applied across three separate scales of ecological organization:

1. Landscapes
2. Ecosystems
3. Species and populations

The Yale Framework is evaluated by assessment teams across the country. The assessments address a variety of climate and adaptation-related issues, including renewable energy siting, identification of migration corridors and other critical habitats, and broad-scale land-use change. The Geos Institute, NatureServe, EcoAdapt, Florida Natural Areas Inventory / Florida State University, the Conservation Biology Institute, and the University of Washington are conducting the current assessments.

More information, and access to online spatial datasets and documentation through Data Basin, is available at http://www.databasin.org/yale.

Projected trends in climate change

Climate models

When looking into potential impacts of climate change, conservation practitioners usually are looking for reliable future climate projections and likely emission scenarios. The various Intergovernmental Panel on Climate Change (IPCC) reports (http://www.ipcc.ch/) have incorporated state-of-the-art information on climate models and projected climate impacts, but the science is changing rapidly as new observations improve current understanding of climate and ecological processes. In the IPCC reports, scientists around the world have compiled published results and summarized current understanding of issues from atmospheric processes and ocean dynamics, to plant and animal responses to changes in their environment.

The climate datasets used for the assessments have been made available for download from various websites (e.g., CMIP3), but their format often is unwieldy for nontechnical groups. Fortunately, other websites have been created to provide the data in more manageable formats or even to make it directly usable on the web (e.g., Climate Wizard and Data Basin). Faced with available projections and maps of potential future climate, land stewards still need to estimate the uncertainty associated with the projections and understand the methods used to create the finer-scale coverage for their area of concern. Because that information is sometimes lacking or difficult to understand, it may not be available to inform the choice of "the best" future scenarios.

It is important to note that although climate projections diverge after 2040, models generally agree for the first half of the century, and the choice of a particular climate model or scenario is less important if the conservation goal is limited to the next two or three decades. Beyond 2040, it becomes critical to rely upon experts who can select climate models based on less-than-perfect criteria. For example, it is common to choose climate models that best simulate past climate dynamics, particularly paying attention to the most important local climate feature—for example, the Arizona Monsoon for the southwestern United States or the Pacific Decadal Oscillation in the Pacific Northwest. However, there is no insurance that a model that reproduces the past well is simulating the processes that will make it project the future reliably.

In addition, it is important to understand that as changes occur and persist (such as ice-free poles, disappearing glaciers, new wind patterns, or changes in ocean currents), the basic assumptions of climate models may become obsolete. Climate scientists update their models as new observations lead to new knowledge and more algorithms to better represent processes.

Finally, the accuracy of the emission scenarios used by the IPCC depends entirely on political decisions and social choices which, by definition, are impossible to predict. These economic projections are used by climate modelers to adjust atmospheric concentrations and are thus subject to constant revisions. Nevertheless, the following paragraphs summarize what most models agree are general trends for future climate.

General climate trends

Temperature. Warming has been recorded around the world and is projected to continue (see figure 13.3), with regional and local patchiness, causing an overall decrease in the longevity, extent, and thickness of glaciers, ice sheets, and snowpacks (Solomon et al. 2007). Observations have shown that land has been warming up at a faster rate than oceans due to the greater thermal inertia of deep oceans. Sea-level rise projections presented the 2007 IPCC report were extremely conservative, and new publications suggest higher levels are likely to be reached by the end of the century if current trends of ice melting and ocean warming continue (e.g., Rahmstorf 2010).

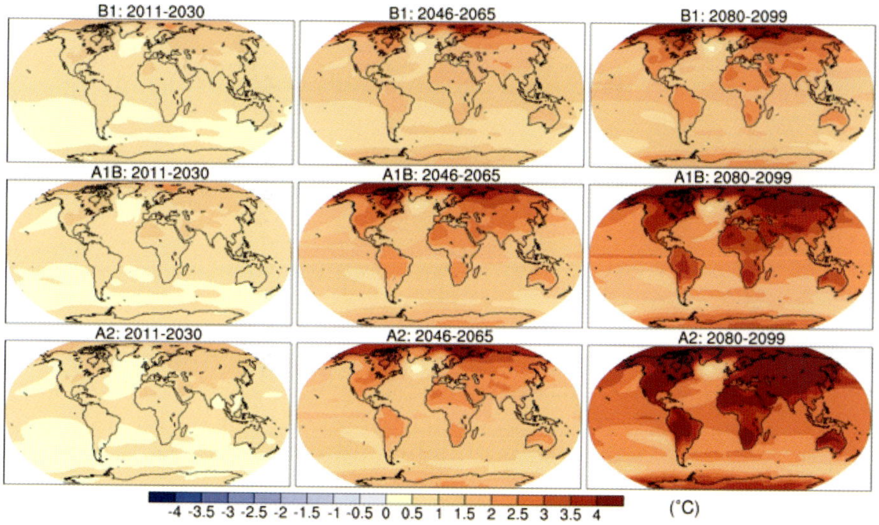

Figure 13.3. Multimodel mean of annual mean surface warming (surface air temperature change, °C) for the scenarios B1 (top), A1B (middle), and A2 (bottom), and three time periods, 2011 to 2030 (left), 2046 to 2065 (middle) and 2080 to 2099 (right). Anomalies are relative to the average of the period 1980 to 1999. From Solomon et al. (2007).

Precipitation. Global warming is likely to drive an increase in global mean precipitation. However, the degree of spatial and seasonal variation remains large even when considering multimodel mean values (see figure 13.4). Historical records of precipitation show large natural variability and sensitivity to teleconnections. Because there is no reason this might change in the future, patterns of change will be complex. All simulations point to increases in precipitation at high latitudes, where more rain than snow has been observed recently, a trend that will likely continue as winter temperatures increase. Models agree to a lesser extent concerning an increase of precipitation in the tropics. There is also general agreement about precipitation decreases in the subtropics. However, cloud formation and wind patterns are elements of uncertainty in model structure because current understanding remains limited.

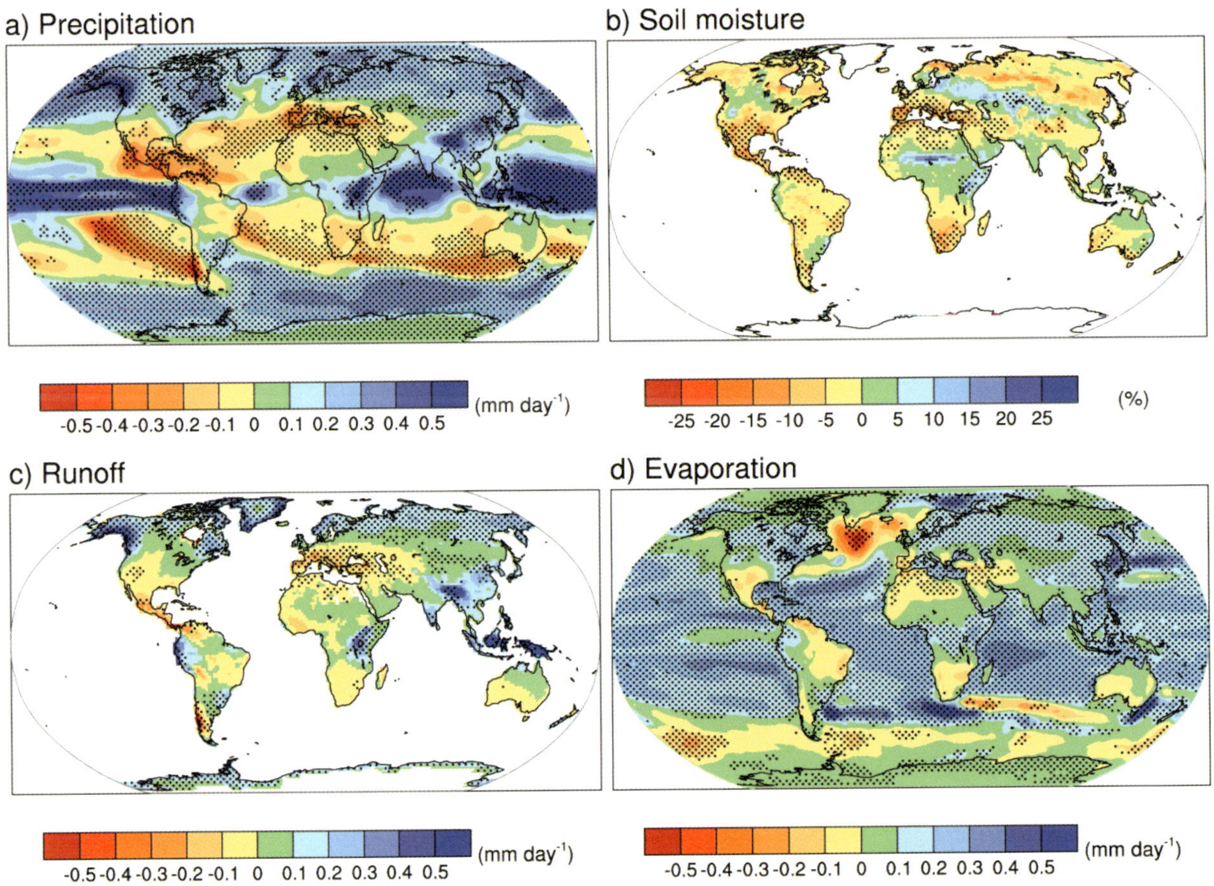

Figure 13.4. Multimodel mean changes in (a) precipitation (mm/day), (b) soil moisture content (%), (c) runoff (mm/day) and (d) evaporation (mm/day). To indicate consistency in the direction of change, regions are stippled where at least 80% of models agree on the direction of the mean change. Changes are annual means for the SRES A1B scenario for the period 2080 to 2099 relative to 1980 to 1999. Soil moisture and runoff changes are shown at land points with valid data from at least 10 models. From Solomon et al. (2007).

Natural climate variability (e.g., the El Niño Southern Oscillation [ENSO], the Pacific Decadal Oscillation [PDO], or the Atlantic Multidecadal Oscillation) and its impacts have been well documented for many regions of the world (e.g., Kitzberger et al. 2001), but the limited understanding of the causes of shifts in teleconnections renders them difficult to include in climate models.

Extreme events (intense droughts, floods, hurricanes/typhoons) are also difficult to predict by general circulation models (see figure 13.5). The 2007 report from the IPCC warns about the increased risk of more intense, more frequent, and longer-lasting heat waves, as exemplified by the European heat wave of 2003 that killed several thousands of people and caused widespread forest mortality (Ciais et al. 2005). Along with a greater risk of drought, there is an increased chance of intense precipitation and flooding due to the greater water-holding capacity of a warmer atmosphere, such that both wet and dry extremes are likely to become more severe. Several modeling studies project less numerous but more severe future cyclones, with greater wind speeds, more intense precipitation, and higher ocean waves.

Such extreme events, although unpredictable, are often what shape our landscapes. Past extreme events, such as the drought of the 1930s that caused the Dust Bowl in the United States, or the 1998 floods in China caused by heavy rainfall that affected 240 million people, certainly affected natural ecosystems and human land use. Recently, records of extreme events have been increasing. For example, drought in the summer of 2010 caused crop failure and huge fires in Russia, while record rainfall caused extensive flooding and loss of lives in both China and Pakistan. These concurrent extreme events might be evidence of climate destabilization; they certainly are consistent with climate scientists' expectations. Extremes pose a challenge to land managers more comfortable thinking about chronic linear change rather than abrupt and unpredictable change. Yet these events might be what they most need to account for when they plan for change. In the past, the reliability of models was tested in part by simulating large disturbances and observing the simulated system's response. It may be of interest to practitioners to focus on disturbance simulation to explore fully the resilience of their system.

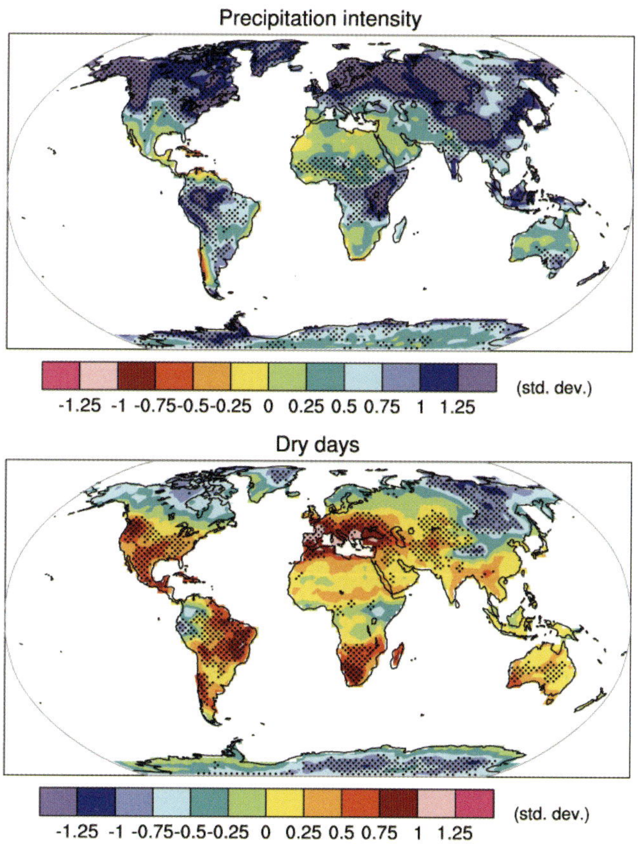

Figure 13.5. Changes in extremes based on multimodel simulations from nine global coupled climate models, adapted from Tebaldi et al. (2006). Top: Changes in spatial patterns of simulated precipitation intensity (defined as the annual total precipitation divided by the number of wet days) between two 20-year means (2080–2099 minus 1980–1999) for the A1B scenario. Bottom: Changes in spatial patterns of simulated dry days (defined as the annual maximum number of consecutive dry days) between two 20-year means (2080–2099 minus 1980–1999) for the A1B scenario. From Solomon et al. (2007).

Nonclimate events—for example, volcanic eruptions ejecting sulfur dioxide into the stratosphere; plankton mortality (caused by oil spills or anoxic zones created by excess fertilizers washed downstream) affecting the carbon sequestration (mitigation) potential of oceans—can directly or indirectly affect the climate without being predictable. Volcanic dust and ash fall out quickly and have no climate impacts. But sulfur dioxide gas can be injected into the stratosphere and combine with water to form sulfuric acid and sulfate particles, producing stratospheric haze that scatters solar radiation back to space, causing cooling. Studies have shown that a major eruption can cool the planet by several degrees (e.g., the 1815 Mount Tambora eruption was followed by crop failure in New England and Europe during the "year without a summer"). Pollutants and changes in conditions (temperature or acidity) of seawater affect phytoplankton growth and the large carbon absorption potential of oceans that might mitigate greenhouse gas emissions. As the concentration of CO_2 increases, water pH decreases and affects the growth and survival of marine organisms that fix carbon, thus eventually affecting climate.

Projections of future global climate conditions are available on the web from the IPCC (http://www.ipcc-data.org/) and the World Climate Research Programme's Coupled Model Intercomparison Project phase 3 (CMIP3; http://www-pcmdi.llnl.gov/ipcc/about_ipcc.php/). Seasonal means from all the models downscaled globally to a 10-km spatial grain are available by decades on the Data Basin site (http://bit.ly/wUzWaV). The Nature Conservancy's Climate Wizard (http://www.climatewizard.org/) provides average temperature and precipitation maps for historical and future periods for various emission scenarios and general circulation models using CRU (Climate Research Unit, UK Meteorological Office) observed global data as well as PRISM (parameter-elevation regressions on independent slopes model) historical climate data (http://www.prism.oregonstate.edu/) and the CMIP3 multimodel dataset. ClimaScope (http://www.elanadapt.net/climascope), a website where multiple climate annual variables from the most up-to-date scenarios used by the IPCC are made available, was developed through a collaboration of World Wildlife Fund with the Community Integrated Assessment System at the Tyndall Center (UK).

Limitations—Model uncertainty

Model verification is obviously impossible for future projections. Confidence in climate models often is based on the ability of the models to reproduce past climatic changes. However, there is no guarantee that a model that reproduces the past well will simulate the future accurately. Current models incorporate our current understanding of past climate dynamics, but future dynamics may change drastically with changes in tropospheric and stratospheric composition as well as in the planet's albedo.

General circulation models (GCMs; also known as global climate models) were designed to simulate the planet's climate, and their results compare well to global-scale climate observations. Their skills decline at the local scale due to their inherent coarse spatial resolution, averaging the diverse vegetation cover and complex topography important to conservation practitioners. Downscaling techniques (statistical or dynamic) bring GCM results to the scale of concern, but their accuracy is limited to that of the original projection, and the GCM bias is carried over in the boundary conditions for regional climate models. Furthermore, feedbacks from the biosphere to the atmosphere continue to be woefully underrepresented in global models, and regional model feedbacks to GCMs have not been developed yet. The uncertainty of GCM projections result from the imperfect knowledge of (1) initial conditions, such as sea-surface temperatures, that are difficult to measure, (2) the levels of future anthropogenic emissions, which are unknowable because they depend on current and future political decisions and social choices, and not on physical laws of nature, and (3) general system behavior (such as clouds or ice-sheet melt) that continues to be the subject of basic climate research and constitutes the “known unknowns” (see figure 13.6) of the climate system (see, for example, Cox and Stephenson 2007). Finally, although surprises have been projected for the biological world as the climate changes, “unknown unknowns” such as the unexpectedly rapid Larsen B ice shelf collapse bring climate scientists back to the drawing board to improve existing models.

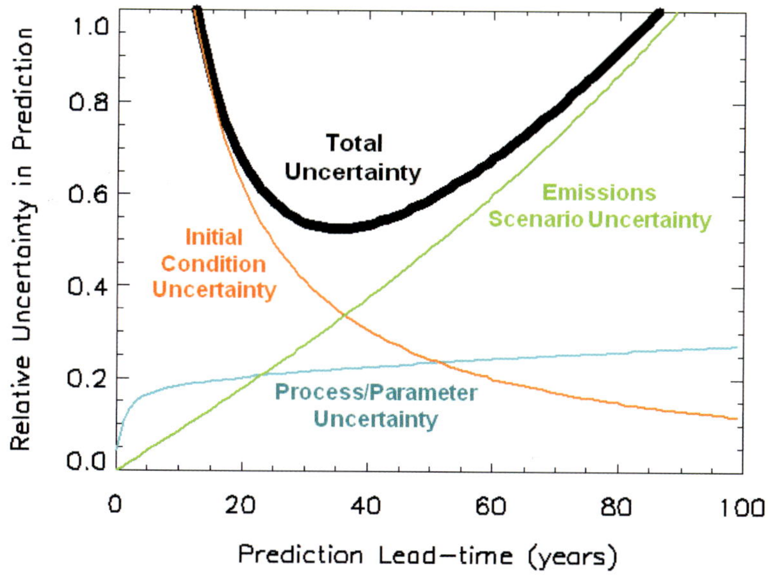

Figure 13.6. Uncertainty in climate change predictions is primarily the result of underlying uncertainty in initial conditions, future emissions levels, and climate process behavior. From Cox and Stephenson (2007).

It is important for conservation practitioners to ask relevant questions of providers of climate scenarios so that decisions are influenced by applicable climate information. A brief summary table (table 13.1), inspired by Daly et al. (2008), includes various themes conservation biologists could start exploring with their climatologist colleagues.

Finally, practitioners should ask themselves: What would happen if future projections were correct but not taken into account? If negative consequences and opportunities are inconsequential, then model outcomes can be ignored. However, if negative consequences are potentially large, and they could alter the future of a particular ecosystem, then the best models should be used to as a basis to prepare for change. Preparedness and awareness of possible changes can buffer negative consequences even when unpredictable extreme events occur (e.g., decade-long drought, catastrophic fire, or hurricane) and the conservation landscape is transformed.

Table 13.1. General uncertainty criteria about climate projections

Themes	Assumptions	Questions to ask
Reliability of interpolated climate	Future climate projections usually are based on anomalies modifying a historical baseline created from interpolated station records. Few stations and short records reduce the reliability of the climate baseline.	How many meteorological stations are close to your site? How long are the records from those stations? How much infilling has occurred to complete the records?
Influence of topography	Complex terrain often means decoupling from regional climate trends through such processes as cold air drainage or inversions. Climate models assume elevation homogeneity in all grid cells.	What is the spatial resolution of the climate projections used? How complex is the terrain around your site?
Proximity to relief	Large relief features such as mountain ranges can have shadow effects that affect local climate and that are not simulated well by climate models.	How far from a relief feature (e.g., a mountain range) is your site?
Proximity to a water body	Proximity to water can affect atmospheric moisture. Climate model scale may preclude the inclusion of such influence; even downscaled climate may miss this.	How far is your site from a water body such as a coast (fog and wind), lake (size imports), or stream (healthy riparian corridor—shade, moisture, and groundwater)?
Human influence	Land use affects local climate: urbanization, agriculture, and plantations affect temperature (e.g., urban heat islands), atmospheric moisture, cloud cover (e.g., the role of cloud condensation nuclei from pollution), albedo, wind direction, turbulence, and fire ignition.	How much is your site affected by humans? How close is it to an urban center? How close is it to irrigated crops? How many roads are nearby?
Natural climate variability	Seasonal or interannual climate variability may have conferred some resilience to change to local populations. Teleconnections such as PDO or ENSO can exacerbate or mitigate climate change.	How variable is the climate at your site? What magnitude of change would be required (e.g., 10% or 40% decrease in rainfall) to affect organisms? Is it close to the projected change? Would the timing of change (e.g., winter versus summer temperature) affect organisms?
Records from past extreme events	Using the knowledge of ecosystem responses to past extreme events helps to define the potential vulnerability of the system and prioritize strategies.	Are there known thresholds or tipping points (e.g., hydraulic failure for specific drought conditions) in local communities/species, and how close to such conditions are future climate projections?

Inspired by Daly et al. (2008).

Observed climate change impacts

Many climate-related impacts continue to emerge in the scientific literature. I will only cite a few to illustrate the point that climate change is occurring now and is not an event that will begin at some future time. The most visible secondary effects of climate change include changes in fire regimes and outbreaks of pests and pathogens. As environmental conditions change, scientists discover that thresholds are passed that trigger large-scale impacts, such as mountain pine beetle (*Dendroctonus ponderosae* Hopkins) epidemics in British

Columbia (Kurz et al. 2008; Carroll et al. 2004) and the Rocky Mountain region (e.g., Logan and Powell 2001; Logan et al. 2010), associated with warmer winters at higher latitudes and elevations. Westerling et al. (2006) have been documenting an increase in fire occurrence in recent decades in the western United States. Both wildfires and insect outbreaks in western coniferous forests have been attributed not only to fire suppression and reduced thinning but also to warming conditions (McCullough et al. 1998; Ayres and Lombardero 2000; Flannigan et al. 2005; Westerling et al. 2006; Kurz et al. 2008; Raffa et al. 2008; Waring et al. 2009).

In the last 40 years, natural rates of tree mortality in old-growth forests have increased while seedling recruitment has decreased across western North America (Van Mantgem et al. 2009). Increased evaporative demand, causing hydraulic failure or carbon starvation, has been shown to increase tree vulnerability to drought and cause tree mortality (Breshears et al. 2009), often mediated by pest outbreaks (Allen et al. 2010).

Many scientists and citizen-science networks have reported widespread changes in plant phenology in England and North America (e.g., Beaubien and Freeland 2000; Conover and Pelikan 2011; Wolkovich et al. 2012). Bird migration seasonality and routes are shifting (e.g., Booth et al. 2006) and animals are found at higher elevations (Moritz et al. 2008) as tree lines generally move higher on mountainsides and alpine zones constrict (Baker and Moseley 2007). Migration routes of a variety of animals are shifting by latitudes (e.g., Parmesan 2006; Rijnsdorp et al. 2009; La Sorte and Jetz 2012). Furthermore, while dead zones develop along the coasts, coastal wetland marshes are suffering salt-water intrusion as sea levels creep upward (Gitay et al. 2001).

Whether the climate models are reliable or not, changes in climate are causing environmental changes that are being documented by the few monitoring networks in existence today. Long-term monitoring and careful interpretation of observations are key to understanding what the future might bring. Tipping points can be detected, and ecologists now are charged with explaining these thresholds so that management strategies can be designed to address them.

Fazey et al. (2010) warn practitioners that action plans often focus on short-term coping strategies that can engender maladaptive vulnerability to unforeseen future changes—for example, the building of levees in New Orleans that increased the vulnerability of the city to hurricanes. Complex interactions are difficult to disentangle, and technofixes unfortunately can be quick and easy to implement. Richard Stone (2010) discusses the argument between conservationists and ecologists over assisted migration and cites examples of past experiments gone awry (e.g., the introduction of cane toads in Australia). The tolerance of species to climate variation is very much debated. Little is known about acclimation, adaptation potential, or even simple physiological thresholds. Or, as Taleb (2010) puts it, would you let a pilot using a map of Atlanta's airport land in New York City "because there is nothing else" to inform him? Doing nothing can sometimes be preferable to doing something potentially harmful that could reduce system resilience to future, yet-unknown changes. Specific criteria are being drawn up for considering managed relocation (e.g., Richardson et al. 2009) as a function of the confidence in ecological understanding and the level of perceived risk associated with both the lack of and assisted migration (McLachlan et al. 2008). However, gathering data to better understand the extent of species responses to ongoing changes in climate is the first order of business.

Projected climate change impacts

Faced with a changing world, the response of successful organisms include persisting in suitable (micro) refugia (e.g., microsites, deep valleys with cold air drainage and inversions, or foggy coasts), adapting to novel conditions (e.g., adaptive behavior of pikas or genetic adaptation of birds), or dispersing to new sites (northward or upward shifts). Although uncertainties have been much discussed that are inherent to climate models and to the levels of anthropogenic emissions our future world will produce, species responses to changes in climate are difficult to estimate, given our limited knowledge of their natural adaptive capacity, acclimation potential, and physiological thresholds and tipping points. It has been pointed out that although we currently know their "realized niche," their true niche will only be discovered as conditions for establishment, growth, and reproduction change. I have chosen a few examples to illustrate the challenges of trying to forecast how species will be affected by future environmental conditions and to briefly discuss niche models.

Lower snowpacks, lynx, and wolverines

Declared threatened and endangered in the conterminous United States under the Endangered Species Act in 2000 by the US Fish and Wildlife Service, the lynx is becoming the poster animal for the effects of climate change. The lynx is a specialist of the deep powdery snow of northern forests, where its massive paws keep it afloat in open forest and boreal habitat and give it a competitive advantage over other predators. A future with shorter winters and more rain than snow defeats the purpose of its snow-adapted morphology. The snowshoe hare, its favorite prey, is also perfectly adapted to snowy landscapes, with wide feet and white winter fur, and but easily falls victim to other predators as the snowpack dwindles. Furthermore, warming-enhanced mountain pine beetle outbreaks are affecting vast expanses of lodgepole pines, the main winter food for the hare. Projections of lynx survival under future climate conditions are not optimistic. However, it is important to remember that the main causes of lynx mortality in the twentieth century are linked to such human activities as trapping, habitat destruction (old-growth logging), landscape fragmentation, and killing by motor vehicle. Separating the impacts of future climate from those of future human activities and land use as a result of climate change may be presumptuous.

Copeland et al. (2010) showed that snow is a key component of wolverine habitat and that spring snow cover is essential for the animal, who uses it for denning, both as protection from predators and as insulation from cold temperatures. They concluded their study by stating that it was likely that future reductions in spring snow cover due to warming would reduce wolverine habitat. However, McKelvey et al. (2010) noted that teasing a climate-change signature from available observations was impossible to do with any rigor at this point.

Conservation practitioners are faced with the challenge of using modeling tools that can enable them to move toward a better understanding of climate change impacts on natural habitat as well as species' vulnerability to change, but the complex web of interactions with other species, and the critical importance of human actions affecting habitat quality and extent, may render that effort irrelevant.

Warming temperatures and pika behavior

The American pika lives on rocky slopes in the western mountains of North America. Because most climate-change impact models project the disappearance of alpine tundra areas, scientists have speculated that the range of suitable pika habitat might contract upward in elevation as lower-elevation site temperatures increase. However, in the Sierra Nevada and southwestern Great Basin, pikas appear to be thriving and tolerating a wide range of thermal environments, showing a greater distribution than previously thought. Millar and Westfall (2010) found that 12% of the occupied pika sites were lower in elevation than the historical (early 20th century) records indicate, revealing that pika live 500 meters (1,640 feet) below what was previously known for this region. Rock-ice features provide pikas with insulation and thermal refugia, and account for 83% of the 420 sites studied. In other words, pikas use talus adaptively to avoid extreme heat and severe cold.

So in this case, a cold-adapted species has been found to develop a behavior that provides a mechanism to avoid temperature fluctuations in its habitat. The quality of the data collected by scientists to support their case of vulnerability or resilience of a species is critical because it informs managers, who will base their decisions to dedicate resources for the protection of a species on its interpretation. Moreover, temporal and spatial scales of data collection are crucial to the understanding of the current state of habitat quality and viability of populations. For example, nesting bird populations have been known to migrate along the Pacific coast of the United States as a function of sea-surface temperatures, and their presence in a particular locale can be linked to the phase of teleconnections such as ENSO or PDO. Similarly, salmon fisheries have long documented temporal variation in population levels linked to the PDO phase in the Pacific Northwest and Alaska. Careful consideration of a climate change signal, rather than natural climate variability, is important to develop effective conservation strategies.

Species models and their limitations

To simulate a particular species response, niche models (also called bioclimatic-envelope or species-distribution models) have been used (Wiens et al. 2009). These correlative models vary substantially in their ability to predict current species distributions (Wiens et al. 2009) and produce highly variable predictions of range shifts in response to changing climate (Araújo and Rahbek 2006; Elith et al. 2006; Pearson et al. 2006). Moreover, they do not take into account biotic interactions, which may be critical. Because current species distributions depend on both climate and competitive interactions, the validity of envelope models relies on these interactions remaining constant in the future (Rehfeldt et al. 2006). However, paleoecologic records have demonstrated, for example, that plant associations do change (e.g., Ackerly 2003). Consequently, projections of changes in vegetation, and thus habitat characteristics, are subject to uncertainty and should be regarded as suggestions based on the current state of models rather than predictions.

Conservation planning strategies

Taking into account the various aspects of climate change, the following conservation strategies are proposed to land managers to preserve species or ecosystems of interest, and to continue to enjoy the valuable services they render to society. A mutually supportive set of strategies, such as those described below, which span a suite of ecological and societal approaches, is far more likely to ensure the persistence of valuable ecosystems or communities than those that focus solely on only one of these factors.

Promote heterogeneity

Following Taleb's (2010) advice, it is important to remember "nothing is ever too big to fail." Application of this principle to conservation indicates that even large preserves may not be sufficient to sustain a particular endangered species or ecosystem function if a specific threshold is passed. Concentrating on the projected shifts of leading or trailing edges of a population range or of ecotones between biomes, and creating large buffer zones around preserves, may underestimate the impacts of change because "optimal" conditions at the geographic center of that population may change and cause widespread decline there rather than at the edges.

A diversity of communities or ecosystems in the landscape can facilitate smoother transitions from a dominant community that might be less adapted to the new set of environmental conditions to a better-adapted community already existing in the landscape. For example, in the Puget Sound lowlands, Frelich and Reich (2010) note that remnant grasslands and savannas, even if transformed into no-analog communities of native species by climate warming and the northward expansion of southern species, could constitute an important seed source that would allow "forests to make a graceful transition to prairie and/or savanna" when climate conditions become too extreme to sustain forests.

Although it is economically important for the timber industry to produce even-aged forest stands, this type of focused management increases the vulnerability of the system to disturbance. Homogeneous forests are vulnerable to fires and insect or pathogen outbreaks, both of which are finding continuous optimal conditions for their spread: the forest exposure to disturbance is maximized. Dunwiddie et al. (2009) use as an example the Tieton River Canyon, in eastern Washington, where The Nature Conservancy (TNC) is trying to reverse the homogenization of dry forests caused by fire exclusion, grazing, wildlife management, and timber harvesting. These past human activities have increased the system's vulnerability to the large stand-replacing fires projected as a result of climate change (Lenihan et al. 2008; Littell et al. 2010). Manipulating stand structure and surface fuels with a combination of mechanical treatments and prescribed fire should help increase the diversity of age, structure, and size, and decrease catastrophic fire risk. The goal of this approach is to facilitate the gradual adaptation of the forests to warmer summer temperatures by creating a more natural mosaic of habitats with diverse potential responses to climate-driven disturbances, including not only fire but also pest outbreaks. The focus is on keeping the identity of the system intact despite, or rather thanks to, disturbances. Forest managers want to maintain dry forests and prevent a shift to a more open vegetation type. This strategy will work well if climate changes are gradual, so that managers have enough time to achieve site heterogeneity before a catastrophic disturbance occurs. In other words, the success of this strategy depends on the timing and amount of resources allocated to restoring a more diverse habitat.

Reduce stress but also rely on stress

Organisms that live in marginal areas where the stress level is already high and conditions are suboptimal might be less affected by a change in climate that would only represent one more level of stress. Seed sources or genetic material from marginal populations may become the most valuable tool for conservation of these organisms. "What is fragile will break early" (Taleb 2010), and no buffer size is likely protect a fragile (vulnerable) system from rapid disturbance. Thus, conserving a diversity of habitats—including marginal habitats—might be much more productive that creating large buffers of optimal habitat because diverse habitats likely include valuable refugia that conservation practitioners may not consider as optimal.

When populations of a particular species become small or disjunct, they may find refuge in local micro-habitats that buffer them from altered disturbance regimes, climate change, or competitive exclusion in the surrounding landscape. Potential refugia may include such features as rock outcrops that offer a diversity of micro-environmental conditions along gradients of light, moisture, and temperature, or riparian corridors where conditions are buffered from warming (through shading and high relative humidity) and drought (groundwater availability) or even disturbances such as fires. Refugia can also include stressful environments, such as a particular soil type that buffers species from competitive exclusion by more generalist species. The recent history of lowland Pacific Northwest prairies and savannas provide a case in point. Following the loss of historic fire regimes, shallow soils and harsh exposures reduced or prevented conifer invasion on small areas of historic prairie and savanna, allowing the drought-tolerant oaks, along with many associated prairie grasses and forbs, to persist in small pockets, often surrounded by dense conifer forest (Murphy 2008).

However, certain species or ecosystems will benefit from a reduction in the number of stress factors affecting their life cycle or their function. As an example, salmon require diverse habitats throughout their life cycle, and stress factors add up along the route from river to ocean. To address salmon issues, it is necessary to look from the spawning gravels up in the watershed, pools and side channels for rearing and overwintering, to estuaries and shorelines for rearing. Degraded riparian habitat, dams, overfishing, and hatcheries are often mentioned as the major threats to salmon along this continuum. Climate is a stress factor because it has the potential to greatly affect salmon phenology. Despite the fact that Pacific Northwest salmon are already adapted to wide fluctuations in climate, from one phase of the PDO to another (Mantua et al. 2010), climate change projections may bring more severe conditions. Decreased snowpacks and increased winter rains would generate high winter flows and earlier spring runoff (Milly et al. 2005), while decreases in summer rains and longer warmer growing seasons would cause low summer stream flows, affecting the timing of spawning and out-migration. As a result, strategies to improve or preserve salmon habitat vary in both time and space. The following example focuses on the work of The Nature Conservancy on the northern coast of the state of Washington, where sea-level rise is occurring due to geologic processes (isostatic rebound for Vancouver Island) and is projected to increase in the future (Glick et al. 2007).

Simulation results suggest that the removal of a sea dike at the mouth of the Stillaguamish River could increase freshwater residence time in the estuary, which might increase resistance to salinity intrusion imposed by sea-level rise, and thereby increase the functional resilience of freshwater tidal habitats (Dunwiddie et al. 2009). In contrast, a mid-delta project, removing a sea dike in an area where riverine connectivity has not yet been restored, might accelerate salinity intrusion and cause the loss of existing tidal marshes. In the latter case, productive salmon-rearing habitat would be restored in the short term but, despite the best intentions, the project could accelerate habitat loss in the long term. The timing of the restoration project becomes crucial because it depends on first restoring connectivity with rivers to ensure processes that will sustain the site as sea level rises.

Create redundancy

What Dunwiddie et al. (2009) calls "the Boeing approach" builds system resilience to change through the replication of the pieces that compose the system. The goal is to sustain species and their communities at multiple locations at or near the sites where they currently exist, assuming that if mortality occurs, it may not affect every instance of a dispersed population. Redundancy becomes insurance against the negative impacts of a localized disturbance. One of the recommendations of Millar et al. (2007) is to enable systems to "respond to change," but it is also a strategy to resist change.

Landscape fragmentation and habitat loss in the Puget Sound area have caused the decline of once thriving populations of Mazama (western) pocket gophers (*Thomomys mazama*), which play an important role in promoting soil aeration and germination of nondominant species, contributing to the prevention of invasive encroachment (Knudsen 2003). The gopher was listed as threatened in 2006 and only exists today in the few prairie remnants, where high mortality rates are due to roads, poisoning, trapping, and pets in the suburban environment. Local groups have been moving gopher populations to nearby military land that acts as preserves in this fragmented urban landscape. As the climate warms and invasive species get introduced that are better adapted to future conditions, it is important for local conservation groups to maintain healthy populations of gophers and their influence on native grass germination and growth.

This strategy will work well when change is geographically limited, and no general physiological threshold is reached that affects all of the redundant populations in the same way at the same time. It is an effective strategy if an extreme event (e.g., flooding, fire, or storm surge) affects a fraction of the landscape.

Preserve the function if not the composition

What Dunwiddie et al. (2009) call "the Microsoft approach" is the idea that as the climate changes, managers of restoration projects or natural systems can introduce new species, or novel associations of species, that are functionally similar to the existing ones but can eventually replace them as conditions become less suitable for the original biota. It corresponds to the recommendation of Millar et al. (2007) to enable forests to "respond" to change, encouraging transitions. An example is the establishment of warm-water corals that are transported to areas where cold-adapted corals did not survive bleaching events (Hoegh-Guldberg et al. 2007). Another example is the Ellsworth Creek Preserve in southwest Washington, where timber production over the last 100 years has shaped existing forests as even-aged stands of Douglas fir and western hemlock (Dunwiddie et al. 2009). Old-growth forest restoration at the site includes the option to bring back the native western red cedar, alder, and Sitka spruce that originally populated these forests. Western red cedar is one species that actually may expand northward and increase in frequency in its current range under future climate conditions (Hamann and Wang 2006). Managers will continue to preserve Douglas fir because it is the dominant species of the fire-adapted landscapes in the drier southern and eastern portion of the region

(Agee 1993) and because Chen et al. (2011) concluded from their modeling study that "despite local variations, climate change impacts are likely to be overall neutral for the coastal Douglas fir variety and interior populations of the 'inland rain forest' ecosystems." However, consideration of future climate projections and potential increases in wildfires is compelling TNC managers to also consider introducing fire-adapted California coastal redwoods.

Because summer fog decline reported along the California coast (Johnstone and Dawson 2010) may cause drought stress among marine fog-adapted redwoods, its introduction to the Pacific Northwest Coast might also serve to protect that species. This introduction of a new species could be interpreted as an example of the "resistance" strategy proposed by Millar et al. (2007). But the newly introduced trees would also increase functional redundancy and conserve important native forest features, such as canopy platforms for nesting marbled murrelets, rich epiphytic vegetation (including nitrogen fixers), and large woody debris that create critical habitats in streams and on the forest floor. The introduction of redwoods would allow the forests to transition smoothly from one community type to another (hopefully) better adapted to new climate conditions, changing composition but not function. The large trees would maintain the cool, moist microclimatic conditions in the understory capable of effectively buffering regional climate variations and creating climate refugia for animals that may need it, as existing old-growth forests already do. In this case, the focus is on system function rather than on system identity, and—just like Microsoft does not care if one uses a pull-down menu or an icon as long as one reaches one's goal—forest managers want to keep big trees and the environment they create to preserve the communities that depend on them as temperatures rise. This approach will work well if projections of future warming are correct, and southern species such as redwoods are viable under the new conditions of the Pacific coast. In case projections are not realized, mortality of patches of ill-suited, newly introduced species will only contribute to the diversity of habitats in old-growth forests.

Understand the limits of optimization strategies

Ecosystem resilience will be enhanced through slack and redundancy, not optimization. Optimization models are useful to provide guidance and information, and provide additional options, but not to dictate a course of action as discussed by Williams (chapter 12). As climate models continue to be developed, and the scale of the projections becomes finer and more relevant to local issues, it is possible that better tools might become available to help solve problems more rigorously. However, models will never become the infallible tools some decision makers expect. The future will continue to bring climate surprises, and intricate webs of interactions among system components will continue to be the subject of research and controversies. While scientists continue to learn by monitoring ecosystem responses, managers cannot wait for a crystal ball that will never be invented for them. Uncertainties about cloud formation, future global energy uses, and adaptability potential of common species, to name a few issues, do not support the optimization of proposed strategies.

Genetic engineering of species that focuses on future conditions that may or not materialize, depending on the surprises or "unknown unknowns" in the climate system, chances losing that species. In the Sahel region of Africa, one of the keys to local population survival in the drought cycles of the 1970s and 1980s was to have conserved natural plant biodiversity. Mortimore (2010) found in one community 121 nondomesticated plant species of economic value, including "famine foods" for which identification and preparation has been taught by women to their daughters for generations. The maintenance of a range of varieties of staple crops (agrodiversity) was critical in sustaining populations dependent on uncertain rainfall, a "resilient but unstable system" (Holling 1973). The existence of rare varieties in the Sahel depends entirely on elderly farmers who continue breeding them despite the wide spread diffusion of "miracle" varieties developed on research stations (Mortimore 2010). The reliance of a handful of genetically modified staple crops for our global economy is a sign of the fragility of our agricultural system if superbugs or superpathogens were to emerge and decimate the fields. Similarly, in natural systems, the reliance on "better adapted" species overestimates the accuracy of current projections.

Allow migration by creating working landscapes

"The Starbucks approach" (Dunwiddie et al. 2009) ensures that suitable habitats are always within easy reach of one another. This conservation approach provides linkages, corridors, or other mechanisms to facilitate the movement of species as they respond to climate changes. Just as Starbucks intends to keep coffee drinkers within easy range of their next cup, conservation practitioners want to ensure that migration routes remain functional and that organisms can escape to better habitats whenever possible. Approximately 650,000 ha previously converted to agriculture are currently enrolled in the Conservation Reserve Program (CRP) in eastern Washington (Farm Service Agency 2009) to protect and restore native sagebrush. Ensuring that CRP lands remain in permanent vegetation cover is an important strategy to conserve biodiversity in the Moses Coulee Conservation Area in north-central Washington (Schroeder and Vander Haegen 2006). In the past, the focus has been on CRP fields adjacent to shrub-steppe systems so as to help the ecosystem expand into recovering land areas. But potential climate change impacts cause parcels providing connections between native shrub-steppe patches, even if surrounded by cropland, to become increasingly important as dispersal corridors. It is essential that conservation efforts involve private landowners to develop "working" landscapes, including pastures, farmland, and rural residential properties. Although dedicated reserves will always be central to conserving ecosystems and biodiversity, viable (even if transitional) habitats may exist, be restored, or be created on private property. It will be difficult, if not impossible, to maintain ecosystems and associated species at the landscape scale if private landowners do not engage in the effort. Such efforts become particularly important as human populations migrate on the landscape in response to climate change impacts on sources of food, water, and energy. As Margules and Pressey (2000) noted more than a decade ago, "the realization of conservation goals requires strategies for managing whole landscapes including areas allocated to both production and protection," which is the very definition of a working landscape.

Develop new expectations for future communities

Invasive species are becoming widely distributed in many countries. They represent 10%–20% of any US state's flora. On islands around the world, floras can include more than 50% invasives. Tom Stohlgren (US Geological Survey, personal communication) talks about the "tradecene," part of Paul Crutzen's anthropocene, as a period of homogenization. Thanks to ships, planes, trains, cars, and trucks, one can find the same species anywhere one goes. Cheat grass is a well-known example. Native to Eurasia and accidentally introduced in the United States in the mid-1880s, it is now present throughout North America, and is a widespread problem species in arid regions across much of the western United States. One option is to start developing the landscapes of the future, preserves or otherwise, by integrating invasives that are here to stay. Rather than ignoring their ubiquitous presence and their resilience to eradication techniques, they should be located and mapped, and their future location should be projected so that land-use decisions can take them into account. Just like humans, invasive species are active participants in current ecosystems. Today's ecosystems include them, and the no-analog communities of Williams and Jackson (2007) will certainly do so.

Conclusions

Conservation planning has developed as the world has been changing. Climate change often is associated with negative impacts and rarely is considered as a set of opportunities that land managers could exploit. Today's projections of future climate change actually give valuable early warnings of likely challenges ahead. Although large uncertainties are associated with climate model results, there is consensus that our planet is getting warmer, precipitation is increasing near the poles, and weather extremes are becoming more frequent. Every day brings new information about the various changes occurring around the planet. Although the challenges are numerous, positive results should also be discussed: complex terrain may buffer regional climate variation (cold air drainage), behavioral (alpine mammals) or morphological (birds) adaptations will allow certain species to thrive, and valuable ecosystem services (e.g., more crops in Canada) will become available in new areas.

The impacts of climate change on ecosystem processes and species survival are difficult to predict because (1) the rate of climate change is uncertain and depends on human choices, (2) the timing and magnitude of extreme weather events that might completely transform landscapes are unpredictable, (3) the complexity of biotic interactions that might allow acclimation, adaptation, or mitigation are mostly unknown, and (4) physiological and ecological thresholds continue to be explored and should be monitored more extensively. Although technology has helped shape our environment to meet our needs, it might behoove us to revisit what has caused problems (pollution, invasives, system homogenization, and landscape fragmentation) in the past. Our partial understanding of the natural world should make us pause before basing adaptation strategies on new technology. Taleb reminds us that we should develop our respect for time and nondemonstrative knowledge, which means we should first observe change and learn about processes without interfering, thus likely avoiding worse surprises.

Basic principles that have been used in the past to deal with various disturbances continue to apply. I have described a few in this chapter: promoting redundancy as insurance against localized disturbance, restoring heterogeneity to allow a wide range of responses to stress, focusing on function rather than composition to prevent the loss of valuable ecosystem services for our society, allowing movement on the landscape by creating working landscapes that include human land-use, reducing avoidable stress that limits systems' or species' natural response to disturbance, and developing new expectations for future communities as climate and human activities result in new species assemblages. Common sense might be more valuable than stress-testing and scenario analyses when the future is uncertain and the boundaries of the extent of change are unknown. Chronic gradual change, although more easily understandable and predictable, is less likely to have drastic consequences than abrupt extreme events resulting directly or indirectly from changes in climate. Maintaining flexibility of response in natural and managed systems, and the flexibility to modify expectations and tolerate change in the minds of conservation practitioners (West et al. 2009), should be the most important objective of integrating climate change trends into conservation planning.

References

Ackerly, D. D. 2003. "Community Assembly, Niche Conservatism and Adaptive Evolution in Changing Environments." *International Journal of Plant Sciences* 164: S165–S184.

Agee, J. K. 1993. "Fire Ecology of Pacific Northwest Forests." Covelo, CA: Island Press.

Allen, C. D., A. K. Macalady, H. Chenchouni, D. Bachelet, N. McDowell, M. Vennetier, T. Kitzberger, A. Rigling, D. D. Breshears, E. H. Hogg, P. Gonzalez, R. Fensham, Z. Zhang, J. Castro, N. Demidova, J-H. Lim, G. Allard, S.W. Running, A. Semerci, and N. Cobb. 2010. "A Global Overview of Drought and Heat-Induced Tree Mortality Reveals Emerging Climate Change Risks for Forests." *Forest Ecology and Management* 259: 660–84.

Araújo, M. B., and C. Rahbek. 2006. "How Does Climate Change Affect Biodiversity?" *Science* 313: 1,396–97.

Aschmann, H. 1973. "Distribution and Peculiarity of Mediterranean Ecosystems." In *Mediterranean Type Ecosystems: Origin and Structure,* edited by F. Di Castri and H. A. Mooney, 11–20. Berlin and New York: Springer-Verlag.

Ashcroft, M. B. 2010. "Identifying Refugia from Climate Change." *Journal of Biogeography* 37: 1,407–13.

Ayres, M. P., and M. J. Lombardero. 2000. "Assessing the Consequences of Global Change for Forest Disturbance from Herbivores and Pathogens." *The Science of the Total Environment* 262: 263–86.

Baker, B. B., and R. K. Moseley. 2007. "Advancing Treeline and Retreating Glaciers: Implications for Conservation in Yunnan, P. R. China." *Arctic, Antarctic, and Alpine Research* 39: 200–9.

Beaubien, E., and H. Freeland. 2000. "Spring Phenology Trends in Alberta, Canada: Links to Ocean Temperature." *International Journal of Biometeorology* 44: 53–59.

Booth, C., S. Bouwhuis, C. M. Lessells, and M. E. Visser. 2006. "Climate Change and Population Declines in a Long-Distance Migratory Bird." *Nature* 441: 81–83.

Breshears, D. D., O. B. Myers, C. W. Meyer, F. J. Barnes, C. B. Zhou, C. D. Allen, N. G. McDowell, and W. T. Pockman. 2009. "Tree Die-Off in Response to Global Change-Type Drought: Mortality Insights from a Decade of Plant Water Potential Measurements." *Frontiers in Ecology and Environment.* 7: 185–89.

Carroll, A. L., S. W. Taylor, J. Regniere, and L. Safranyik. 2004. In *Mountain Pine Beetle Symposium: Challenges and Solutions,* Report number BC-X-399, edited by T. L. Shore, J. E. Brooks, and J. E. Stone, 223–32. Victoria, BC: Natural Resources Canada, Canadian Forest Service.

Chen, B., N. C. Coops, T. A. Black, R. Jassal, J. M. Chen, and M. S. Johnson. 2011. "Modeling to Discern Nitrogen Fertilization Impacts on Carbon Sequestration in a Pacific Northwest Douglas-Fir Forest in the First-Post Fertilization Year." *Global Change Biology* 17: 1,442–60, doi:10.1111/j.1365-2486.2010.02298.x.

Ciais, P., M. Reichstein, N. Viovy, A. Granier, J. Ogée, V. Allard, M. Aubinet, N. Buchmann, C. Bernhofer, A. Carrara, F. Chevallier, N. de Noblet, A. D. Friend, P. Friedlingstein, T. Grünwald, B. Heinesch, P. Keronen, A. Knohl, G. Krinner, D. Loustau, G. Manca, G. Matteucci, F. Miglietta, J. M. Ourcival, D. Papale, K. Pilegaard, S. Rambal, G. Seufert, J. F. Soussana, M. J. Sanz, E. D. Schulze, T. Vesala, and R. Valentini. 2005. "Europe-wide Reduction in Primary Productivity Caused by the Heat and Drought in 2003." *Nature* 437: 529–33.

Conover, D., and S. Pelikan. 2011. "Earlier Flowering in a Restored Wetland–Prairie Correlated with Warmer Temperatures (Ohio)." *Ecological Restoration* 28 (4): 428–30.

Copeland, J. P., K. S. McKelvey, K. B. Aubry, A. Landa, J. Persson, R. M. Inman, J. Krebs, E. Lofroth, H. Golden, J. R. Squires, A. Magoun, M. K. Schwartz, J. Wilmot, C. L. Copeland, R. E. Yates, I. Kojola, and R. May. 2010. "The Bioclimatic Envelope of the Wolverine (*Gulo gulo*): Do Climatic Constraints Limit its Geographic Distribution?" *Canadian Journal of Zoology* 88: 233–46.

Cox, P., and D. Stephenson. 2007. "A Changing Climate for Prediction." *Science* 317: 207–8.

Cross, M. S., E. S. Zavaleta, D. Bachelet, M. L. Brooks, C. A. F. Enquist, E. Fleishman, L. Graumlich, C. R. Groves, L. Hannah, L. Hansen, G. Hayward, M. Koopman, J. J. Lawler, J. Malcolm, J. Nordgren, B. Petersen, D. Scott, S. L. Shafer, M. R. Shaw, and G. M. Tabor. 2012. "The Adaptation for Conservation Targets (ACT) Framework: A Tool for Incorporating Climate Change into Natural Resource Management." *Environmental Management* doi 10.1007/s00267-012-9893-7.

Daly, C., M. Halbleib, J. I. Smith, W. P. Gibson, M. K. Doggett, G. H. Taylor, J. Curtis, and P. A. Pasteris. 2008. "Physiographically-Sensitive Mapping of Temperature and Precipitation across the Conterminous United States." *International Journal of Climatology* 28: 2,031–64.

Dunwiddie, P. W., S. A. Hall, M. W. Ingraham, J. D. Bakker, K. S. Nelson, R. Fuller, and E. Gray. 2009. "Rethinking Conservation Practice in Light of Climate Change." *Ecological Restoration* 27 (3): 320–29, doi:10.3368/er.27.3.320.

Elith J., C. Graham, R. Anderson, M. Dudik, S. Ferrier, A. Guisan, R. Hijmans, F. Huettmann, J. Leathwick, A. Lehmann, J. Li, L. Lohmann, B. A. Loiselle, G. Manion, C. Moritz, M. Nakamura, Y. Nakazawa, J. M. C. M. Overton, A. T. Peterson, S. Phillips, K. Richardson, R. Scachetti-Pereira, R. Schapire, J. Soberón, S. Williams, M. Wisz, and N. Zimmermann. 2006. "Novel Methods Improve Prediction of Species' Distributions from Occurrence Data." *Ecography* 29: 129–51.

Farm Service Agency. 2009. "Summary of Active and Expiring CRP Cropland Acres by County." US Department of Agriculture Report No. MEPEII-R1.

Fazey, I., J. Gamarra, J. Fischer, M. S. Reed, L. C. Stringer, and M. Christie. 2010. "Adaptation Strategies for Reducing Vulnerability to Future Environmental Change." *Frontiers in Ecolology and the Environment* 8 (8): 414–22.

Flannigan, M. D., K. A. Logan, B. D. Amiro, W. R. Skinner, and B. J. Stocks. 2005. "Future Area Burned in Canada." *Climatic Change* 72: 1–16.

Frelich, L. E., and P. B. Reich. 2010. "Will Environmental Changes Reinforce the Impact of Global Warming on the Prairie-Forest Border of Central North America?" *Frontiers in Ecology and the Environment* 8: 371–78.

Gitay, H., S. Brown, W. Easterling, and B. Jallow. 2001. "Ecosystems and Their Goods and Services." In *Climate Change 2001: Impacts, Adaptation and Vulnerability,* edited by J. J. McCarthy, O. F. Canziani, N. A. Leary, D. J. Dokken, and K. S. White, 735–800. Cambridge: Cambridge University Press.

Glick, P., J. Clough, and B. Nunley. 2007. *Sea-Level Rise and Coastal Habitats in the Pacific Northwest: An Analysis for Puget Sound, Southwestern Washington, and Northwestern Oregon.* Reston, VA: National Wildlife Federation.

Griffith, B., J. M. Scott, R. S. Adamcik, D. M. Ashe, B. Czech, R. Fischman, P. Gonzalez, J. J. Lawler, A. D. McGuire, and A. Pidgorna. 2009. "Climate Change Adaptation Options for the U.S. National Wildlife Refuge System." *Environmental Management* 44: 1,043–52.

Hamann, A., and T. L. Wang. 2006. "Potential Effects of Climate Change on Tree Species and Ecosystem Distribution in British Columbia." *Ecology* 87: 2,773–86.

Hannah, L., G. Midgley, S. Andelman, M. Araújo, G. Hughes, E. Martinez-Meyer, R. Pearson, and P. Williams. 2007. "Protected Area Needs in a Changing Climate." *Frontiers in Ecology and the Environment* 5: 131–38.

Hoegh-Guldberg, O., P. J. Mumby, A. J. Hooten, R. S. Steneck, P. Greenfield, E. Gomez, D. R. Harvell, P. F. Sale, A. J. Edwards, K. Caldeira, N. Knowlton, C. M. Eakin, R. Iglesias-Prieto, N. Muthinga, R. H. Bradbury, A. Dubi, and M. E. Hatziolos. 2007. "Coral Reefs under Rapid Climate Change and Ocean Acidification." *Science* 318: 1,737–42.

Holling, C. S. 1973. "Resilience and Stability of Ecological Systems." *Annual Review In Ecological Systems* 4: 1–23.

Intergovernmental Panel for Climate Change. 2007. *Climate Change 2007: The Physical Science Basis. Contribution of Working Group I to the Fourth Assessment Report of Intergovernmental Panel on Climate Change,* edited by S. Solomon, D. Qin, and M. Manning. Cambridge and New York: Cambridge University Press. http://www.ipcc.ch/publications_and_data/ar4/wg1/en/contents.html.

Johnstone, J. A., and T. E. Dawson. 2010. "Climatic Context and Ecological Implications of Summer Fog Decline in the Coast Redwood Region." *Proceedings of the National Academy of Sciences* 107 (10): 4,533–38.

Kitzberger, T., T. W. Swetnam, and T. T. Veblen. 2001. "Inter-hemispheric Synchrony of Forest Fires and the El Niño–Southern Oscillation." *Global Ecology and Biogeography* 10: 315–26.

Klausmeyer, K. R., and M. R. Shaw. 2009. "Climate Change, Habitat Loss, Protected Areas and the Climate Adaptation Potential of Species in Mediterranean Ecosystems Worldwide." *PLoS ONE* 4 (7): e6392, doi:10.1371/journal.pone.0006392.

Knudsen, C. 2003. "The *Thomomys mazama* Pocket Gopher in Washington Prairies: A Contemporary View for Management." Master's thesis, Evergreen State College Archives. QL737.R654 K68 2003.

Kurz, W. A., C. C. Dymond, G. Stinson, G. J. Rampley, E. T. Neilson, A. L. Carroll, T. Ebata, and L. Safranyik. 2008. "Mountain Pine Beetle and Forest Carbon Feedback to Climate Change." *Nature* 452: 987–90.

La Sorte, F. A., and W. Jetz. 2012. "Tracking of Climatic Niche Boundaries under Recent Climate Change." *Journal of Animal Ecology,* doi: 10.1111/j.1365-2656.2012.01958.x.

Lenihan, J. M., D. Bachelet, R. P. Neilson, and R. J. Drapek. 2008. "Simulated Response of Conterminous United States Ecosystems to Climate Change at Different Levels of Fire Suppression, CO^2 Emissions Rate, and Growth Response to CO^2." *Global and Planetary Change* 64: 16–25.

Littell, J. S., E. E. O'Neil, D. McKenzie, J. A. Hicke, J. A. Lutz, R. A. Norheim, and M. M. Elsner. 2010. "Forest Ecosystems, Disturbance, and Climatic Change in Washington State, USA." *Climatic Change* 102: 129–58.

Logan, J. A., W. W. Macfarlane, and L. Willcox. 2010. "Whitebark Pine Vulnerability to Climate Change Induced Mountain Pine Beetle Disturbance in the Greater Yellowstone Ecosystem." *Ecological Applications* 20: 895–902.

Logan, J. A., and J. A. Powell. 2001. "Ghost Forests, Global Warming, and the Mountain Pine Beetle." *American Entomologist* 47: 160–73.

Mantua, N., I. M. Tohver, and A. F. Hamlet. 2010. "Climate Change Impacts on Streamflow Extremes and Summertime Stream Temperature and Their Possible Consequences for Freshwater Salmon Habitat in Washington State." *Climatic Change* 102: 187, doi:10.1007/s10584-010-9845-2.

Margules, C. R., and R. L. Pressey. 2000. "Systematic Conservation Planning." *Nature* 405 (6783): 243–53.

McCullough, D. G., R. A. Werner, and D. Neumann. 1998. "Fire and Insects in Northern and Boreal Forest Ecosystems of North America." *Annual Review of Entomology* 43: 107–27.

McKelvey, K. S., E. C. Lofroth, J. P. Copeland, K. B. Aubry, and A. J. Magoun. 2010. "Comments on Brodie and Post: Climate-Driven Declines in Wolverine Populations: Causal Connection or Spurious Correlation?" *Population Ecology* 53: 263–66, doi 10.1007/s10144-010-0242-5.

McLachlan, J. S., J. J. Hellman, and M. W. Schwartz. 2008. "A Framework for Debate of Assisted Migration in an Era of Climate Change." *Conservation Biology* 21: 297–302.

Meehl, G. A., C. Covey, T. Delworth, M. Latif, B. McAvaney, J. F. B. Mitchell, R. J. Stouffer, and K. E. Taylor. 2007. "The WCRP CMIP3 Multi-Model Dataset: A New Era in Climate Change Research." *Bulletin of the American Meteorological Society* 88: 1,383–94.

Millar, C. I., N. L. Stephenson, and S. L. Stephens. 2007. "Climate Change and Forests of the Future: Managing in the Face of Uncertainty." *Ecological Applications* 17 (8): 2,145–51.

Millar, C. I., and R. D. Westfall. 2010. "Distribution and Climatic Relationships of the American Pika (*Ochotona princeps*) in the Sierra Nevada and Western Great Basin, U.S.A.: Periglacial Landforms as Refugia in Warming Climates." *Arctic, Antarctic, and Alpine Research* 42 (1): 76–88.

Milly, P. C. D., K. A. Dunne, and A. V. Vecchia. 2005. "Global Pattern of Trends in Streamflow and Water Availability in a Changing Climate." *Nature* 438: 347–50.

Moritz, C., J. L. Patton, C. J. Conroy, J. L. Parra, G. C. White, and S. R. Beissinger. 2008. "Impact of a Century of Climate Change on Small-Mammal Communities in Yosemite National Park, USA." *Science* 322: 261–64.

Mortimore, M. 2010. "Adapting to Drought in the Sahel: Lessons for Climate Change." *WIREs Climate Change* 1: 134–43.

Murphy, M. S. 2008. "Edaphic Controls over Succession in Former Oak Savanna." Willamette Valley, OR. MS Thesis, University of Oregon, Eugene.

Parmesan, C. 2006. "Ecological and Evolutionary Responses to Recent Climate Change." *Annual Review of Ecology and Systematics* 37: 637–69.

Pearson, R.G., W. Thuiller, M. B. Araújo, E. Martinez-Meyer, L. Brotons, C. McClean, L. Miles, P. Segurado, T. P. Dawson, and D. C. Lees. 2006. "Model-Based Uncertainty in Species Range Prediction." *Journal of Biogeography* 33: 1,704–11.

Pressey, R. L., M. Cabeza, M. E. Watts, R. M. Cowling, and K. A. Wilson. 2005. "Conservation Planning in a Changing World." *TRENDS in Ecology and Evolution* 22 (11): 583–92.

Raffa, K. F., B. H. Aukema, B. J. Bentz, A. L. Carroll, J. A. Hicke, M. G. Turner, and W. H. Romme. 2008. "Cross-Scale Drivers of Natural Disturbance Prone to Anthropogenic Amplification: The Dynamics of Bark Beetle Eruptions." *BioScience* 580: 501–17.

Rahmstorf, S. 2010. "A New View on Sea Level Rise." *Nature Reports Climate Change,* doi:10.1038/climate.2010.29.

Rehfeldt, G. E., N. L. Crookston, M. V. Warwell, and J. S. Evans. 2006. "Empirical Analyses of Plant-Climate Relationships for the Western United States." *International Journal of Plant Sciences* 167 (6): 1,123–50. http://www.fs.fed.us/rm/pubs_other/rmrs_2006_rehfeldt_g001.pdf.

Richardson, D. M., J. J. Hellmann, J. S. McLachlan, D. F. Sax, M. W. Schwartz, P. Gonzalez, E. J. Brennan, A. Camacho, T. L. Root, O. E. Sala, S. H. Schneider, D. M. Ashe, J. Rappaport Clark, R. Early, J. R. Etterson, E. D. Fielder, J. L. Gill, B. A. Minteer, S. Polasky, H. D. Safford, A. R. Thompson, and M. Vellend. 2009. "Multidimensional Evaluation of Managed Relocation." *Proceedings of the National Academy of Sciences of the United States of America* 106: 9,721–24, doi: 10.1073/pnas.0902327106.

Rijnsdorp, A. D., M. A. Peck, G. H. Engelhard, C. Möllmann, and J. K. Pinnegar. 2009. "Resolving the Effect of Climate Change on Fish Populations." *ICES Journal of Marine Science* 66: 1,570–83.

Schroeder, M. A., and W. M. Vander Haegen. 2006. "Use of Conservation Reserve Program Fields by Greater Sage-Grouse and Other Shrub Steppe-Associated Wildlife in Washington State." Final report prepared for USDA Farm Service Agency. Olympia: Washington Department of Fish and Wildlife.

Solomon, S., D. Qin, M. Manning, R. B. Alley, T. Berntsen, N. L. Bindoff, Z. Chen, A. Chidthaisong, J. M. Gregory, G. C. Hegerl, M. Heimann, B. Hewitson, B. J. Hoskins, F. Joos, J. Jouzel, V. Kattsov, U. Lohmann, T. Matsuno, M. Molina, N. Nicholls, J. Overpeck, G. Raga, V. Ramaswamy, J. Ren, M. Rusticucci, R. Somerville, T. F. Stocker, P. Whetton, R. A. Wood, and D. Wratt. 2007. "Technical Summary." In *Climate Change 2007: The Physical Science Basis. Contribution of Working Group I to the Fourth Assessment Report of the Intergovernmental Panel on Climate Change,* edited by S. Solomon, D. Qin, M. Manning, Z. Chen, M. Marquis, K. B. Averyt, M. Tignor, and H. L. Miller. Cambridge: Cambridge University Press.

Stone, R. 2010. "Home, Home Outside the Range?" *Science* 329: 1,592–94.

Taleb, N. N. 2010. *The Black Swan: The Impact of the Highly Improbable.* New York: Random House.

Tebaldi C., K. Hayhoe, J. M. Arblaster, and G. A. Meehl. 2006. "Going to the Extremes: An Intercomparison of Model-Simulated Historical and Future Changes in Extreme Events." *Climatic Change* 79: 185–211.

Van Mantgem, P. J., N. L. Stephenson, J. C. Byrne, L. D. Daniels, J. F. Franklin, P. Z. Fulé, M. E. Harmon, A. J. Larson, J. M. Smith, A. H. Taylor, and T. T. Veblen. 2009. "Widespread Increase of Tree Mortality Rates in the Western United States." *Science* 323: 521–24.

Waring, K. M., D. M. Reboletti, L. A. Mork, C. H. Huang, R. W. Hofstetter, A. M. Garcia, P. Z. Fulé, and T. S. Davis. 2009. "Modeling the Impacts of Two Bark Beetle Species under a Warming Climate in the Southwestern USA: Ecological and Economic Consequences." *Earth & Environmental Science* 44: 824–35.

West, J. M., S. H. Julius, P. Kareiva, C. Enquist, A. E. Johnson, J. J. Lawler, B. Petersen, and E. R. Shaw. 2009. "U.S. Natural Resources and Climate Change: Concepts and Approaches for Management Adaptation." *Environmental Management* 44: 1,001–21.

Westerling, A. L., H. G. Hidalgo, D. R. Cayan, and T. W. Swetnam. 2006. "Warming and Earlier Spring Increase Western U.S. Forest Wildfire Activity." *Science* 313: 940–43.

Wiens, J. A., D. Stralberg, D. Jongsomjit, C. A. Howell, and M. A. Snyder. 2009. "Niches, Models, and Climate Change: Assessing the Assumptions and Uncertainties." *Proceedings of the National Academy of Sciences* 106 (2): 19,729–36.

Williams, J. W., and S. T. Jackson. 2007. "Novel Climates, No-Analog Communities, and Ecological Surprises." *Frontiers in Ecology and the Environment* 5 (9): 475–82, doi:10.1890/070037.

Wilson, K., R. L. Pressey, A. Newton, M. Burgman, H. Possingham, and C. Weston. 2005. "Measuring and Incorporating Vulnerability into Conservation Planning." *Environmental Management* 35 (5): 527–43.

Wolkovich, E.M., B. I. Cook, J. M. Allen, T. M. Crimmins, J. L. Betancourt, S. E. Travers, S. Pau, J. Regetz, T. J. Davies, N. J. B. Kraft, T. R. Ault, K. Bolmgren, S. J. Mazer, G. J. McCabe, B. J. McGill, C. Parmesan, N. Salamin, M. D. Schwartz, and E. E. Cleland. 2012. "Warming Experiments Underpredict Plant Phenological Responses to Climate Change." *Nature*, doi: 10.1038/nature11014.

14

Mapping biological processes to the appropriate spatial modeling tools

Kevin M. Johnston, PhD; David Western, PhD; and Rodney Jackson, PhD

A completed conservation plan represents decisions made by the conservation planner about the most appropriate land use for a given site at a given time. The goal is to maintain or enhance biodiversity, ecosystem function, and ecosystem structure while accommodating human-oriented land uses that may be detrimental to that goal. The primary tools for achieving this balance are models that represent the biological processes that we are trying to maintain. This chapter is designed to provide additional guidance in using those models.

Many spatial tools are available to model biological processes; this book has provided an introduction to some of them. Underlying each tool or series of tools are mathematical or spatial algorithms and myriad assumptions. These spatial modeling tools can be misused inadvertently because the modeler may not be aware of, or have full understanding of, the biological process, the underlying algorithms, or the assumptions associated with the tool. Often, the modeler may also allow the spatial modeling tool to drive the modeling efforts rather than having the biological process lead in the selection and implementation of the tools. A biological process as defined in this chapter is a series of events generally occurring over time between living species interacting with one another and with the landscape. Examples of biological processes include elephants selecting the best locations to live, snow leopards moving through a landscape, trees reacting to a fire burning through an area, and wolverines adapting to changes caused by a warming climate. To model a biological process accurately, the spatial relationships of the biological process must be defined, and then the appropriate tools must be identified, with the appropriate algorithm and assumptions to capture the process.

If the biological process is not clearly defined and matched with the appropriate spatial tools (the algorithm and assumptions), the result will be an incorrect portrayal of the phenomena being modeled. For instance, Cost Path, a common corridor modeling tool, has been used to model ant as well as elephant movement; for migration, seasonal, and daily movement; for defining movement of a species reintroduced into a new area; for movement of species within established territories; and for movement of species into new territories. Clearly, the decision making for the species in each of these situations is not the same; therefore, the same tool with identical algorithms and assumptions cannot be used to capture all of these various biological processes accurately.

This chapter explores some of the most common biological processes relevant to conservation biology, and it discusses the algorithms and assumptions underlying specific tools that can be used to address those processes to identify and avoid common mistakes. This chapter does not provide specific answers, but it raises questions and considerations that should be addressed when modeling biological processes.

Each modeling approach is presented by identifying the biological process, the algorithm to model the process, and the assumptions behind the model. In many cases, the process is used in an example, with a discussion of the scenario behind the example.

Models and their use

Modeling is a process, and generally not an end in itself. A model allows you to explore what you know and what you do not know. Models can possibly provide additional insight into the biological process being addressed, which may lead to better management decisions. In this chapter, when we present the biological process with the algorithm and the underlying assumptions, we are not suggesting the process is necessarily applicable for your situation or that the tools capture the full range of biological interactions. No model captures the entire story. Each modeling tool provides a limited insight into the biological process. By understanding what portion of the biological process the model is capturing, and the limitations and assumptions associated with the modeling tool, you can understand how to refine the analysis using other models, or you can use your knowledge of the biological process to adjust the more important parameters or submodels.

There is rarely a "right" answer in modeling, but there are correct questions to ask. One goal of this chapter is to emphasize the fact that biological processes are very complex. It is our hope that this chapter will provide you with additional insight into modeling biological processes, so that you have a better understanding of what information can be derived from the models so that you can make more informed decisions. All models have strengths and limitations, and it is difficult for any single model to fully capture the complexities of any biological process. Models are almost always useful. In some cases, such as in climate change analysis, models are the only means to estimate the potential effects of predicted climate change on biological processes.

Mapping models to biological processes

Because in this chapter we mainly are modeling biological processes, the model and its results should be related back to the biology with care. For example, you might create a complex model that identifies usage of areas by species. This model identifies the areas with the greatest biological flow, or use over time by the greatest number of species. A model identifying biological flow can be powerful and may suggest that, by protecting the heavily used areas, you will save the greatest number of species. However, that might not necessarily be the case. For example, an escape route for bighorn sheep or a calving area for elk are, by definition, not heavily used, and thus will not be identified as being critical by this model. Without including areas that are crucial for the species, protecting the highest-use areas alone will be useless and can even be harmful for the populations in question.

To further understand this model and how to use the results in decision making, it is useful to know the kind of flow the model captures: seasonal, migration, or yearly flow. If the model captures yearly flow, then lower elevations in certain biological processes will be identified as most critical for the winter habitat of grazing animals that herd together at that time, but the summer range, where the animals are more dispersed, will be deemed less critical. In another example, the highest biological flow for monarch butterflies will be found on their mating grounds, but if the butterflies cannot reach the area due to fragmentation of their migration routes, protecting the mating grounds will be pointless. If this model is applied to a source/sink population, it will evaluate the breeding and rich source habitat as a high-flow area. The sink habitat, where the species is just able to survive, and the population is declining, will be evaluated as low flow. However, in some cases, without both the source and sink habitat, the species cannot survive. There is some evidence that elephants occasionally need a trace element found in marginal habitats that are low-flow areas. Without all the habitats necessary to maintain all its biological requirements, the species may not be able to survive. By understanding this model, you can accentuate the output based on these biological needs to make the results more realistic, which will lead to better decisions.

Understanding what aspect of the biology of a species a model captures provides insight into interpreting model results. For example, if you are given a corridor design of connected habitat, you might first inquire as to what the design is capturing and for which species. Are the corridors used for seasonal movement, annual movement, or longer-term dispersal to allow for genetic movement? How is the core habitat identified: is it a combination of the requirements of several species? Which core habitats should be connected and why? Which species the design is for are important (core and matrix for one species may be matrix and core for another)? What happens in a multispecies corridor network if one core is a stepping stone to reach another core, and the stepping stone is inhospitable for one of the species but not the other? Are the same criteria used to define the core habitat and the cost surfaces for the connectivity model? This is an approach that is often taken, and it assumes that the same habitat is used for living as well as traveling. However, the decision making of a species when it is moving is often much different from when it is foraging or resting in its core habitat, so why would the same surface be used in a connectivity model? This is not to say the design is poor, but it may be based upon incomplete knowledge, or it may be the only design possible with the available data. Not all these questions can be answered. In summary, all models are limited, and it is necessary to understand the limitations to ensure the model is useful.

Based upon our experience, we recommend running as many different models on a problem as possible. Each model provides a different perspective and insight into a biological process. Rarely can the biological process be fully understood, and even if it is, there is even less likelihood that a single model can fully capture it. It is therefore extremely useful for a conservation planner to have broad experience with many types of models. In the rest of this chapter, we discuss the most useful models for conservation planning that we have had the opportunity to use.

Chapter outline

This chapter is divided into six sections. The first section covers suitability models and how to incorporate a time component into the suitability models. The second section discusses the derivation of rules and parameters from field data, which can be used to reclassify and weight values in the model. The third section explores explicit space and time models, focusing on deterministic and stochastic processes. The fourth section discusses the basic decision making behind movement models—particularly, the movement of species. The fifth section explores causality using agent-based modeling (ABM). The final section identifies a variety of common misuses of spatial modeling tools, specifically, density mapping and interpolation that are applied to field data (particularly radio-location data).

Habitat and suitability modeling

Habitat and suitability models identify the most preferred locations for a species in terms of environmental covariates (see Dickson et al., chapter 7). Suitability models are more broadly used and can range from identifying the most favorable locations used by deer, to the best location for a house, to the areas with the highest potential for crime, or to the areas most susceptible to a spreading chemical spill. Habitat suitability index models are one form of suitability models that describe habitat quality. For the purposes of this chapter, "habitat and suitability modeling" refers to any model that is used to identify the most preferred habitats for a species, sometimes leading to the label of "core habitats" (see Church, chapter 9) that may be used when developing a corridor design (see Singleton and McCrae, chapter 10).

The most preferred, or core, habitat, can be derived from expert-based models or from empirical field data. The general suitability model is one of the more commonly used methodologies for habitat modeling. When a suitability model is implemented in the raster environment, each cell is assigned a value based on the preference of the phenomenon (such as winter use preference) to the features contained at the cell location (such as vegetation type). In this section, we explore the most important features of basic suitability models. If data is available, more advanced methods of modeling habitat can be employed as discussed by Dickson et al. (chapter 7). In many cases, especially in remote environments or in developing countries, expert opinion can be used as a substitute for detailed location data.

The basic steps of a suitability model

Whether the reclassification and weight values are derived from expert opinion, empirical data, statistical models, data mining, or any other technique, the basic steps are similar when creating suitability models. However, the various approaches may emphasize (or eliminate) certain steps over others. The general steps common to developing suitability models are the following:

1. *Define the goal.* The goal identifies the overall objective of the phenomenon or biological process being modeled. Every component in the model should relate to this goal (e.g., identify the best habitat for snow leopard).
2. *Establish how to evaluate the model.* The evaluation technique identifies the success of the model. Each output from a model must be evaluated by objective, measurable criteria generally relating back to the overall goal. If the model cannot be evaluated, it has limited usefulness, and another model should be chosen.
3. *Identify the criteria and attributes that are important to the biological process.* The interaction of these criteria and their attributes within the biological process must be clearly understood and defined. Only the criteria and attributes that are relevant to the biological process should be entered into the model (e.g., water is important as a positive component of elephant habitat, and roads are important as a negative component of elk habitat).
4. *Reclassify the features.* There are a variety of numbering systems: ratio (e.g., distance), interval (e.g., pH), ordinal (e.g., 1st, 2nd), and nominal (e.g., phone book). The numbering systems cannot be directly compared. A common scale is generally established to reclassify the attributes of the significant criteria. Each input attribute for each criterion is assigned to the scale—for example, a 1 to 10 scale, from least to most preferred. The more preferred the attribute, the higher the reclassified value that is assigned to the cell location. This reclassification process is performed for each value for each criterion (e.g., slope, aspect, and distance to water). The scaling of the reclassified values is relative to each other within a criterion and between criteria.
5. *Weight the criteria.* Certain criteria are more important than others and must be weighted appropriately before they are combined.
6. *Combine the reclassified values of all criteria.* A variety of methods can be used to combine the different criteria. Because all the values are on a common scale, the easiest method is simply to add them together. The cell locations with the highest values are assumed to contain more features that are more preferred in terms of the biological process for the species in question.
7. *Analyze the results.* The most preferred locations (the locations with the highest values) are identified and verified. Conservation decisions about those locations can then be made.
8. *Validate the results.* The outcome of the model should be ground-truthed if possible. Expert opinion can be used to determine whether the results are logical. The reclassified values and weights should be altered (sensitivity analysis) to determine the robustness of the parameters, and an error analysis should be performed.

A simplified suitability model workflow

The following is a simplified example to demonstrate how the above steps are realized in a suitability model. The goal of the model is to identify the best habitat in a study area for elephants. Three criteria have been identified that are important to the biology of elephants: food, security, and water. A land-use raster dataset is used that identifies the land-use types for each cell location: 1 = fields, 2 = deciduous forest, . . . 7 = single family residential, 8 = multifamily residential, . . . and 13 = urban. Starting with the food criterion, the value for each land-use type is reclassified to a 1 to 10 scale, with 10 being the most preferred class. In this example, land use 1 (fields) will be reclassified to 8 (high value as food); 2 (deciduous forest) to 10 (highest food value); 7 (low-use residential) to 5; 8 (high-use residential) to 3; and 13 (urban) to 1. Values for security will be reclassified in the same way. Areas near roads and villages that are open are given low security values. To reclassify the water criterion, a raster is created identifying the Euclidean distance of each cell location to the closest stream. The closer areas are then reclassified as being most preferred. For example, using 200-meter increments, a cell 0 to 200 meters from streams is reclassified as 10, cells 200–400 meters distant are classified as 9, and so forth.

As a result of the reclassification, three rasters are produced, each identifying the preference for each cell relative to the criteria—food, security, and distance from water (with 10 being the most preferred on each raster). In this particular model, each criterion is considered to be of equal importance, so no weights will be applied to the values. A final value for each cell is then found by taking the sum of the three rasters. The resulting values range from 3 to 30. Those areas that are assigned to 30 are the most preferred areas; they will have the most favorable elephant food, security, and proximity to water. These areas are considered to be the most important to meet the biological needs of the elephant.

Adding additional considerations

More complex suitability models can have many more criteria and many input layers for each criterion. A variety of approaches and techniques can be used to implement the general steps within the suitability modeling process, which include the additive approach identified above, fuzzy logic, value/utility functions, direct scaling, linear transformation, analytical hierarchy process, pair-wise weighting, the trade-off method, simple additive weighting, and ideal point methods. For more information on suitability and habitat modeling, see Malczewski (1999).

The desired result of a suitability model is the identification and ranking of the most important areas of habitat for a species or population. Populations in different areas may utilize slightly different habitats. Preferred habitat can be identified empirically from field observations or telemetry location data. Models using these types of data are referred to as empirically based or data-driven models (see also Dickson et al., chapter 7). The basic assumption is that animals spend their time in preferred locations. One caveat to keep in mind is that animals may move through less desirable habitat and may be observed there. Another caveat is that individuals may vary in their habitat preferences, and the locations observed are only a sample of the habitat used by the entire population. Ideally, observations of all sex and age groups should be included unless the goal is to model the habitat used by a specific demographic.

A wide variety of other knowledge-based and empirically based habitat modeling approaches exist. Because the biological needs of different species vary greatly, the full complement of habitats in any area is required to maintain all the species that exist there. A key problem for conservation planners is to determine where the best of each type of habitat is located and how to preserve it. The most preferred habitat for a zebra is much different than for a reptile living in the same environment. Therefore, it is difficult to identify a single set of core habitat that meets the needs for all species present. To make this problem more tractable, Brock and Atkinson (chapter 6) discuss how focal species are selected so that the full range of habitats are represented.

To illustrate the modeling process, consider the following series of suitability models. The sequence begins with a basic suitability model that captures a single season. Each subsequent model presented adds complexity. First, the reclassification values are defined in alternate ways, and then the dimension of time is added. Time is first introduced by modeling different seasons and then combining the seasons together. The final suitability model attempts to capture a critical event that might influence the survival of the species. Depending on the needs of the species you are modeling, multiple approaches may be necessary. We discuss how the identified algorithm and assumptions for each of the models presented result in limitations to the modeling approach. In certain situations, the limitations may not be important, but in other cases they may be critical. Keep in mind, if the assumptions underlying the algorithm that represents the biological process are not realistic, then this modeling approach is probably not suitable to capture the process of interest. Determining how realistic the assumptions are can often be difficult. This can often be evaluated by sensitivity analysis and should improve with experience.

Model 1: The single-season weighted overlay suitability model

This section explores the biological principles, model algorithm, and model assumptions for a single-season weighted overlay suitability model. The concepts are demonstrated through an example elephant suitability model.

Biological principles addressed by the model

The species is reacting to the environmental attributes at a particular location. The more of the preferred attributes found at a location, the higher the habitat value for the species. Use of this location should confer higher fitness on individuals. Generally, a suitability model emphasizes the environmental conditions and the season that are most stressful to the species. A major assumption is that if the biological needs are met for the most critical season, the biological needs of the other seasons will also be met. This model refines the simplified elephant suitability model workflow discussed earlier and focuses it to a single season.

Model algorithm

Each cell location (when performed in the raster environment) is assigned a value to each attribute (or cell value) for each criterion the species is reacting to. The total preference value represents the relative preference of that species for the attributes at that location for all the criteria. This value is arrived at through the weighting and additive processes identified above. The higher the value, the more preferred the cell location is assumed to be. A number of basic assumptions underlie this model approach.

Assumptions underlying the model

Developing a model requires assumptions to be made about the data, the process being modeled, and the environment in which that process occurs. Almost all assumptions are violated at some time or other. To understand what the results mean, and how to interpret them, the modeler needs to understand which assumptions are not met and how that failure can be taken into account in the final analysis. The basic assumptions underlying a simple suitability model include the following:

- The habitat value (or species preference) at a specified location can be determined from the environmental attributes that exist there.
- All, or the most significant, criteria (e.g., the need for water, staying secure, and food) influencing species preference can be identified so that preference can be determined.
- All criteria that have been identified and included in the model are important to the species and are used when determining preference for each location.
- The biological process being modeled has reached an equilibrium. The process is not transitioning into a more mature or stable state.
- Individuals of the species have had access to all locations in the study area and have selected those identified as the most preferred solely because of the attributes at each location. The individuals are not utilizing those locations for other reasons (e.g., they were closest).
- The greater the presence of good attributes or features, the better, in a weighted sum combination. The attributes within a criterion are identified through specific values (e.g., its distance from a water source, its security level, or its food type). Being close to water, in a secure location, with good food might all be favorable attributes.
- Habitat value or preference rank is the result of an independent evaluation. The evaluation is based solely on the attributes found at a particular location, independent of surrounding locations (with the exception of neighborhood and distance calculations). That is, there is no adjustment to a particular cell location if it is surrounded by highly preferred cells; the preference evaluation is only based on what is at the location.
- The habitat value is not a measure of the fitness of individuals of the species. It does not constitute proof that the species can live there. It is only a relative weighting of the multiple criteria at each location.
- The values assigned to each cell have no history. The reclassification values assigned to each attribute for each criterion, and the attribute values themselves, are lost in each subsequent step. Only the final accumulative score is known; the modeler does not know what criterion values went into creating that final value.
- The habitat values are sensitive to the reclassification values assigned to the attributes at each location for each criterion and to the weights assigned to the criteria itself.
- The model is dependent on data that already exists and is available.
- The value assigned to each attribute in a criterion, and the weight given to the criterion, are often ambiguous. Many times, there are conflicting opinions and research identifying what is preferred by a species and how much it is preferred.
- Often, continuous data, such as slope, distance, and so on, is reclassified using a linear decay function. That is the preference for continuous criteria that decrease by equal intervals (e.g., 0 to 200 is assigned a 10; > 200 to 400, a 9; > 400 to 600, an 8; and so forth).

Not all of these assumptions are true in any given case. Equal access to all locations is assumed for all of the suitability models presented in this chapter: that is, individuals of the species have full knowledge of all locations within the study area, can reach any location at any point in time, and select the most preferred habitat within the area. In many cases, this assumption is violated. For example, the Serengeti wildebeest migrate long distances, and this travel is costly in terms of energy expenditure and other personal resources. To avoid additional costs, individuals may limit movement (other than migration) to find high-quality habitat, thus reducing their choices to habitat near the migration route. Thus, the base assumption of equal access to all cells is violated, and this modeling approach may not adequately identify preferred habitat.

Note that all the models described in the following scenarios are at the early stages of the modeling process. An iterative process, working with domain experts, statistics, and focused field data gathering, will ensue to refine the models to more accurately capture the biological process.

A model scenario: The single-season weighted overlay suitability model

In this example, the Maasai tribal leaders wish to ensure that elephants migrating from highland forest to lowland habitats during both the wet and dry seasons have sufficient favorable habitat to sustain the herds and attract tourists. By identifying the best locations for the elephants, the Maasai can also identify other locations to graze their livestock to avoid grazing conflicts with the elephants. By separating the cattle and elephant herds both spatially and temporally, the Maasai hope to coexist with the elephants (see figure 14.1).

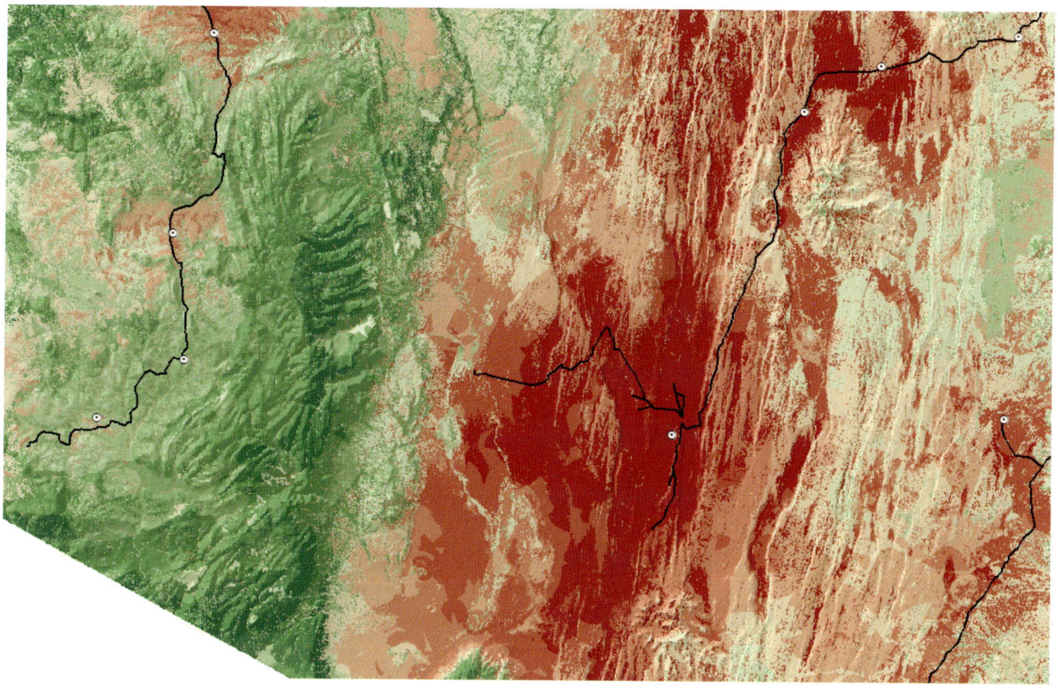

Figure 14.1. A habitat linear suitability or preference map for the forest elephant in Kenya modeled for the wet season. Darker reds indicate least preferred locations. Brighter greens are more preferred. Roads (in black) and towns (white points) are displayed for context. The criteria entered into the suitability model evaluation include land cover, elevation, and slope, as well as distance from water, towns, villages, and roads. Data from the African Conservation Center.

Model 2: Nonlinear suitability models

Expanding the concepts presented in the single-season suitability model described in the previous section, this section discusses how to use nonlinear functions to assign the reclassification values within a criterion to more accurately capture the biology of a species.

Biological principles addressed by the model

In a basic suitability model, most cell values generally are divided into 10 equal intervals (or another convenient number of "bins" for the reclassification scale). This is often the case when using continuous metrics (e.g., distance from water or a road). However, most biological processes of interest are generally nonlinear processes, and transforming them into preference values should be done using nonlinear functions. Often, as animal locations get farther from a feature (e.g., a stream), the preference for the distant locations exponentially declines—with farther locations being much less preferred than closer locations. In certain cases, the preference can be defined by a mathematical function that precisely calculates the reclassification value. For example, the cost of providing water to a species in time of drought can be precisely calculated by the cost of pumping water per cubic meter. That cost function can be used to reclassify the features within a study area. Watering holes farther from the water source will be less preferred in the suitability model because of the greater cost to pump water to them, and therefore, they will be less likely to receive water.

Model algorithm

The basic algorithm is the same as for the linear suitability model discussed above, but nonlinear reclassification functions are applied to the attributes within a criterion (see sidebar "How to reclassify data using nonlinear formulas").

Assumptions of the basic nonlinear suitability model

Assumptions are the same as for the basic suitability model, with one additional assumption: the species preference for the attribute in question is not a linear relationship but varies by some nonlinear function. For example, the preference for cells farther from water exponentially decreases with distance.

A model scenario: The nonlinear suitability model

The model scenario for this nonlinear suitability model is the same as for the basic suitability model, except that the preference for the continuous features is not treated as a linear decay, but rather is assumed to be an exponential decay (see figure 14.3). That is, an elephant has a high preference for areas close to water which decreases rapidly for areas that are farther from water.

How to reclassify data using nonlinear formulas

A nonlinear function can be implemented by applying a precise, known formula to each of the input criteria, using ArcGIS Spatial Analyst Map Algebra.

When the exact mathematical formula is not known, a second approach for applying a nonlinear classification is to change the ranges for each reclassification category. This can be done by typing the desired ranges in the Reclassification table in the Reclassify tool or by clicking the Classify . . . button in the Reclassify tool and manually moving the reclassification bars.

The reclassification images below were applied to reclassify distance to water. Figure 14.2A demonstrates a linear function being applied to the criterion. Notice that there is an equal distance between the ranges (the blue lines) in the histogram; therefore, each of the reclassification ranges will be equal (e.g., 0–816, 816–1633, . . .).

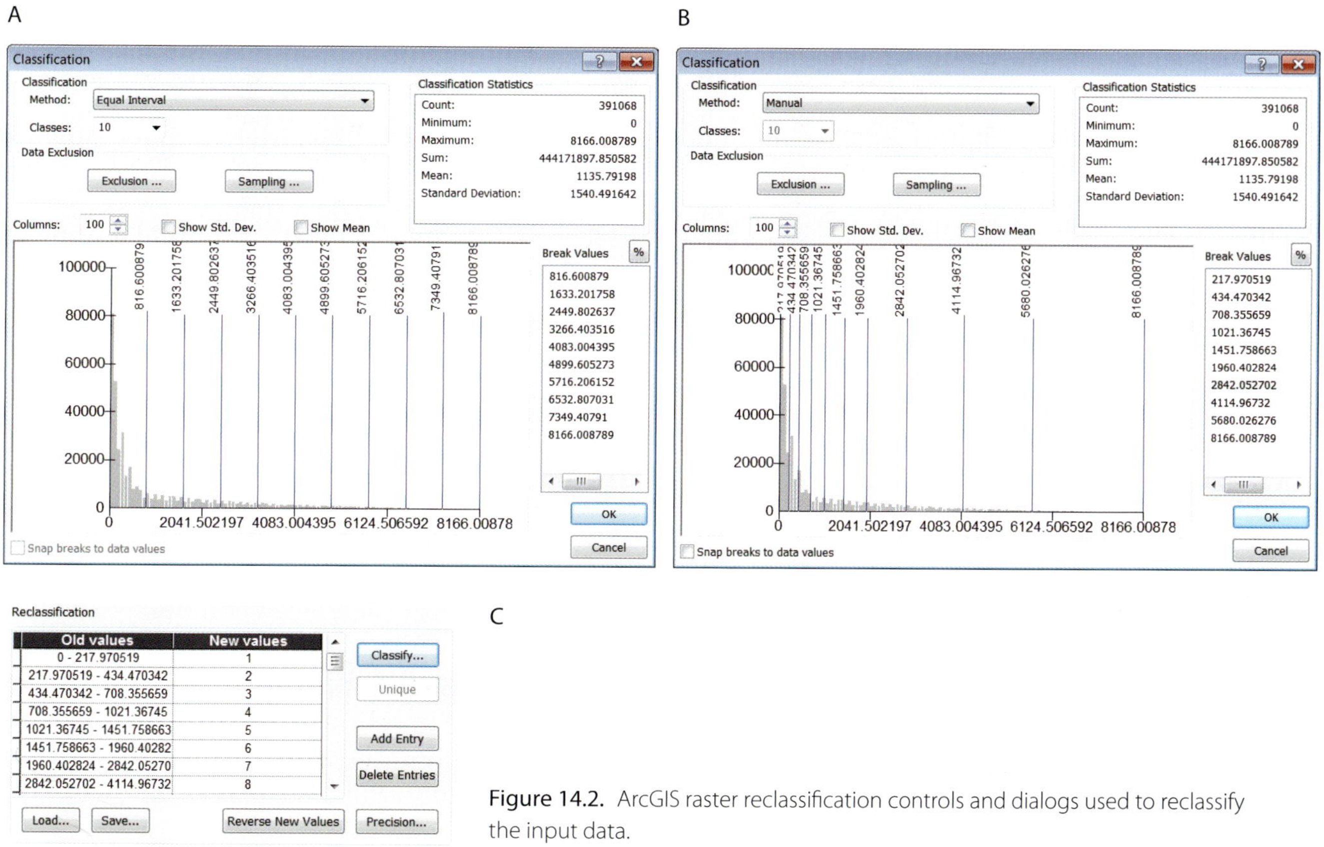

Figure 14.2. ArcGIS raster reclassification controls and dialogs used to reclassify the input data.

Figure 14.2B applies a nonlinear function: exponential decay. Notice that the classes assigned to the closer values are closer together; therefore, more classes are assigned to cells nearer to water. By reversing the reclassification scale (see the "Reverse New Values" button in figure 14.2C), the cell locations that are closer to water are given higher preference than those farther away.

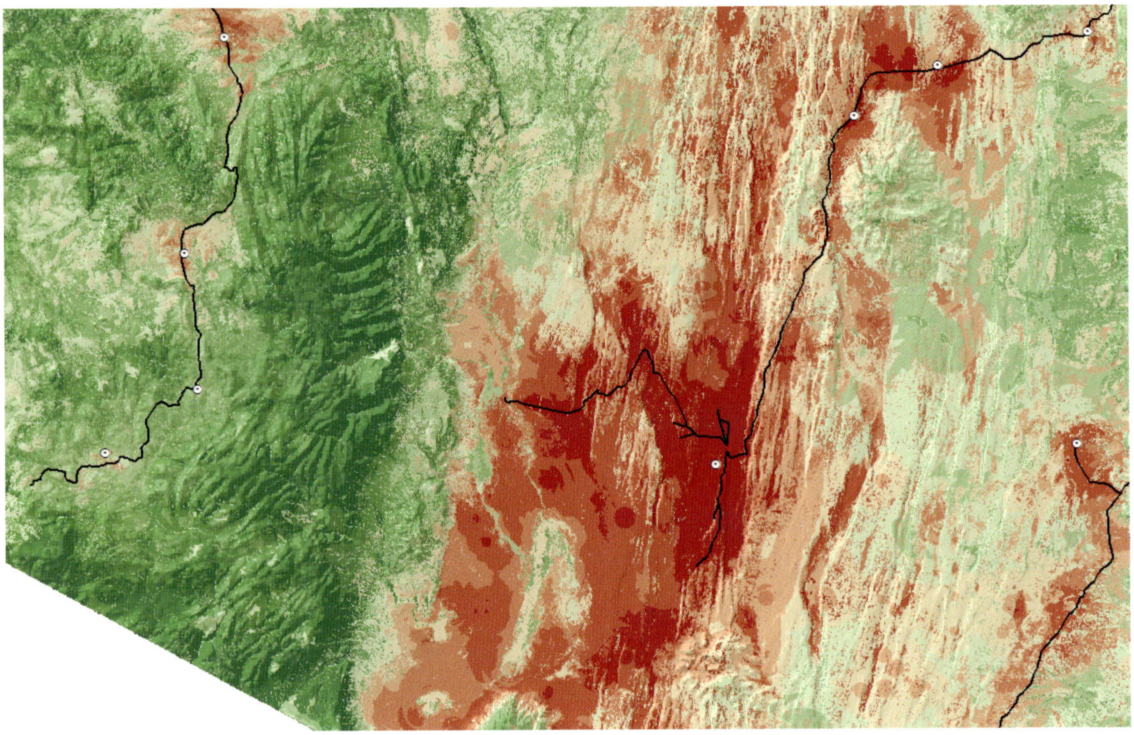

Figure 14.3. A nonlinear habitat suitability or preference map for the forest elephant in Kenya, modeled for the wet season. Darker reds indicate least-preferred locations. Brighter greens are more preferred. Roads (in black) and towns (white points) are displayed for context. Distance from roads, towns, and water have been reclassified using an exponential decay function (the interval distances are not equal). The nonlinear reclassification can better capture the fact that the towns are relatively small and the roads are not heavily used. Notice that the roads and towns do not have as much negative influence on elephants in farther locations as they did in the linear suitability model above. The roads and towns influence is more localized around the features (there is much less red as you move away from the features). At a certain distance, the towns and roads have little, if any effect, on the elephants. The resulting map more accurately captures the biology of the elephant relative to the features. Data from the African Conservation Center.

Model 3: Adding time–seasonal suitability modeling

In the previous two suitability models time was treated as a static snap shot. In this section seasons are introduced into the suitability modeling process.

Biological principles addressed by the model

Temporal elements are added to the model. The biological needs for the species vary uniquely based on the season. An individual's needs for each and every season must be met for the survival of the animal and the population.

Model algorithm

Individual suitability models are created for each season, and the seasonal models are combined to create a suitability model incorporating the full yearly biological needs of the species.

Assumptions underlying the model

Assumptions are the same as for the suitability models, plus two additional assumptions:

- Capturing the most critical season does not capture the full biology of the species. Species need different, specific habitat conditions during each season that may not be captured in modeling for the critical season alone.
- Seasonal suitability models are not necessarily additive. Caution must be taken when adding the different seasonal suitability surfaces together because they can offset each other. If different areas are important in different seasons, there may not be any differentiation in the additive result. Also, areas that are important in one season but unimportant in other seasons may end up with low overall value despite being crucial to the survival of the population.

A model scenario: Incorporating seasonal time with suitability models

This model scenario adds complexity to the two scenario cases above—the unique biological needs for the elephants for each season are incorporated. To capture the full biology of the elephant more accurately, the biological needs for each season are modeled separately (see figure 14.4).

Model 4: Adding connectivity to the suitability model

This section discusses the biological principles, model algorithm, and model assumptions related to a common corridor modeling approach used to maintain connectivity between the best habitat—cost path analysis. An example of a seasonal corridor connectivity network is presented.

Biological principles addressed by the model

The most favorable habitats within each season need to have a functional biological connection through the intervening habitat so that animals can successfully travel between them to meet their biological needs. An example using cost path analysis is used to illustrate this concept. Additional connectivity models are discussed in detail by Singleton and McRae (chapter 10).

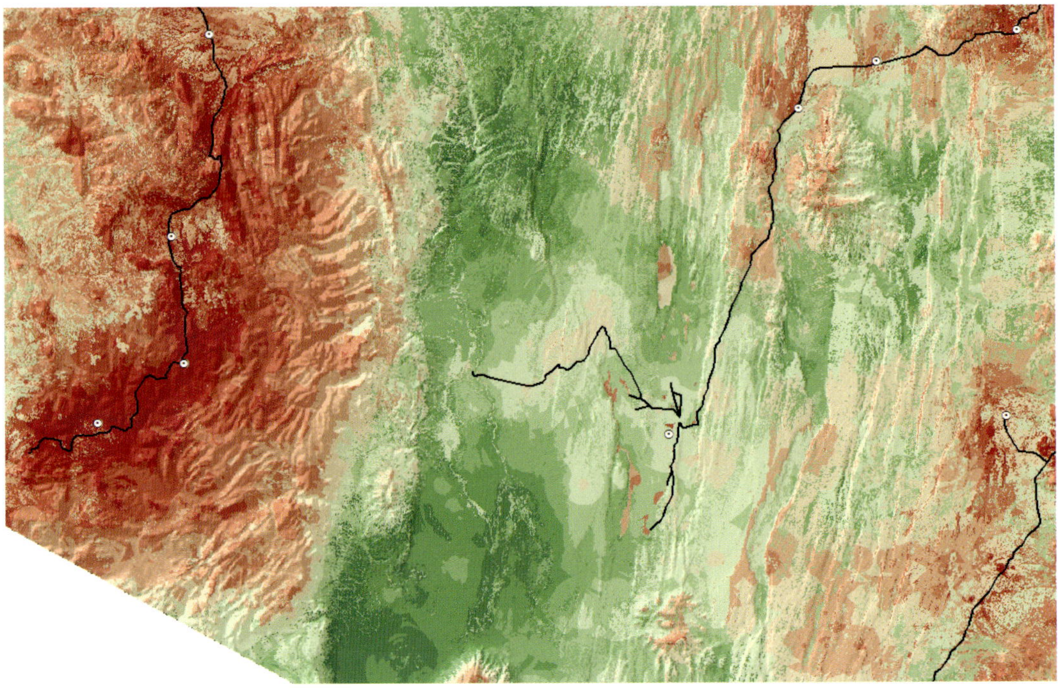

Figure 14.4. A nonlinear habitat suitability or preference map for the forest elephant in Kenya, modeled for the dry season (the wet season was modeled earlier, see figure 14.3). Darker reds indicate least preferred locations. Brighter greens are more preferred. Data from the African Conservation Center.

Model algorithm

Many times, modelers will use the suitability surface (the result of suitability model) as the cost surface for a connectivity model using a cost path analysis (see the sidebar "Performing a cost path analysis" later in this chapter). However, this assumes that the decision making (the biological process) for the species is identical for selecting highly preferred (core) habitats and for moving between cores. This is often a false assumption. When traveling between core habitats, an animal is often more concerned about remaining safe than finding food. To reduce the exposure to risk, an animal will often move much more rapidly than during normal foraging activities. Once again, the key biological criteria must be identified and incorporated into the algorithm to accurately capture the process. The assumptions must be clearly understood so as to adjust model inputs and interpret the results. To model the decision making of the species when it is moving, a separate cost surface should be created that weights each cell according to preference for movement rather than for favorable habitat, and that surface should be used as the cost surface in the cost path analysis.

Assumptions underlying the model

Assumptions are the same as for the suitability models, plus three additional assumptions:

- Animal habitat selection for movement may use different criteria than for foraging or other activities. Various animals may have different preference criteria for movement (e.g., cover may be more important for species that avoid predation through concealment; open habitat may be more important for species that avoid predation by flight).

- The animal knows the desired endpoint of its movement.
- The animal chooses the next cell along the corridor with knowledge of the overall path, rather than simply moving into the closest best habitat.

A model scenario: Identifying the best habitat for connectivity within seasonal preferred habitats

Corridors are needed to allow the elephants to move between core habitats during each season (see figures 14.5 and 14.6).

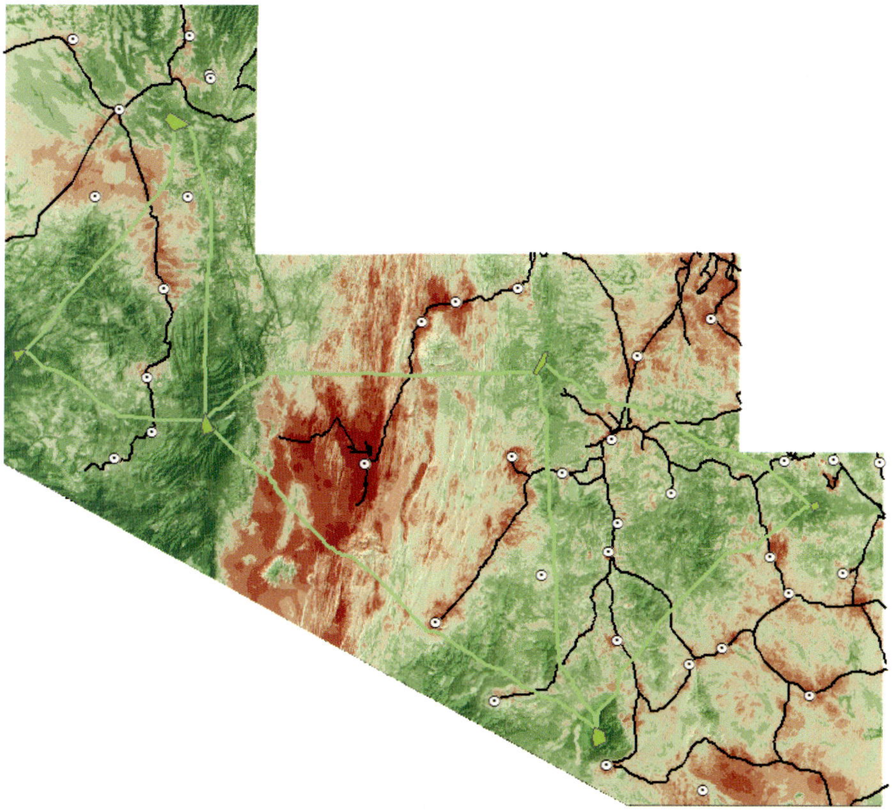

Figure 14.5. A proposed corridor network connecting the most preferred habitats for forest elephant in Kenya for the wet season. The nonlinear suitability surface is displayed for context. The light green polygons are the most preferred core areas for the wet season. The core habitats are connected with corridors (the light green lines) using cost path analysis. These corridors allow the elephants to travel between core habitats in the wet season and should be maintained. Data from the African Conservation Center.

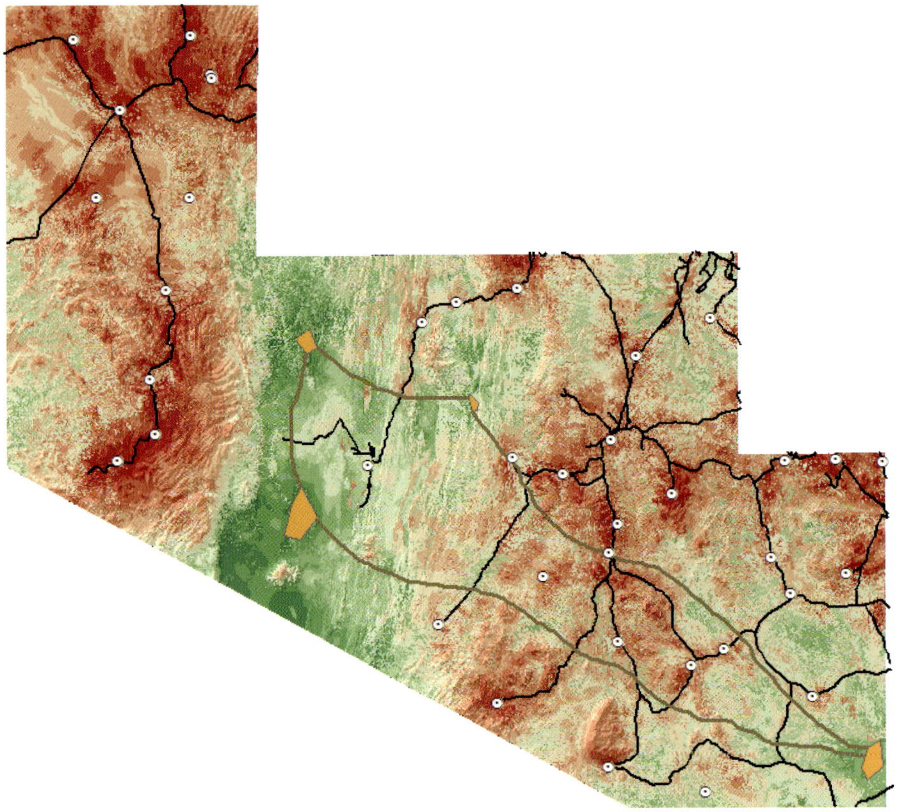

Figure 14.6. A proposed corridor network connecting the most preferred habitats for the dry season. The nonlinear suitability surface is displayed for context. The tan polygons are the most preferred core areas for the dry season. The core habitats are connected with corridors (the light brown lines) using cost path analysis. These corridors allow the elephants to travel between core habitats in the dry season and should be maintained. Data from the African Conservation Center.

This corridor connectivity process identifies the best habitat for movement between the preferred (core) habitats within a season. These are areas where individuals of the species can move most efficiently between the core habitats. The corridor connectivity process may also need to "connect" the habitats between seasons identified by each of the seasonal suitability models (see figure 14.7). This connectivity process may appear obvious in terms of island biogeographic principles, metapopulation concepts, landscape ecology theories, and source/sink concepts (see Singleton and McRae, chapter 10), but from a modeling perspective it raises a series of practical questions: Which patches do you connect? How many corridors are needed? How wide should the corridors be? What species is the corridor designed for? How do you evaluate the quality and success of the corridors? How does the distance between habitat patches affect the corridors?

When addressing which core habitat patches should be connected and how many corridors are needed, we tried to connect the closest patches, and we had at least two corridors going to each core habitat in case one of the corridors is severed.

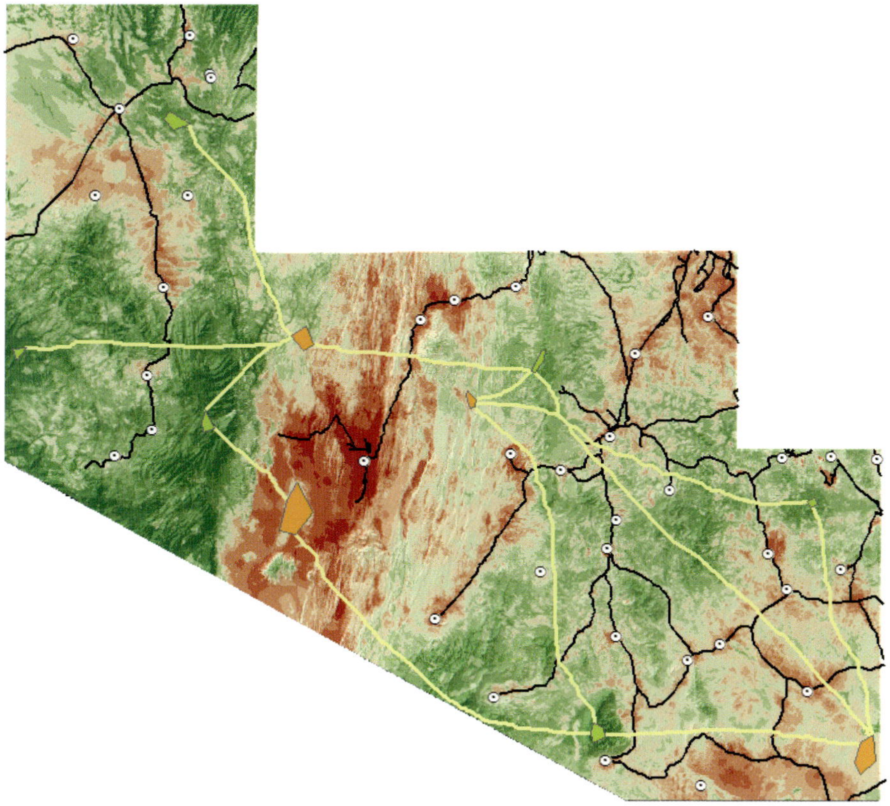

Figure 14.7. Corridors to connect the wet to dry seasons. The core wet season habitat (light green polygons) and the core dry season habitat (tan polygons) are connected with corridors using cost path analysis. Corridors were also created to connect the dry to wet seasons. The resulting corridors between wet to dry and dry to wet generally followed similar paths, but in certain circumstances the wet to dry and dry to wet corridors connecting the same core habitat patches differed considerably. Data from the African Conservation Center.

Model 5: Suitability model for a stress event

Not only is it important that individuals of the species are able to move within seasons and between seasons, but they must also endure specific events. This section discusses the biological principles, model algorithm, and model assumptions when individuals encounter a stress event. An example of a drought in Kenya is presented as a stress event and how the event can be addressed within the suitability modeling framework.

Biological principles addressed by the model

Every so often, a unique stressful event occurs, such as drought, heavy snowfall, or climate change. If a population cannot survive these events, the species will become locally extinct. In designing a conservation plan, it is therefore not only critical to meet the biological seasonal needs of a species, but also to ensure its survival during unique stressful events.

Model algorithm

Suitability models are created for each season and combined as described previously. Additional models are created to identify the areas that are critical for individual survival during stress events. These stress suitability models are added to the seasonal models to capture the full biology of the species.

Assumptions underlying the model

Assumptions are the same as for the suitability models, with one additional assumption: the needs during the unique stress events must be met to ensure survival of the population.

A model scenario: A suitability model capturing a stress event

For elephants, a severe drought is a stress event. The occurrence of a severe drought is unpredictable, and it can devastate a population. Human activity can greatly amplify the impact of a drought on elephants.

In Amboseli National Park in Kenya in the 1970s, elephants stopped migrating out of the park with other herbivores during the wet season due to heavy poaching outside the park boundaries. As a result, elephant densities within the park increased, causing overutilization of foraging habitat, eventual conversion to other habitat types, and depletion of the late season forage in permanent swamp areas. The loss of habitat and forage forced herds out of the late season foraging areas that were secure refuges, reversing the wet season migration pattern. The steady depletion of the late season forage left progressively less of this reserve area with each drought. The swamp pastures became so depleted in the 2009 drought that more than 75% of the wildebeest, zebra, and buffalo died of starvation, and 400 of the 1,600 elephants perished.

During a drought, water and forage persist in distant areas. For the population to survive, elephant corridors must be maintained to allow movement between the various dry season forage and water reserves (see figure 14.8).

All the corridors—within a season and stress events, and between seasons and stress events—need to be combined to create a corridor network to meet the long-term survival of the elephants (see figure 14.9).

Climate change can be considered a special case of a stress event. As the habitat configurations are altered with climate change, the preferred movement pathways will shift temporally (e.g., over a 100-year time frame). Such temporal corridors will become increasingly important for species survival because they allow populations the flexibility to adjust to shifting habitats over decades and centuries.

Figure 14.8. The combined corridors for within the wet and dry seasons and stress events, and between seasons and stress events. The green polygons are the core habitats for the wet season, the tan polygons are the core habitats for the dry season, and the red polygons are the core habitat for the stress events. Certain corridors within a season or stress events and between seasons and stress events can be combined into single corridors that are used for multiple purposes. The light green corridors are used within the wet season, and the light brown corridors are used within the dry season. The red corridors are used during the stress event. The light-yellow corridors are used only between the wet and dry seasons, the brown corridors are used within the dry season and between seasons, and the khaki corridors are used within the wet season and between seasons. The purple corridors are used within the dry and wet seasons as well as between seasons and the stress event. The corridor network will meet the full long-term biology of the elephants. Data from the African Conservation Center.

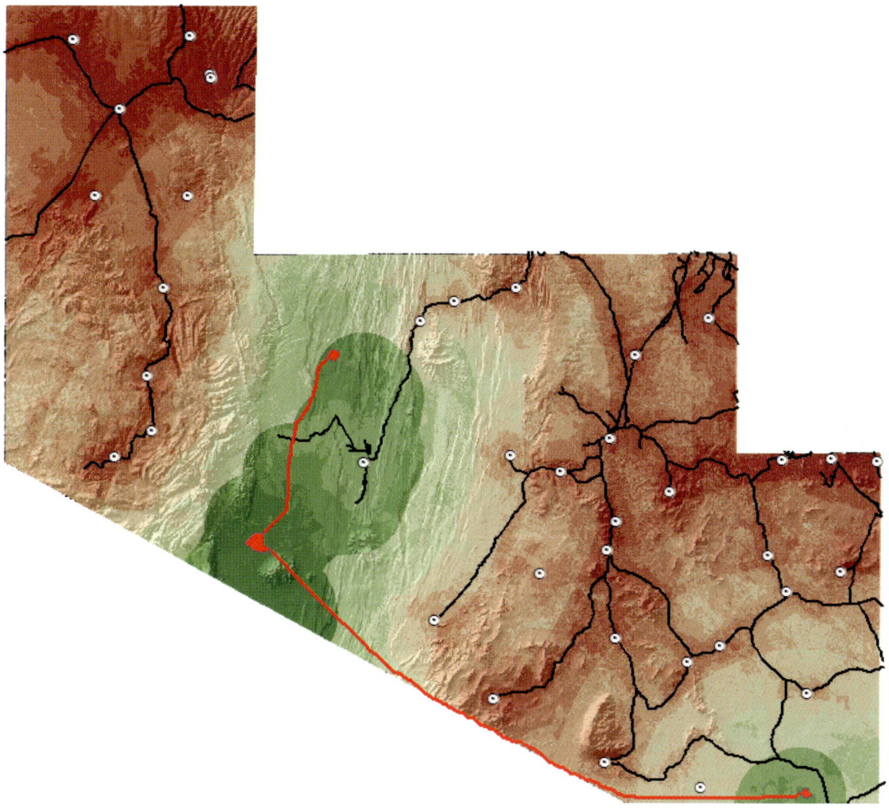

Figure 14.9. A nonlinear habitat suitability or preference map for the forest elephant in Kenya, modeled for an extreme drought event with least-cost movement paths—that is, corridors (red lines). During a severe drought, corridors between water sources must be maintained so that elephants are able to safely move between them. Data from the African Conservation Center.

A model scenario: Snow leopard modeling that integrates the principles presented above

Even though elephants and snow leopards have very different biological requirements and use different decision making, the general suitability-modeling framework can be applied to both species. In this model, we try to understand the biological needs for snow leopards in Sagarmatha Park in Nepal (Mount Everest) and develop a conservation plan based on protecting the highest-quality habitat. The biological process in question, habitat use of snow leopards, is a dynamic process, similar to the elephant examples. The snow leopard can exist between the snow line (the upper bound) and habitat occupied by the common leopard (the lower bound). The lower bound can be identified by the tree line (upper limit of tree growth). The upper bound shifts up and down depending on the season, but the lower bound is stable from year to year; thus, snow leopard habitat becomes compressed during winter. Between the upper bound and the lower bound, the snow leopards must share their possible habitat with people and livestock that also move up and down in elevation seasonally. The people move up the valleys to grazing pastures in the summer and move down the valley in the winter to avoid the cold and snow. Detailed human and livestock population data has been collected for each season and each area (see figure 14.10).

Snow leopards and people move up with the upper bound (snow line) in the summer and move down in the winter. To capture this seasonal movement, four seasonal suitability models were created (winter, spring, summer, and fall). The upper bound for each suitability model varied, based on the season. Upper snow lines were defined for the north- and south-facing aspects for each season. A linear relationship was assumed between all intermediate aspects. Within the zone of possible habitat for each season, snow leopards prefer more rugged terrain for hunting. These areas also offer protection because they are away from human development and less preferred by livestock. Other criteria that went into each suitability model are land use, vegetation, slope, elevation, aspect, distance from human activity (which include towns, villages, hotels, and schools), and distance from roads.

The summer and winter suitability surfaces are displayed below (spring and fall were also created). Notice that the upper bound, the snow line, moves up the valleys in the summer.

The habitat patches within each season are connected (see figures 14.11 and 14.12 for the summer and winter corridor connections). The cost surface for determining movement included cover, ruggedness, and distance from humans.

The four seasonal corridor/patch networks are combined to create an overall yearly connected corridor network to meet the yearly biological needs for the snow leopards (see figure 14.13).

A suitability model for the stress event (a winter with very heavy snow) must be added to the combined season models to allow the snow leopards to survive through years of high stress. During winters with heavy snow, leopards will be pushed close to their lower boundary and in greater proximity to humans and livestock. An opposite type of stress may be caused by climate change—the upper limit may move farther up in elevation rapidly while the lower limit moves upward gradually as the tree line expands in altitude. Eventually, humans may have full access to snow leopard potential habitat all year long, and conflicts will increase.

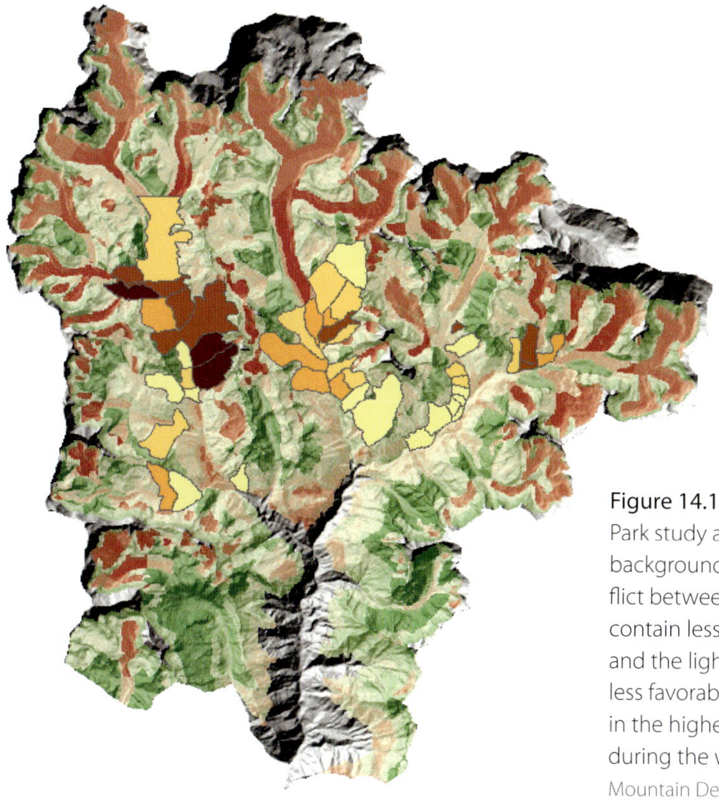

Figure 14.10. Livestock density during summer in the Sagarmatha Park study area, with the summer season suitability map in the background. Higher density of livestock increases the risk of conflict between snow leopards and humans. The dark tan polygons contain less livestock and are more favorable to snow leopards, and the light tan polygons have higher density of livestock and are less favorable. In the summer, there is a higher density of livestock in the higher elevations, and that density shifts down in elevation during the winter. Data from the International Centre for Integrated Mountain Development (ICIMOD).

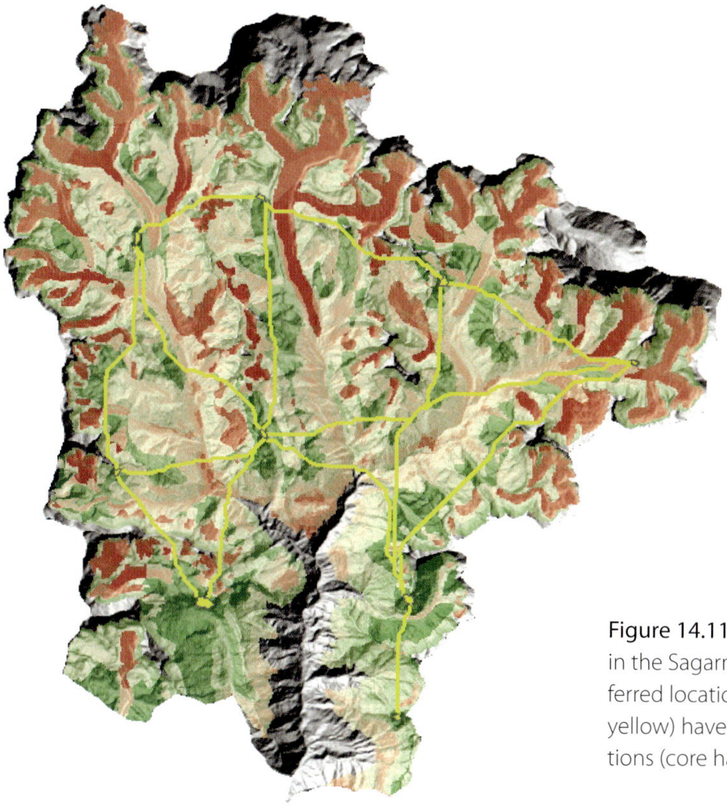

Figure 14.11. The suitability map for snow leopard summer habitat in the Sagarmatha Park study area. Darker reds indicate least preferred locations. Brighter greens are more preferred. Corridors (in yellow) have been identified connecting the most preferred locations (core habitats) within the summer season. Data from ICIMOD.

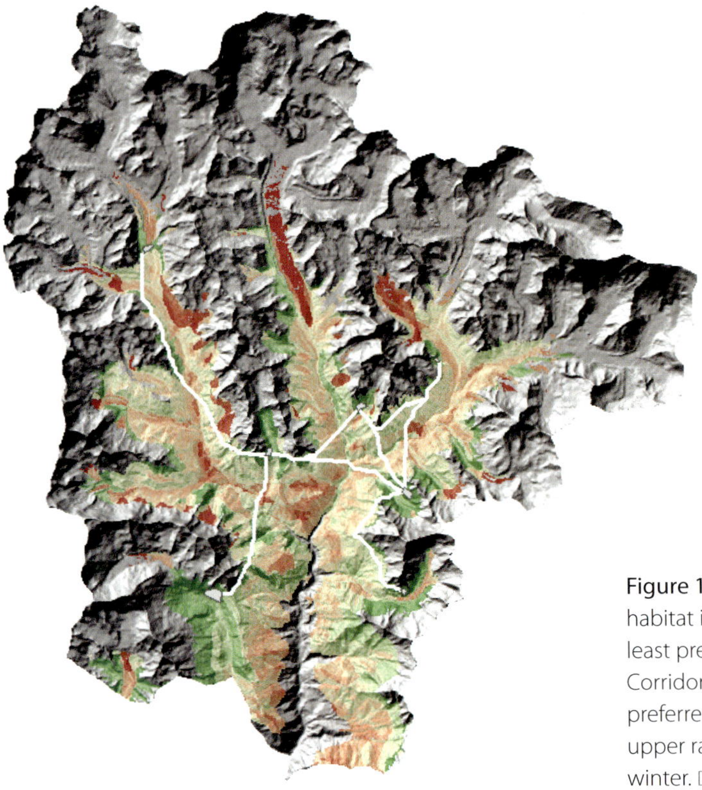

Figure 14.12. The suitability map for snow leopard winter habitat in the Sagarmatha Park study area. Darker reds indicate least preferred locations. Brighter greens are more preferred. Corridors (in white) have been identified connecting the most preferred locations within the winter season. Notice that the upper range for the snow leopard is greatly reduced in the winter. Data from ICIMOD.

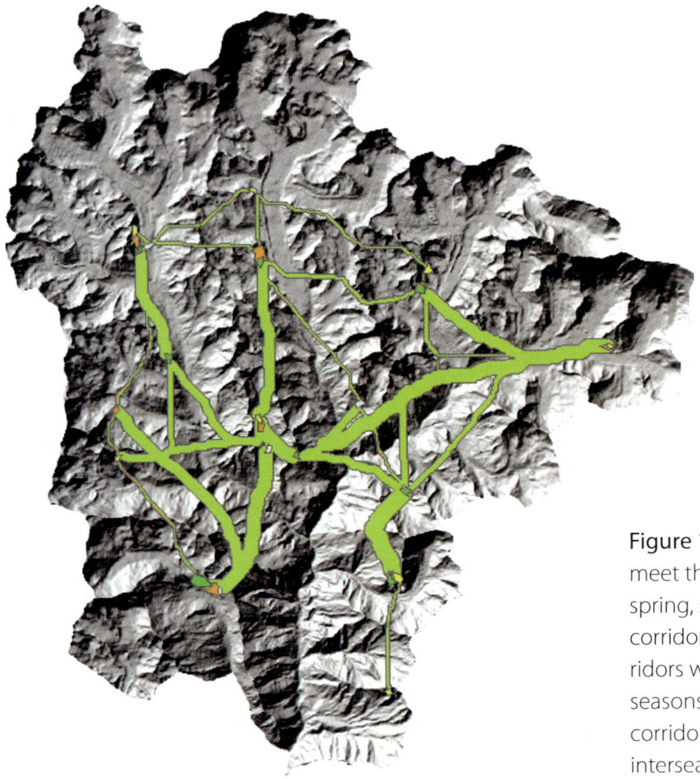

Figure 14.13. The four season combined corridor network to meet the yearly biological needs of the snow leopard. Winter, spring, summer, and fall suitability models were created, with corridors connecting the core habitat within each season. Corridors were also developed to allow for movement between seasons. Many of the corridors overlapped. The width of the corridors vary to encompass several of the seasonal and interseasonal corridors into a single corridor. Data from ICIMOD.

Empirically deriving reclassification and weight value

The previous section discussed how the reclassification and weight values in a suitability model are subjectively established by expert opinion. In this section, we explore spatial modeling tools that can be used to derive empirical reclassification and weight values that are more objectively based on where the species is observed and how much it prefers the locations. The results from these models can be used for defining the reclassification and weight values for a subsequent suitability model for the species being analyzed.

Model 6: Linear regression

In this section, linear regression is used to define the features that are significant to a species and quantifies their importance (weight) to create a preference surface.

Biological principles addressed by the model

A species responds to certain attributes in the landscape. What the species is responding to can be identified. By knowing where a species currently exists, how much the species prefers the locations, and the quantities of the attributes the species prefers or dislikes, the preference for any location in the study area can be determined.

Model algorithm

From sample points identifying locations of where the species is observed and where it is not, and the magnitude of the preference (the dependent variable), linear regression can be used to quantify through coefficients the relationship of the independent variables to the species preference for the locations. Location data is commonly obtained through direct observation; indirect observations such as scat, track, or hair samples; remote camera photos; or telemetry data. This data can be used to develop an occupancy model (see Dickson et al., chapter 7) to create a probability surface for modeling suitability.

Assumptions underlying the model

Assumptions underlying a linear regression model are the following:

- Regression is a classic statistic that requires independent observations from its sample points. If the input is derived from spatial data (in particular from a raster), the sample points generally are spatially autocorrelated and therefore will not meet the independent observation assumption. To reduce the spatial autocorrelation, sample points must be identified (e.g., randomly or systematically [every fifth cell of every fourth row]) and the raster values for each of the dependent and independent variables for those locations are assigned to the points. The points are then passed to the regression analysis.

- In a linear regression, there is a magnitude of preference by the species (the dependent variable) for each sample point.
- The current locations of individuals of the species can define a preference for the attributes at those locations. Once again, equal access to all locations is assumed, and individuals have selected these locations because they prefer them.
- Individuals of the species will move to other areas with similar favorable features.
- The preference for the location can be predicted from an analysis of the attributes at the locations of the sample points.
- The greater the magnitude of preference of the species at the sample point, the more preferred the features are at that location. If the species is not located at a sample point, it is assumed the species does not prefer the attributes at the location.
- All locations with similar attributes will have the same preference values.
- The individuals will first move to the most preferred locations. Having depleted the resources at the most preferred areas, they will move to those areas that are next highest preferred.
- Selection of subsequent preferred areas are chosen independently, with no spatial consideration. The individuals will not move into a neighboring location that is less preferred simply because it is closer.
- The more of the favorable attributes a location has, the more preferred the location will be.
- Individuals of the species have equal access to all locations, and the system is stable, not in transition. It is assumed that the individuals are at their current locations by choice, not because they got to those locations first. Distances from certain features (e.g., distance from water) are surrogates representing the time it takes an individual to move to that feature.

A model scenario: A linear regression identifying the reclassification values for a budworm suitability model

The locations of budworm (an insect that kills trees) in a study area (see figure 14.14) are known by observation. The percent canopy cover damage by the insect can be determined. It is assumed that locations having the highest amount of canopy damage contain attributes that are most preferred by the budworm. Budworm preference values can be predicted for all locations in the study area from the existing locations of the budworms and how much damage they have caused.

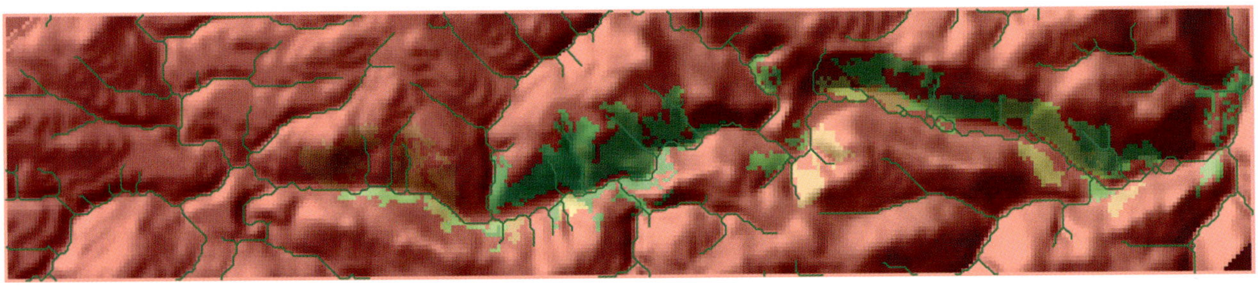

Figure 14.14. The distribution of budworm. The brighter the green, the more damage to the canopy cover by the insect. Figure from Kevin M. Johnston.

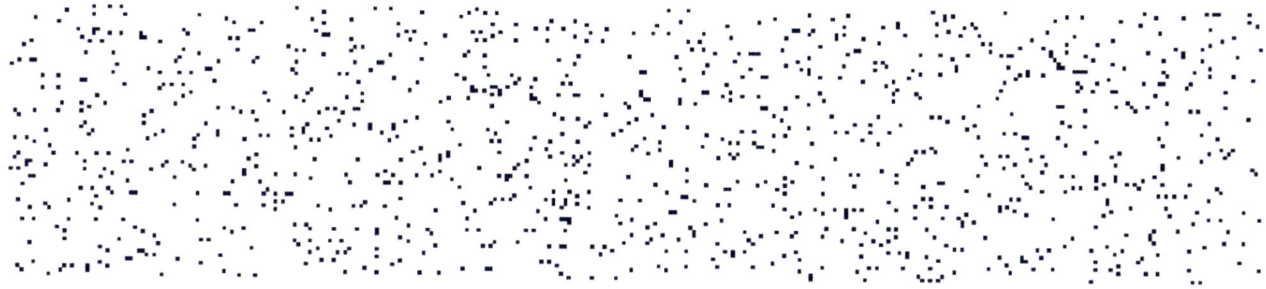

Figure 14.15. The sample points, taken randomly, to which the raster values for each criterion will be assigned. Figure from Kevin M. Johnston.

Regression is a classic statistical method that requires the observations to be independent, but spatial data by nature is spatially autocorrelated. To reduce this spatial bias, sample points are extracted. For each input criterion (raster), the cell value for that sample location is identified. Sampling can be done systematically, such as every fifth cell of every fourth row, or sampling can be done randomly (see figure 14.15).

In this model, a classical linear regression was run on the sample points, and coefficients were calculated. Each coefficient for each criterion was multiplied by the cell value for each attribute. The following is an ArcGIS Spatial Analyst Map Algebra statement used to create an output continuous suitability surface. The layer names for each criterion are used in the statement. Each cell for each criterion (the raster) is multiplied by the appropriate coefficient.

outputSurface = 9.09753 + 3.03041 * spruovma + –0.00049821 * eucfroad + –0.00417 * eucsstream + 1.68426 * pine + 0.91464 * firoverm + 0.07131 * slope + –3.3 * deadsoft + 0.9845 * spruimma + 0.01431 * aspect + 1.39141 * firmatur + –0.00181 * eucpstream

A continuous budworm suitability surface results (see figure 14.16).

A spatial regression analysis was also run. Spatial regression uses spatial autocorrelation in its calculations. That is, not only are the coefficients determined from the attributes of the dependent and independent variables, but also the relative positions of the sample points are accounted for. As a result, not only are the values at the sample locations critical, but also their relative positions in space are important (see figure 14.17). Sample points that contain positive or negative attributes near other sample points that contain positive or negative attributes adjust their coefficients. Notice how the resultant surface is smoother than the results from the classical linear regression (which independently calculates the coefficients for each sample point) because the spatial relationship between the points is taken into consideration.

Figure 14.16. The resulting suitability surface from a classical linear regression analysis. The brighter the green, the more preferred the location should be. Figure from Kevin M. Johnston.

Figure 14.17. The resulting suitability surface from a spatial regression, using the same independent variables as the classical linear regression model. Figure from Kevin M. Johnston.

Model 7: Logistics regression

Unlike the linear regression presented in the previous section, there are situations where the magnitude of the preference where the species is located is not known. In this section a logistics regression is used to create a preference surface for those situations.

Biological principles addressed by the model

The biological principles are the same as for classical linear regression; however, only presence and absence is known for the sample points. That is, we know where the species is (presence) and we know where it is not (absence). The magnitude of the preference of individuals of the species for the attributes at the location is not known. Presence and absence data is commonly misused (see discussion of model 17 below). An observation of an individual at a location is a valid data point, but lack of an observation does not signify that individuals are not present (absence), or that the location is not used.

Model algorithm

The algorithm is the same as above for linear regression, except the logistics regression is applied to presence and absence data to produce a surface of the probability of occurrence.

Assumptions underlying the model

Assumptions are the same as for the classical linear regression, plus the following:

- The samples must be true presence and true absence, not simply that the species has been spotted in some locations and not seen in others. The algorithm evaluates the attributes at the occupied (true presence) locations (attributes it prefers) with the attributes at the true absence locations (attributes it avoids).
- The samples must not be pseudoreplicates. One way to ensure this is that no two samples should be taken within the distance of the coarsest cell size.

- Logistics regression creates a probability surface. The assumption is, the higher the probability, the more preferred the location is to the species.

A model scenario: Analyzing how climate change may affect the distribution of wolverine using logistics regression

The potential effects of climate change on the distribution of an animal species can be predicted from known current locations derived from the existing distribution of the species. A series of independent variables have been identified (the criteria the species is responding to), which include vegetation type as dummy variables (one layer per vegetation type, with 1 being assigned to cells if the vegetation type is present, 0 if not), elevation, slope, aspect, and distance from roads and cities. The Hadley climate model can be used to predict the effects of climate change on vegetation, precipitation, and temperature. The model has two scenarios: a moderate mid-level carbon scenario (A1B), and a higher, more extreme, carbon scenario (A2). Each scenario has three time periods for this century: averaged 2020–2024, averaged 2050–2054, and averaged 2095–2099.

Sample points are created from the raster data to reduce the effects of spatial autocorrelation on the observations, as discussed for the previous model (see figure 14.18).

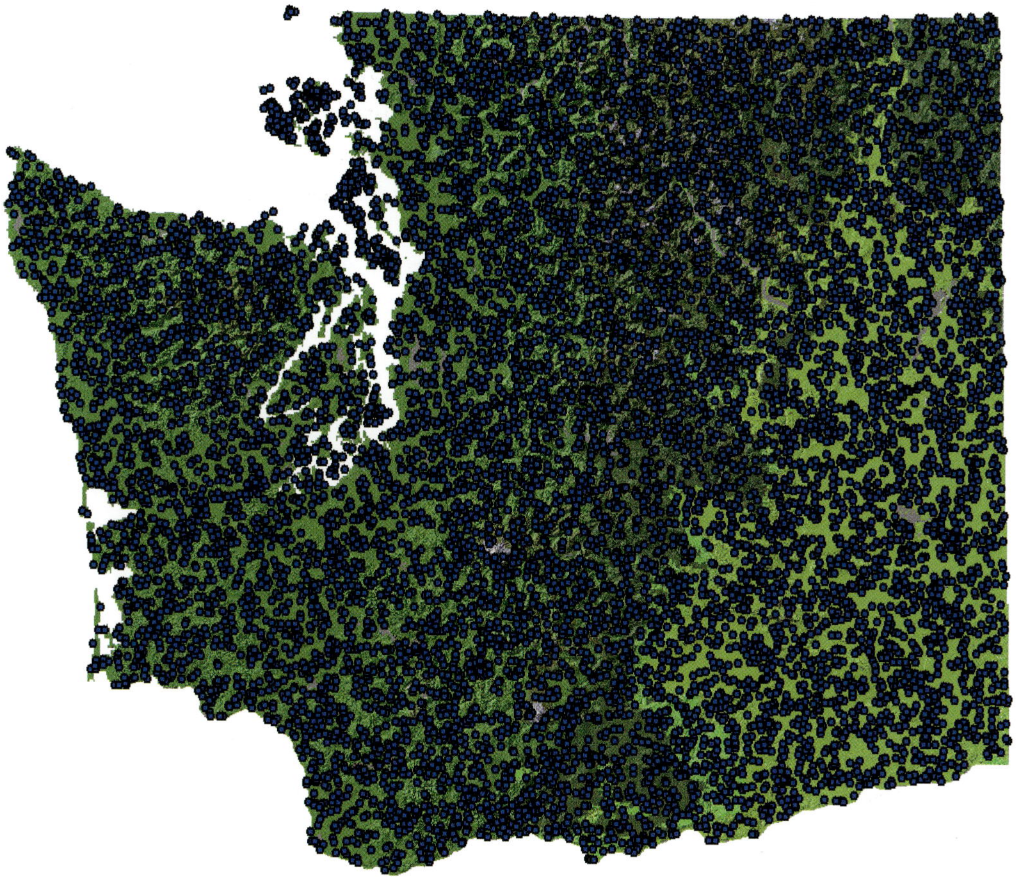

Figure 14.18. The sample points for the logistics regression analysis. Data from US Department of the Interior, Bureau of Land Management.

A prediction surface is created from the coefficients using ArcGIS Map Algebra. The expression for the prediction surface for modeling the current distribution is below (note this is the formula for the logistics regression, which differs from the linear regression discussed above):

outputRaster = 1 / (1 + exp(–1 * (9.595857 + (–1.28212 * tmp1991) + (–0.003687 * ppt1991) + (0.426121 * veg8_10) + (–0.560821 * veg7_10) + (–2.077026 * veg6_10) + (–2.941375 * veg2_10) + (–0.496024 * veg17_10) + (–1.740473 * veg16_10) + (0.557113 * veg12_10) + (–7.103907 * veg10_10) + (0.016223 * slope) + (–0.000674 * elevation) + (–0.000555 * aspect) + (–0.000062 * disthigh) + (0.000049 * distcity))))

This output surface represents the probability of finding a species at any given location. All cells with a probability of 0.5 or greater (having 50% or greater probability of finding the species at the location) are selected first on the current climate conditions to allow the model to predict the current distribution of the species (see figure 14.19). Other probability thresholds can be used to obtain a better fit to the current species distribution, but those probabilities are dependent on a priori information—the known species distribution. Since this a priori information is not known for the future predictions, a 0.5 probability is the most logical threshold to use. When the predicted distribution for the species for the current climate conditions is similar to the actual distribution, the model coefficients can be applied to future predictions caused by climate change with some certainty. Diagnostics statistics such as AIC and spatial comparisons of the actual distribution to the predicted distribution can be used to evaluate the fit of the model.

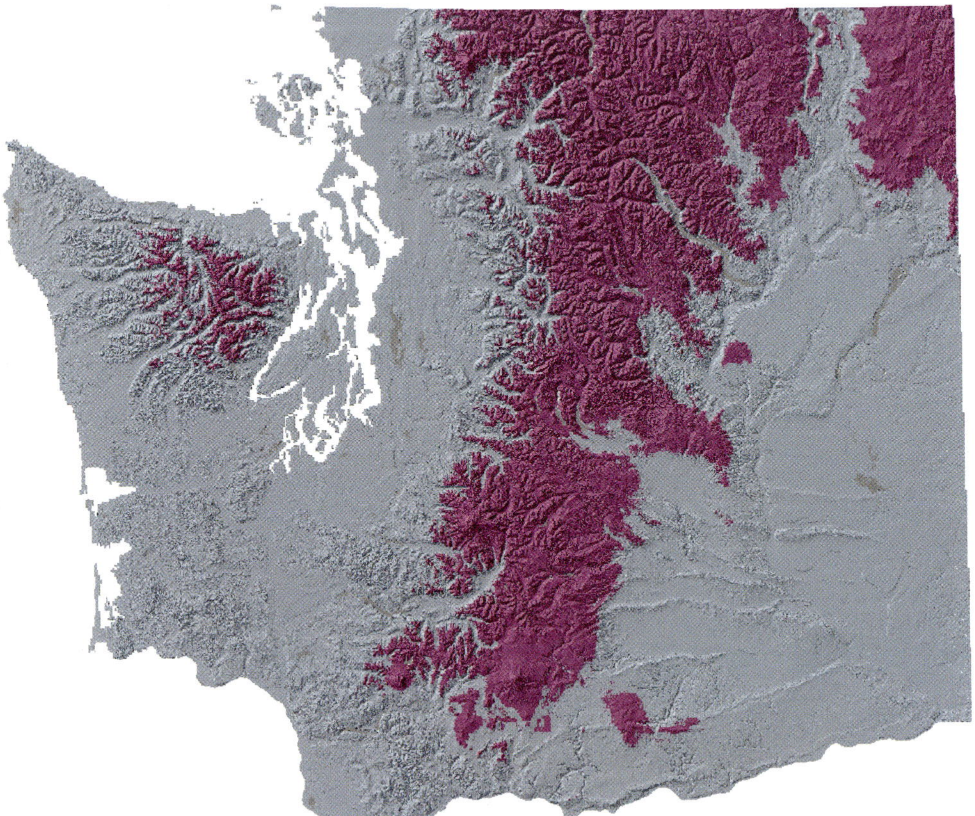

Figure 14.19. The predicted distribution surface developed from the current climate conditions for wolverine. Purple areas identify the predicted current distribution of the wolverine. Data from US Department of the Interior, Bureau of Land Management.

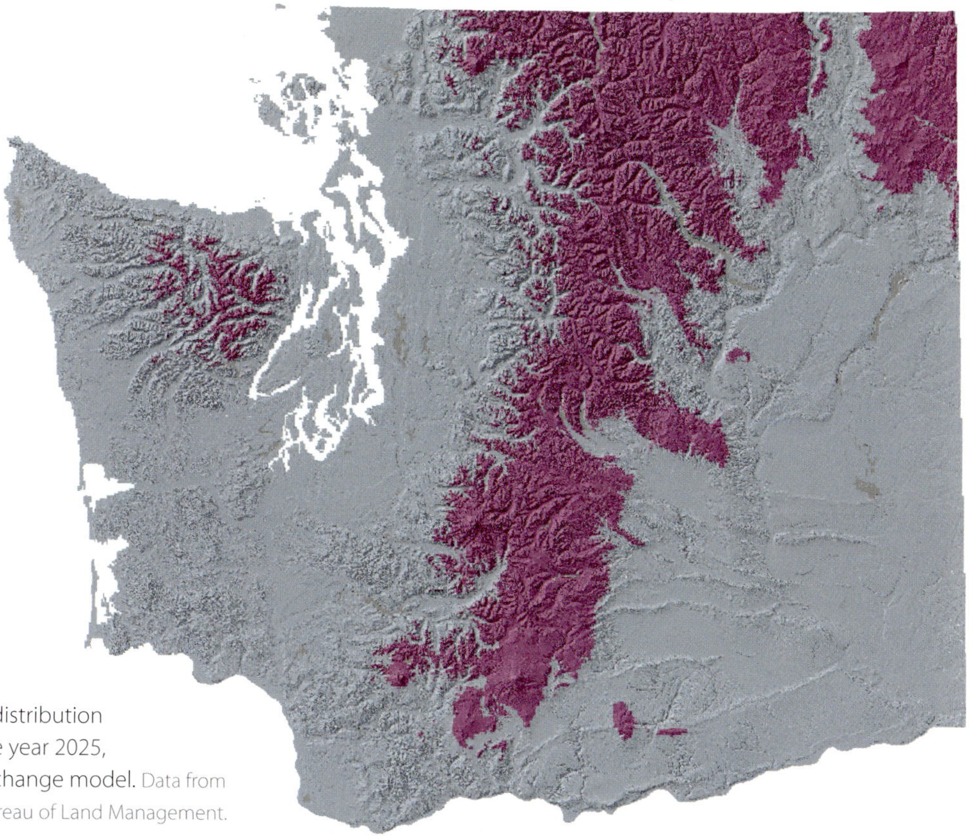

Figure 14.20. The predicted distribution surface of the wolverine in the year 2025, based on the Hadley climate change model. Data from US Department of the Interior, Bureau of Land Management.

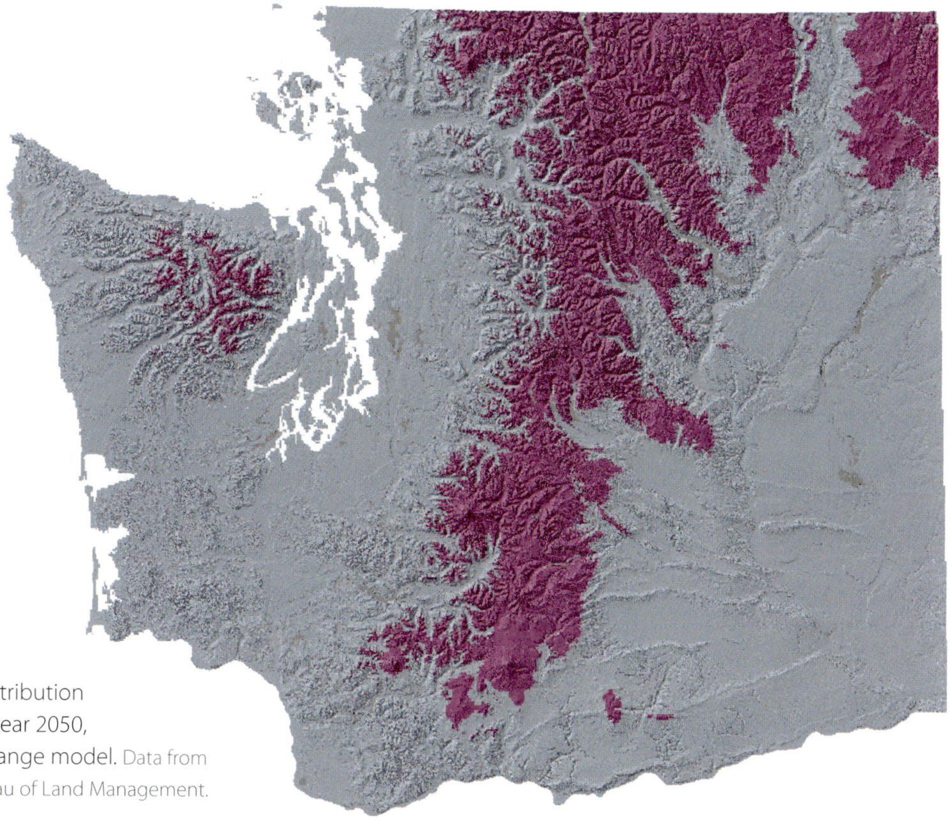

Figure 14.21. The predicted distribution surface of the wolverine in the year 2050, based on the Hadley climate change model. Data from US Department of the Interior, Bureau of Land Management.

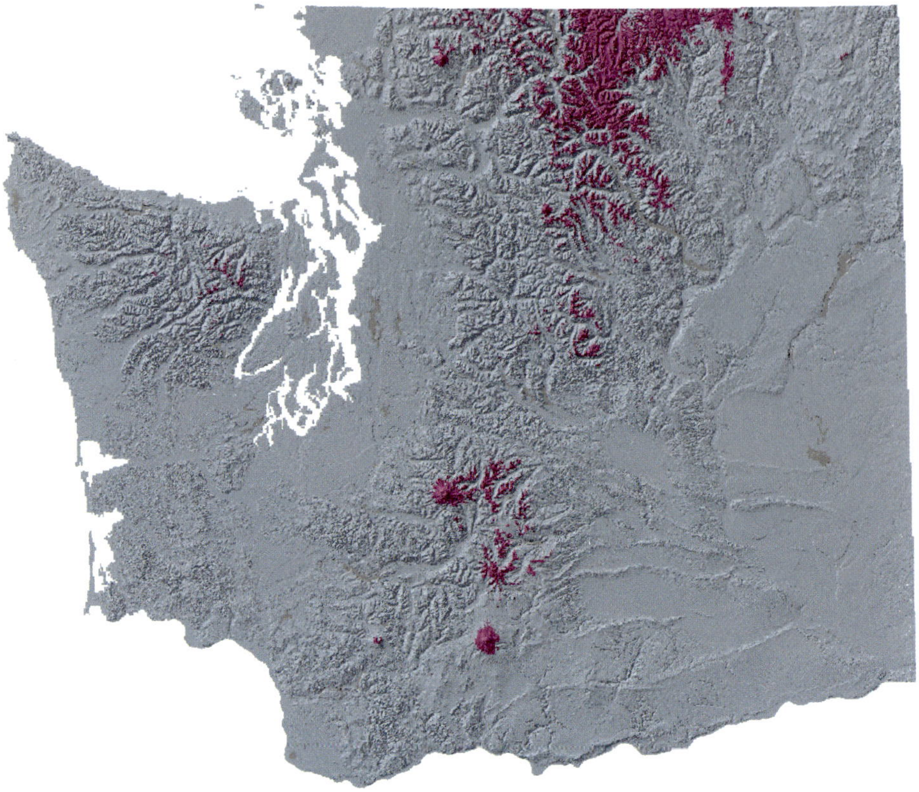

Figure 14.22 The predicted distribution of the wolverine in the year 2100, based on the Hadley climate change model. Data from US Department of the Interior, Bureau of Land Management.

The coefficients are then applied to the significant variables defined by the climate change models to predict how the climate change may affect the future distributions of the species (see figures 14.20, 14.21, and 14.22).

The study identifies several modeling points and questions:

1. How can the model be validated to ensure its accuracy before applying it to future conditions? If the model cannot be proven accurate, how can it be applied to future predictions with confidence? If a cell-to-cell comparison is used for validation, a cell that is said to contain the species may actually not contain it. However, it may be located next to a cell that does contain the species. This may not be as unfavorable a misrepresentation as a cell identified as containing the species that is actually far from cells that currently contain it. An objective quantitative assessment should be established to measure the accuracy of the model.
2. What level of comfort does the modeler have in the climate change models for depicting what will actually occur?
3. The farther the prediction is in the future, the less confidence there is in the output.
4. This particular model uses current species distributions and looks at the attributes found at these locations. The wolverine is not currently found in the Olympic Range (in the northwest section of the state), but they were found there historically. This model identifies the wolverine's historic range because conditions that are favorable for the species still exist in the Olympic Range. However, they were extirpated and have not returned. Is the model wrong?

Explicitly modeling space and time—Dynamic modeling

In the previous sections, the evolution of static descriptive suitability models were described. Seasonal as well as rare events were added to capture more of the full biological requirements of the species. In all these models, time was treated as a static event by modeling it in discrete slices.

In this section, we explore how to explicitly incorporate time into the model to replicate the interactions of biological processes.

Generally, in a dynamic model there is an action (or a series of actions) associated with each phenomenon in each time step. Movement of some sort is usually involved in the action in the time steps. Depending on the movement, the state of the phenomenon changes (for example, a wildfire expands, or an animal eats and becomes less hungry) or the conditions of the landscape can change in the time step (for example, the leading edge of the fire dries out the vegetation, or the animal uses up the forage at a location). Based on the new status of the phenomenon and the new configuration of the landscape, in the subsequent time step, the phenomenon may move again. This process continues for a specified number of time steps.

When developing a dynamic model, the initial location of the phenomenon is identified, the time interval for each step is established, the general movement rules are defined, and the number of time steps to run the model is determined. The output from one time step is the input to the subsequent step. Time and space are explicit in a dynamic model.

Model 8: Deterministic dynamic models

The biological principles, model algorithm, and model assumptions for a deterministic dynamic model are discussed in this section. An example of a fire growth demonstrates the underlying concepts.

Biological principles addressed by the model

The biological process moves explicitly in time and space. All interactions associated with the biological process are understood and can be determined and defined.

Model algorithm

The input starting conditions and locations of the phenomenon are entered into a series of defined rules. Based on the rules, the phenomenon may change its state and its location and may also change the conditions of the landscape and other phenomena that it encounters. The new state and location of the phenomenon and the state and locations of other phenomena are input into the defined rule set.

Assumptions underlying the model

The assumptions underlying a deterministic dynamic model are the following:

- The precise outcome can be predicted because full knowledge of the process and its relationships are understood (e.g., wildfire growth model).
- There is no randomness in the biological process, and all interactions are understood.
- Given the same starting conditions, the output will always be identical.
- The model environment requires feedback and looping capabilities.
- Time is dictated by the modeler's knowledge of the phenomenon, what aspects of the phenomenon are being modeled, how far the phenomenon can move in a time step, and the resolution of the data. It may not be possible to model fire movement every second, or geological plate tectonic movement every hour, because of a lack of understanding of the process at these timescales. It would not be possible to model the movement of an animal every minute if the resolution of the data is 10 kilometers.

A model scenario: A simplified deterministic fire model

This model simulates the spread of a fire in a forest landscape (see figure 14.23). Common characteristics of a fire are the following:

- Fire either burns a location or it does not. So, fire moves in an all-or-nothing way, based on a series of criteria.
- It moves based on a series of environmental factors—wind speed, wind direction, rain, and temperature.

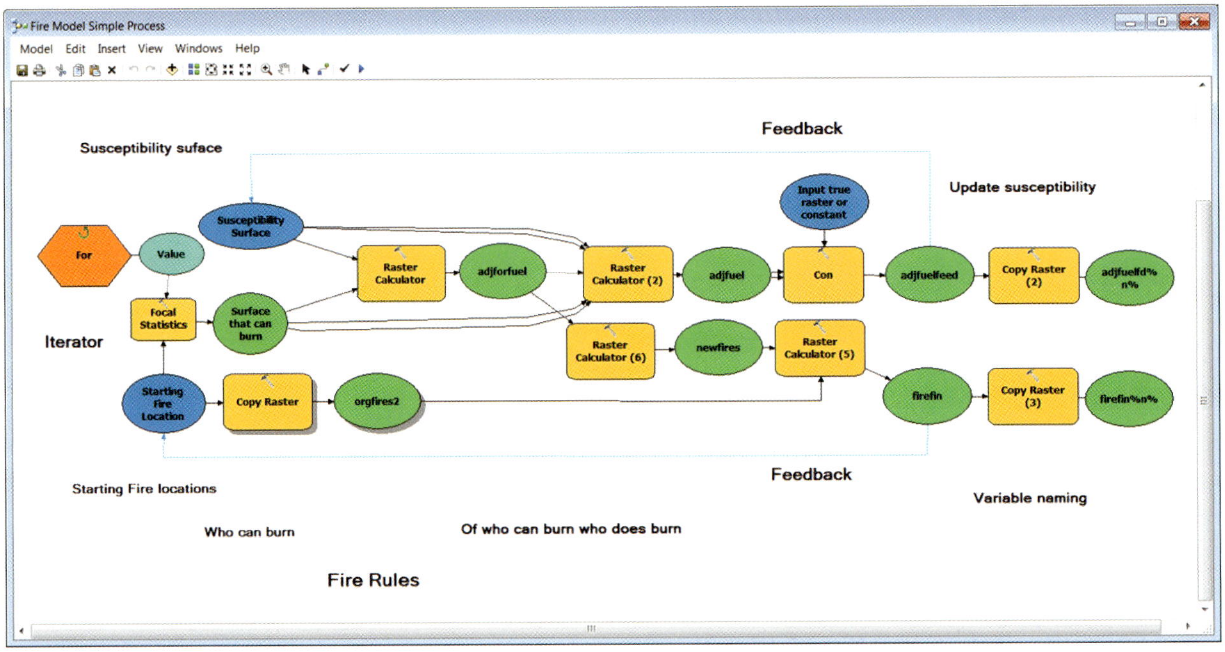

Figure 14.23. The rule set for a simple fire growth model to deterministically spread a fire explicitly in time (dynamically) in ArcGIS ModelBuilder. Figure from Kevin M. Johnston.

- The locations a fire may burn are determined by the characteristics of the landscape—fuel load, aspect, and slope.
- The locations a fire may burn are also determined by a characteristic of the fire—temperature of the fire.

Model 9: A stochastic dynamic model

Building on the previous deterministic dynamic model, randomness is integrated in a stochastic dynamic model in this section, to better capture the phenomenon.

Biological principles addressed by the model

Some of the events in the biological process may appear random. The events may be random, or they may appear random due to incomplete information about the process. In the latter situation, the process is still deterministic (as for model 8) but the modeler does not have full understanding of the rules defining the biological process. In either case in the modeling environment, randomness is included in the model.

Model algorithm

This approach uses the same methodology described in the deterministic dynamic model but uses random numbers to add stochasticity. A stream of random numbers is generated using one of a number of standard random number generators, such as standard C, ACM599, or Mersenne twister. The random numbers are transformed into a distribution such as uniform, normal, gamma, and so on. Each distribution captures characteristics of the stochasticity of the biological process. A random number is selected from the distribution. Based on the number selected, the value can be added to existing input data, used to determine what action should or should not be performed, or can alter existing values within a criterion affecting the decision making of the phenomenon.

Assumptions underlying the model

The assumptions underlying a stochastic dynamic model are the following:

- The random events in the biological process being modeled can be defined by a specified distribution.
- As with most dynamic models, and more so with dynamic models with stochastic processes, it is virtually impossible to model the precise movement of a phenomenon every time step. Therefore, the models are run many times, and the results are combined to create probability surfaces identifying the likelihood of where the phenomenon will be.
- More than likely, each run of the model will produce different results.

A model scenario: Adding stochasticity to a fire model

This approach introduces unpredictable events into the fire model. The model attempts to simulate fire spotting, which appears to be a random event. Also, the wind direction and fire temperature for a time step is modeled as random processes (neither can be precisely determined).

Each of these unpredictable processes is modeled using random numbers that are assigned to these phenomena, using appropriate distributions: fire spotting is assigned a random number from a gamma distribution, the wind direction from a normal distribution (with the mean based on the prevailing wind), and fire temperature from a uniform distribution (see figure 14.24).

Incorporating temporal scale

In many dynamic modeling environments, such as ModelBuilder in ArcGIS, a model can be created within another, container model. These two models can vary in temporal scales: the container model could be running once per day while the inner model repeats every hour.

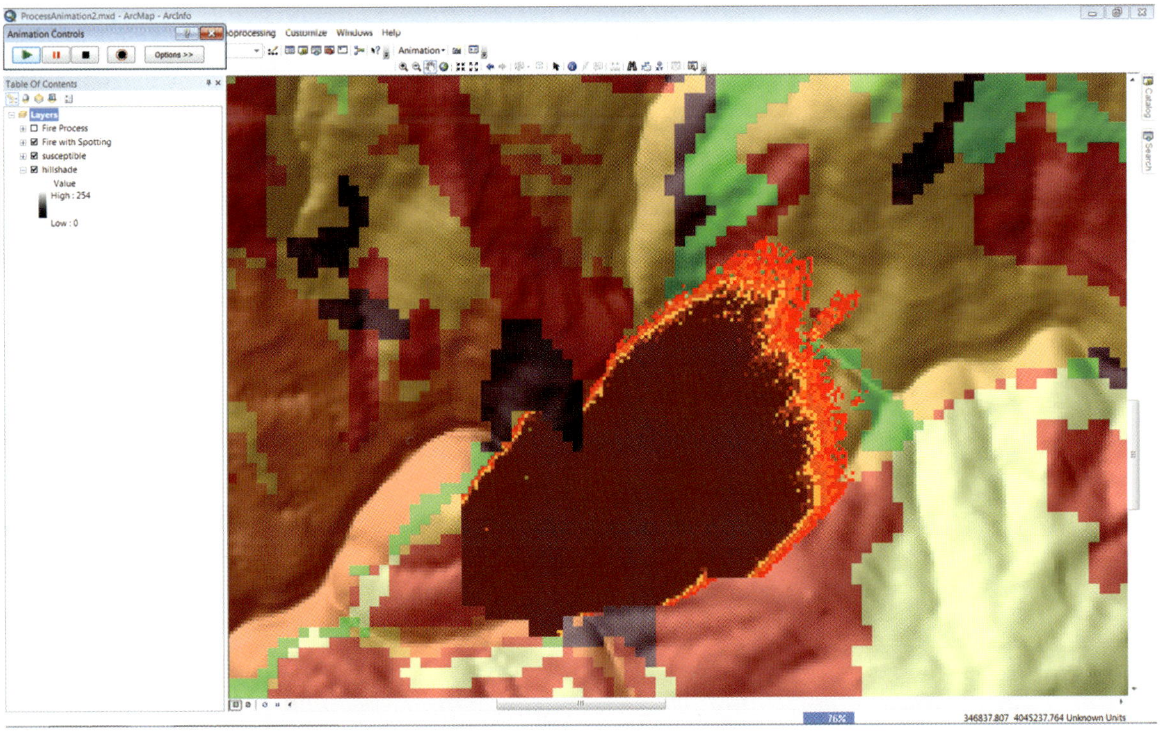

Figure 14.24. A fire growth model incorporating stochastic events. In this example, fire spotting is the stochastic event. The burning fire is displayed over a fire resistance surface (a suitability surface), with the brighter red areas being more resistant to fire and the green areas being more susceptible. Dark red areas indicate where the fire has burned itself out. Notice the spot at the leading edge of the fire (the brighter red). Figure from Kevin M. Johnston.

Error and sensitivity analysis

The tools used in dynamic models (using iterations, feedback, and random numbers) can also be used to perform error and sensitivity analysis through simulation modeling. The general concept underlying error analysis using simulation modeling assumes that there is error in the input data. Using a priori information about the form of the error distribution, the modeler adds random numbers from this distribution to the input data.

For example, in creating a hydrologic network from an existing elevation surface, the modeler may know that the elevation values are accurate to within 2 feet. This known error (defined by a random number, based on a normal distribution with a mean of 0 and possibly a standard deviation of 2) is added (plus or minus) to each elevation value to create another possible reality of the stream network that uses this new elevation surface. This process can be repeated for 100 iterations, and the results of each output can be added together to produce a probability surface of stream locations. In this example, generally where the variation in elevation is greatest, variation errors have less impact. Error analysis identifies those areas that might need more investigation before making a decision. It can also provide some level of confidence when making a decision.

To perform sensitivity analysis using dynamic modeling tools, the modeler systematically changes one parameter slightly and examines how the change affects the output. A series of systematic changes are repeated for the parameter until a reasonable level of comfort is reached. For example, in a suitability model, distance from water may be an input criterion (e.g., 0 to 100 meters initially are assigned a 10, with 10 being the most favorable). A 10% change in the ranges of the reclassified values can be used to assess the model's sensitivity to variations in this distance. In sensitivity analysis, the modeler may want to alter the most favorable class, assigning the range of 0–110 meters to a 10, and then 1–120 meters to a 10, and examine the effects on the output. If there is little change in the outcome, the model is robust to the parameter, and the modeler can comfortably make a decision relative to the identified parameter.

Modeling decision rules that determine movement

In the previous section, we discussed how time and space can be explicitly incorporated into a model. Each of the models discussed can be used to identify habitat, and core areas, for conservation planning. In this section, we will continue with the explicit time–space environment, but we will explore various techniques of modeling the decision-making process of moving phenomena to capture different circumstances. The decision making of a moving animal is a particular focus of this section. These models can be used to identify connectivity habitat and movement routes for conservation planning

The model scenario for each of the situations described in this section involves stepwise choices made by an individual of the species that is at a particular cell location each time step. The animal moves into one of the eight surrounding cells or remains in the cell it currently occupies for each time step. Each cell in the raster has been assigned a number based on the preference of the species for the attributes in the cell location.

Model 10: Random walk

In this section, the biological principles, model algorithm, and model assumptions for a random walk are introduced.

Biological principles addressed by the model

The model for this biological process (movement) assumes the species has no preference for the features or location of the surrounding cells. There is no perceivable logic in the animal's decision making. The random walk usually represents the null hypothesis.

Model algorithm

Each of the eight surrounding cells, plus the cell in which the animal currently resides, has equal probability of being selected in the next time step (see sidebar "Creating a random walk").

Creating a random walk

Each of the eight surrounding cells in the raster environment and the current location have equal probability of being selected for the animal to move into. To implement the algorithm, each cell can be assigned a value from 1 to 9, with each cell receiving a unique value.

A random value from a uniform integer distribution ranging from 1 to 9 is determined. Whichever cell the random value corresponds to will be the new location for the species. This process is repeated for each time step.

Assumption underlying the model

A random walk model assumes that where the cell is relative to the location of the animal, and what that cell contains, has no influence on which cell is selected (see figure 14.25).

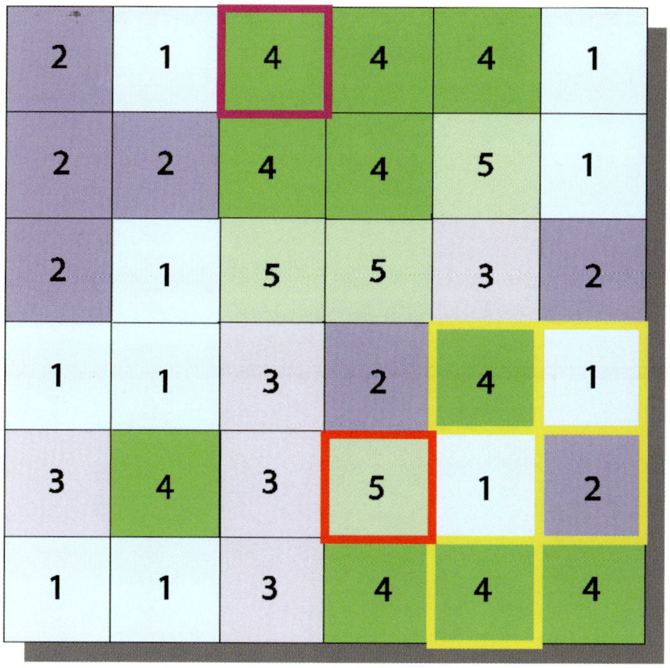

Figure 14.25. A random walk on a least-cost surface. The starting location is the red cell. The number assigned to each cell identifies the preference for movement per distance unit within the cell. A lower value indicates less cost, and therefore is more preferred. Because the species is selecting one of the eight surrounding cells (or its current location) randomly, these movement preference values are irrelevant. In this example, the species moves first to the cell with the yellow outline to its lower right and then moves to the next cell with a yellow outline (the cell with a value of 2), and so forth. The cell with a magenta outline is a destination. It will be used in the subsequent examples, but in the random walk, it has no influence in the decision making of the animal movement. Figure from Kevin M. Johnston.

Model 11: Preferred random walk

Building on the random walk discussed in the previous section, in this section a random walk in which some preference logic is associated with the surrounding cells—a preferred random walk—is presented.

Biological principles addressed by the model

In the preferred random walk model, the animal is aware of its surroundings and moves to the most highly preferred neighboring location (see figure 14.26). The preferred random walk simulates an animal introduced into a new location with an unfamiliar landscape.

Model algorithm

The species moves into the neighboring cell that has the highest preference value (see sidebar "Creating a preferred random walk model").

Creating a preferred random walk model

Each cell in a raster is assigned a value based on the preference by the species for the attributes at the location.

The neighboring cell that is most preferred is the cell the species will move into. If multiple cells contain the same preference value, and that value is the most preferred, either additional criteria need to be evaluated to distinguish between the values, or one of the cells is randomly selected.

Assumptions underlying the model

Assumptions underlying a preferred random walk model are the following:

- The animal has no long-term goal; it will only respond to what is immediately surrounding it.
- The animal always makes the local optimum decision and selects the most preferred location. It makes no wrong decisions.
- Species preference is fully understood and is represented by the preference value in each cell.
- The current location of the animal is not considered in the movement decision.
- If two or more cells contain the most preferred value, then the algorithm must randomly choose between them, or some other logic must be added.

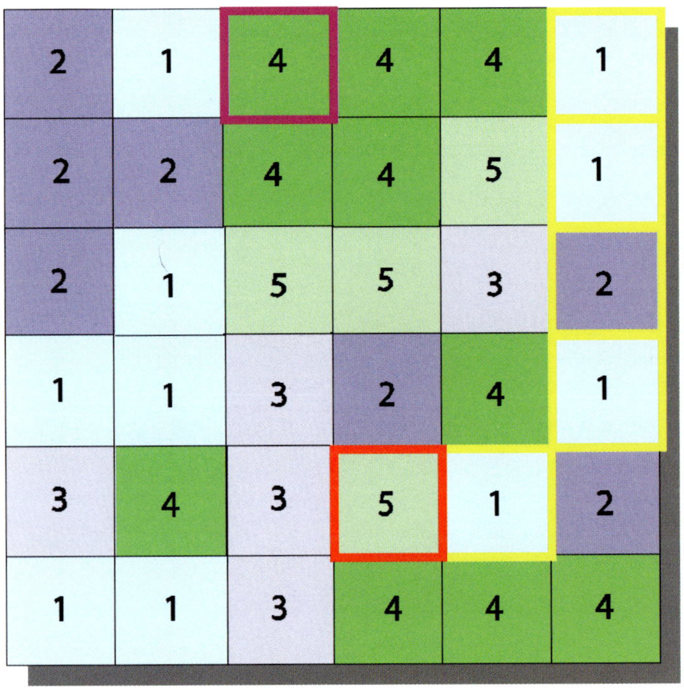

Figure 14.26. The preferred random walk on a least-cost surface. The starting location for the animal is the cell with the red outline. It will move to the cell with the least cost (highest preference value). The first move is to the right, into the cell with a value of 1, and then to the subsequent cells outlined in yellow. The purple cell is a destination, and it has no influence in this model. Figure from Kevin M. Johnston.

Model 12: Preferred random walk with probability

Continuing the progression of the random walk and preferred random walk models identified in the previous sections, probability is added to the preferred random walk to capture uncertainty in the decision-making process of a species in this section.

Biological principles addressed by the model

This model uses the same algorithm as the preferred random walk, but randomness is incorporated into decisions. This approach can be effective if the modeler does not have full understanding of what the species prefers, and therefore, it is difficult to accurately assign a single number to each cell to capture the animal's preference. It can also model cases in which the animal does not always make the optimum decision (maybe it does not have full knowledge of its surroundings before it makes a move).

Model algorithm

A distribution is created from the values assigned to each of the surrounding eight cells, plus the current cell location. A random number is selected from the distribution. The random number that is selected defines the location selected. The more preferred the value of the cell (in this example, the higher the value), the more likely that cell will be selected (see the sidebar "Creating a random walk model with probability using multiple criteria").

Assumptions underlying the model

Assumptions underlying a preferred random walk model with probability are the following:

- The modeler cannot know everything that the animal is reacting to when it makes a decision to move.
- The animal may make a suboptimal decision.
- The animal generally will move into locations that contain more favorable attributes.
- The animal can (with less probability) move into the least preferred location.
- There is randomness or stochasticity in the decision-making process.

Creating a random walk model with probability using multiple criteria

The marble algorithm commonly is used to add randomness to decision making. Conceptually, it works as follows: a distribution is created from the eight surrounding cells, plus the current cell, for a single criterion. To visualize this, imagine that each of the cells within the distribution contains a bucket. In the bucket are placed the same number of marbles as are assigned to the preference for the attribute in the cell for the defined criterion. Each marble from each bucket contains information on which cell it comes from. All the marbles from the nine cells are dumped into a larger bucket. One marble is selected from the larger bucket. The origin (the cell) of the marble that was selected is the location the species will move into. The cells that are most preferred will contribute more marbles to the larger bucket, and therefore, cells with higher preference values will have a greater chance of being selected.

This process can be extended for multiple criteria. A distribution is created as above, but from all the criteria. For each of the nine cells, the cell is assigned a value for the preference of the species for the attributes at the location relative to each criterion. For each criterion, the marbles are added into the corresponding bucket for each cell. The marbles in each of the nine buckets for each of the criteria are added to the larger bucket, and one marble is selected from the aggregation of the marbles.

During a time step, one criterion can be more important than another. For instance, if the animal is hungry, the prey or food availability criterion can be weighted more than the other, competing criteria. To capture this changing decision-making process, the number of marbles in each cell's bucket can be multiplied by a weight. This weight will increase the number of marbles in each bucket for the criterion, making marbles from this criterion more likely to be selected when all the marbles are added together.

Many factors can influence the weighting of each criterion. For example, the cell's relative distance from a destination, or to some attractor, may influence an animal's choices. If the cell is the continuation of some physical feature (for example, snow leopards prefer to move along a ridge), it may be given a higher weight. The state of an agent during a particular time step may also alter the preference for a cell (e.g., as previously mentioned, if the animal is hungry, then the cells with the most food may be weighted more).

Model 13: Preferred random walk with probability and a goal

In the random walks discussed in the previous sections individuals of the species had no long-term destination influencing their decision making. In this section, a preferred random walk with probability is presented but the animal also wants to reach a distant destination.

Biological principles addressed by the model

This model represents movement toward a known goal. The direction to the distant goal influences the immediate decision making of the animal. The attributes at each of the surrounding cells affect the decision making of the species, as well as the relative position of the cell relative to the distant goal.

The animal has some reason it wants to get to the distant goal. Maybe it has been to the goal and wants to return to it (this simulates memory), or it has some other way of detecting the distant goal (e.g., via sight, smell, or hearing).

Model algorithm

The preferred random walk with probability and a goal has the same algorithm as the others, except the distribution created by the values of the surrounding cells is adjusted based on the distance from each cell to the distant goal. Generally, reaching the goal is implemented as another criterion, just as for hunting or staying secure. For weighting the cells within the criterion, the closer the cell is to the distant goal, the more preference is added to the location. Conceptually, more marbles are added to the cell's bucket (see the sidebar "Creating a random walk model with probability using multiple criteria").

Assumptions underlying the model

Assumptions underlying a preferred random walk model with probability and a goal are the following:

- The values and locations of the surrounding cells influence the decision making of the animal. The influence of each cell on the decision making of the animal can be predicted. The influence can change at each time step based on a number of factors (e.g., how hungry the animal is at that time step).
- The animal may make suboptimal local or immediate decisions when moving toward the distant goal. The overall path will probably not be a global optimum (see figure 14.27).

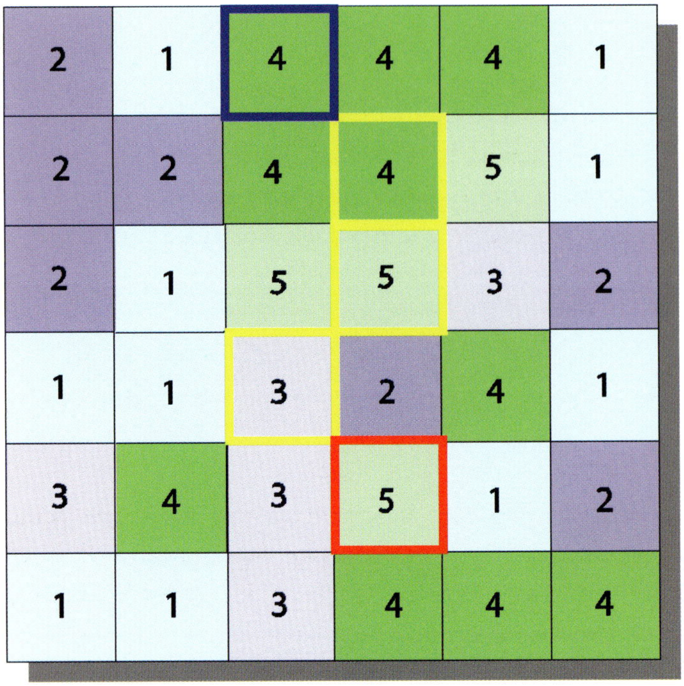

Figure 14.27. A preferred random walk moving toward a destination. The animal starts at the cell outlined in red and is trying to move toward the cell outlined in blue. In this example, it moves generally toward the goal by attempting to move into lower-cost cells, but because there is some randomness in the decision making, the cell with the lowest value is not always selected. The influence or weight of the distant goal can vary with each time step. The cost assigned to each cell is cost per unit, so diagonal movement is multiplied by 1.414214. Figure from Kevin M. Johnston.

Model 14: Escape or pursuing

In this section, a deviation is taken from the progression of the random models presented in the previous sections; a specific event dominates the decision making of individuals.

Biological principles addressed by the model

In this model, the animal is escaping or pursuing another animal. While doing either action, its typical decision-making process changes, based on fleeing its pursuer or catching its prey. The decisions that will allow the animal to flee or capture its prey may override all other priorities, and thus its decision making may be reckless in these pursuits.

Model algorithm

The relative location of the predator or prey strongly influences the weighting of the neighboring cells, generally making any cells farther from the predator or closer to the prey more favorable.

Assumptions underlying the model

Assumptions underlying an escape or pursuing model are the following:

- The animal may be more reckless in movement, with escape and pursuit taking precedence.
- The animal is trying to gain a position of advantage when fleeing from a predator or pursuing prey.
- Other criteria may take lower priority (e.g., security and food) when in pursuit or being pursued.

Model 15: Moving with long-term intent and with memory (a least-cost path)

In the previous random walk models, an animal reacted to its immediate surroundings (maybe being influenced by a distant goal) resulting, possibly, in local optimization. In this section, the decision making of an individual is dominated by optimally reaching a distant goal, which assumes the individual has knowledge of intervening habitat.

Biological principles addressed by the model

In this model, the animal is at one location and knows it wants to get to a distant location. It may or may not even detect the distant location, but it has knowledge of the goal and has familiarity with the entire area and the features within it. The species has long-term memory.

Model algorithm

The algorithm guarantees to identify the least-cost path to travel between the source and the destination (see sidebar "Performing a cost path analysis"). This modeling approach is generally solved with cost path analysis (see figure 14.28) or circuit theory.

Assumptions underlying the model

Assumptions underlying a least-cost path model are the following:

- The cost of the cells immediately around the animal influences the decision making, but the overall cost to travel from a source to a destination dictates the path the animal should take.
- The animal makes decisions that will achieve the global optimum.
- The animal has full knowledge of its environment.
- Time is frozen, and the algorithm does not account for the time to travel between cells. As a result, there is the assumption nothing changes in the environment between the source and destination once the calculation is started.
- The animal generally makes suboptimal local decisions to obtain the global optimum.

Performing a cost path analysis

A source, destination, and a cost surface are input into the ArcGIS Cost Distance tools.

The cost surface identifies the cost the animal pays or gains per unit when traveling through a cell. The cost surface can be based on energetics gained, the level of safety, or any other series of preference criteria.

Cost path is a two-step process. First, the source and the cost surface are entered into the Cost Distance tool. This tool produces two output surfaces. The output cost distance raster identifies for each cell in the study area the total least cost for the species to return to the cheapest source over the cost surface. The output back-link identifies for each cell which of the eight surrounding cells the species should move into once leaving the cell.

The second step is defining the optimum path, which is accomplished by running the Cost Path tool. This tool utilizes the cost-distance and the back-link rasters that are output from the Cost Distance tool and the destination. The output from Cost Path tool produces the global optimum path to move from the source and destination.

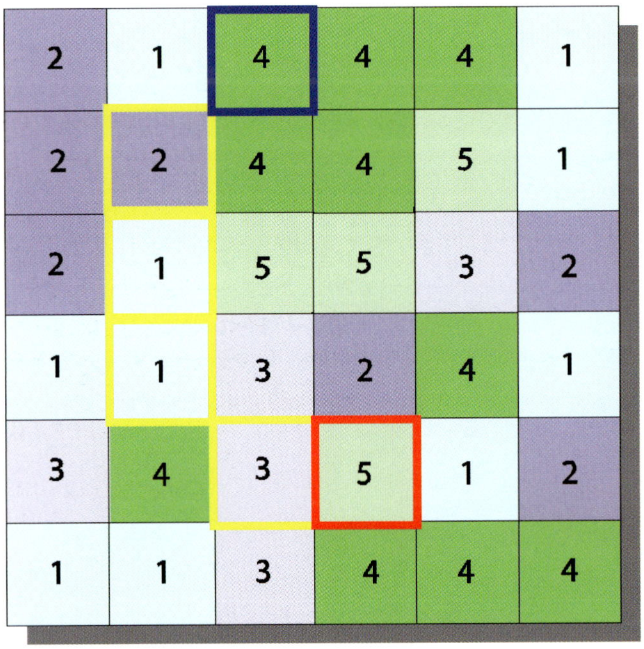

Figure 14.28. A cost path movement from the source (the red cell) to the destination (the blue cell). The path is the least-cost path to travel. The cost surface (defining the per unit cost of travel) in this model ranges from 1 to 5, with 1 being the least costly or the most preferred. Notice the first move is to the immediate left, toward a cell with a value of 3, as opposed to moving to the upper cell with a value of 2. The species will make a suboptimal local decision to obtain the global optimum. Figure from Kevin M. Johnston.

Cost path analysis was used in the elephant and snow leopard models discussed above that connect the most preferred habitat patches within and between seasons.

General movement types in models

There are almost as many different movement types as there are phenomena. The underlying principles of the movement of the phenomenon being modeled must be understood before it can be captured. A single movement tool cannot capture all types of movement. Not only is it important to identify the differences in movement, but it is also important to look for commonalities in the movement and the underlying causality.

Different species may respond broadly in similar ways. Macroecology analyzes the relationship of species evolving in a similar manner in spatially separate locations, while allometry studies the relationship between size and weight. Size affects the rate of a species' life-history properties and its ecological functions. These similar facets of evolution shape the species requirements (such as energetics, nutrition, and so on), thus accounting for why species often respond in a similar manner but with different intensities (e.g., density, passage rates, and so on) and timing.

The commonality is that individuals of all species are trying to better their own position, to survive, or to obtain a goal, or they are driven by some natural, predictable, external forces. The challenge for the modeler and planner is to find the simplest explanation and to use a model that produces realistic results.

Integrating causality in models

The progression of descriptive suitability models discussed above attempts to represent a slice of time, or a series of time slices, that captures critical components of a species' biology. Different methods were used to derive model rules based on the current distributions of the species. Explicit space–time models were discussed to model the biological process more realistically. And finally, in the explicit space–time modeling environment, the different decision-making processes that could be used to capture movement were explored.

However, all of these approaches have limitations. One limitation is that all individuals are assumed to behave in the same fashion. To more realistically model biological processes, advanced techniques have been developed for simulating the intelligence in individual animals or agents.

Model 16: Agent-based modeling

The biological principles, model algorithm, and model assumptions for an agent-based model used to capture causality are discussed in this section.

Biological principles addressed by the model

Biological processes (in the sense of this chapter) are composed of individuals making decisions, and these decisions generally result in some action by the individuals (agents). In many cases, an action involves movement. The movement can be based on the surrounding landscape features, and the movement can change the landscape with which the agent interacts. From the aggregation of many individual actions, patterns emerge. Agent-based modeling (ABM) explores the causality of the decision-making process that produces patterns; it does not simply quantify existing patterns. For detailed information, refer to Johnston (2013), as well as chapter 10 in this text.

Model algorithm

A series of fields and actions are created for each agent that simulate actions, reactions, and conditional variables in movement. The actions are scheduled and the model is run.

Assumptions underlying the model

Assumptions underlying agent-based modeling are the following:

- Patterns are created based on the decisions of individuals. These patterns can be quantified using standard statistical modeling techniques that are applied to any biological process.
- Many of the assumptions in the models above also apply to ABM. They include movement and decision making by the species is well understood, the movement is determined each time step based on needs and general conditions, and there is intent in the decision making of the species.

A model scenario: A cougar movement ABM model

There has been increasing conflict between cougars and humans near Flagstaff, Arizona. The goal is to model the individual movement of the cougars in the area, based on observed decision making by the cougars, to understand the biological process of movement and gain insight into the conflicts.

Each male cougar agent during each time step will prioritize or make trade-offs using the following criteria:

1. Remaining secure
2. Remaining with a kill once a kill is made
3. Pursuing a female cougar agent (for a male cougar agent)
4. Mating
5. Staying within its home range
6. Moving with intent toward good habitat/hunting

Each of the cells surrounding the current cougar agent location is evaluated based on the attributes at the locations relative to each criterion. Each criterion is weighted relative to one another, based on the state of the cougar agent during that time step. The physical state of the cougar is based on its energetic state.

Certain criteria, such as pursuing a female and remaining with a kill, are modeled based on attractors. That is, when a kill is made, the location of that kill becomes a strong attractor, and the weight for that criterion will be high until the kill is consumed.

Five different agent types exist in this particular model, two primary and three supporting. Male cougar agents and female cougar agents are the primary agents. Home range boundary, cells (the eight cells surrounding a particular cougar agent), and prey are three supporting agents.

Each agent type is assigned specific rules on how to interact with the landscape and with other agents, given various states and situations. Those actions or interactions are scheduled. When the model is run, each agent makes decisions about whether to move (or not) during each time step. The agent's state may change, and possibly the landscape may alter during the time step.

This particular model was created using the Agent Analyst extension to ArcGIS (see figure 14.29). The basis for multicriteria decision making of the agents is the marbles algorithm (discussed in the sidebar "Creating a random walk with probability using multiple criteria"). For a full discussion of this model, see chapters 1, 2, 5, and 8 of Johnston (2013).

ABM is considered one of the more realistic approaches for modeling biological processes. It incorporates a series of individuals making decisions, and from those decisions, patterns are created. The process of creating an agent-based model is as important in the modeling process as the model itself. The model development forces the modeler to fully understand, in detail, the biological process and to represent it in the model environment.

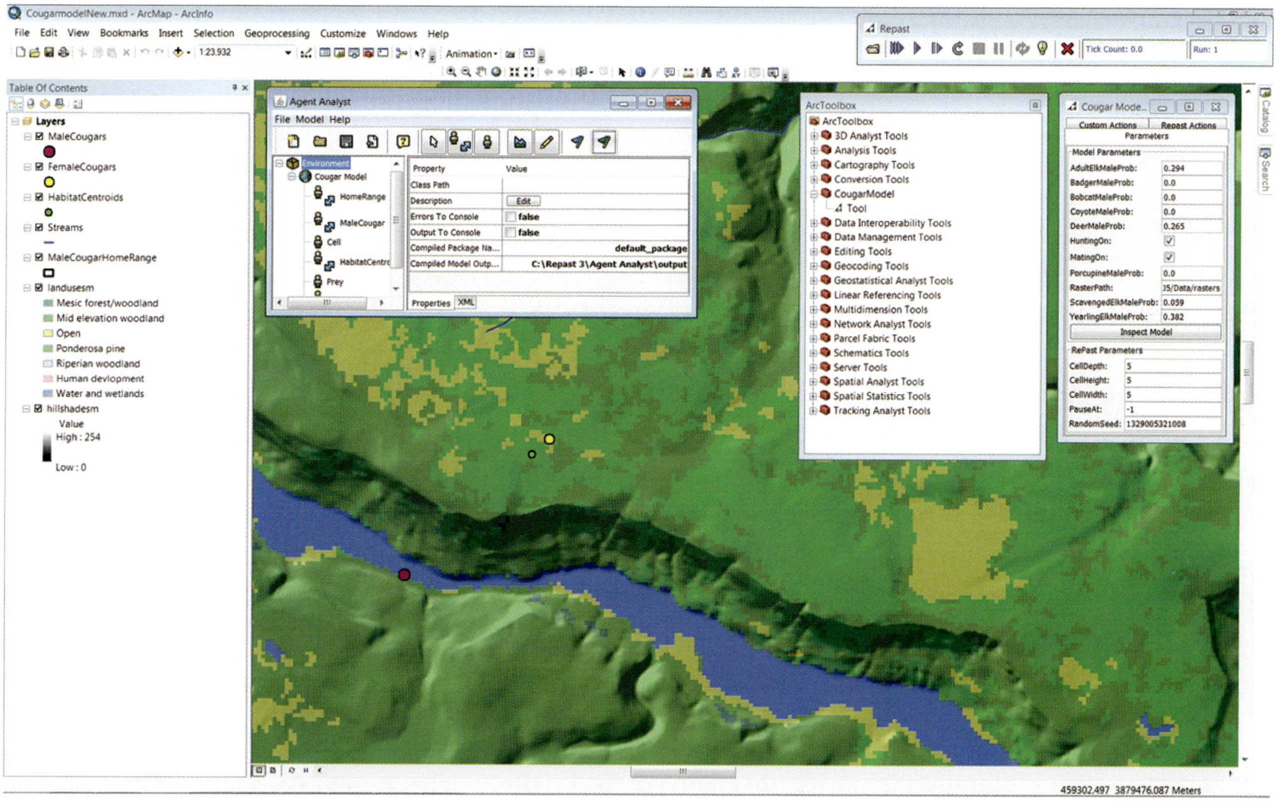

Figure 14.29. The Agent Analyst interface used to create agent-based models. The model displayed is the cougar model. The purple dot is a male cougar, and the yellow dot is a female cougar. Notice in the Agent Analyst dialog box (center) the different agent types. The top right toolbar controls running the model. The right dialog box allows the user to specify certain parameters within the model. Data from the US Geological Survey.

Common misuses of spatial modeling that incorrectly capture biological processes

In this section, we explore the common misuse of two spatial modeling tools when capturing biological processes from existing field data: density mapping and interpolation.

Model 17: Density mapping

By understanding the biological principles, model algorithm, and model assumptions of a density model, it is less likely the approach will be misused.

Biological principles addressed by the model

The modeler acquires data aggregated at a point or linear feature. For example, some census data is aggregated with the number of people existing in a certain area being assigned to a point, or there is a count of the number of animals visiting a water hole. The modeler knows that the population of the commodity (i.e., people or animals) does not in fact exist at that point or at the water hole; the goal is to spread the known commodity across the landscape as realistically as possible, using the principles underlying spatial autocorrelation.

Model algorithm

The magnitude of commodity at each point (or line) and the relative positions of the points to one another is known. Locations closer to the point are more alike (spatial autocorrelation). Therefore, if a location is near a point with a high level of the commodity, then it is more likely that when that commodity is spread across the landscape, that location will receive a higher percent of the commodity identified by that point. With these principles, the commodity for all the points are spread across the landscape, creating a continuous surface (a raster) to represent the most likely distribution of that commodity.

Assumptions underlying the model

Assumptions underlying a density-mapping model are the following:

- From the locations and the magnitude of the samples, the modeler can approximate the values of the commodity for all locations in the study area.
- It is assumed the commodity is spread as a continuous phenomenon across the landscape.

- The sum of the commodity assigned to each cell should be similar to the sum of the commodity at the sample points.
- The output values are number per square unit.
- Care must be taken when changing the output square units to a coarser unit. In doing so, it is assumed the same density identified at the finer scale can be extrapolated to the coarser scale.
- The input units and the per-unit output must make sense. If the input units are in degrees, generally the results are difficult to interpret.
- Generally, in density mapping, the area outside the search radius around the sample points is assigned to zero.
- In kernel density (a specific form of density mapping), the resulting values are generally highest at the location of the point, and diminish with increasing distance from the point.

A model scenario: Using density on elephant collar data

The following example is a common misuse of density mapping: using density mapping on radio-location data. In this example, data has been acquired from a radio collar on an elephant. Each point in the data represents a reading from the elephant's collar. From this elephant's locations, the modeler tries to predict the distribution of elephants across the landscape by predicting the density of the elephant population at each location. Initially, it appears logical that elephants will more likely be near locations where other elephants occur.

There are problems with using density mapping for this situation. The first issue is that radio collar data is simply presence data with a magnitude of one. That is, only one elephant is known for certain at the data location. Everywhere else is assumed to be zero. Using density mapping in such cases results in density values of less than one elephant at a location. Such values are only meaningful in a probability-of-presence map, not a density map.

Density modeling assumes that the underlying process driving the system is continuous. In the case of elephants, generally the underlying process dictating the locations of an elephant is the availability of food. The elephant in this model is located in Kenya, where food is not continuous. Just because an elephant is at a particular location does not mean it will be at nearby locations, especially if there is no food there.

In density mapping, the magnitude and spatial relationship of the sample points establish the density of the phenomenon. However, the underlying process (the availability of food) is not continuous, and there is no magnitude associated with the samples. Hence, the output map may look nice, but the assumptions are violated, and the density results are meaningless.

To address the issue of having a single elephant at each telemetry location, the modeler can alter the example so that each data point represents the number of elephants that were observed from that data point (within a specified distance). This creates magnitude (the number of elephants around the collared animal) at each sample point, and a seemingly realistic map can be created because there is some variation; however, the above assumptions were again violated, and the numbers are meaningless (see figure 14.30).

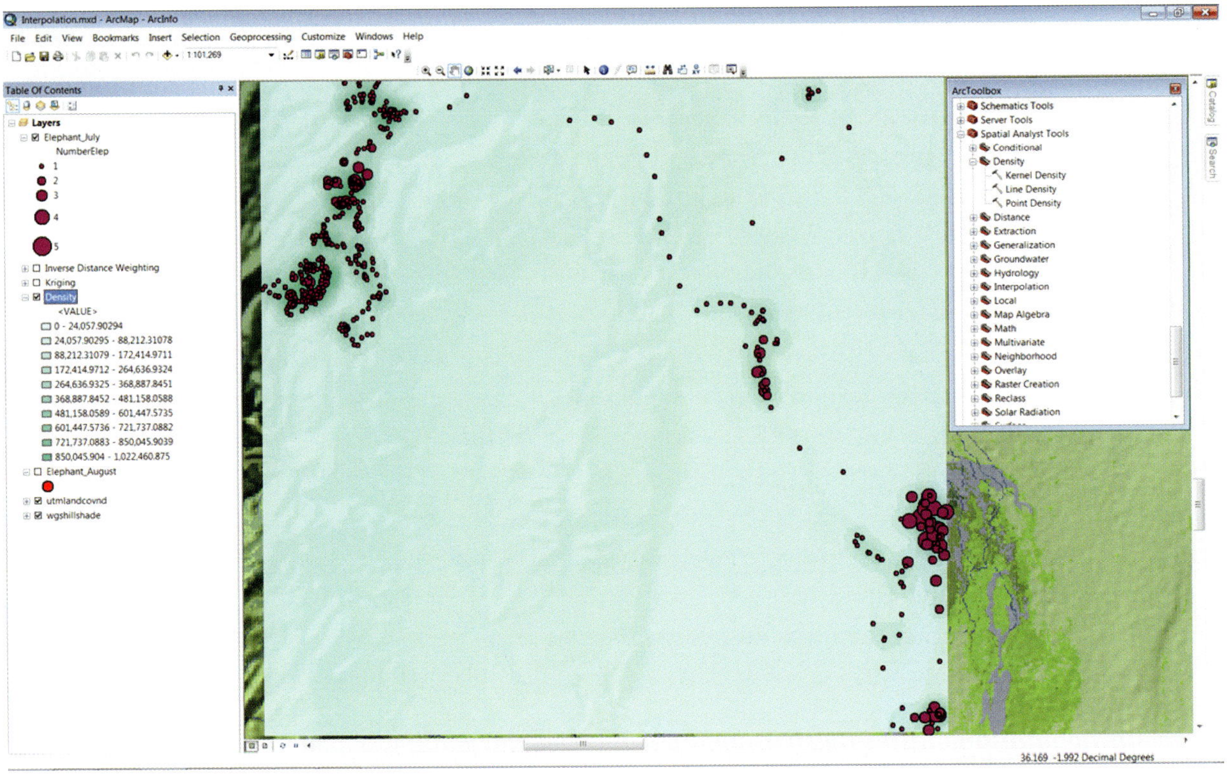

Figure 14.30. A density map created from sample points identifying the number of elephants observed at each point (in purple). Notice in the table of contents that the numbers are floating-point values, resulting in locations with densities of portions of elephants. The density values are high because the input unit was in meters, and the output density is in square kilometers. An understanding of the interactions of the input units and the search radius must be exercised when specifying the output units. Data from the African Conservation Center.

Model 18: Interpolation

Similar to density modeling described in the previous section, interpolation from point observations is commonly misused to depict how a species might be distributed through the landscape.

Biological principles addressed by the model

The distribution of individuals of the species across the landscape can be predicted using data on location and the number of individuals present at sample points (the same as described in density mapping discussed above).

Model algorithm

Based on the spatial distribution and the magnitude of the phenomenon at specified sample points, a continuous surface can be derived. As with density mapping, interpolation derives the output values based on principles of spatial autocorrelation. Where interpolation differs is that it does not spread a known commodity across the landscape but derives values based on the values of nearby sample points. The closer the prediction location is to a sample point, the more likely the output value will be similar to the value associated with that sample point.

There are several interpolation methods for creating a surface, including inverse distance weighted, global polynomial, local polynomial, spline, and kriging algorithms.

Assumptions underlying the model

Assumptions underlying an interpolation model are similar to those of density mapping, plus the following:

- All interpolators are based to some degree on spatial autocorrelation—a basic principle of geography that things that are closer together are more alike.
- The underlying phenomenon driving the system must be continuous.
- Defining the driving force for the system is critical. For example, as was the case for the location of elephants in the density mapping example, the location of the elephants is driven by the distribution of food. Food was not a continuous phenomenon in that example.
- The system contains stationarity, that is, the covariance between two points is similar for two sample points that are the same direction and distance apart at another location in the study area.
- The distribution of the sample points is normal.
- The predictions within the minimum-bounding rectangle have greater certainty.
- If presence data is only denoted as 1 (as in telemetry data), then the modeler lacks variance, and every location will have a value of 1.

A model scenario: Using interpolation to create a population distribution map of elephants

The following model scenario is an example of a common misuse of interpolation when applied to field data. This example uses the same scenario as for density mapping, described above. A collared elephant provides presence (location) data. Similar assumptions to density mapping apply to interpolation. The underlying process should be continuous, and the spatial relationships between the sample points and their magnitude are central to creating a continuous surface from the sample points. Once again, both of these assumptions are violated in this example.

In fact, it is not possible to even create a surface from the sample points with a magnitude of 1—because there is no variance in the sample points, there is no spatial structure in the sample points. No matter how far or close two sample points are to other sample points, all points have a value of 1; therefore, the only thing that can be concluded from the data is that everywhere has a value of 1. Where there are no samples does not mean those locations have a value of zero. Conclusions can only be derived from known samples.

To address the presence-only issue (similar to the density mapping example), the sample points are altered to represent the number of elephants observed from each sample point, thus giving magnitude and variation to the points so that some spatial structure can be established. Once again, an interesting map is produced, but because the underlying assumptions for interpolation are violated, the resulting output values are meaningless. Also notice in figure 14.31 that no locations have less than one elephant. This is because there are no sample values with a zero. In this example, a kriging algorithm was used. Because there is no measurement error (which would add some randomness to the predictions), kriging is an exact interpolator. Therefore, no prediction value on the resulting surface will be above the maximum or below the minimum values. If samples containing zero are added, then only the location of the sample point will contain zero. As the locations move away from the zero sample point, they increase in magnitude. But in reality, most of the locations will be zero and not some percent of an elephant.

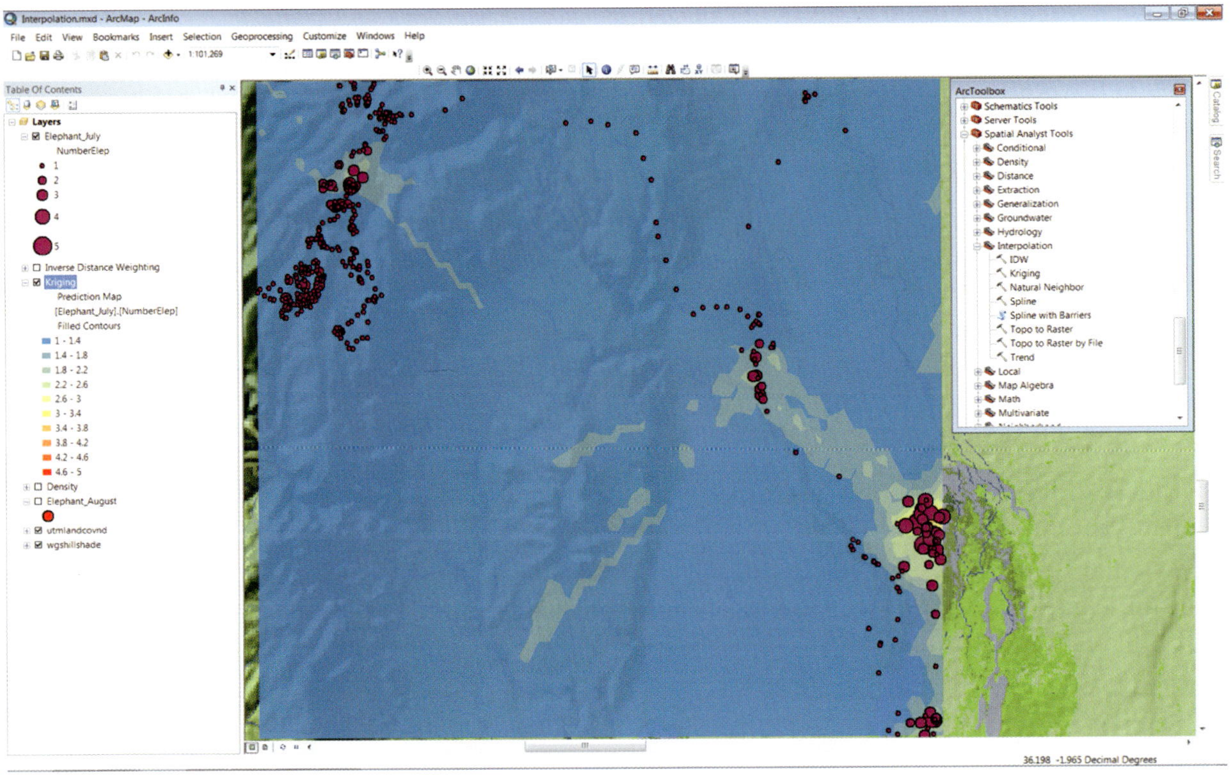

Figure 14.31. A kriging surface created from the number of elephants observed at each sample point. Most locations contain one elephant because there are no samples with zero values. Notice in the table of contents that in the legend for the "Kriging Prediction Map," some cell locations have portions of elephants (e.g., 1.4 to 1.8 elephants). Data from the African Conservation Center.

Conclusion

Most biological processes are complex and difficult to fully understand. If the biological process is not precisely defined, it cannot be modeled. Every model is defined by the algorithm and a set of assumptions underlying the algorithm. If the biological process is not defined clearly, the wrong modeling tools (algorithm) may be applied. If the biological process is defined clearly but the appropriate modeling tools (based on the correct algorithm and assumptions) are not applied, the process will not be captured. In either case, invalid conclusions and wrong decisions may be made because the biological process was not captured.

Modelers occasionally try to capture a biological process with generalized rules on small datasets. Often, no validation is done on the model output. Sometimes, modelers will then make generalizations of the biological process from the results of these simplified models. It is critical that the correct modeling tools are applied to the appropriate biological process and the models be validated before generalizations and conclusions can be drawn with any confidence.

Each biological process is unique, and so is each spatial modeling tool. All models are limited: there is no single model that can capture each biological process in all situations. It is important to understand what aspect of the biological process a particular model captures so as to interpret its results and understand its limitations. Field data, other models, and expert knowledge can be used to supplement a particular model.

Generally, it is a good practice to apply a series of different models when exploring a biological process. With each model application, a deeper understanding of the biological process is gained, and more informed decisions can be made.

Acknowledgments

We would like to thank Michelle A. Gudorf, Matthew D. Lisk, and Sarah Lamb for their tremendous input into the creation of the data and the development of the models used in this chapter and for their comments on the text. We would also like to thank ICIMOD in Nepal for the use of their data for the snow leopard models, Brandon Rogers and Ronald P. Neilson of the USDA Forest Service for providing the climate change prediction models for the Pacific Northwest, and Dr. David Mattson and Terrence R. Arundel from the US Geological Survey Southwest Biological Science Center for the data used in the cougar agent-based model.

References

Johnston, K. M., ed. 2013. *Agent Analyst: Agent-Based Modeling in ArcGIS.* Redlands, CA: Esri Press.

Malczewski, J. 1999. *GIS and Multicriteria Decision Analysis.* New York: John Wiley & Sons.

15

Summary: Building a broader base for conservation planning

Frank W. Davis, PhD

This book is a cornucopia of technical information and guidance on how to produce a biodiversity conservation plan. The editors and authors expertly summarize and illustrate the growing sophistication, intellectual breadth, and technological capacity of the field of conservation planning. Rather than trying to summarize the book's wide-ranging and dense content, I will briefly highlight three recurrent themes that caught my attention: thinking like an ecologist, the plan versus the planning process, and the new era of data-intensive networked conservation science. I will conclude by contrasting conservation planning to traditional environmental planning, and suggest that conservation planners now have an unprecedented opportunity to build a broader base for conservation planning by bringing their science, methods, and tools into the everyday arena of urban, landscape, and regional planning.

Theme 1: To be an effective conservation planner you have to think like an ecologist (or at least collaborate closely with one)

Editors Lance Craighead and Charles Convis establish this theme in the first paragraph of chapter 1 when they assert that conservation planners have to *understand the processes* (italics added) responsible for the distribution and composition of plant and animal communities that an ecosystem supports. Perhaps this theme seems obvious. Just as transportation planners must understand transportation systems and users, and water resource planners must understand processes underlying water supply and use, conservation planners must understand biodiversity processes.

Thinking like an ecologist is an important conservation planning principle, but unless you are a trained conservation biologist or ecologist, it seems a tall order. Early systematic conservation planning, although informed by ecological reasoning, was mainly pattern-based and dominated by mapping, map overlays, gap analysis, and reserve siting. By comparison, most of the contemporary tools and approaches provided in this book demand working familiarity with, if not explicit modeling of, spatial ecological processes of habitat selection, dispersal and migration, population regulation, disturbance regimes, and ecosystem dynamics. Careful attention to ecological processes informs the conservation planner's decisions about goals, conservation targets, planning approaches, spatiotemporal scales, data sources, and model structures.

Attention to ecological processes puts conservation planning on firmer scientific footing and better integrates conservation planning with conservation biology, biogeography, and landscape ecology. But the tension is palpable between the scientist's desire for realism and completeness versus the conservation planner's need for expedience and accessibility of the planning process to nonexperts.

My parenthetical comment about collaboration is added to the first theme because I hope that many readers of this book will be planners whose main expertise is not ecology or conservation biology. As underscored by several of the authors, conservation planning is ultimately a team sport. Disciplinary expertise from geography, urban and regional planning, economics, behavioral sciences, law, and other disciplines is needed to fully address the spatial, ecological, and social dimensions of conservation planning.

The interdisciplinary character of conservation planning has been noted repeatedly in the recent conservation literature (e.g., Knight et al. 2010) and nicely demonstrated here by Chan and coauthors (chapter 2), who point to Ostrom's (2009) social-ecological systems approach—multiscale analysis of complex, interacting resource systems (e.g., fisheries), resource units (e.g., fish), users, and governance systems—as a beacon for conservation planning research. Approaching conservation planning from this perspective requires a broadly interdisciplinary framing of the planning problem and a willingness to "harness complexity, rather than eliminate it from such systems" (Ostrom 2009, 420). If we follow Ostrom's advice, conservation planning will migrate from its current position at the outskirts of mainstream land planning to the leading edge of a more comprehensive environmental planning process, a point I will return to below.

Theme 2: The conservation planning process is ultimately more important than the conservation plan (or is it?)

In their opening chapter, the editors make clear that the book is about producing a conservation plan, not plan implementation. I consider this a wise decision because in my experience, implementation takes many forms (Pressey and Bottrill 2008). Conservation plans are rarely implemented fully, or as initially envisioned, unless the plan is motivated and supported by an implementing authority—for example, the authority under California's Marine Life Protection Act to implement a network of marine protected areas (Gleason et al. 2010) or under the US Endangered Species Act to implement a single- or multispecies habitat conservation plan to mitigate take of listed species (Wilhere 2009). Clearly, plan adoption is not the only measure of success. A conservation plan is implemented to the extent that it leads to better conservation investments and management decisions. Based on this view of implementation, which is probably the rule, not the exception, success depends more on the planning process than on the plan itself.

This book intentionally focuses mainly on the "nuts and bolts" (chapter 1) of producing a plan, but the editors and authors are certainly attentive to stakeholder processes and the value of *collaborative planning*. Indeed, the editors describe stakeholder use of computers and software tools as "the new frontier of conservation planning" (chapter 1). In moving toward more collaborative planning processes, conservation planning is following the same trajectory as the broader field of land planning (Ahern 2006). Klosterman (1997, 2008) describes this evolution as a progression through three different planning ideals that include planning as design, planning as applied science, and planning as reasoning together.

Planning as design. This approach gained importance in planning in the United States during the first half of the twentieth century. Rooted in architecture and landscape architecture, planning by design aspired to go beyond simple blueprints to produce comprehensive plans that were the planners' rendering of what was putatively the public's shared view of a desired future. Ultimately, the fatal problem with this approach proved to be the planner's inability to craft a desired future that the public would agree on.

Planning as applied science. This approach emerged with the rise of computers in the 1950s and 1960s and focused on the use of models (e.g., urban growth models) and quantitative techniques from social sciences and operations research to develop scientifically objective plans. But rational, quantitative land planning approaches have proven problematic given the social and information challenges that prevail in most cases. As neatly summarized by Couclelis (2005, 1,355), "Land-use planning is a hopelessly complex human endeavor. It involves actions taken by some to affect the use of land controlled by others, following decisions taken by third parties based on values not shared by all concerned, regarding issues no one fully comprehends, in an attempt to guide events and processes that very likely will not unfold in the time, place, and manner anticipated."

Much of the literature in conservation planning would seem to subscribe to planning as applied science. Justin Williams (chapter 12) accurately portrays the heavy reliance on optimization models by many conservation planners. He also points out that these models have in practice mainly been used to provide guidance and insight for planners and stakeholders into very complex spatial problems, not for decision making.

Planning as reasoning together. This approach responds to the limitations of top-down rational planning by emphasizing broad-based collaborative planning in which stakeholders are not simply kept informed but are meaningfully engaged in identifying problems and solutions, supplying data and information, and evaluating alternative futures (Brail and Klosterman 2001; Klosterman 2008). This ideal of planning has promoted the concept of "planning support systems" (as opposed to "decision support systems") that combine computer-based models into an integrated system to support a planning process in which stakeholders and professionals engage in mutual learning, spatial understanding, and plan development (Klosterman 1997). Conservation planning increasingly entails assembling diverse groups to reason together about strategies and tactics to maintain native biological diversity. For this kind of planning, the "who, when, where, and how" of the planning process are more important than any particular plan realization.

The spatial data sources, tools, and techniques described in this book could be viewed as the components of conservation planning support systems to support collaborative planning. Chan et al. (chapter 2) and Ban et al. (chapter 8) provide several examples of meaningful stakeholder involvement in plan development. In chapter 6, Brent Brock and Eric Atkinson state the premise of collaborative planning succinctly when they assert that "local stakeholders need to be empowered to make biologically appropriate conservation decisions."

As I suggest above, the ideal of conservation planning as reasoning together poses a conundrum for those of us developing and espousing tools and methods for data-driven, science-based conservation planning. If you can read and understand all of the material in this book, then you are in a small minority of the planners and stakeholders involved in land and water planning. Most land planners and most stakeholders engaged in participatory land planning are not trained as scientists or quantitative geographical analysts. How can we as conservation planners satisfy our urge to inject increasingly complex ecological reasoning and increasingly sophisticated tools and technologies into the conservation planning process, while at the same time promoting inclusive, collaborative, bottom-up planning processes? The onus is on conservation planners to understand our clientele, to keep our models as simple as we can without sacrificing scientific credibility, and to communicate our science clearly and succinctly. As we do so, we will broaden the participant base for successful conservation planning.

Theme 3: Open, data-driven conservation planning is feasible and desirable almost everywhere

Conservation science and conservation planning, as with science in general (Nielsen 2012), have benefitted from the breathtaking increase in network access to open data and computing software. In their review of conservation planning tools, Sarkar et al. (2006, 127) claimed that "there is now no terrestrial region for which there is not enough data for planning tools to improve policy formulation." This is a bold claim, but this book (notably chapters 3, 4, 5, and 14) provides strong supporting evidence. The authors also make clear that good conservation planning does not depend on having complete data, only on making appropriate use of the best available data, whether it be data about biodiversity, threats, land use and land management, or conservation costs. Dave Theobald (chapter 5) provides good advice when he admonishes conservation planners to aim for surrogates and indicators that are appropriately comprehensive, readily interpretable, parsimonious, and based on solid theory.

Interestingly, Peter Singelton and Brad McRae (chapter 10) are the only authors who explicitly provide a small bit of guidance on the curation and distribution of the products of conservation planning. Many such products could serve as important inputs for other analyses. Unfortunately, conservation biologists and conservation planners are avid consumers but undependable providers of public data. As a result, much useful data and information on biodiversity patterns and trends, as well as on conservation priorities and opportunities, is not readily discoverable or publicly available.

Enormous potential exists to improve conservation planning through large-scale data synthesis (e.g., see Kelling et al. [2009] and the National Center for Ecological Analysis and Synthesis website, http://www.nceas.ucsb.edu, for many other examples), but this depends on access to the kinds of data, tools, and products described throughout this book. Rapid progress has been made on the informatics challenges, such as standards and tools for describing data, tracking data provenance, and data discovery and online retrieval, especially for map and image data (Jones et al. 2006; Reichman et al. 2011). The issues are now more cultural than technological in nature. Most conservation planners lack the resources, formal training, or sufficiently strong incentives to employ best data management practices and to publish their data and information in a way that facilitates reuse. I will simply lodge a plea for allocating more resources to training and open sharing of data and information to continue broadening the collaborative network of conservation planners.

Connecting conservation planning to the everyday world of urban, landscape, and regional planning

In chapter 1, the editors state that this book "should be most valuable for planners in nongovernmental conservation organizations and in city, county, and state agencies; government professionals in land-management agencies; and students interested in conservation or planning. Although applicable anywhere in the world, this book may be particularly useful in developing countries."

I cannot help but wonder how receptive professional planners in city, county, and state agencies, who comprise the key target audience, will be to the material provided here. Most planners may have had some training in environmental planning but few have any familiarity with conservation planning.

To highlight the divide between conservation planning and environmental planning, compare figure 15.1, a word cloud generated from the chapter 1 overview of this book using Wordle (http://www.wordle.net), to figure 15.2, a word cloud generated from an expanded table of contents of a widely used environmental planning handbook (Daniels and Daniels 2003).

Figure 15.1. Word cloud of chapter 1, which provides an overview of the contents of this book. The size of each word is proportional to its frequency of use. Figure from Frank W. Davis, using Wordle.net.

Figure 15.2. Word cloud of the expanded table of contents of the *Handbook of Environmental Planning* by Daniels and Daniels (2003). Figure from Frank W. Davis, using Wordle.net.

The lack of overlap between topics of conservation planning and environmental planning is striking but perhaps not surprising. Conservation planners identify places to protect. Most planners have the complementary job of determining where and how development should occur, in compliance with laws that assure public health, protect farmlands and endangered species, and regulate air and water quality. The two cultures have intersected mainly over endangered species and, to a lesser extent, over migratory species and wetlands.

Other systemic barriers are present to broader integration of conservation planning and environmental planning. For example, the term "conservation planning" is essentially absent from the curriculum of most planning schools and programs, at least in the United States. Conversely, students in conservation biology and conservation planning usually have little or no exposure to urban and regional planning or landscape architecture. As a result, they are unaware of a venerable literature on landscape and regional planning theory and practice, and usually have only superficial familiarity with the local, state, and federal statutes that establish currently operating planning frameworks.

Let me quickly add that there are some notable exceptions to this intellectual divide between land planning and conservation planning. For example, both Richard Forman and Carl Steinitz at the Harvard School of Design have creatively integrated theory and practice from landscape planning and conservation planning (Forman and Collinge 1997; Steinitz 1997). Recent literature on ecological landscape planning is also helping to bridge the divide between traditional land and water resource planning and conservation planning (Leitao and Ahern 2002; Ahern 2006).

In the United States, conservation planning has not been in the mainstream of planning policy, professional practice, or education. Fortunately, this situation appears to be changing, making this book especially timely. For example, in the United States over the past decade, all states have engaged in the production of comprehensive state wildlife conservation strategies (or "wildlife action plans"). For 37 out of 50 states, these plans represent the first comprehensive statewide synthesis of information on wildlife species, habitats, threats, conservation priorities, and opportunities (Davis et al. 2008). Many were produced and are being updated through new collaborations between conservation scientists, state wildlife agencies, federal agencies, conservation NGOs (notably The Nature Conservancy and the Audubon Society) and local planning agencies (http://www.biogeog.ucsb.edu/SWAP/SWAP-home2.html). The recently established Landscape Conservation Cooperatives Program in the US Fish and Wildlife Service (http://www.fws.gov/science/shc/lcc.html), which is promoting science-based conservation planning across public and private lands, presents another opportunity to bring conservation planning together with local and regional environmental planning processes.

Strong forces are now bringing the approaches presented in this book into established urban and regional planning processes. Planning for ecosystem services (chapter 2) integrates naturally with traditional local planning activities concerned with flood control, water supply, coastal protection, green infrastructure planning, and sustainable landscape planning. Climate change adaptation presents another important nexus between traditional planning and modern conservation planning. As Dominique Bachelet describes in chapter 13, climate change adaptation is a major concern for local planning departments looking for cost-effective strategies to cope with changing climate and weather patterns, sea-level rise, higher frequency of extreme events, vulnerable water supplies, changing fire regimes, and new biotic assemblages.

An example of conservation planning informing climate change adaptation from my own backyard is The Nature Conservancy's (TNC) Santa Clara River Project, just north of Los Angeles. For more than a decade, TNC has systematically prioritized and begun to acquire and restore floodplain properties that simultaneously benefit native riparian species, reduce flood risk to urban populations and prime farmland, and increase resilience to sea-level rise. What I find especially encouraging is that TNC staff is doing this work in collaboration with the local flood control district, farm bureau, county planners, and other non-traditional conservation partners. In doing so, they are mainstreaming the modern theory and practice of conservation planning.

Clearly, conservation planners can and must play an expanded role in addressing pressing societal problems related to environmental degradation, biodiversity loss, declining ecosystem services, and risks to society due to rapid climate change. We offer understanding of spatial ecological processes and spatial analytical planning methods that are currently lacking in local and regional planning departments. And the opportunities for integrating conservation planning into mainstream planning processes are expanding rapidly.

References

Ahern, J. 2006. "Theories, Methods and Strategies for Sustainable Landscape Planning." In *From Landscape Research to Landscape Planning: Aspects of Integration, Education and Application,* edited by B. Tress, G. Tress, G. Fry, and P. Opdam, 119–32. Dordrecht, The Netherlands: Springer.

Brail, R. K., and R. E. Klosterman, eds. 2001. *Planning Support Systems: Integrating Geographic Information Systems, Models, and Visualization Tools.* Redlands, CA: Esri Press.

Couclelis, H. 2005. "Where Has the Future Gone? Rethinking the Role of Integrated Land-Use Models in Spatial Planning." *Environment and Planning A* 37: 1,353–71.

Daniels, T., and K. Daniels. 2003. *Environmental Planning Handbook: For Sustainable Communities and Regions.* Chicago, IL: American Planning Association Planners Press.

Davis, F., B. Griffith, S. Henke, L. Maguire, V. Meretsky, J. M. Scott, D. Goble, J. Vaughn, S. Yaffee, and D. Stoms. 2008. *Initial Implementation of the State Wildlife Action Plans: Conservation Impacts, Challenges and Enabling Mechanisms.* Final report to the National Council for Science and the Environment's Wildlife Habitat Policy Research Program. http://www.ncseonline.org/sites/default/files/1A_fdavis_WHPRP_final_report%281%29.pdf.

Forman, R. T. T., and S. K. Collinge. 1997. "Nature Conserved in Changing Landscapes with and without Spatial Planning." *Landscape and Urban Planning* 37: 129–35.

Gleason, M., S. McCreary, M. Miller-Henson, J. Ugoretz, E. Fox, M. Merrifield, W. McClintock, P. Serpa, and K. Hoffman. 2010. "Science-Based and Stakeholder-Driven Marine Protected Area Network Planning: A Successful Case Study from North Central California." *Ocean & Coastal Management* 53: 52–68.

Jones, M. B., M. P. Schildhauer, O. J. Reichman, and S. Bowers. 2006. "The New Bioinformatics: Integrating Ecological Data from the Gene to the Biosphere." *Ecology, Evolution, and Systematics* 37: 519–44, doi: 10.1146/annurev.ecolsys.37.091305.110031.

Kelling, S., W. M. Hochachka, D. Fink, M. Riedewald, R. Caruana, G. Ballard, and G. Hooker. 2009. "Data-Intensive Science: A New Paradigm for Biodiversity Studies." *Bioscience* 59: 613–20.

Klosterman, R. E. 1997. "Planning Support Systems: A New Perspective on Computer-Aided Planning." *Journal of Planning Education and Research* 17: 45–54.

———. 2008. "A New Tool for a New Planning: The What If Planning Support System." In *Planning Support Systems for Cities and Regions,* edited by R. K. Brail, 85–100. Cambridge, MA: Lincoln Institute of Land Policy.

Knight, A. T., R. M. Cowling, M. Difford, and B. M. Campbell. 2010. "Mapping Human and Social Dimensions of Conservation Opportunity for the Scheduling of Conservation Action on Private Land." *Conservation Biology* 24: 1,348–58.

Leitao, A. B., and J. Ahern. 2002. "Applying Landscape Ecological Concepts and Metrics in Sustainable Landscape Planning." *Landscape and Urban Planning* 59: 65–93.

Nielsen, M. 2012. *Reinventing Discovery: The New Era of Networked Science.* Princeton, NJ: Princeton University Press.

Ostrom, E. 2009. "A General Framework for Analyzing Sustainability of Social-Ecological Systems." *Science* 325: 419–22.

Pressey, R. L., and M. C. Bottrill. 2008. "Opportunism, Threats, and the Evolution of Systematic Conservation Planning." *Conservation Biology* 22: 1,340–45.

Reichman, O. J., M. B. Jones, and M. P. Schildhauer. 2011. "Challenges and Opportunities of Open Data in Ecology." *Science* 331: 703–5.

Sarkar, S., R. L. Pressey, D. P. Faith, C. R. Margules, T. Fuller, D. M. Stoms, A. Moffett, K. A. Wilson, K. J. Williams, P. H. Williams, and S. Andelman. 2006. "Biodiversity Conservation Planning Tools: Present Status and Challenges for the Future." *Annual Review of Environment and Resources* 31: 123–59.

Steinitz, C. 1997. *An Alternative Future for the Region of Camp Pendleton, California.* Cambridge, MA: Harvard University Graduate School of Design.

Wilhere, G. F. 2009. "Three Paradoxes of Habitat Conservation Plans." *Environmental Management* 44: 1,089–98.

Afterword

A book like this is necessarily a series of compromises, or a balancing act between the most rigorous science and the most practical approaches, forged in the heat of deadlines, time constraints, and budgets. Frank Davis captured this inherent stress in chapter 15: " . . . the tension is palpable between the scientist's desire for realism and completeness versus the conservation planner's need for expedience and accessibility of the planning process to nonexperts."

Although we have strived to provide the best available science for conservation planning, we realize that this is a moving target: better science is continually being developed. Also, the best available science does not mean "better than nothing," as one reviewer reminded us; but it does not always mean "best possible science" either. Using only the best possible science might require years of data gathering and involve costs that are not generally practical in the time remaining before serious land-use decisions are made.

The rapid pace of climate change and anthropogenic habitat alteration adds a great deal of uncertainty to even the most carefully researched and rigorously analyzed planning decisions. In other words, there is still a lot of art, as well as science, involved, as Reed Noss and colleagues have noted (see chapter 1). Some of the art includes choosing tools and approaches that are most appropriate in a given location at a given time. To that end, we have tried to provide some information and resources on alternate approaches beyond the ones emphasized in each chapter.

This book is just a sample of the information currently available. To access greater detail, and to keep up with new advances, the Internet is essential. Modern planning tools such as ArcGIS software are moving to the web, and cloud computing may soon make desktop software obsolete. The role of hard-copy books, like this one, is limited; even the URLs printed here need to be retyped into a browser. Online, web-based content is the future of conservation planning texts, and the texts themselves will be dynamic. Guides to technological approaches will continually be works in progress.

With the support of Esri, we have established a web-based source for further information, updates, new tools, and new approaches. The success of this online approach depends on the network: the conservation community that is using and developing these tools. Other sites as well can offer complementary, and sometimes overlapping, information. Our hope is that the functionality of this site, http://www.esri.com/conservationbook, based on the ArcGIS platform, will provide tools and ideas that are not available anywhere else. But again, it is up to you, the readers and practitioners, to help make this happen.

Finally, to put conservation planning into perspective, we can contemplate that the human imagination is one of the special gifts that sets humankind apart from the rest of the animal kingdom. The ability to imagine, and especially to imagine the future, is an attribute like being a toolmaker (as Benjamin Franklin declared), the possession of articulate speech (as Thomas Huxley believed), the ability to form civilizations, or the capacity for compassion; that appear (at least with our imperfect knowledge) to be uniquely human. Our ability to imagine, and to shape, the future is the foundation for this book. This ability requires intelligence, tools, dexterity, communication, and compassion—the best of humankind's gifts—but reaches beyond them into the realms of probability and possibility. Shaping the future, by using tools and ideas, for the common good of all life on Earth, is what we call conservation planning. We hope that this book will provide conservation planners at all levels of expertise with the tools and information to help make the world a better place—or at least a healthy place—for our children and grandchildren.

Index